D0026405

# Organic Chemistry Laboratory Manual

## CHEM 3105 & 3106

Authors

Mayo

Cover Image © R. Gino Santa Maria / Shutterstock

Copyright © 2012 by John Wiley & Sons, Inc.

All rights reserved.

No part of this publication may be reproduced, stored in a retrieval system or transmitted in any form or by any means, electronic, mechanical, photocopying, recording, scanning or otherwise, except as permitted under Sections 107 or 108 of the 1976 United States Copyright Act, without either the prior written permission of the Publisher, or authorization through payment of the appropriate per-copy fee to the Copyright Clearance Center, Inc., 222 Rosewood Drive, Danvers, MA 01923, website www.copyright.com. Requests to the Publisher for permission should be addressed to the Permissions Department, John Wiley & Sons, Inc., 111 River Street, Hoboken, NJ 07030-5774, (201)748-6011, fax (201)748-6008, website http://www.wiley.com/go/permissions.

To order books or for customer service, please call 1(800)-CALL-WILEY (225-5945).

Printed in the United States of America.

ISBN 978-1-118-46261-4

# Contents

# Organic Chemistry Laboratory Manual

Special thanks to:

**David M. Birney**
**Ron Erickson**
**Gary Miracle**
**Yanfei Yang**
**Amanda L. Boston**
Texas Tech University

# Periodic Table of the Elements

Atomic number→ 6
Symbol → C
Name (IUPAC) → Carbon
Atomic mass → 12.011

IUPAC recommendations→
Chemical Abstracts Service group notation →

| 1<br>IA | 2<br>IIA | 3<br>IIIB | 4<br>IVB | 5<br>VB | 6<br>VIB | 7<br>VIIB | 8<br>VIIIB | 9<br>VIIIB | 10<br>VIIIB | 11<br>IB | 12<br>IIB | 13<br>IIIA | 14<br>IVA | 15<br>VA | 16<br>VIA | 17<br>VIIA | 18<br>VIIIA |
|---|---|---|---|---|---|---|---|---|---|---|---|---|---|---|---|---|---|
| 1<br>H<br>Hydrogen<br>1.0079 | | | | | | | | | | | | | | | | | 2<br>He<br>Helium<br>4.0026 |
| 3<br>LI<br>Lithium<br>6.941 | 4<br>Be<br>Berylium<br>9.0122 | | | | | | | | | | | 5<br>B<br>Boron<br>10.811 | 6<br>C<br>Carbon<br>12.011 | 7<br>N<br>Nitrogen<br>14.007 | 8<br>O<br>Oxygen<br>15.999 | 9<br>F<br>Fluorine<br>18.998 | 10<br>Ne<br>Neon<br>20.180 |
| 11<br>Na<br>Sodium<br>22,990 | 12<br>Mg<br>Magnesium<br>24.305 | | | | | | | | | | | 13<br>Al<br>Aluminum<br>26.982 | 14<br>Si<br>Silicon<br>28.086 | 15<br>P<br>Phosphorus<br>30.974 | 16<br>S<br>Sulfur<br>32.065 | 17<br>Cl<br>Chlorine<br>35.453 | 18<br>Ar<br>Argon<br>39.948 |
| 19<br>K<br>Potassium<br>39.098 | 20<br>Ca<br>Calcium<br>40.078 | 21<br>Sc<br>Scandium<br>44.956 | 22<br>Ti<br>Titanium<br>47.867 | 23<br>V<br>Vanadium<br>50.942 | 24<br>Cr<br>Chromium<br>51.996 | 25<br>Mn<br>Manganese<br>54.938 | 26<br>Fe<br>Iron<br>55.845 | 27<br>Co<br>Cobalt<br>58.933 | 28<br>Ni<br>Nickel<br>58.693 | 29<br>Cu<br>Copper<br>63.546 | 30<br>Zn<br>Zinc<br>65.409 | 31<br>Ga<br>Gallium<br>69.723 | 32<br>Ge<br>Germanium<br>72.64 | 33<br>As<br>Arsenic<br>74.922 | 34<br>Se<br>Selenium<br>78.96 | 35<br>Br<br>Bromine<br>79.904 | 36<br>Kr<br>Krypton<br>83.798 |
| 37<br>Rb<br>Rubidium<br>85.468 | 38<br>Sr<br>Strontium<br>87.62 | 39<br>Y<br>Yttrium<br>88.906 | 40<br>Zr<br>Zirconium<br>91.224 | 41<br>Nb<br>Niobium<br>92.906 | 42<br>Mo<br>Molybdenum<br>95.94 | 43<br>Tc<br>Technetium<br>(98) | 44<br>Ru<br>Ruthenium<br>101.07 | 45<br>Rh<br>Rhodium<br>102.91 | 46<br>Pd<br>Palladium<br>106.42 | 47<br>Ag<br>Silver<br>107.87 | 48<br>Cd<br>Cadmium<br>112.41 | 49<br>In<br>Indium<br>114.82 | 50<br>Sn<br>Tin<br>118.71 | 51<br>Sb<br>Antimony<br>121.76 | 52<br>Te<br>Tellurium<br>127.60 | 53<br>I<br>Iodine<br>126.90 | 54<br>Xe<br>Xeno<br>131.29 |
| 55<br>Cs<br>Caesium<br>132.91 | 56<br>Ba<br>Barium<br>137.33 | 57<br>*La<br>Lanthanum<br>138.91 | 72<br>Hf<br>Hafnium<br>178.49 | 73<br>Ta<br>Tantalum<br>180.95 | 74<br>W<br>Tungsten<br>183.84 | 75<br>Re<br>Rhenium<br>186.21 | 76<br>Os<br>Osmium<br>190.23 | 77<br>Ir<br>Iridium<br>192.22 | 78<br>Pt<br>Platinum<br>195.08 | 79<br>Au<br>Gold<br>196.97 | 80<br>Hg<br>Mercury<br>200.59 | 81<br>Tl<br>Thallium<br>204.38 | 82<br>Pb<br>Lead<br>207.2 | 83<br>Bi<br>Bismuth<br>208.98 | 84<br>Po<br>Polonium<br>(209) | 85<br>At<br>Astatine<br>(210) | 86<br>Rn<br>Radon<br>(222) |
| 87<br>Fr<br>Francium<br>(223) | 88<br>Ra<br>Radium<br>(226) | 89<br>#Ac<br>Actinium<br>(227) | 104<br>Rf<br>Rutherfordium<br>(261) | 105<br>Db<br>Dubnium<br>(262) | 106<br>Sg<br>Seaborgium<br>(266) | 107<br>Bh<br>Bohrium<br>(264) | 108<br>Hs<br>Hassium<br>(277) | 109<br>Mt<br>Meitnerium<br>(268) | 110<br>Uun<br>(281) | 111<br>Uuu<br>(272) | 112<br>Uub<br>(285) | | 114<br>Uuq<br>(289) | | | | |

| | | | | | | | | | | | | | | |
|---|---|---|---|---|---|---|---|---|---|---|---|---|---|---|
| *Lanthanide Series | 58<br>Ce<br>Cerium<br>140.12 | 59<br>Pr<br>Praseodymium<br>140.91 | 60<br>Nd<br>Neodymium<br>144.24 | 61<br>Pm<br>Promethium<br>(145) | 62<br>Sm<br>Samarium<br>150.36 | 63<br>Eu<br>Europium<br>151.96 | 64<br>Gd<br>Gadolinium<br>157.25 | 65<br>Tb<br>Terbium<br>158.93 | 66<br>Dy<br>Dysprosium<br>162.50 | 67<br>Ho<br>Holmium<br>164.93 | 68<br>Er<br>Erbium<br>167.26 | 69<br>Tm<br>Thulium<br>168.93 | 70<br>Yb<br>Ytterbium<br>173.04 | 71<br>Lu<br>Lutetium<br>174.97 |
| # Actinide Series | 90<br>Th<br>Thorium<br>232.04 | 91<br>Pa<br>Protactinium<br>231.04 | 92<br>U<br>Uranium<br>238.03 | 93<br>Np<br>Neptunium<br>(237) | 94<br>Pu<br>Plutonium<br>(244) | 95<br>Am<br>Americium<br>(243) | 96<br>Cm<br>Curium<br>(247) | 97<br>Bk<br>Berkelium<br>(247) | 98<br>Cf<br>Californium<br>(251) | 99<br>Es<br>Einsteinium<br>(252) | 100<br>Fm<br>Fermium<br>(257) | 101<br>Md<br>Mendelevium<br>(258) | 102<br>No<br>Nobelium<br>(259) | 103<br>Lr<br>Lawrencium<br>(262) |

## Common Organic Solvents: Table of Properties

| Solvent | formula | MW | boiling point (°C) | melting point (°C) | density (g/mL) | solubility in water (g/100g) | Dielectric Constant | flash point (°C) |
|---|---|---|---|---|---|---|---|---|
| acetic acid | $C_2H_4O_2$ | 60.05 | 118 | 16.6 | 1.049 | Miscible | 6.15 | 39 |
| acetone | $C_3H_6O$ | 58.08 | 56.2 | −94.3 | 0.786 | Miscible | 20.7(25) | −18 |
| acetonitrile | $C_2H_3N$ | 41.05 | 81.6 | −46 | 0.786 | Miscible | 37.5 | 6 |
| benzene | $C_6H_6$ | 78.11 | 80.1 | 5.5 | 0.879 | 0.18 | 2.28 | −11 |
| 1-butanol | $C_4H_{10}O$ | 74.12 | 117.6 | −89.5 | 0.81 | 6.3 | 17.8 | 35 |
| 2-butanol | $C_4H_{10}O$ | 74.12 | 98 | −115 | 0.808 | 15 | 15.8(25) | 26 |
| 2-butanone | $C_4H_8O$ | 72.11 | 79.6 | −86.3 | 0.805 | 25.6 | 18.5 | −7 |
| *t*-butyl alcohol | $C_4H_{10}O$ | 74.12 | 82.2 | 25.5 | 0.786 | Miscible | 12.5 | 11 |
| carbon tetrachloride | $CCl_4$ | 153.82 | 76.7 | −22.4 | 1.594 | 0.08 | 2.24 | — |
| chlorobenzene | $C_6H_5Cl$ | 112.56 | 131.7 | −45.6 | 1.1066 | 0.05 | 5.69 | 29 |
| chloroform | $CHCl_3$ | 119.38 | 61.7 | −63.7 | 1.498 | 0.795 | 4.81 | — |
| cyclohexane | $C_6H_{12}$ | 84.16 | 80.7 | 6.6 | 0.779 | <0.1 | 2.02 | −20 |
| 1,2-dichloroethane | $C_2H_4Cl_2$ | 98.96 | 83.5 | −35.3 | 1.245 | 0.861 | 10.42 | 13 |
| diethyl ether | $C_4H_{10}O$ | 74.12 | 34.6 | −116.3 | 0.713 | 7.5 | 4.34 | −45 |
| diethylene glycol | $C_4H_{10}O_3$ | 106.12 | 245 | −10 | 1.118 | 10 | 31.7 | 143 |
| diglyme (diethylene glycol dimethyl ether) | $C_6H_{14}O_3$ | 134.17 | 162 | −68 | 0.943 | Miscible | 7.23 | 67 |
| 1,2-dimethoxy-ethane (glyme, DME) | $C_4H_{10}O_2$ | 90.12 | 85 | −58 | 0.868 | Miscible | 7.2 | −6 |
| dimethylether | $C_2H_6O$ | 46.07 | −22 | −138.5 | NA | NA | NA | −41 |
| dimethyl-formamide (DMF) | $C_3H_7NO$ | 73.09 | 153 | −61 | 0.944 | Miscible | 36.7 | 58 |
| dimethyl sulfoxide (DMSO) | $C_2H_6OS$ | 78.13 | 189 | 18.4 | 1.092 | 25.3 | 47 | 95 |
| dioxane | $C_4H_8O_2$ | 88.11 | 101.1 | 11.8 | 1.033 | Miscible | 2.21(25) | 12 |
| ethanol | $C_2H_6O$ | 46.07 | 78.5 | −114.1 | 0.789 | Miscible | 24.6 | 13 |
| ethyl acetate | $C_4H_8O_2$ | 88.11 | 77 | −83.6 | 0.895 | 8.7 | 6(25) | −4 |
| ethylene glycol | $C_2H_6O_2$ | 62.07 | 195 | −13 | 1.115 | Miscible | 37.7 | 111 |
| glycerin | $C_3H_8O_3$ | 92.09 | 290 | 17.8 | 1.261 | Miscible | 42.5 | 160 |
| heptane | $C_7H_{16}$ | 100.20 | 98 | −90.6 | 0.684 | 0.01 | 1.92 | −4 |
| Hexamethylphosphoramide (HMPA) | $C_6H_{18}N_3OP$ | 179.20 | 232.5 | 7.2 | 1.03 | Miscible | 31.3 | 105 |
| Hexamethylphosphorous triamide (HMPT) | $C_6H_{18}N_3P$ | 163.20 | 150 | −44 | 0.898 | Miscible | ?? | 26 |
| hexane | $C_6H_{14}$ | 86.18 | 69 | −95 | 0.659 | 0.014 | 1.89 | −22 |
| methanol | $CH_4O$ | 32.04 | 64.6 | −98 | 0.791 | Miscible | 32.6(25) | 12 |
| methyl *t*-butyl ether (MTBE) | $C_5H_{12}O$ | 88.15 | 55.2 | −109 | 0.741 | 5.1 | ?? | −28 |
| methylene chloride | $CH_2Cl_2$ | 84.93 | 39.8 | −96.7 | 1.326 | 1.32 | 9.08 | 1.6 |
| *N*-methyl-2-pyrrolidinone (NMP) | $CH_5H_9NO$ | 99.13 | 202 | −24 | 1.033 | 10 | 32 | 91 |
| nitromethane | $CH_3NO_2$ | 61.04 | 101.2 | −29 | 1.382 | 9.50 | 35.9 | 35 |
| pentane | $C_5H_{12}$ | 72.15 | 36.1 | −129.7 | 0.626 | 0.04 | 1.84 | −49 |
| Petroleum ether (ligroine) | — | — | 30–60 | −40 | 0.656 | — | — | −30 |
| 1-propanol | $C_3H_8O$ | 88.15 | 97 | −126 | 0.803 | Miscible | 20.1(25) | 15 |
| 2-propanol | $C_3H_8O$ | 88.15 | 82.4 | −88.5 | 0.785 | Miscible | 18.3(25) | 12 |
| pyridine | $C_5H_5N$ | 79.10 | 115.2 | −41.6 | 0.982 | Miscible | 12.3(25) | 17 |
| tetrahydrofuran (THF) | $C_4H_8O$ | 72.11 | 66 | −108.4 | 0.886 | 30 | 7.6 | −21 |
| toluene | $C_7H_8$ | 92.14 | 110.6 | −93 | 0.867 | 0.05 | 2.38(25) | 4 |
| triethyl amine | $C_6H_{15}N$ | 101.19 | 88.9 | −114.7 | 0.728 | 0.02 | 2.4 | −11 |
| water | $H_2O$ | 18.02 | 100.00 | 0.00 | 0.998 | — | 78.54 | — |
| water, heavy | $D_2O$ | 20.03 | 101.3 | 4 | 1.107 | Miscible | ?? | — |
| *o*-xylene | $C_8H_{10}$ | 106.17 | 144 | −25.2 | 0.897 | Insoluble | 2.57 | 32 |
| *m*-xylene | $C_8H_{10}$ | 106.17 | 139.1 | −47.8 | 0.868 | Insoluble | 2.37 | 27 |
| *p*-xylene | $C_8H_{10}$ | 106.17 | 138.4 | 13.3 | 0.861 | Insoluble | 2.27 | 27 |

T = 20 °C unless specified otherwise.
Source: http://virtual.yosemite.cc.ca.us/smurov/orgsoltab.htm

# MICROSCALE ORGANIC LABORATORY

## with Multistep and Multiscale Syntheses

FIFTH EDITION

**Dana W. Mayo**
*Charles Weston Pickard*
*Professor of Chemistry, Emeritus*
Bowdoin College

**Ronald M. Pike**
*Professor of Chemistry, Emeritus*
Merrimack College

**David C. Forbes**
*Professor of Chemistry*
University of South Alabama

**John Wiley & Sons, Inc.**

| | |
|---|---|
| VICE PRESIDENT AND PUBLISHER | Kaye Pace |
| ASSOCIATE PUBLISHER | Petra Recter |
| PROJECT EDITOR | Jennifer Yee |
| EDITORIAL PROGRAM ASSISTANT | Catherine Donovan |
| PRODUCTION SERVICES MANAGER | Dorothy Sinclair |
| PRODUCTION EDITOR | Janet Foxman |
| EXECUTIVE MARKETING MANAGER | Christine Kushner |
| MARKETING MANAGER | Kristine Ruff |
| CREATIVE DIRECTOR | Harry Nolan |
| TEXT AND COVER DESIGNER | Madelyn Lesure |
| MEDIA EDITOR | Marc Wezdecki |
| PRODUCTION SERVICES | Sherrill Redd/Aptara |

Cover images courtesy of David C. Forbes, University of South Alabama.

This book was set in 11/13 Palatino Light by Aptara and printed and bound by Courier/Kendallville. The cover was printed by Courier/Kendallville.

This book is printed on acid-free paper. ∞

Copyright © 2011, 2000, 1994, 1989 John Wiley & Sons, Inc. All rights reserved. No part of this publication may be reproduced, stored in a retrieval system or transmitted in any form or by any means, electronic, mechanical, photocopying, recording, scanning or otherwise, except as permitted under Sections 107 or 108 of the 1976 United States Copyright Act, without either the prior written permission of the Publisher, or authorization through payment of the appropriate per-copy fee to the Copyright Clearance Center, Inc., 222 Rosewood Drive, Danvers, MA 01923, website *www.copyright.com.* Requests to the Publisher for permission should be addressed to the Permissions Department, John Wiley & Sons, Inc., 111 River Street, Hoboken, NJ 07030-5774, (201) 748-6011, fax (201) 748-6008, website *www.wiley.com/go/permissions.*

Evaluation copies are provided to qualified academics and professionals for review purposes only, for use in their courses during the next academic year. These copies are licensed and may not be sold or transferred to a third party. Upon completion of the review period, please return the evaluation copy to Wiley. Return instructions and a free of charge return shipping label are available at *www.wiley.com/go/returnlabel.* Outside of the United States, please contact your local representative.

**Library of Congress Cataloging-in-Publication Data**

Mayo, Dana W.
Microscale organic laboratory : with multistep and multiscale syntheses / Dana W. Mayo, Ronald M. Pike & David C. Forbes. — 5th ed.
p. cm.
ISBN 978-0-471-21502-8 (cloth)
1. Chemistry, Organic—Laboratory manuals. I. Pike, Ronald M. II. Forbes, David C. III. Title.
QD261.M38 2010
547.0078—dc22

2009048001

ISBN 978-0-471-21502-8 (Main Book)

Printed in the United States of America

TO JEANNE D’ARC, MARILYN, AND CAROL

# PREFACE

Twenty-five years ago, in 1985, when *Microscale Organic Laboratory (MOL)* was first published (as paperback Xerox copies of an unproofed manuscript!), it was the only microscale organic laboratory text available. In the February 1999 *Book Buyers Guide Supplement to the Journal of Chemical Education,* however, there were seventeen laboratory manuals (of a total of thirty-nine) containing miniaturized, fully microscale, or a mixture of micro and macro experiments. Fast forward ten years and without any doubt, microscale techniques have solidly established their place in chemical education. The number of lab manuals currently in print reflects the growing number of students being introduced to organic chemistry through microscale techniques. While the conversion may not yet be quite as high as the eighty percent predicted by David Brooks back in 1985, a conservative estimate would be that a solid two-thirds majority of sophomore students now work with miniaturized experiments compared with the amounts of material employed in these laboratories in the late 1970s.

The major changes that were made to *MOL* in the fourth edition were very well received by our readers. Indeed, we are now nearing the fine-tuning stage in the evolution of this laboratory text. Hence, *MOL5* on the surface will look very much like *MOL4. MOL5,* however, has undergone further significant internal reorganization and rewriting. Many helpful suggestions have been received from reviewers and from instructors who have used previous editions of this text. As a result, some major changes have been made for this new edition:

- A key change to the 5th edition is the modification of the procedural sections to allow for inquiry-based experimentation. Reaction times have been replaced with guidelines, and options on how to best monitor reactions and gauge product purity are left to the discretion of the student or instructor. Many ideas and new approaches can stem from this change—for example, instructional sections can be split into small groups and each group can approach the monitoring of a reaction differently. Using a completely separate set of experiments, discussions can be pursued which focus on reaction purity and evidence which offers the experimentalist sufficient data about what was prepared. Students can then compare notes at the end of the lab period and discuss the various approaches and end results. We hope that this change will allow the lab to become a more interactive experience between groups of students, should that be the wish of the instructor. The opportunity to monitor a reaction rather than assume reaction completion by simply following a time-based instruction and to allow for students to gather additional evidence of product purity empowers the student and adds an element of excitement to the lab experience. Optional inquiry-based guidelines have been added to experiments 5A, 5B, 7, 19B, 24A, and 32. Experiments 11A, 16, and 28 have been modified in a way which focuses on validation of product purity. References to inquiry-based guidelines **?** and validation experiences **V** are noted in the text.
- The use of microwave heating as a tool in synthetic organic chemistry is fast-growing and is becoming an enabling technology. Optional instructions have been added to experiments to allow for the integration of microwave heating as a tool for performing reactions. Since reaction times are shorter than when conventional heating methods are used,

students have the opportunity to supplement these activities with traditional techniques and as stated above engage in discussions comparing the two. Optional microwave heating instructions have been added to experiments 7, 8, 15, 22, and 30. References to microwave use are noted in the text by the use of this icon M.

- A rich collection of end of chapter exercises and the addition of pre and post lab questions provides students with the valuable opportunity to test and practice their own understanding of each laboratory experiment.
- Discussion sections that appear at the beginning of each Experiment have been added, revised, and expanded upon. These discussion sections provide chemical context/background for each experiment, and provide more information regarding the chemical principles involved in each experimental procedure.
- **The Refractive Index material, which was formally in chapter 4, has been moved to the book companion web site: http://www.wiley.com/college/mayo.**
- **Chapter 10W, "Advanced Microscale Organic Laboratory Experiments," (formerly chapter 7) has been moved to the book companion web site: http://www.wiley.com/college/mayo.**

**Additional Resources**

**Text web site**-http://www.wiley.com/college/mayo

As with the previous edition, a major portion of the background theoretical discussions have been moved to the text web site, without affecting the operational part of the text. Likewise, the web site has allowed us to move a number of more advanced discussions out of the printed text. Wherever the shift of this material has occurred the move is flagged by reference call-outs using an icon www.

These **web reference discussions** include information on the following topics:

- Microscale lab equipment and techniques
- Semimicroscale distillation
- Reduced pressure distillations with microspinning band columns
- Vacuum pumps and pressure regulation
- Crystallization
- Measurement of Specific Rotation
- Introduction to Infrared Spectroscopy—Introduction to Theory
- Group Frequencies of the Hydrocarbons
- Characteristic Frequencies of the Heteroatom Functional Groups
- Instrumentation—the Infrared Interferometer
- Tables of Derivatives

The majority of the background infrared spectra and the associated discussions used to develop the use of group frequencies from these spectra are also found on the web site, while the text still contains the essential tables of characteristic frequencies that are in every day use in the laboratory. The many compound data tables, used primarily in the chapter on qualitative identification, also reside on the web site. The Classification of Experiments Based on Mechanism is also available on the web site.

The **Instructor's Manual,** also available on the web site, provides a list of chemicals for each experiment, setup suggestions, and anticipated outcomes. The Instructor's Manual has a separate listing for each experiment developed in the text, which often includes tips for avoiding potential trouble spots and adds considerable information and important references.

### Wiley Custom Select

Wiley's custom publishing program, "Wiley Custom Select" (http://customselect.wiley.com/) gives you the freedom to build your course materials exactly the way you want them. Through a simple, on-line three step process, Wiley Custom Select allows instructors to select content from a vast database of experiments to create a customized laboratory text that meets the needs of their particular course. Each book can be fully customized—instructors can select their own output method, create a cover, arrange the sequence of content, and upload their own materials. At any time, instructors can preview a full version of what the customized book will look like, before the final order is placed.

### Acknowledgements

We continue to acknowledge the outstanding contributions of the early pioneers of instructional microscale programs and techniques, such as F. Emich and F. Pregl in Austria; N. D. Cheronis (who first defined 100 mg of starting substrate in an organic reaction as a microscale transformation), L. Craig, R. C. Fuson, E. H. Huntress, T. S. Ma, A. A. Morton, F. L. Schneider, and R. L. Shriner, in the United States; and J. T. Stock in both England and the United States. These educators laid the foundation on which we were able to fashion much of the current introductory program.

In addition, we are grateful to the colleagues listed below whose careful reviews, helpful suggestions, comments, and thoughtful criticisms of the manuscript have been of such great value to us in developing the final version of this fifth edition of *MOL*.

John Bonte, *Clinton Community College*
Debra Dolliver, *Southeastern Louisiana University*
Maryam Foroozesh, *Xavier University of Louisiana*
Dale Ledford, *University of Southern Mississippi*
A. L. Findeisen, *Temple University*
Neil Freeman, *St. Johns River Community College*
Valerie Keller, *University of Chicago*
Dennis Kevill, *Northern Illinois University*
Deborah Lieberman, *University of Cincinnati*
Barbara Mayer, *California State University—Fresno*
Cynthia McClure, *Montana State University*
Mark C. McMills, *Ohio University*
Richard Narske, *Augustana Collage*
Manoj Patil, *Western Iowa Technical Community College*
Robert Rosenberg, *Transylvania University*
Mark Ruane, *University of Wisconsin—Platteville*
Daniel J. Sandman, *University of Massachusetts-Lowell*

James D. Stickler, *Allegany College of Maryland*
Luise E. Strange de Soria, *Georgia Perimeter College*
Richard T. Taylor, *Miami University*
Richard E. Tompson, *Louisiana State University—Shreveport*
Eric Trump, *Emporia State University*
Nanette Wachter, *Hofstra University*
Don B. Weser, *Frostburg State University*
James Zubricky, *University of Toledo*

We are as well grateful to those who revisited each experiment in MOL4 which was the central theme of this project. Taking ownership and seeing its completion was nothing short of a herculean effort. The team who undertook this task are listed below. They were all under the guidance of Brian Finnigan who deserves much of the credit as he served as point person for the revision project.

Stephen E. Arnold, University of South Alabama
Sampada V. Bettigeri, Mathoshri Prathisthan College of Engineering
Amanda C. Brewton, University of South Alabama
Sarah S. Dolbear, University of South Alabama
Brian P. Finnigan, University of South Alabama

We also appreciate the support from Wiley which allowed us to revisit each exeriment in MOL4 so that we could properly identify which experiments were best served to be modified. And finally, we would like to extend our gratitude to Petra Recter, Sherrill Redd, and Jennifer Yee who shepherded this projected from conception to press.

We continue to applaud the widespread development of affordable glassware for use in microscale instructional laboratories. We are particularly pleased to note that the particular style of equipment (cap-seal connectors) that we developed for this program at Bowdoin College has accomplished an outstanding record of survival on the battleground of the sophomore laboratory bench. Much of the credit for the granitelike character of this equipment goes to J. Ryan and Larry Riley of the ACE Glass Company. Several contributors have played long-term roles in the successful evolution of the microscale organic laboratory program, and we are happy to acknowledge them: Janet Hotham, Judy Foster, Henry Horner, Lauren Bartlett, Robert Stevens, and Samuel Butcher have all made vital contributions along the way.

We are particularly indebted to our colleagues Nicholas Leadbeater, Cynthia McGowan, and Elizabeth Stemmler. Their willingness to contribute to this project is gratefully appreciated. Cynthia and Nicholas provided in its entirety the microwave contribution to MOL5. The discussion section, experimentation, and safety contribution truly adds to the wealth of this edition and the excitement of a comprehensive introductory laboratory experience. As it was with MOL4, Elizabeth's contribution of an introductory discussion on the *Application of Mass Spectrometry to Organic Chemistry* continues to offer the reader a diverse experience using this powerful technique to the introductory laboratory experience.

The development of our kinetics experiment fell on the strong shoulders of Paulette Messier, Laboratory Instructor, and adds just one more accomplishment to her unending contributions to the development of the microscale program at Bowdoin College. Paulette is rapidly closing in on three decades of continuous laboratory instruction at the microscale level, a unique record of experience in microscale anywhere in the world of chemical education. Paulette, more than any other person, has made this program a success in the trenches between the lab benches where it really counts. The thousands of students who have dealt directly with her and gained her respect are a tribute to Paulette's quiet, confident way of instilling enthusiasm and excitement into the microscale experience. Paulette Messier is indelibly linked to the Microscale Organic Laboratory at Bowdoin College.

It is with deep regret that the authors note the passing of Peter K. Trumper at far too young an age. Peter made many contributions to the development of the microscale program at Bowdoin. In particular, in the early days of its evolution Peter's expertise with NMR theory and experimental practice greatly enhanced the integration of this technique into the undergraduate microscale laboratory starting as early as in the second edition. His engaging nature and wit is sorely missed.

With the publication of the Fifth Edition, *Microscale Organic Laboratory* might be considered to have reached a mature state. In our opinion, however, chemical education is as dynamic as the subject itself. For on our drawing boards are thoughts almost as outrageous as the idea that occurred in the early winter of 1980 to 1981— to run an introductory organic laboratory program on a milligram scale!

DANA W. MAYO
RONALD M. PIKE
DAVID C. FORBES
***January 2010***

## CH$_4$apter 1

# INTRODUCTION

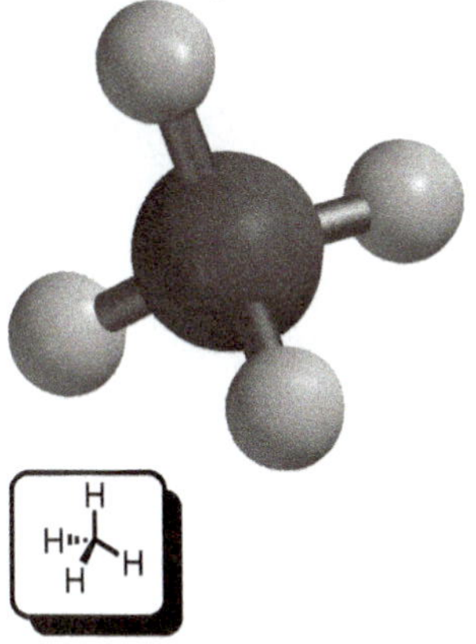


*You are about to embark on a challenging adventure—the microscale organic chemistry laboratory!*

Your course is going to be quite different from the conventional manner in which this laboratory has been taught in past decades. You will be learning the experimental side of organic chemistry from the microscale level. Although you will be working with very small amounts of materials, you will be able to observe and learn more organic chemistry in one year than many of your predecessors did in nearly two years of laboratory work. You will find this laboratory an exciting and interesting place to be. While we cannot guarantee it for you individually, the majority of students who went through the program during its development found the microscale organic laboratory to be a surprisingly pleasant adventure.

This textbook is centered on helping you develop skills in microscale organic laboratory techniques. Its focus is twofold. For those of you in the academic environment and involved with the introductory organic laboratory, it allows the flexibility of developing your own scaling sequence without being tied to a prescribed set of quantities. For those of you working in a research environment at the advanced undergraduate or graduate level or in the industrial area, this text will provide the foundation from which you can develop a solid expertise in microscale techniques directly applicable to your work. Working at the microscale level is substantially different from using conventional operations in the organic laboratory with multigram quantities of materials.

During the last two decades, the experimental side of organic chemistry has moved ever closer to the microscale level. This conversion started in earnest nearly thirty years ago and has been spurred on by the rapidly accelerating cost of chemical waste disposal. As we have said, you will be working with very small amounts of materials, but the techniques that you will learn and experience you will gain will allow you to accomplish more organic chemistry in the long run than many of your predecessors.

First, we want to acquaint you with the organization and contents of the text. ***With the fifth edition, a continued effort has been made to streamline the basic reference material from the text using our accompanying website (www.wiley.com/college/MOL5). Accordingly, Chapter 10W (formerly Chapter 7 of the fourth edition) in its entirety has been placed online. Throughout this edition, Chapter 10W is identified with a "W" (e.g., Chapter 10W), indicating its location online. Furthermore, an icon will be used in the margin to indicate website material that will be of interest to the user. We hope this treatment of the laboratory will make the more important aspects of the basic text easier to access and will speed your laboratory work along.*** We then give you a few words of advice, which, if they are heeded, will allow you to avoid many of the sand traps you will find as you develop microscale laboratory techniques. Finally, we wax philosophical and attempt to describe what we think you should derive from this experience.

WWW

WWW

WWW

After this brief introduction, the second chapter is concerned with safety in the laboratory. This chapter supplies information that will allow you to estimate

**Chapter 1: $CH_4$, Methane**
a substance of natural origin, known as Marsh Gas to the alchemists.

your maximum possible exposure to volatile chemicals used in the microscale laboratory. Chapter 2 also discusses general safety protocol for the laboratory. It is vitally important that you become familiar with the details of the material contained in this chapter; your health and safety depend on this knowledge.

The next three chapters are concerned primarily with the development of experimental techniques. Chapter 3 describes in detail the glassware employed in microscale organic chemistry: the logic behind its construction, tips on its usage, the common arrangements of equipment, and various other laboratory manipulations, including techniques for transferring microquantities of materials. Suggestions for the organization of your laboratory notebook are presented at the end of this chapter.

Chapter 4 deals with equipment and techniques for determining a number of physical properties of microscale samples. Chapter 5 is divided into nine technique sections. Detailed discussions develop the major areas of experimental technique that are used in the microscale organic laboratory.

WWW→ Chapters 6, 7, and 10W contain the main experimental sections of this text. Chapter 6 is focused primarily on preparative organic chemistry at the microscale level and consists of 35 experiments. While the number of experiments has not changed with this edition, there have been changes to how the reactions are monitored and conducted. Six experiments (Experiments 5A, 5B, 7, 19B, 24A, and 32) in Chapter 6 have been modified in a way which replaces the posting of a reaction time with the task of monitoring the reaction by TLC until complete. The TLC technique is asked of the experimentalist in three more experiments (Experiments 11A, 16, and 28) in order to provide additional evidence of reaction purity upon recrystallization of the crude reaction mixture. And finally, five experiments (Experiments 7, 8, 15, 22, and 30) in Chapter 6 now have optional exercises which utilize microwave technologies. Additional selections of individual experiments can be drawn from those experiments presented in Chapter 7. WWW→ Chapter 10W, which is now located online, contains a series of seven experiments of a more sophisticated nature. A number of the experiments WWW→ contained in Chapters 6 and 10W are of optional scale so that you may also have the opportunity to gain some experience with experimentation at larger scales. Chapter 7 consists of a set of six sequential experiments that are essentially identical to the type of problems tackled by research chemists involved in synthetic organic chemistry. A number of these multistep procedures begin the first step in the experiment with large-scale, multigram quantities of starting material, but require microscale techniques to complete the final step or two. The use of this chapter is most appropriate in the final stages of the course, for example, the latter part of the second semester of a two-semester sequence.

Chapter 8 develops the characterization of organic materials at the microscale level by spectroscopic techniques. The chapter starts with a brief discussion of the interpretation of infrared (IR) group frequencies and is followed by a more detailed treatment of nuclear magnetic resonance (NMR) spectral data, a brief discussion of ultraviolet-visible (UV–vis) spectroscopy, and a brief introduction to the theory, experimental techniques, and applications of mass spectrometry to organic chemistry. A more detailed introduction to the theoretical basis for these spectroscopic techniques is also presented on the accompanying website.

Chapter 9 develops the characterization of organic materials at the microscale level by the use of classical organic reactions to form solid derivatives. Tables of derivative data for use in compound identification by these techniques are discussed and are included on the website as Appendix A.

A list of all the experiments grouped by reaction mechanism is given on the web-site as Appendix B.

The organization of the experimental procedures given in Chapters 6, 7, and 10W is arranged in the following fashion. A short opening statement describing the reaction to be studied is followed by the reaction scheme.

WWW

Generally, a brief discussion of the reaction follows, including a mechanistic interpretation. In a few cases of particularly important reactions, or where the experiment is likely to precede presentation of the topic in the classroom, a more detailed description is given. The estimated time needed to complete the work, and a table of reactant data come next. For ease in organizing your laboratory time, the experimental section is divided into four subsections: *reagents and equipment, reaction conditions, isolation of product, and purification and characterization.*

We then introduce a series of questions and problems designed to enhance and focus your understanding of the chemistry and the experimental procedures involved in a particular laboratory exercise. Finally, a bibliography offering a list of literature references is given. Although this list comes at the end of the experimental section, we view it as a very important part of the text. The discussion of the chemistry involved in each experiment is necessarily brief. We hope that you will take time to read and expand your knowledge about the particular experiment that you are conducting. You may, in fact, find that some of these references become assigned reading.

A prompt (➡) in the text indicates that experimental apparatus involved with that stage of the experiment are shown in the margin. Important comments are italicized in the text, and **Warnings** and **Cautions** are given in boxes and also indicated in the margins.

*In an effort to streamline our treatment of the laboratory we have moved a considerable quantity of material from the previous editions, MOL3 and MOL4, and placed it in easily accessible form on our website (www.wiley.com/college/MOL5). An icon lets you know that supplemental material is available on the website. New to this edition is a detailed listing within the table of contents of all materials available online. We hope this format will make the more important aspects of the basic text easier to access and speed your laboratory work along.*

WWW

## GENERAL RULES FOR THE MICROSCALE LABORATORY

**1. Study the experiment before you come to lab.** This rule is a historical plea from all laboratory instructors. In the microscale laboratory it takes on a more important meaning. You will not survive if you do not prepare ahead of time. In microscale experiments, operations happen much more quickly than in the macroscale laboratory. Your laboratory time will be overflowing with many more events. If you are not familiar with the sequences you are to follow, you will be in deep trouble. Although the techniques employed at the microscale level are not particularly difficult to acquire, they do demand a significant amount of attention. For you to reach a successful and happy conclusion, you cannot afford to have the focus of your concentration broken by having to constantly refer to the text during the experiment. Disaster is ever present for the unprepared.

**2. ALWAYS work with clean equipment.** You must take the time to scrupulously clean your equipment before you start any experiment. Contaminated glass-ware will ultimately cost you additional time, and you will experience the frustration of inconsistent results and lower yields. Dirty equipment is the primary cause of reaction failure at the microscale level.

**3. CAREFULLY measure the quantities of materials to be used in the experiments.** A little extra time at the beginning of the laboratory can speed you on your way at the end of the session. A great deal of time has been spent optimizing the conditions employed in these experiments in order to maximize yields. Many organic reactions are very sensitive to the relative quantities of substrate (the material on which the reaction is taking place) and reagent (the reactive substance or substances that bring about the change in the substrate). After equipment contamination, the second-largest cause of failed reactions is attempting to run a reaction with incorrect quantities of the reactants present. Do not be hurried or careless at the balance.

**4. Clean means DRY.** Water or cleaning solution can be as detrimental to the success of a reaction as dirt or sludge in the system. You often will be working with very small quantities of moisture-sensitive reagents. The glass surface areas with which these reagents come in contact, however, are relatively large. A slightly damp piece of glassware can rapidly deactivate a critical reagent and result in reaction failure. *This rule must be strictly followed.*

**5. ALWAYS work on a clean laboratory bench surface, preferably glass!**

**6. ALWAYS protect the reaction product that you are working with from a disastrous spill by carrying out all solution or solvent transfers over a crystallizing dish.**

**7. ALWAYS place reaction vials or flasks in a clean beaker when standing them on the laboratory bench.** Then, when a spill occurs the material is more likely to be contained in the beaker and less likely to be found on the laboratory bench or floor.

**8. NEVER use cork rings to support round-bottom flasks, particularly if they contain liquids.** You are inviting disaster to be a guest at your laboratory bench.

**9. ALWAYS think through the next step you are going to perform *before* starting it.** Once you have added the wrong reagent, it is back to square one.

**10. ALWAYS save everything you have generated in an experiment until it is successfully completed.** You can retrieve a mislabeled chromatographic fraction from your locker, but not from the waste container!

## THE ORGANIC CHEMISTRY LABORATORY

The confidence gained by mastering the microscale techniques described here will pay big dividends as you progress into modern-day experimental chemistry. The organic laboratory has had a reputation of being smelly, long, tedious, and pockmarked with fires and explosions; but present-day organic chemistry is undergoing a revolution at the laboratory bench. New techniques are sweeping away many of the old complaints, as an increasing fraction of industrial and academic research is being carried out at the microscale level.

This book allows the interested participant to rapidly develop the skills needed to slice more deeply into organic chemistry than ever before. The attendant benefits are greater confidence and independence in acquired laboratory techniques. The happy result is that in a microscale-based organic chemistry laboratory, you are more likely to have a satisfying encounter with the experimental side of this fascinating field of knowledge.

$C_2H_4$apter 2

# SAFETY

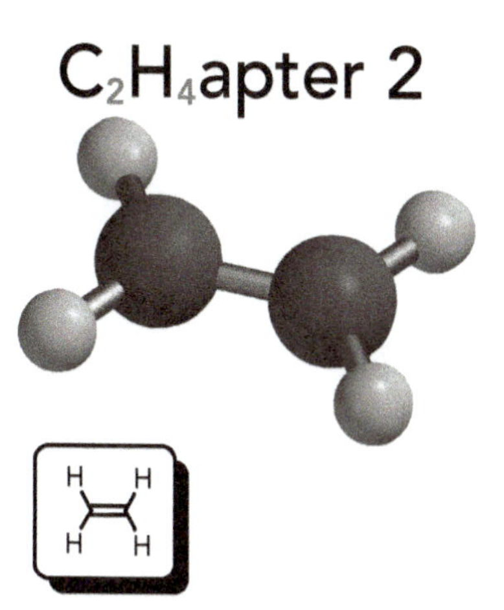


Research laboratories vary widely with respect to facilities and support given to safety. Large laboratories may have several hundred chemists and an extensive network of co-workers, supervisors, safety officers, and hazardous-waste managers. They also, according to government regulations, have an extensive set of safety procedures and detailed practices for the storage and disposal of hazardous wastes. In small laboratories, the individual chemist may have to take care of all these aspects of safety. Some laboratories may routinely deal with very hazardous materials and may run all reactions in hoods. Others may deal mainly with relatively innocuous compounds and have very limited hood facilities.

Our approach is to raise some questions to think about and to suggest places to look for further information. In this chapter, we do not present a large list of safety precautions for use in all situations; rather, we present a list of very basic precautionary measures. A bibliography at the end of the chapter offers a list of selected references. *We urge you to consult these references concerning specific safety regulations.* Many laboratories may have safety guidelines that will supercede this very cursory treatment. This chapter is no more than a starting point.

## MAKING THE LABORATORY A SAFER PLACE

Murphy's law states in brief, "If anything can go wrong, it will." Although it is often taken to be a silly law, it is not. Murphy's law means that if sparking switches are present in areas that contain flammable vapors, sooner or later there will be a fire. If the glass container can move to the edge of the shelf as items are moved around or because the building vibrates, at some time it will come crashing to the floor. If the pipet can become contaminated, then the mouth pipetter will eventually ingest a contaminant.

We cannot revoke Murphy's law, but we can do a lot to minimize the damage. We can reduce the incidence of sparks and flames and flammable vapors. We can make sure that if the accident does occur, we have the means to contain the damage and to take care of any injuries that result. All of this means thinking about the laboratory environment. Does your laboratory have or enforce regulations related to important items such as eye, face, and foot protection, safety clothing, respiratory equipment, first aid supplies, fire equipment, spill kits, hoods, and compliance regulations? *Think ahead* about what could go wrong and then *plan* and *prepare* to minimize the chance of an accident and be prepared to respond when one does occur.

## NATURE OF HAZARDS

The chemistry laboratory presents a wide assortment of risks. These risks are outlined briefly here so that you can begin to think about the steps necessary to make the laboratory safer:

1. **Physical hazards.** Injuries resulting from flames, explosions, and equipment (cuts from glass, electrical shock from faulty instrumentation, or improper use of instruments).

---

**Chapter 2: $C_2H_4$, Ethylene**
a substance of natural origin, released by ripening fruit.

2. **External exposure to chemicals.** Injuries to skin and eyes resulting from contact with chemicals that have spilled, splashed, or been left on the bench top or on equipment.
3. **Internal exposure.** Longer term (usually) health effects resulting from breathing hazardous vapors or ingesting chemicals.

## REDUCTION OF RISKS

Many things can be done to reduce risks. The rules below may be absolute in some laboratories. In others, the nature of the materials and apparatus used may justify the relaxation of some of these rules or the addition of others.

1. **Stick to the procedures described by your supervisor.** This attention to detail is particularly important for the chemist with limited experience. In other cases, variation of the reagents and techniques may be part of the work.
2. **Wear approved safety goggles.** We can often recover quickly from injuries affecting only a few square millimeters on our bodies, unless that area happens to be in our eyes. Larger industrial laboratories often require that laboratory work clothes and safety shoes be worn. Wear them, if requested.
3. **Do not put anything in your mouth under any circumstances while in the laboratory.** This includes food, drinks, chemicals, and pipets. There are countless ways that surfaces can become contaminated in the laboratory. Since there are substances that must never be pipetted by mouth, one must get into the habit of *never* mouth pipetting anything.
4. **Be cautious with flames and flammable solvents.** Remember that the flame at one end of the bench can ignite the flammable liquid at the other end in the event of a spill or improper disposal. Flames must never be used when certain liquids are present in the laboratory, and flames must always be used with care. Check the *fire diamond* hazard symbol, if available.
5. **Be sure that you have the proper chemicals for your reaction.** Check labels carefully, and return unused chemicals to the proper place for storage. Be sure to replace caps on containers immediately after use. An open container is an invitation for a spill. Furthermore, some reagents are very sensitive to moisture, and may decompose if left open.
6. **Minimize the loss of chemicals to air or water and dispose of waste properly.** Some water-soluble materials may be safely disposed of in the water drains. Other wastes should go into special receptacles. Pay attention to the labels on these receptacles. Recent government regulations have placed stringent rules on industrial and academic laboratories for proper disposal of chemicals. *Severe penalties are levied on those who do not follow proper procedures.* We recommend that you consult general safety references nos. 3 and 4 at the end of the chapter.
7. **Minimize skin contact with any chemicals.** Use impermeable gloves when necessary, and promptly wash any chemical off your body. If you have to wash something off with water, use lots of it. Be sure that you know where the nearest water spray device is located.

*NOTE. Do not use latex gloves. They are permeable to many chemicals and some people are allergic to them. A recommended substitute are the various grades of nitrile gloves.*

8. **Do not inhale vapors from volatile materials.** Severe illness or internal injury can result.
9. **Tie back or confine long hair and loose items of clothing.** You do not want them falling into a reagent or getting near flames.
10. **Do not work alone.** Too many things can happen to a person working alone that might leave him or her unable to obtain assistance. As in swimming, the "buddy system" is safest.
11. **Exercise care in assembling glass and electrical apparatus.** All operations with glass, such as separating standard taper glassware, involve the risk that the glass may break and that lacerations or punctures may result. Seek help or advice with glassware, if necessary. Special containers should be provided for the disposal of broken glass. Electrical shock can occur in many ways. When making electrical connections, make sure that your hands, the laboratory bench, and the floor are all dry and that *you* do not complete an electrical path to ground. Be sure that electrical equipment is properly grounded and insulated.
12. **Report any injury or accident to the appropriate person.** Reporting injuries and accidents is important so that medical assistance can be obtained if necessary. It also allows others to be made aware of any safety problems; these problems may be correctable.
13. **Keep things clean.** Put unused apparatus away. Immediately wipe up or care for spills on the bench top or floor. This also pertains to the balance area and to where chemicals are dispensed.
14. **Never heat a closed system.** Always provide a vent to avoid an explosion. Provide a suitable trap for any toxic gases generated in a given reaction.
15. **Learn the correct use of gas cylinders.** Even a small gas cylinder can become a lethal bomb if not properly used.
16. **Attend safety programs.** Many industrial laboratories offer excellent seminars and lectures on a wide variety of safety topics. Pay careful attention to the advice and counsel of the safety officer.
17. **Above all, use your common sense.** Think before you act.

## PRECAUTIONARY MEASURES

*Know the location and operation of safety equipment in the laboratory. Locate the nearest*

- Fire extinguisher
- Telephone
- Exit
- First aid kit
- Emergency shower
- Fire blanket
- Eye wash

*Know where to call (have the numbers posted) for*

- Fire
- Medical emergency
- Spill or accidental release of corrosive or toxic chemicals.

*Know where to go*

- In case of injury
- To evacuate the building

## THINKING ABOUT THE RISKS IN USING CHEMICALS

The smaller quantities used in the microscale laboratory carry with them a reduction in hazards caused by fires and explosions; hazards associated with skin contact are also reduced. However, care must be exercised when working with even the small quantities involved.

There is great potential for reducing the exposure to chemical vapors, but these reductions will be realized only if everyone in the laboratory is careful. One characteristic of vapors emitted outside hoods is that they mix rapidly throughout the lab and will quickly reach the person on the other side of the room. In some laboratories, the majority of reactions may be carried out in hoods. When reactions are carried out in the open laboratory, each experimenter becomes a polluter whose emissions affect nearby people the most, but these emissions become added to the laboratory air and to the burden each of us must bear.

The concentration of vapor in the general laboratory air space depends on the vapor pressure of the liquids, the area of the solid or liquid exposed, the nature of air currents near the sources, and the ventilation characteristics of the laboratory. One factor over which each individual has control is evaporation, which can be reduced by the following practices:

- Certain liquids must remain in hoods.
- Reagent bottles must be recapped when not in use.
- Spills must be quickly cleaned up and the waste discarded.

Chemicals must be properly stored when not in use. Some balance must be struck between the convenience of having the compound in the laboratory where you can easily put your hands on it and the safety of having the compound in a properly ventilated and fire-safe storage room. Policies for storing chemicals will vary from place to place. There are limits to the amounts of flammable liquids that should be stored in glass containers, and fire-resistant cabinets must be used for storage of large amounts of flammable liquids. Chemicals that react with one another should not be stored in close proximity. There are plans for sorting chemicals by general reactivity classes in storerooms; for instance, Flinn Scientific Company includes a description of a storage system in their (2009) chemical catalog.

## DISPOSAL OF CHEMICALS

Chemicals must also be segregated into categories for disposal. The categories used will depend on the disposal service available and upon federal, state, and local regulations. For example, some organic wastes are readily incinerated, while those containing chlorine may require much more costly treatment. Other wastes may have to be buried. For safety and economic reasons, it is important to place waste material in the appropriate container. In today's world, it often costs more to dispose of a chemical than to purchase it in the first place! The economic impact of waste generation and disposal is gigantic. Based upon the toxic release inventory from the EPA for the year of 2005, do you realize that the chemical industry in the United States released more than 4 billion pounds of on-site and off-site chemical waste?[1] Each year hundreds of billions of dollars is spent per year in waste treatment, control, and disposal costs!

It is our obligation as chemists to decrease the impact that hazardous chemicals have on our environment. A movement is currently underway (referred to as "Green Chemistry" or "Benign by Design") to accomplish this goal by focusing on the design, manufacture, and use of chemicals and chemical processes that have little or no pollution potential or environmental risk.

## MATERIAL SAFETY DATA SHEETS

Although risks are associated with the use of most chemicals, the magnitudes of these risks vary greatly. A short description of the risks is provided by a Material Safety Data Sheet, commonly referred to as an MSDS. All participants of a laboratory experience are strongly encouraged to educate themselves on the risks, large or small, of the chemicals they are scheduled to work with while in lab. The information contained on these sheets can be obtained from a number of locations. While they are normally provided by the manufacturer or vendor of the chemical, and users are required to keep on file the MSDS of each material stored or used, data sheets can be easily obtained online.

As an example, the 1985 MSDS for acetone is shown here. This sheet was provided by the J. T. Baker Chemical Company. Sheets from other sources, especially those online[2], do provide a very illuminating comparison of what has transpired over 20+ years. Much of the information on these sheets is self-explanatory, but let's review the major sections of the acetone example.

www

Section I provides identification numbers and codes for the compound and includes a summary of the risks associated with the use of acetone. Because these sheets are available for many thousands of compounds and mixtures, there must be a means of unambiguously identifying the substance. A standard reference number for chemists is the Chemical Abstracts Service Number (CAS No.).

A quick review of the degree of risks is given by the numerical scale under Precautionary Labeling. This particular scale is a proprietary scale that ranges from 0 (very little or nonexistent risk) to 4 (extremely high risk). The National Fire Protection Association (NFPA) uses a similar scale but the risks considered are different. Other systems may use different scales, and there are some that represent low risks by the highest number! Be sure that you understand the scale being used. Perhaps some day one scale will become standard.

Section II covers risks from mixtures. Because a mixture is not considered here, the section is empty. Selected physical data are described in Section III. Section IV contains fire and explosion data, including a description of the toxic gases produced when acetone is exposed to a fire. The MSDSs are routinely made available to fire departments that may be faced with fighting a fire in a building where large amounts of chemicals are stored.

[1]For online access to the EPA Toxic Release Inventory, see: www.epa.gov/tri/
[2]For an online MSDS for acetone, see: www.jtbaker.com/msds/englishhtml/A0446.htm

J. T. BAKER CHEMICAL CO. 222 RED SCHOOL LANE, PHILLIPSBURG, NJ 08865
MATERIAL SAFETY DATA SHEET
24-HOUR EMERGENCY TELEPHONE — (201) 859-2151
CHEMTREC # (800) 424-9300 — NATIONAL RESPONSE CENTER # (800) 424-8802

A0446 -01 ACETONE PAGE: 1
EFFECTIVE: 10/11/85 ISSUED: 01/23/86

SECTION I – PRODUCT IDENTIFICATION

PRODUCT NAME: ACETONE
FORMULA: (CH3)2CO
FORMULA WT: 58.08
CAS NO.: 00067-64-1
NIOSH/RTECS NO.: AL3150000
COMMON SYNONYMS: DIMETHYL KETONE; METHYL KETONE; 2-PROPANONE
PRODUCT CODES: 9010,9006,9002,9254,9009,9001,9004,5356,A134,9007,9005,9008

PRECAUTIONARY LABELLING

BAKER SAF-T-DATA(TM) SYSTEM

HEALTH - 1
FLAMMABILITY - 3 (FLAMMABLE)
REACTIVITY - 2
CONTACT - 1

LABORATORY PROTECTIVE EQUIPMENT

SAFETY GLASSES; LAB COAT; VENT HOOD; PROPER GLOVES; CLASS B EXTINGUISHER

PRECAUTIONARY LABEL STATEMENTS

DANGER
EXTREMELY FLAMMABLE
HARMFUL IF SWALLOWED OR INHALED
CAUSES IRRITATION

KEEP AWAY FROM HEAT, SPARKS, FLAME. AVOID CONTACT WITH EYES, SKIN, CLOTHING. AVOID BREATHING VAPOR. KEEP IN TIGHTLY CLOSED CONTAINER. USE WITH ADEQUATE VENTILATION. WASH THOROUGHLY AFTER HANDLING. IN CASE OF FIRE, USE WATER SPRAY, ALCOHOL FOAM, DRY CHEMICAL, OR CARBON DIOXIDE. FLUSH SPILL AREA WITH WATER SPRAY.

SECTION II – HAZARDOUS COMPONENTS

| COMPONENT | % | CAS NO. |
|---|---|---|
| ACETONE | 90-100 | 67-64-1 |

SECTION III – PHYSICAL DATA

BOILING POINT: 56 C ( 133 F) VAPOR PRESSURE(MM HG): 181
MELTING POINT: −95 C ( −139 F) VAPOR DENSITY(AIR=1): 2
SPECIFIC GRAVITY: 0.79 (H2O=1) EVAPORATION RATE: 5.6 (BUTYL ACETATE=1)

SOLUBILITY(H2O): COMPLETE (IN ALL PROPORTIONS) % VOLATILES BY VOLUME: 100
APPEARANCE & ODOR: CLEAR, COLORLESS LIQUID WITH FRAGRANT SWEET ODOR.

SECTION IV – FIRE AND EXPLOSION HAZARD DATA

FLASH POINT: −18 C ( 0 F) NFPA 704M RATING: 1-3-0
FLAMMABLE LIMITS: UPPER – 13 % LOWER – 2 %

FIRE EXTINGUISHING MEDIA
USE ALCOHOL FOAM, DRY CHEMICAL OR CARBON DIOXIDE.
(WATER MAY BE INEFFECTIVE.)

SPECIAL FIRE-FIGHTING PROCEDURES
FIREFIGHTERS SHOULD WEAR PROPER PROTECTIVE EQUIPMENT AND SELF-CONTAINED (POSITIVE PRESSURE IF AVAILABLE) BREATHING APPARATUS WITH FULL FACEPIECE.
MOVE EXPOSED CONTAINERS FROM FIRE AREA IF IT CAN BE DONE WITHOUT RISK.
USE WATER TO KEEP FIRE-EXPOSED CONTAINERS COOL.

UNUSUAL FIRE & EXPLOSION HAZARDS
VAPORS MAY FLOW ALONG SURFACES TO DISTANT IGNITION SOURCES AND FLASH BACK.
CLOSED CONTAINERS EXPOSED TO HEAT MAY EXPLODE. CONTACT WITH STRONG OXIDIZERS MAY CAUSE FIRE.

SECTION V – HEALTH HAZARD DATA

THRESHOLD LIMIT VALUE (TLV/TWA): 1780 MG/M3 ( 750 PPM)
SHORT-TERM EXPOSURE LIMIT (STEL): 2375 MG/M3 ( 1000 PPM)

TOXICITY: LD50 (ORAL-RAT)(MG/KG) - 9750
LD50 (IPR-MOUSE)(G/KG) - 1297

EFFECTS OF OVEREXPOSURE
CONTACT WITH SKIN HAS A DEFATTING EFFECT, CAUSING DRYING AND IRRITATION.
OVEREXPOSURE TO VAPORS MAY CAUSE IRRITATION OF MUCOUS MEMBRANES, DRYNESS OF MOUTH AND THROAT, HEADACHE, NAUSEA AND DIZZINESS.

EMERGENCY AND FIRST AID PROCEDURES
CALL A PHYSICIAN.
IF SWALLOWED, IF CONSCIOUS, IMMEDIATELY INDUCE VOMITING.
IF INHALED, REMOVE TO FRESH AIR. IF NOT BREATHING, GIVE ARTIFICIAL RESPIRATION. IF BREATHING IS DIFFICULT, GIVE OXYGEN.
IN CASE OF CONTACT, IMMEDIATELY FLUSH EYES WITH PLENTY OF WATER FOR AT LEAST 15 MINUTES. FLUSH SKIN WITH WATER.

SECTION VI – REACTIVITY DATA

STABILITY: STABLE HAZARDOUS POLYMERIZATION: WILL NOT OCCUR

CONDITIONS TO AVOID: HEAT, FLAME, SOURCES OF IGNITION

INCOMPATIBLES: SULFURIC ACID, NITRIC ACID, STRONG OXIDIZING AGENTS

SECTION VII – SPILL AND DISPOSAL PROCEDURES

STEPS TO BE TAKEN IN THE EVENT OF A SPILL OR DISCHARGE
WEAR SUITABLE PROTECTIVE CLOTHING. SHUT OFF IGNITION SOURCES; NO FLARES, SMOKING, OR FLAMES IN AREA. STOP LEAK IF YOU CAN DO SO WITHOUT RISK. USE WATER SPRAY TO REDUCE VAPORS. TAKE UP WITH SAND OR OTHER NON-COMBUSTIBLE ABSORBENT MATERIAL AND PLACE INTO CONTAINER FOR LATER DISPOSAL. FLUSH AREA WITH WATER.

J. T. BAKER SOLUSORB(R) SOLVENT ADSORBENT IS RECOMMENDED FOR SPILLS OF THIS PRODUCT.

DISPOSAL PROCEDURE
DISPOSE IN ACCORDANCE WITH ALL APPLICABLE FEDERAL, STATE, AND LOCAL ENVIRONMENTAL REGULATIONS.

EPA HAZARDOUS WASTE NUMBER: U002 (TOXIC WASTE)

SECTION VIII – PROTECTIVE EQUIPMENT

VENTILATION: USE GENERAL OR LOCAL EXHAUST VENTILATION TO MEET TLV REQUIREMENTS.

RESPIRATORY PROTECTION: RESPIRATORY PROTECTION REQUIRED IF AIRBORNE CONCENTRATION EXCEEDS TLV. AT CONCENTRATIONS UP TO 5000 PPM, A GAS MASK WITH ORGANIC VAPOR CANNISTER IS RECOMMENDED. ABOVE THIS LEVEL, A SELF-CONTAINED BREATHING APPARATUS WITH FULL FACE SHIELD IS ADVISED.

EYE/SKIN PROTECTION: SAFETY GLASSES WITH SIDESHIELDS, POLYVINYL ACETATE GLOVES ARE RECOMMENDED.

SECTION IX – STORAGE AND HANDLING PRECAUTIONS

SAF-T-DATA(TM) STORAGE COLOR CODE: RED

SPECIAL PRECAUTIONS
BOND AND GROUND CONTAINERS WHEN TRANSFERRING LIQUID. KEEP CONTAINER TIGHTLY CLOSED. STORE IN A COOL, DRY, WELL-VENTILATED, FLAMMABLE LIQUID STORAGE AREA.

SECTION X – TRANSPORTATION DATA AND ADDITIONAL INFORMATION

DOMESTIC (D.O.T.)

PROPER SHIPPING NAME ACETONE
HAZARD CLASS FLAMMABLE LIQUID
UN/NA UN1090
LABELS FLAMMABLE LIQUID

INTERNATIONAL (I.M.O.)

PROPER SHIPPING NAME ACETONE
HAZARD CLASS 3.1
UN/NA UN1090
LABELS FLAMMABLE LIQUID

(TM) AND (R) DESIGNATE TRADEMARKS.
N/A = NOT APPLICABLE OR NOT AVAILABLE

THE INFORMATION PUBLISHED IN THIS MATERIAL SAFETY DATA SHEET HAS BEEN COMPILED FROM OUR EXPERIENCE AND DATA PRESENTED IN VARIOUS TECHNICAL PUBLICATIONS. IT IS THE USER'S RESPONSIBILITY TO DETERMINE THE SUITABILITY OF THIS INFORMATION FOR THE ADOPTION OF NECESSARY SAFETY PRECAUTIONS. WE RESERVE THE RIGHT TO REVISE MATERIAL SAFETY DATA SHEETS PERIODICALLY AS NEW INFORMATION BECOMES AVAILABLE.

Health hazards are described in Section V. The entries of most significance for evaluating risks from vapors are the Threshold Limit Value (or TLV) and the Short-Term Exposure Limit (STEL). The TLV is a term used by the American Conference of Governmental Industrial Hygienists (ACGIH). This organization examines the toxicity literature for a compound and establishes the TLV. This standard is designed to protect the health of workers exposed to the vapor 8 hours a day, five days a week. The Occupational Safety and Health Administration (OSHA) adopts a value to protect the safety of workplaces in the United States. Their value is termed the Time-Weighted Average (TWA) and in many cases is numerically equal to the TLV. The STEL is a value not to be exceeded for even a 15-minute averaging time. TLV, TWA, and STEL values for many chemicals are summarized in a small handbook available from the ACGIH (2000); they are also collected in the *CRC Handbook of Chemistry and Physics*.

The toxicity of acetone is also described in terms of the toxic oral dose. In this case, the $LD_{50}$ is the dose that will cause the death of 50% of the mice or rats given that dose. The dose is expressed as milligrams of chemical per kilogram of body weight of the subject animal. The figures for small animals are often used to estimate the effects on humans. If, for example, we used the mouse figure of 1297 mg/kg and applied it to a 60-kg chemist, a dose of 77,820 mg (~98.5 mL) would kill 50% of the subjects receiving that dose. As a further example, chloroform has an $LD_{50}$ of 80 mg/kg. For our 60-kg chemist, a dose of 4800 mg (~3 mL) would be fatal for 50% of these cases. The effects of exposure of skin to the liquid and vapor are also described.

Section VI describes the reactivity of acetone and the classes of compounds with which it should not come in contact. For example, sodium metal reacts violently with a number of substances (including water) and should not come in contact with them. Strong oxidizing agents (such as nitric acid) should not be mixed with organic compounds (among other things). The final sections (Sections VII–X) are self-explanatory.

## ALTERNATE SOURCES OF INFORMATION

Similar information in a more compact form can be found in the *Merck Index* (Merck). This basic reference work provides information on the toxicity of many chemicals. It often refers one to the *NIOSH Pocket Guide to Chemical Hazards* (National Institute for Occupational Safety and Health). The *Merck Index* also supplies interesting information about the common uses of the chemicals listed, particularly related to the medical area. References to the chemical literature are also provided. The *CRC Handbook of Chemistry and Physics*, which is updated each year, contains a wide range of data (located in tables) in the area of health, safety, and environmental protection. It also includes directions for the handling and disposal of laboratory chemicals. Your laboratory should have a copy of this work. Most chemical supply houses now label their containers with data showing not only the usual package size, physical properties, and chemical formula, but also pictures or codes showing hazard information. Some include a pictogram (for example, see the newer Aldrich Chemical labels on their bottles). The J. T. Baker Company uses the Baker SAF-T-DATA System.

## ESTIMATING RISKS FROM VAPORS

Other things (availability, suitability) being equal, one would, of course, choose the least toxic chemical for a given reaction. Some very toxic chemicals play very important roles in synthetic organic chemistry, and the toxicity of the chemicals in common use varies greatly. Bromine and benzene have TLVs of 0.7 and 30 mg/m$^3$, respectively, and are at the more toxic end of the spectrum of chemicals routinely used. Acetone has a TLV of 1780 mg/m$^3$. These representative figures do not mean that acetone is harmless or that bromine cannot be used. In general, one should exercise care at all times (make a habit of good laboratory practice) and should take special precautions when working with highly toxic materials.

The TLV provides a simple means to evaluate the relative risk of exposure to the vapor of any substance used in the laboratory. If the quantity of the material evaporated is represented by $m$ (in milligrams/hour) and the TLV is expressed by $L$ (milligrams per cubic meter), a measure of relative risk to the vapor is given by $m/L$. This quantity represents the volume of clean air required to dilute the emissions to the TLV. As an example, the emission of 1 g of bromine and 10 g of acetone in one hour leads to the values of $m/L$ of 1400 m$^3$/hour (h) for the bromine and 5.6 m$^3$/h for acetone. These numbers provide a direct handle on the *relative* risks from these two vapors. It is difficult to assess the absolute risk to these vapors without a lot of information about the ventilation characteristics of the laboratory. If these releases occur within a properly operated hood, the threat to the worker in the laboratory is probably very small. (However, consideration must be given to the hood exhaust.)

Exposure in the general laboratory environment can be assessed if we assume that the emissions are reasonably well mixed before they are inhaled and if we know something about the room ventilation rate. The ventilation rate of the room can be measured by a number of ways.[3] Given the ventilation rate, it might be safe to assume that only 30% of that air is available for diluting the emissions. (This accounts for imperfect mixing in the room.) The effective amount of air available for dilution can then be compared with the amount of air required to dilute the chemical to the TLV.

Let us continue our example. Suppose that the laboratory has a volume of 75 m$^3$ and an air exchange rate of 2 air changes per hour. This value means that $(75\ \text{m}^3)(2/\text{h})(0.3) = 45\ \text{m}^3/\text{h}$ are available to dilute the pollutants. There may be enough margin for error to reduce the acetone concentration to a low level (5.6 m$^3$/h is required to reach the TLV), but use of bromine should be restricted to the hood. An assessment of the accumulative risk of several chemicals is obtained by adding the individual $m/L$ $(\frac{\text{mg/h}}{\text{mg/m}^3})$ values.

The $m/L$ figures may also be used to assess the relative risk of performing the experiment outside a hood. Since $m/L$ represents the volume of air for each student, this may be compared with the volume of air actually available for each student. If the ventilation rate for the entire laboratory is $Q$ (in cubic meters per minute) for a section of $n$ students meeting for $t$ minutes, the volume for each student is $kQt/n$ cubic meters. Here $k$ is a mixing factor that allows for the fact that the ventilation air will not be perfectly mixed in the laboratory before it is exhausted. In a reasonable worst-case mixing situation a $k$ value of 0.3 seems reasonable. Laboratories with modest ventilation rates supplied by

[3]Butcher, S. S.; Mayo, D. W.; Hebert, S. M.; Pike, R. M.,"Laboratory Air Quality, Part I"; *J. Chem. Educ.* **1985,** *62,* A238; and"Laboratory Air Quality, Part II"; *J. Chem. Educ.* **1985,** *62,* A261.

15–20 linear feet of hoods can be expected to provide 30–100 $m^3$ per student over a 3-h laboratory period if the hoods are working properly. Let us take the figure of 50 $m^3$ per student as an illustration. If the value of *m/L* for a compound (or a group of compounds in a reaction) is substantially less than 50 $m^3$, it may be safe to do that series of operations in the open laboratory. If *m/L* is comparable to or greater than 50 $m^3$, a number of options are available: (1) Steps using that compound may be restricted to a hood. (2) The instructional staff may satisfy themselves that much less than the assumed value is actually evaporated under conditions present in the laboratory. (3) The number of individual repetitions of this experiment may be reduced. The size of the laboratory section can be reduced or the experiment may be done in pairs or trios.

Conducting reactions in a hood does not automatically convey a stamp of safety. Hoods are designed to keep evaporating chemicals from entering the general laboratory space. For hoods to do their job, there must be an adequate flow of air into the hood, and this air flow must not be disturbed by turbulence at the hood face. A frequently used figure of merit for hood operation is the face velocity of 100 ft/min. This is an average velocity of air entering the hood opening. (In the event that your lab does not have monitoring systems already housed within the hoods, instruments for measuring flow rate are available and can be purchased from most major equipment suppliers.) Even with a face velocity of 100 ft/min, vapors can be drawn out of an improperly designed hood simply by people walking by the opening, or by drafts from open windows.

Hood performance should be checked at regular intervals. The face velocity will increase as the front hood opening is decreased. If an adequate face velocity cannot be maintained with a front opening height of 15 cm, use of the hood for carrying out reactions will be limited. A low face velocity may indicate that the fans and ductwork need cleaning, that the exhaust system leaks (if it operates under lower than ambient pressure), or that the supply of makeup air is not adequate. When the hood system is properly maintained, the height of the hood opening required to provide an adequate face velocity is often indicated with a sticker.

Hoods are often used for storage of volatile compounds. A danger in this practice is that the hood space can become quickly cluttered, making work in the hood difficult, and the air flow may be disturbed. Of course, hoods being used for storage must never be turned off.

## MICROWAVE SAFETY*

Scientific microwave apparatus is designed for preparative chemistry and is built with safety in mind. *Domestic (household) microwave ovens should not be used for preparative chemistry.* When employing microwave heating, all the safety precautions that are taken when performing a reaction using conventional heating should be adhered to, particularly the fact that reaction vessels can be hot and, when sealed, residual pressure needs to be released carefully at the end of the reaction. There are also some safety precautions that are specific to reactions using microwave heating:

- Adherence to the microwave manufacturer's user manual and guidelines is essential.

*This section has been written by Dr. Nicholas E. Leadbeater from the Department of Chemistry at the University of Connecticut, and Dr. Cynthia B. McGowan from the Department of Chemistry at Merrimack College, MA.

- Before running reactions in sealed vessels, it is prudent to check the reaction vessels for cracks or any other signs of damage prior to use.
- Only fill the reaction vessels to manufacturer's specifications; do not overfill the reaction vessels. An approximate "rule of thumb" is to fill the vessel to no further than half its capacity.
- Only seal a closed reaction vessel with the manufacturer's recommended cap. These caps are designed to vent and reseal in the case of an overpressurization during a reaction.
- If the cap is a twist-on type, be sure to use the appropriate tool to tighten the cap to the manufacturer-specified torque.
- It is important to monitor temperature and pressure profiles during the course of a reaction and to set safety limits before starting. A reaction mixture can be heated gradually to the set temperature or, in some cases, chemists prefer to heat up the contents as rapidly as possible. In the case of the latter, care needs to be taken to ensure that the temperature and pressure do not rise uncontrollably. As a result, it is best to use a low initial microwave power.
- Before performing a reaction at elevated temperatures, chemists should consider carefully the stability of the reagents and solvents they use at these temperatures.
- It is important to ensure efficient stirring, especially when using heterogeneous reaction mixtures and metal catalysts or reagents since localized heating can occur, resulting in some cases in melting of the vessel walls and, if under pressure, failure of the vessel.
- The stir bars used for agitation should not be of exactly 3 cm in length since this equates to ¼ wavelength of a microwave at 2450 MHz and thus acts as an antenna.
- Upon completion of a microwave run, the microwave unit will start a cooling process. In the case of the monomode microwave, the pressure sensor will release when the tube is cool enough to handle (50 °C). With multimode microwave units, the apparatus is set to cool for a period of time but, at the end of this, the reaction vessels may still be hot. Check the temperature before removing the reaction vessels.
- When opening sealed vessels at the end of a reaction, be sure to point the vessel away from your face and any other person, preferably doing so in a hood. Any remaining pressure will release as soon as the cap is removed. If the tube is very warm, cool it in an ice bath before removing the cap.

## CONCLUDING THOUGHTS

This brief chapter touches only a few of the important points concerning laboratory safety. The risk from vapor exposure is discussed in some detail, but other risks are treated briefly. Applications in some laboratories may involve reactions with a risk from radiation or infection or may involve compounds that are unstable with respect to explosion. The chemist must be aware of the potential risks and must be prepared to go to an appropriate and detailed source of information, as needed. The references cited here represent a small fraction of the safety data, texts, and journals available on this subject. It is highly recommended that the library and/or laboratory at your institution have at least this minimal selection. Of course, the selections should be kept up to date!

## QUESTIONS

**2-1.** After bookmarking a reputable MSDS URL, locate a chemical of your choice and print out the data. If the information is not available, go to your stock room and request a copy of the MSDS. Underline on the sheet the CAS No., solubility data, fire and explosion data, reactivity data, and what protective equipment is required when using this chemical. Does your laboratory meet the safety regulations required to use this chemical? Why or why not?

**2-2.** Think and describe what you would do in each of the following situations which could happen in your laboratory.

**(a)** You are working at your station and the 100-mL round-bottom flask in which you are running a reaction in ether solvent suddenly catches fire.

**(b)** The person working across the laboratory bench from you allows hydrogen chloride gas to escape from his or her apparatus.

**(c)** A reagent bottle is dropped, spilling concentrated sulfuric acid.

**(d)** A hot solution"bumps,"splashing your face.

**2-3.** You are working in the laboratory using 3.0 mL of benzene in an extraction procedure. An alternative to benzene is toluene. However, three times more toluene is required to perform the extraction. The isolation of the desired product from the extraction solution requires evaporation of the solvent (benzene or toluene). This takes 0.5 h to complete. Calculate the relative risks of using these two solvents. Which solvent would you use and why?

**2-4.** A laboratory has four hoods; each is 39 in. wide. When the hood door is open to a height of 8 in. and the hoods are operating, the average air velocity through the hood face is 170 ft/min.

**(a)** Evaluate the total ventilation rate for this room, assuming that there are no other exhausts.

**(b)** The laboratory is designed for use by 30 students. Evaluate the air available per student if the mixing factor is 0.3 and the experiments last for 3 h.

**(c)** An experiment is considered in which each student would be required to evaporate 7 mL of methylene chloride ($CH_2Cl_2$). Estimate the average concentration of methylene chloride. Look up the TLV or the TWA for methylene chloride and consider how the evaporation might be performed.

**2-5.** An experiment is considered in which 1 mL of diethylamine would be used by each student. The ventilation rate for the laboratory is 5 $m^3$/min. Look up the TLV (or TWA) for diethylamine, $(C_2H_5)_2NH$. What restrictions might be placed on the laboratory to keep the average concentration, over a 3-h period, less than one-third of the TWA? Assume a mixing factor of 0.3.

## GENERAL SAFETY REFERENCES


1. ACS Committee on Chemical Safety, *Safety in Academic Chemical Laboratories, Vol. 1: Accident Prevention for College and University Students; Vol. 2: Accident Prevention for Faculty and Administrators,* 7th ed.; American Chemical Society: Washington, DC, 2003.
2. *Handbook of Laboratory Safety,* 5th ed.; Furr, A. K. Jr., Ed.; CRC Press: Boca Raton, FL, 2000.
3. Committee on Prudent Practices for Handling, Storage, and Disposal of Chemicals in Laboratories, National Research Council, *Prudent Practices for Handling Chemicals in Laboratories;* National Academy Press: Washington, DC, 1995.
4. Armour, M. A. *Hazardous Laboratory Chemicals Disposal Guide,* 3rd ed.; CRC Press: Boca Raton, FL, 2003.
5. *Working Safely with Chemicals in the Laboratory,* 2nd ed.; Gorman, C. E., Ed.; Genium: Schenectady, NY, 1995.
6. Alaimo, R. J., Ed. *Handbook of Chemical Health and Safety,* 1st ed., Oxford University Press: New York, 2001.
7. *The Sigma–Aldrich Library of Regulatory and Safety Data;* Lenga, R. E.; Votoupal, K. L., Eds.; Aldrich Chemical Co., Milwaukee, WI, 1992.

8. Young, J. A., *Improving Safety in the Chemical Laboratory: A Practical Guide,* 2nd ed.; Wiley: New York, 1991.
9. American Chemical Society, *Less Is Better (Laboratory Chemical Management for Waste Reduction),* American Chemical Society: Washington, DC, 2008.
10. Verschueren, K., *Handbook of Environmental Data on Organic Chemicals,* 3rd ed.; Van Nostrand Reinhold: New York, NY, 1996.
11. Lewis, R. J., Sr., *Hazardous Chemicals Desk Reference,* 6th ed.: Wiley-Interscience: New York, 2008.
12. *NIOSH Pocket Guide to Chemical Hazards,* National Institute for Occupational Safety and Health, U. S. Government Publication Office: Washington, DC, 2008. A CD-ROM version can be obtained from the NIOSH Publications Office (http://wwwn.cdc.gov/pubs/niosh.aspx).
13. *OSHA Regulated Hazardous Substances,* Vols. I and II, Noyes Data Corp.: Park Ridge, NJ, 1990.
14. Lund, G.; Sansone, E. B."Safe Disposal of Highly Reactive Chemicals"; *J. Chem. Educ.* **1994,** *71,* 972.
15. It is also recommended that one refer to the numerous articles on safety that appear regularly in the *Journal of Chemical Education* (http://jchemed.chem.wisc.edu/) and *The Chemical Educator.*


## BIBLIOGRAPHY


ACGIH. *Threshold Limit Values and Biological Exposure Indices.* Available from ACGIH, Kemper Woods Center, 1330 Kemper Meadow Drive, Cincinnati, OH 45240.

*Aldrich, Catalog Handbook of Fine Chemicals,* 1001 W. St. Paul Ave., Milwaukee, WI, 2009–2010.

Anastas, P. T.; Farris, C.A., Eds., *"Benign by Design—Alternate Synthetic Design for Pollution Prevention,"* ACS Symposium Series 577, American Chemical Society; Washington, DC, 1994.

Anastas, P. T.; Williamson, T.C., Eds.,*"Green Chemistry—Designing Chemistry for the Environment,"* ACS Symposium Series 626, American Chemical Society: Washington, DC, 1996.

———, *Green Chemistry: Frontiers in Benign Chemical Synthesis and Processes,* Oxford University Press: New York, 1998.

Mayo, D. W.; Hebert, S. M.; Pike, R. M.,"Laboratory Air Quality, Part I"; *J. Chem. Educ.* **1985,** *62,* A238;"Laboratory Air Quality, Part II"; *J. Chem. Educ.* **1985,** *62,* A261.

Flinn Scientific Company, *Chemical Catalog/Reference Manual* (2009). Available from Flinn Scientific Co., P.O. Box 219, Batavia, IL 60510.

*Handbook of Chemistry and Physics,* 89th ed.; Lide, D. R., Ed.; CRC Press: Boca Raton, FL, 2008–2009.

*The Merck Index,* 14th ed.; Budavari, S., Ed.; Merck Research Laboratories Publications: White-house Station, NJ, 2008. For online access to five different platforms of the Merck Index, see: www.merckbooks.com/mindex/online/html

Mollinelli, R. P.; Reale, M. J.; Freudenthal, R. I., *Material Data Safety Sheets,* Hill & Gernett: Boca Raton, FL, 1992.

$C_3H_4$apter 3

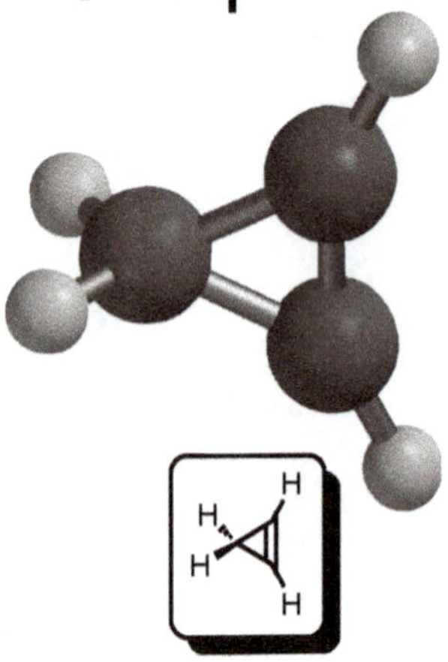

# INTRODUCTION TO MICROSCALE ORGANIC LABORATORY EQUIPMENT AND TECHNIQUES

We begin this chapter with a description of the standard pieces of glassware that are generally employed in a microscale laboratory. Modern standard taper glassware is particularly convenient to use and gives the student a sense of the flavor of the research laboratory. It is not essential, however, for the experimental work in an instructional laboratory, and many courses use glassware with alternative connectors. We describe the standard taper glassware as just one example of microscale equipment that is available. The operations carried out in the laboratory will be very similar or identical if, for example, a plastic connector is used to assemble the experimental setup. We next consider a series of standard experimental apparatus setups that use this equipment, and present a short discussion of the role that they play in the laboratory. We end the chapter with a set of laws that govern how one operates in a microscale laboratory (the rules are a bit different than those for a macroscale laboratory) and a set of guidelines for recording your experimental data. The basic individual pieces of equipment are shown in Figures 3.1 to 3.7.

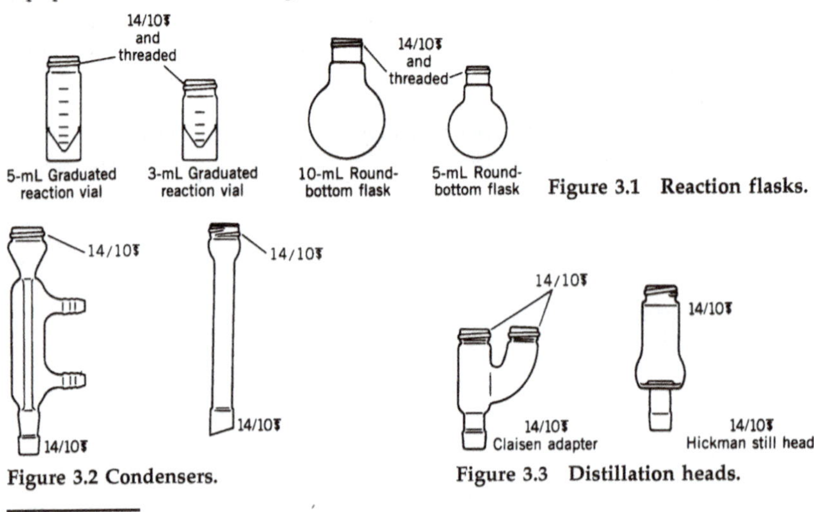


**Figure 3.1 Reaction flasks.**

**Figure 3.2 Condensers.**

**Figure 3.3 Distillation heads.**

---

**Chapter 3: $C_3H_4$, Cyclopropene**
Demyanov and Doyarenko (1922).

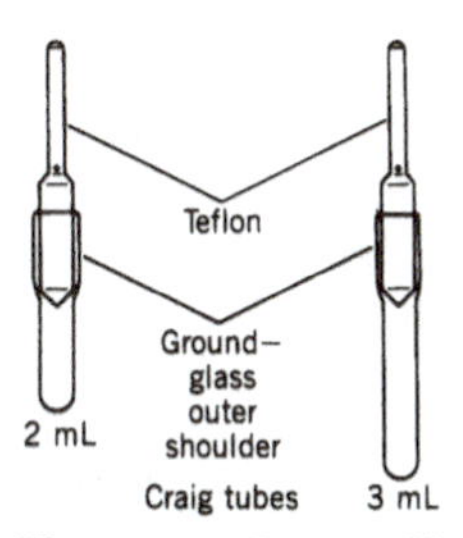

**Figure 3.4 Recrystallization tubes.**

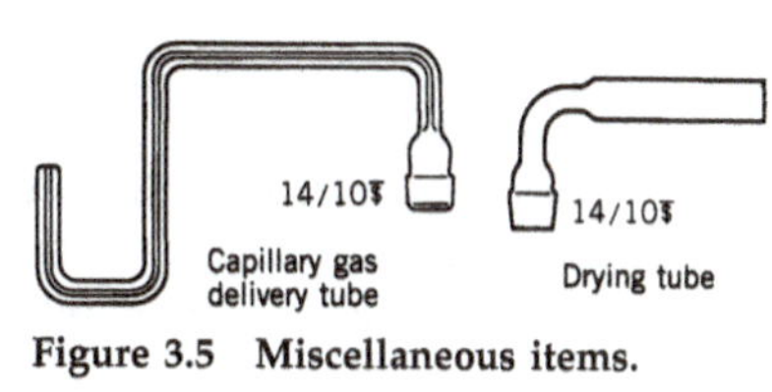


**Figure 3.5 Miscellaneous items.**

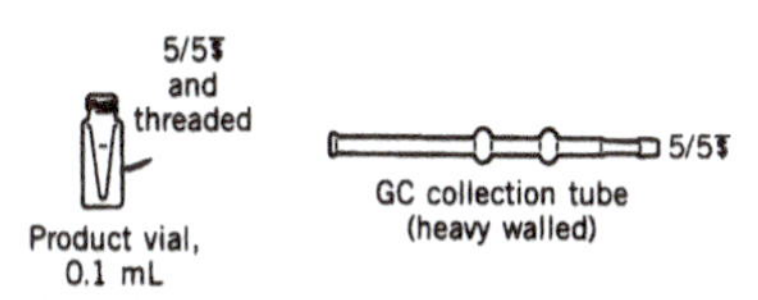


**Figure 3.6 Gas chromatographic fraction collection items.**

14/10₮

Hickman–Hinkle still head with side port

5/5₮ and threaded

14/10₮

**Figure 3.7 Hickman–Hinkle distillation column.**

## MICROGLASSWARE EQUIPMENT

### Standard Taper Joints

Standard taper ground-glass joints are the common mechanism for assembling all conventional research equipment in the organic laboratory. The symbol ₮ is commonly used to indicate the presence of this type of connector. Normally, ₮ is either followed or preceded by #/#. The first # refers to the maximum inside diameter of a female (outer) joint or the maximum outside diameter of a male (inner) joint, measured in millimeters. The second number corresponds to the total length of the ground surface of the joint (Fig. 3.8). The advantage of this type of connection is that if the joint surfaces are lightly greased, a vacuum seal is achieved. One of the drawbacks of using these joints is that contamination of the reacting system readily occurs if the solvents present in the reaction vessel dissolve the grease. In small-scale reactions this contamination can be particularly troublesome.

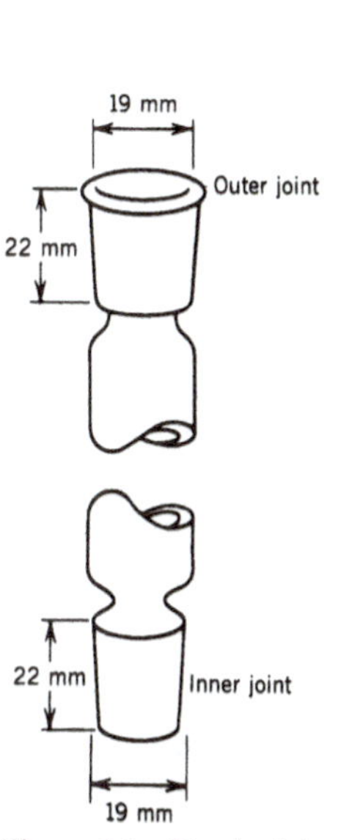


**Figure 3.8 Standard taper joints (₮).** *(From Zubrick, James W.* The Organic Chem Lab Survival Manual, *7th ed.; Wiley: New York, 2008. Reprinted by permission of John Wiley & Sons, Inc., New York.)*

The small joints used in the microscale experimental organic laboratory, however, have the ease of assembly and physical integrity of research-grade, standard taper, ground-glass joints along with a number of important additional features. The joint dimensions are usually ₮ 14/10. The conical vials in which most microscale reactions are carried out use this type of connecting system. Note that in addition to being ground to a standard taper on the inside surface of the throat of the vial, these vials also have a screw thread on the outside surface (Fig. 3.9).

This arrangement allows a standard taper male joint to be sealed to the reaction flask by a septum-type (open) plastic screw cap. The screw cap applies compression to a silicone rubber retaining O-ring positioned on the shoulder of the male joint (Fig. 3.10). The compression of the O-ring thereby achieves a greaseless gas-tight seal on the joint seam, while at the same time clamping the two pieces of equipment together. The ground joint provides both protection from intimate solvent contact with the O-ring and mechanical stability to the connection. The use of this type of connector leads to a further bonus during construction of an experimental setup. Because the individual sections are small, light, and firmly sealed together, the entire arrangement often can be mounted

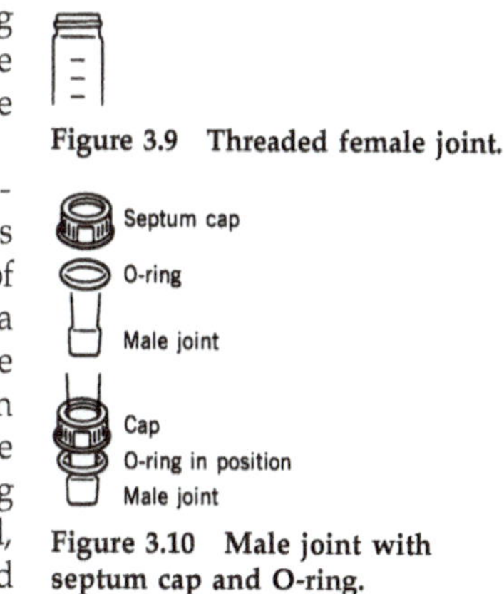


**Figure 3.9 Threaded female joint.**

**Figure 3.10 Male joint with septum cap and O-ring.**

on the support rack by a single clamp. In conventional systems it is often necessary to use at least two clamps. This can easily lead to strain in the glass components unless considerable care is taken in the assembly process. Clamp strain is one of the major sources of experimental glassware breakage. The ability to single-clamp most microscale setups effectively eliminates this problem.

*NOTE. When ground-glass joint surfaces are grease free it is important to disconnect joints soon after use (particularly with basic solutions) or they may become locked or "frozen" together.*

Joints of the size employed in these microscale experiments, however, are seldom a problem to separate if given proper care (*keep them clean!*).

### Conical Vials

Both the conical vials (3 and 5 mL) and the round-bottom flasks are designed to be connected via an O-ring compression cap installed on the male joint of the adjacent part of the system (see Fig. 3.1).

### Condensers

Two types of condensers (air condensers and water-jacketed condensers) are available; in most cases the water-jacketed condenser can work well as an air condenser. Condensers are usually attached to 14/10 ST-jointed reaction flasks. The upper female joints allow connection of the condenser to the 14/10 ST drying tube and the 14/10 ST capillary gas delivery tube (see Fig. 3.2).

### Distillation Heads

The simple Hickman still is used with an O-ring compression cap to carry out semi-micro simple or crude fractional distillations. The Hickman–Hinkle spinning band still uses a 3-cm fractionating column and routinely develops between five and six theoretical plates. The Hickman–Hinkle still is currently available with 14/10 ST joints and can be conveniently operated with the 14/10 ST 3- and 5-mL conical vials (see Figs. 3.1, 3.3, and 3.7). The still head is also available with an optional sidearm collection port.

### Recrystallization Tubes

Craig tubes are a particularly effective method for recrystallizing small quantities of reaction products. These tubes possess a nonuniform ground joint in the outer section. The substitution of Teflon for glass in the head makes these systems quite durable and much less susceptible to breakage during centrifugation (see Fig. 3.4).

### Miscellaneous Items

The Claisen head (see Fig. 3.3) is often used to facilitate the syringe addition of reagents to closed moisture-sensitive systems (such as Grignard reactions) via a septum seal in the vertical upper joint. This joint can also function to position the thermometer (using an adapter) in the well of a Hickman–Hinkle still (see Fig. 3.15). The Claisen adapter is also used to mount the drying tube in a protected position remote from the reaction chamber. The drying tube, in turn, is used to protect moisture-sensitive reaction components from atmospheric water vapor, while allowing a reacting system to remain unsealed. The capillary gas delivery tube is employed in transferring gases formed during reactions to storage containers (see Fig. 3.5 and Chapter 3W, Fig. 3.11W).

### Gas Chromatographic Fraction Collection Items

For fraction collection the gas chromatographic (GC) collection tube is connected directly to the exit port of the GC detector through a stainless steel standard taper adapter. The collected sample is then transferred to a 0.1-mL conical vial for storage. The system is conveniently employed in the resolution and isolation of two-component mixtures (see Fig. 3.6).

## STANDARD EXPERIMENTAL APPARATUS

### Heating and Stirring Arrangements

It is important to be able to carry out microscale experiments at accurately determined temperatures. Very often, transformations are successful, in part, because of the ability to maintain precise temperature control. In addition, many reactions require reactants to be intimately mixed to obtain a substantial yield of product. Therefore, the majority of the reactions you perform in this laboratory will be conducted with rapid stirring of the reaction mixture.

### Sand Bath Technique—Hot Plate Calibration

A most convenient piece of equipment for heating or stirring or for performing both operations simultaneously on a microscale level is the hot-plate–magnetic stirrer. Heat transfer from the hot surface to the reaction flask is generally accomplished with a crystallizing dish containing a *shallow* layer of sand that can conform to the size and shape of the particular vessel employed. The temperature (external) of the system is monitored by a thermometer embedded in the sand near the reaction vessel.

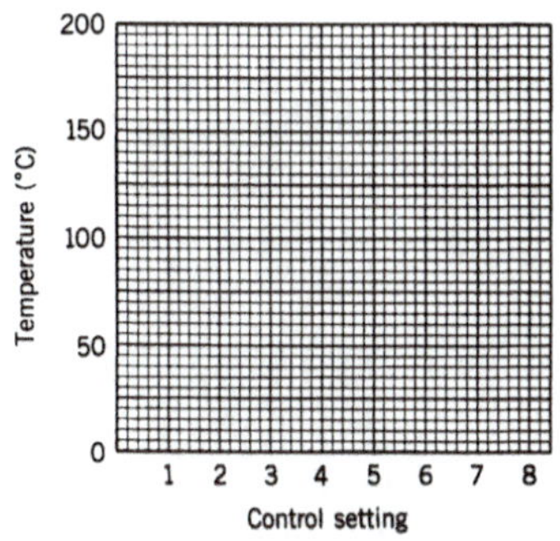


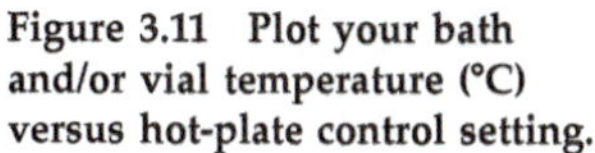
**Figure 3.11 Plot your bath and/or vial temperature (°C) versus hot-plate control setting.**

A successful procedure for determining the temperature inside the vial relative to the bath temperature is to mount a second thermometer in a vial containing 2 mL of high-boiling silicone oil. The vial temperature is then measured at various sand-bath temperatures and the values are entered on graphs of vial temperatures versus hot-plate settings and bath temperatures versus hot-plate settings (see Fig. 3.11 and Chapter 3W, Fig. 3.5W) for your particular hot-plate system (see also section on Metal Heat-Transfer Devices, p. 22). These data will save considerable time when you bring a reaction system to operating temperature. When you first enter the laboratory, it is advisable to adjust the temperature setting on the hot-plate stirrer with the heating device, or bath, in place. The setting is determined from your control setting–temperature calibration curve. This procedure will allow the heated bath to reach a relatively constant temperature by the time it is required. You will then be able to make small final adjustments more quickly, if necessary.

*NOTE. Heavy layers of sand act as an insulator on the hot-plate surface, which can damage the heating element at high temperature settings. When temperatures over 150 °C are required, it is especially important to use the minimum amount of sand.*

Recording the weight of sand used and the size of the crystallizing dish will help to make the graph values more reproducible.

The high sides of the crystallizing dish protect the apparatus from air drafts, and so the dish also operates somewhat as a hot-air bath. Heating can be made even more uniform by covering the crystallizing dish with aluminum foil (see Fig. 3.12 and Chapter 3W, Fig. 3.1W). This procedure works well, but is a bit awkward and is required in only a few instances.

www

The insulating properties of sand provide a readily available variable heat source because the temperature of the sand is higher deeper in the bath; thus, the depth of sand used in the bath is exceedingly important. **The depth should always be kept to a minimum, in the range of 10–15 mm.** Finally, sand baths offer a significant safety advantage over oil baths. Individual grains of sand are so small that they have little heat capacity and thus are less likely to burn the chemist in the event of a spill.

### Metal Heat-Transfer Devices

An alternative to the sand bath is a heat-transfer system that employs copper tube plates or aluminum metal blocks drilled to accommodate the different reaction vials and flasks (Chapter 3W, Fig. 3.3W).

www

### Stirring

Stirring the reaction mixture in a conical vial is carried out with Teflon-coated magnetic spin vanes, and in round-bottom flasks with Teflon-coated magnetic stirring bars (see Fig. 3.12 and Chapter 3W, Fig. 3.1W). It is important to put the reaction flask as close to the center and to the bottom surface of the crystallizing dish as possible when using magnetic stirring. This arrangement is a good practice in general, as it leads to using the minimum amount of sand needed in a sand bath.

www

If the reaction does not require elevated temperatures, but needs only to be stirred, the system can be assembled without the heat-transfer device (sand bath or metal plate). Some stirred reactions, on the other hand, require cooling.

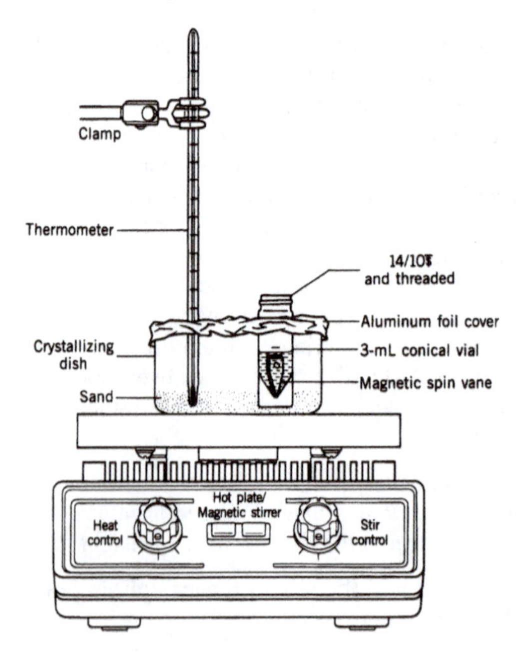


**Figure 3.12 Hot-plate–magnetic stirrer with sand bath and reaction vial.**

In these cases a crystallizing dish filled with ice water, or with ice water and salt, if lower temperatures are called for, will provide the correct environment.

## Reflux Apparatus

To bring about a successful reaction between two substances, it is often necessary to mix the materials together intimately and to maintain a specific temperature. The mixing operation is conveniently achieved by dissolving the materials in a solvent in which they are mutually soluble. If the reaction is carried out in solution under reflux conditions, the choice of solvent can be used to control the temperature of the reaction. Many organic reactions involve the use of a reflux apparatus in one arrangement or another.

What do we mean by *reflux?* The term means to "return," or "run back." This return is exactly how the reflux apparatus functions. When the temperature of the reaction system is raised to the solvent's boiling point (constant temperature), all vapors are condensed and returned to the reaction flask or vial; this operation is not a distillation and the liquid phase remains at a stable maximum temperature. In microscale reactions, two basic types of reflux condensers are utilized: the air-cooled condenser, or air condenser (Chapter 3W, Fig. 3.6W), and the water-jacketed condenser (see Fig. 3.13 and Chapter 3W, Fig. 3.7W). The air condenser condenses solvent vapors on the cool vertical wall of an extended glass tube that dissipates the heat by contact with the laboratory room air. This arrangement functions quite effectively with liquids boiling above 150 °C. Indeed, a simple test tube can act as a reaction chamber and air condenser all in one unit, and many simple reactions can be most easily carried out in test tubes.

WWW

WWW

Air condensers can occasionally be used with lower boiling systems; however, the water-jacketed condenser is more often employed in these situations. The water-jacketed condenser employs flowing cold water to remove heat from the vertical column and thus facilitate vapor condensation. It is highly effective at condensing vapor from low-boiling liquids.

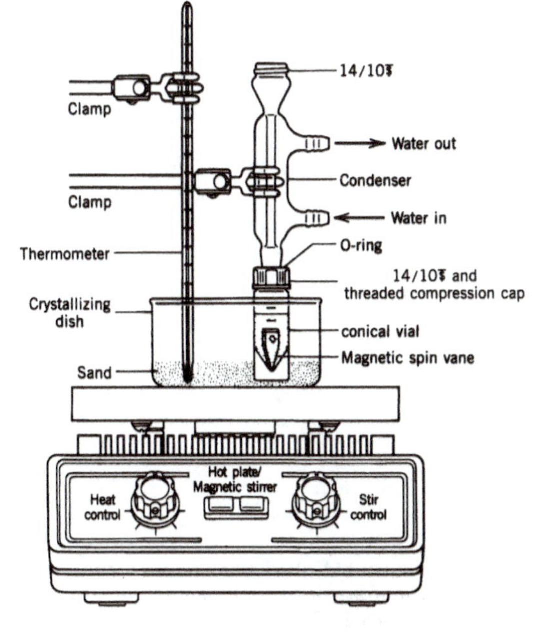


**Figure 3.13 Water-jacketed condenser with conical vial, arranged for heating and magnetic stirring.**

Both styles of condensers accommodate various sizes of reaction flasks and are available with 14/10 $\mathsf{\$}$ standard taper joints. The tops of both condenser columns have a female 14/10 $\mathsf{\$}$ joint.

In refluxing systems that do not require significant mixing or agitation, the stirrer (magnetic spin vane or bar) usually is replaced by a "boiling stone." These sharp-edged stones possess highly fractured surfaces that are very efficient at initiating bubble formation as the reacting medium approaches the boiling point. The boiling stone acts to protect the system from disastrous boilovers and also reduces "bumping."(Boiling stones should be used only once and must **never** be added to a hot solution. In the first case, the vapor cavities become filled with liquid upon cooling, and thus a boiling stone becomes less effective after its first use. In the second case, **adding the boiling stone to the hot solution may suddenly start an uncontrollable boilover**).

## Distillation Apparatus

Distillation is a laboratory operation used to separate substances that have different boiling points. The mixture is heated, vaporized, and then condensed; the early fractions of condensate are enriched in the more volatile component. Unlike the reflux operation, in distillations none, or only a portion, of the condensate is returned to the flask where vaporization is taking place. Many distillation apparatus have been designed to carry out this basic operation. They differ mainly in small features that are used to solve particular types of separation problems. In several of the microscale experiments contained in Chapters 6, 7, and 10W *semi-microscale* distillations are required. In carrying out these distillations the choice of still depends to a large degree on the difficulty of the separation required (generally, how close are the boiling points in the mixture to be separated?).

WWW

The Hickman still head (Fig. 3.14) is ideally suited for simple distillations. This system has a 14/10 $\mathsf{\$}$ male joint for connection to conical vials or round-bottom

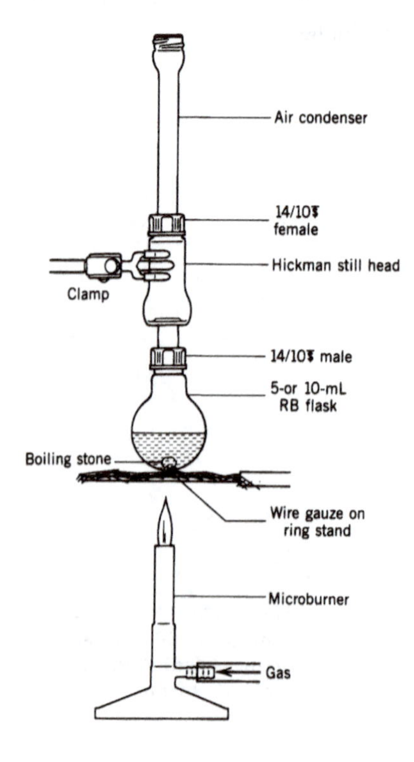


**Figure 3.14 Hickman still head and air condenser with 5-mL round-bottom flask, arranged for microburner heating.**

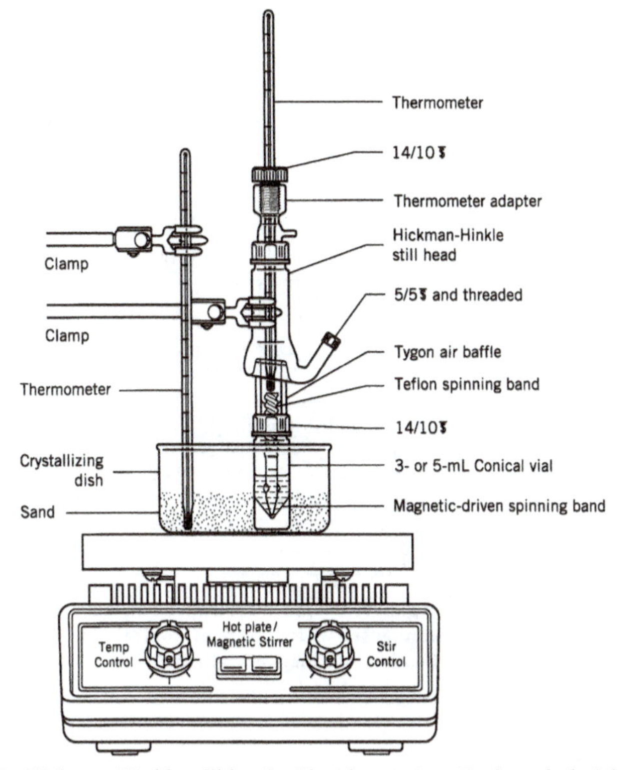


**Figure 3.15 Hickman–Hinkle still head with side-port 3- or 5-mL conical vial, Teflon spinning band, and thermometer adapter and arranged for heating and magnetic stirring.**

flasks. The still head functions as both an air condenser and a condensate trap. For a detailed discussion of this piece of equipment see Experiments [3A] and [3B]. The simple Hickman still has been modified (see Fig. 3.15) with a spinning band. The still continues to function in much the same way as the simple Hickman still, but a tiny Teflon spinning band is now mounted in a slightly extended section between the male joint and the collection collar. When the band is spun at 1500 rpm by a magnetic-stirring hot plate, this still functions as an effective short-path fractional distillation column. In addition, this modified system has a built-in thermometer well that allows reasonably accurate measurement of vapor temperatures plus a sidearm port for removing distillate.

The most powerful microscale distillation system currently available is the 2.5-in. vacuum-jacketed microscale spinning-band distillation column (see Fig. 3.16 for description and details). This still is designed for conventional downward distillate collection and nonstopcock reflux control. The column is rated at ~10 theoretical plates.

For a discussion of reduced pressure distillations see Distillation. WWW

## Moisture-Protected Reaction Apparatus

Many organic reagents react rapidly and preferentially with water. *The success or failure of many experiments depends to a large degree on how well atmospheric moisture is excluded from the reaction system.* The "drying tube," which is packed with a desiccant such as anhydrous calcium chloride, is a handy way to carry out a reaction in an apparatus that is not totally closed to the atmosphere, but

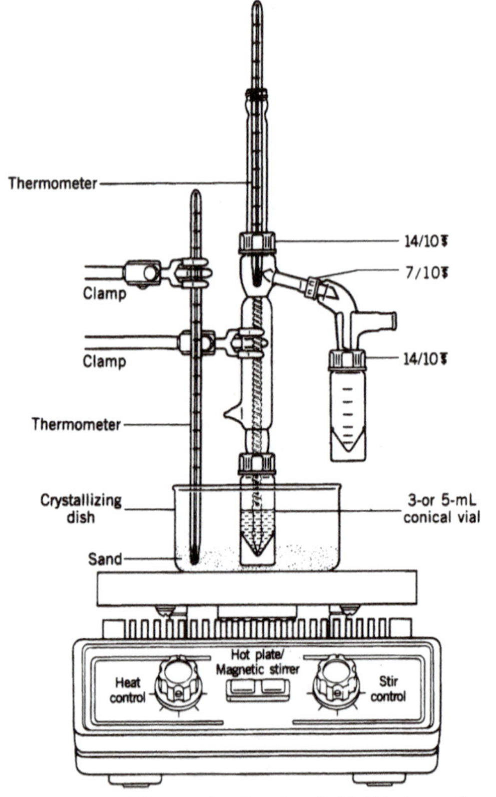


**Figure 3.16 Micro spinning band distillation column (2.5 in).**

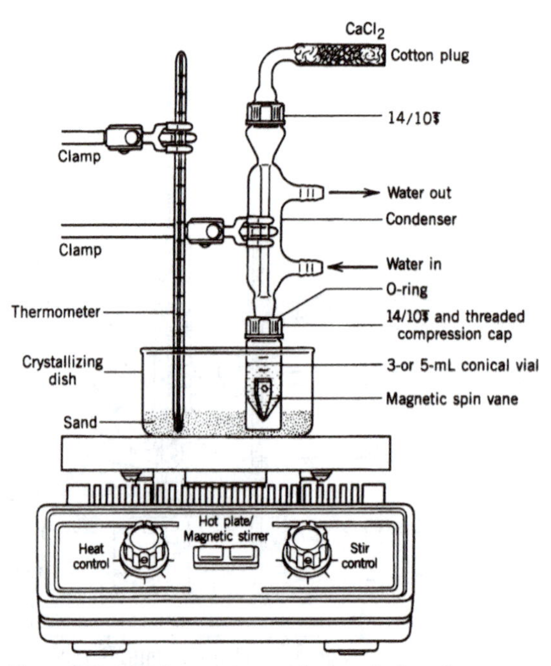


**Figure 3.17 Moisture-protected water-jacketed condenser with 3- or 5-mL conical vial, arranged for heating and magnetic stirring.**

that is reasonably well protected from water vapor. The microscale apparatus described here are designed to be used with the 14/10 $\overline{\mathsf{S}}$ drying tube. The reflux condensers discussed earlier are constructed with female 14/10 $\overline{\mathsf{S}}$ joints at the top of the column, which allows convenient connection of the drying tube if the refluxing system is moisture sensitive (see Fig. 3.17).

Because many reactions are highly sensitive to moisture, successful operation at the microscale level can be rather challenging. If anhydrous reagents are to be added after an apparatus has been dried and assembled, it is important to be able to introduce these reagents without exposing the system to the atmosphere, particularly when operating in a humid atmosphere. In room-temperature reactions that do not need refluxing, adding anhydrous reagents is best accomplished by use of the microscale Claisen head adapter. The adapter has a vertical screw-threaded standard taper joint that will accept a septum cap. The septum seal allows syringe addition of reagents and avoids the necessity of opening the apparatus to the laboratory atmosphere (see Fig. 3.18).

## Specialized Pieces of Equipment

***Collection of Gaseous Products.*** Some experiments lead to gaseous products. The collection, or trapping, of gases is conveniently carried out by using the capillary gas delivery tube. This item is designed to be attached directly to a 1- or 3-mL conical vial (see Fig. 3.19), or to the female 14/10 $\overline{\mathsf{S}}$ joint of a condenser connected to a reaction flask or vial (Chapter 3W, Fig. 3.11W). The tube leads to the collection system, which may be a simple, inverted, graduated cylinder; a blank-threaded septum joint; or an air condenser filled with water

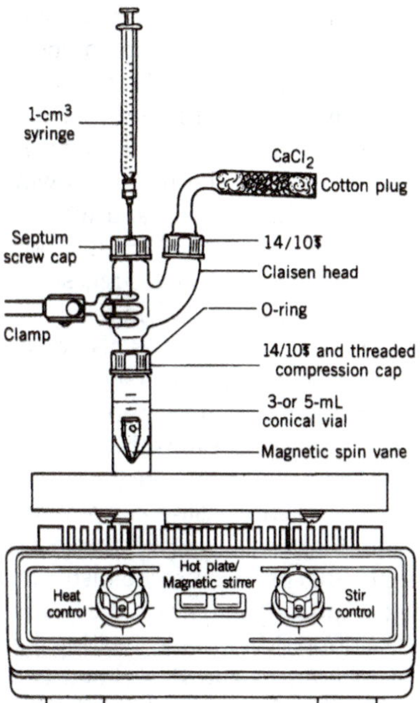


**Figure 3.18 Moisture-protected Claisen head with 3- or 5-mL conical vial, arranged for syringe addition and magnetic stirring.**

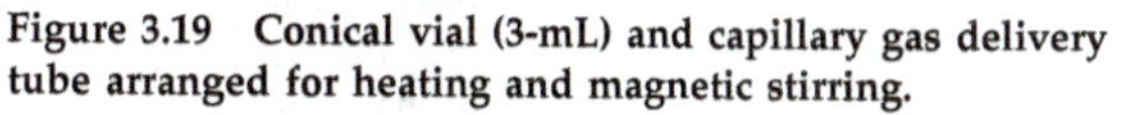


**Figure 3.19 Conical vial (3-mL) and capillary gas delivery tube arranged for heating and magnetic stirring.**

(if the gaseous products are not water-soluble). The 0.1-mm capillary bore considerably reduces dead volume and increases the efficiency of product transfer.

***Collection of Gas Chromatographic Effluents.*** The trapping and collection of gas chromatographic liquid fractions become particularly important exercises in microscale experiments. When yields of a liquid product are less than 100 μL conventional distillation, even using microscale equipment, is impractical. In this case, preparative gas chromatography replaces conventional distillation as the route of choice to product purification. A number of the reaction products in Chapters 6, 7, and 10W depend on this approach for successful purification and isolation. The ease and efficiency of carrying out this operation is greatly facilitated by employing the 5/5 ₮ collection tube and the 0.1-mL 5/5 ₮ conical collection vial ( Chapter 3W, Fig. 3.12W).

WWW

WWW

## MICROWAVE HEATING AS A TOOL FOR ORGANIC CHEMISTRY*

### Introduction

M

An appliance found in almost all homes is a microwave oven. It is possible to heat food much more quickly and easily using a microwave as compared to the stove top. The observation that microwave energy can be used to heat food

*This section has been written by Dr. Nicholas E. Leadbeater from the Department of Chemistry at the University of Connecticut, and Dr. Cynthia B. McGowan from the Department of Chemistry at Merrimack College, MA.

was first made by Percy Spencer, an employee of the Raytheon Corporation.[1] His company was a manufacturer of radar sets and, while working on one he noticed that the candy bar he had in his pocket had melted. Intrigued by this, the next day he brought in some popcorn from home and found that if he placed this near his radar set, it popped. In 1945, Raytheon filed a patent for the microwave cooking process, and in 1947 they built the first microwave oven called the Radarange.[2] It was almost 6 feet (1.8 m) high and weighed over 750 pounds (340 kg) and cost between $2,000 and $3,000.[3] The first popular home model was launched in 1967, and current estimates suggest that over 200 million microwave ovens are in use throughout the world today.[4]

Just as microwave ovens prove so valuable in the kitchen, it is also possible to use similar technology in preparative chemistry. It was in 1986 that the first reports of microwave heating as a tool for organic chemistry appeared in the scientific literature.[5,6] Two research groups published results they had obtained in their laboratories using kitchen microwave ovens. They said that chemistry that usually takes hours to reach completion using conventional heating could be performed in a matter of minutes in a microwave oven. Since these first reports, the use of microwave heating in organic chemistry has grown rapidly. Today, the technology is used in industry and academic laboratories for performing a wide range of reactions. Microwave heating has opened up a range of new areas in organic chemistry, allowing chemists to perform reactions quickly and easily. As an example, in this book Experiment 7, the Cannizzaro reaction, is performed in *one hour* using conventional heating. In the microwave protocol, the reaction is complete *in just one minute*.

The use of microwave heating addresses a number of the green chemistry principles.[7] Since it is often possible to obtain higher yields using microwave heating as opposed to conventional heating, there will be less waste and unused reagents. Also, since microwave heating is fast, there is often not enough time for products to decompose so this makes the product purification cleaner and easier. Chemists have also used the inherent advantages of microwave heating to their advantage for developing cleaner alternatives to known reactions. Take, for example, the use of water as a solvent instead of organics such as dichloromethane and benzene. Work has shown that water is an excellent solvent for organic chemistry, especially when combined with microwave heating. It is possible to heat water well above its boiling point in a sealed reaction vessel very safely and efficiently using microwaves. At these higher temperatures, water behaves more like an organic solvent. While most organic compounds are not soluble in water at room temperature, they can be in this higher temperature water, or at least partly so. This means that reactions can take place and, when the mixture cools down at the end, the product crystallizes out and is easily removed. As well as allowing for a more environmentally friendly solvent to be used, it also makes purification easy.

---

[1] *Reader's Digest*, August 1958, page 114.
[2] Spencer, P. L. 1945. Method of treating foodstuffs. US Patent 2,495,429, filed October 5, 1945, and published January 24, 1950.
[3] Gallawa, J. C. *Complete Microwave Oven Service Handbook: Operation, Maintenance, Troubleshooting, and Repair.* Prentice Hall, 2000.
[4] US Bureau of Labor Statistics.
[5] Gedye, R.; Smith, F.; Westaway, K.; Ali, H.; Baldisera, L.; Laberge, L.; and Rousell, J. "The Use of Microwave Ovens for Rapid Organic Synthesis," *Tetrahedron Lett.* **1986,** *27,* 279.
[6] Giguere, R. J.; Bray, T. L.; Duncan, S. M.; and Majetich, G. "Application of Commercial Microwave Ovens to Organic Sysnthesis," *Tetrahedron Lett.* **1986,** *27,* 4945.
[7] Anastas, P. T.; Warner, J. C. *Green Chemistry: Theory and Practice*; Oxford University Press, New York, 1998.

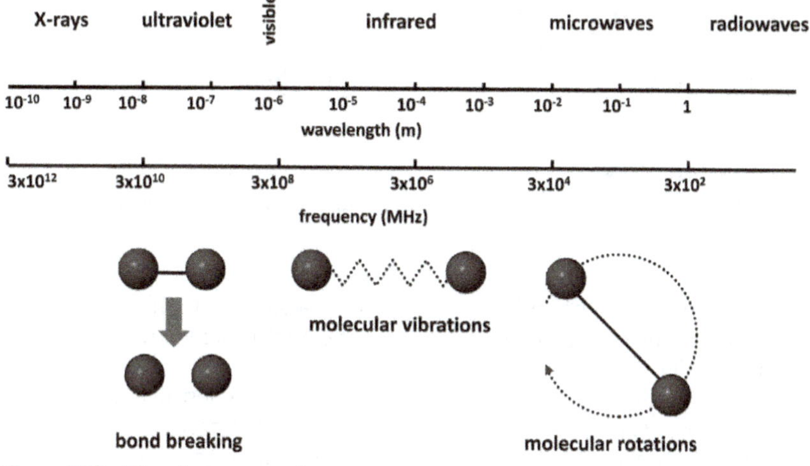


**Figure 3.20 The electromagnetic spectrum.**

Before looking at applications in organic chemistry, it is important to appreciate some of the physical chemistry concepts behind microwave heating. The microwave region of the electromagnetic spectrum (Fig. 3.20) is classified as that between 300 and 300,000 megahertz (MHz). Compared to ultraviolet, infrared, and visible light, microwave irradiation is of relatively low energy. As a result, microwaves are not high enough in energy to break chemical bonds. Instead they can only make molecules rotate. This is very different from the more energetic ultraviolet radiation which, when it interacts with molecules, can break bonds, giving rise to the area of chemistry known as photochemistry.

Both home and scientific microwave equipment operates at 2,450 MHz. Interestingly, the microwave region of the electromagnetic spectrum is also used for navigation, communication, and remote sensing purposes. This includes technologies such as global positioning systems, wireless Internet, and Bluetooth as well as radar. As a result, the frequency used in microwave ovens has to be different from those used for these other applications and so is strictly regulated.

Microwaves, like all electromagnetic energy, move at the speed of light and comprise oscillating electric and magnetic fields (Fig. 3.21). These components oscillate at right angles to each other and to the direction of propagation.

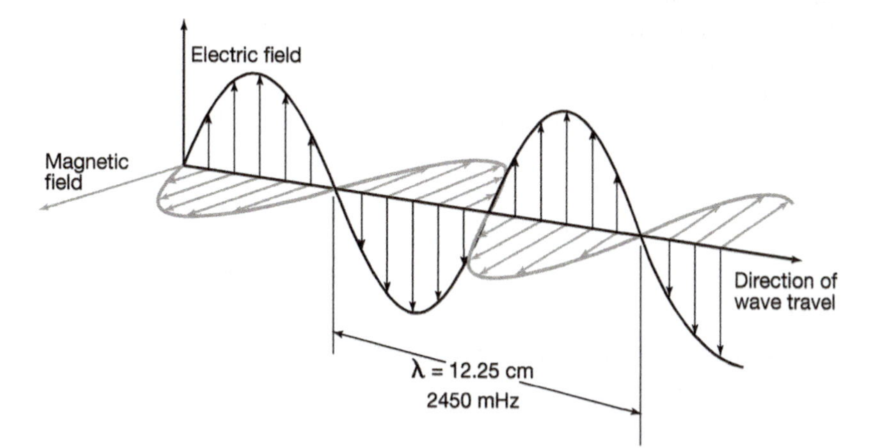


**Figure 3.21 Microwave energy comprises electric and magnetic fields.**

**Figure 3.22 The two mechanisms by which microwave energy leads to heating.**

There are two ways that microwaves can heat a sample, both involving the interaction of molecules in the sample with the electric field of the microwave irradiation (Fig. 3.22).

If a molecule possesses a dipole moment, then when it is exposed to microwave irradiation, the dipole tries to align with the applied electric field. Since the electric field is oscillating, the dipoles constantly try to realign to follow this. Molecules have time to align with the electric field but not to follow the oscillating field exactly. This continual reorientation of the molecules results in friction and thus heat. This heating method is termed dipolar polarization. If a molecule is ionic, then the electric field component of the microwave irradiation moves the ions back and forth through the sample also colliding them into each other. This movement again generates heat and is termed ionic conduction.

Conventionally, in order to heat reaction mixtures, chemists tend to use a hot plate or sand bath. These can be relatively slow and inefficient ways of transferring heat into a sample because they depend on convection currents and the thermal conductivity of the reaction mixture. Also, the walls of the reaction vessel can be hotter than the contents, thus introducing a thermal gradient. This can mean that reagents or products can be decomposed over time because they sit on the walls of the vessel. When using microwave heating, since the energy interacts with the sample directly, heating can be much more effective. In addition, microwave heating is safer than conventional heating; there are no sand baths or hot plates that can burn the chemist.

Each solvent or reagent in a reaction mixture will interact with microwave energy differently. Although not the only factor in determining the absorbance of microwave energy, the polarity of the solvent is a helpful tool for determining how well it will heat when placed into a microwave field; more polar molecules are affected more and nonpolar less. Solvents can be split into three categories; namely, those that absorb microwaves well, moderately, and poorly (Fig. 3.23). High-absorbing solvents will heat up very fast

| LOW ABSORBING | MODERATELY ABSORBING | HIGH ABSORBING |
|---|---|---|
| dichloromethane | water | dimethylsulfoxide (DMSO) |
| toluene | acetonitrile | ethane-1,2-diol (ethylene glycol) |
| hexane | acetone | propanol, ethanol, methanol |

**Figure 3.23 The heating characteristics of common solvents.**

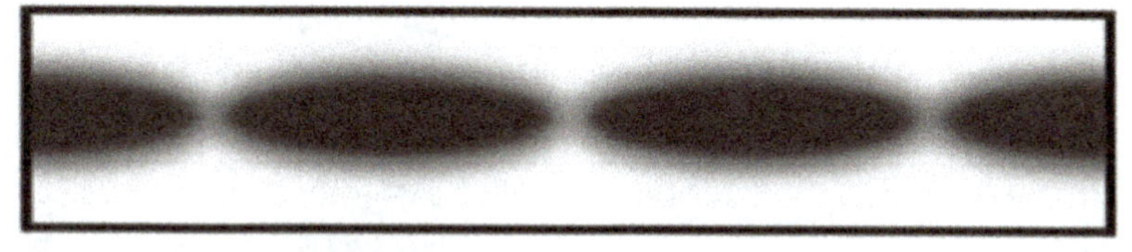

**Figure 3.24 Areas of high and low microwave energy are found in the cavity of a multimode microwave unit.**

upon microwave irradiation. Lower-absorbing solvents can still be used, but work better if one of the reagents in the reaction mixture is itself a good absorber. Interestingly, water absorbs more weakly than methanol while it is considerably more polar. This can be attributed, at least in part, to that fact that the strong, extensive hydrogen bonding in water goes some way to restricting rotation of molecules when irradiated with microwaves.

Electric power is turned into microwave energy using a magnetron, this in essence being a high-voltage tube in which electrons generated from a heated cathode are affected by magnetic and electric fields in such a way as to produce microwave radiation. As the microwaves come into the cavity (heating chamber) of a home microwave unit, they will move around and bounce off the walls. As they do so, they will generate pockets (called modes) of high energy and low energy as the moving waves either reinforce or cancel out each other (Fig. 3.24). This means that the microwave field in the microwave cavity is not uniform. Instead, there will be hot spots and cold spots; these correspond to the pockets of high and low energy, respectively. Domestic microwave ovens are therefore called "multimode" microwave ovens.

While home microwave ovens are useful for heating food, performing chemical reactions using them presents a number of challenges. They have no accurate temperature measurement device; the microwave field inside the oven is not uniform; and they are not safe for containing hot, flammable, organic solvents. These problems have led to the need for scientific microwave apparatus, specifically designed for performing chemical reactions safely and reliably. Scientific multimode microwave units have been developed for use in preparative chemistry (Fig. 3.26*b*). As well as being built to withstand explosions of reaction vessels inside the microwave cavity, temperature and pressure monitoring has been introduced as is the ability to stir reaction mixtures. It is possible to run a number of reactions at the same time in a multimode microwave oven, the samples being placed into tubes and loaded onto a turntable. As the samples move around, because they are large enough to absorb the microwave energy effectively, heating is fairly uniform.

When performing reactions on a small scale, it is sometimes difficult to heat the small volumes of reagents effectively in a multimode microwave apparatus. This is because, with the hot and cold spots that occur in the cavity of a multimode apparatus, it is hard to get constant microwave energy to irradiate the small sample. To overcome these problems, a smaller, single-mode (often called monomode) microwave apparatus has been developed. The cavity of a monomode microwave apparatus is designed for the length of only one wave (mode) (Fig. 3.25). By placing the sample in the middle of the cavity it can be irradiated constantly with microwave energy. Using a monomode apparatus, it is possible to heat samples of as little as 0.2 mL very effectively. The upper volume limit of the monomode apparatus is determined by the size of the microwave cavity and is in the region of 100 mL (Fig. 3.26*a*).

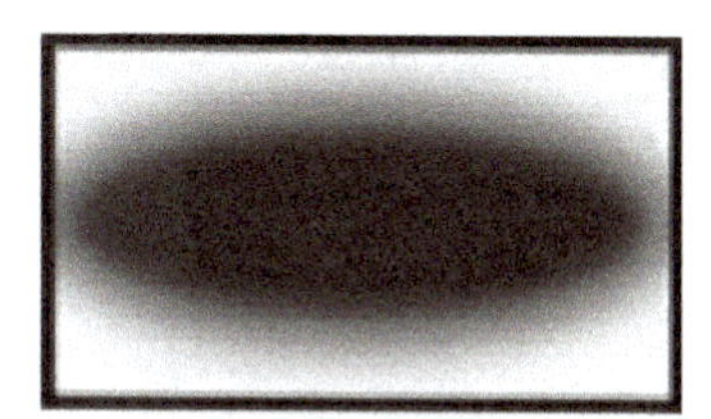

**Figure 3.25 The cavity of a monomode microwave unit is designed to fit just one mode.**

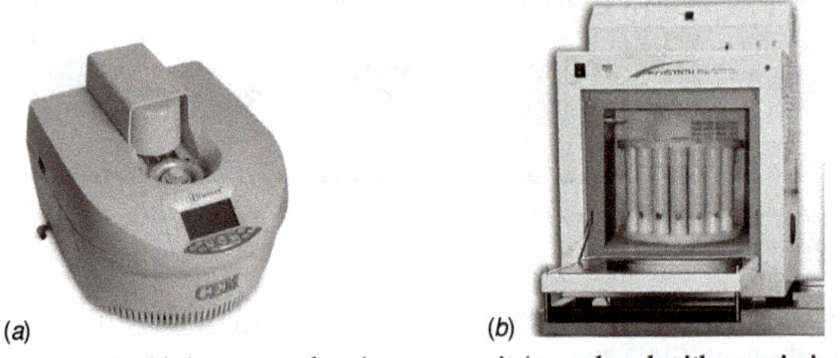


**Figure 3.26 (a) A monomode microwave unit (reproduced with permission from CEM Corporation) and (b) a multimode microwave unit (reproduced with permission from Milestone srl).**

## Applications

Many organic reactions require heat in order to proceed. In the lab, this is traditionally done using a hot plate, steam, oil or sand bath or, before that, a Bunsen burner. For those reactions that do require heat, the problem is that these heating sources are inefficient and reactions can often take a long time to reach completion. By using microwave heating, reaction times can be dramatically reduced and product yields can be higher. Shortening the time of known reactions is not the only advantage that microwave heating is having. It is impacting modern organic chemistry by opening up avenues to compounds that were previously not accessible. It is also a cleaner way to do preparative chemistry. Almost any reaction that needs heat can be performed in a microwave (Fig. 3.27). There are a few exceptions, including those that are known to be highly exothermic.

Microwave heating has proven particularly useful in the pharmaceutical industry where compounds need to be made rapidly so they can be screened for activity as drug candidates. In an interesting experiment undertaken by Boehringer Ingelheim Pharmaceuticals, the exact amount of time saved using microwave as opposed to conventional heating was determined.[8] Two

| | |
|---|---|
| *oxidation* | *rearrangements* |
| *reduction* | *ester and amide synthesis* |
| *substitution* | *ring-forming* |
| *addition* | *heterocycle synthesis* |
| *cycloadditions* | *metal-catalyzed processes* |

**Figure 3.27 Some classes of organic reaction that can be performed using microwave heating.**

[8] "Timesavings associated with microwave-assisted synthesis: A quantitative approach", C. R. Sarko in *Microwave Assisted Organic Synthesis* edited by J. P. Tierney and P. Lidstrom, Blackwell Publishing, Oxford, 2005.

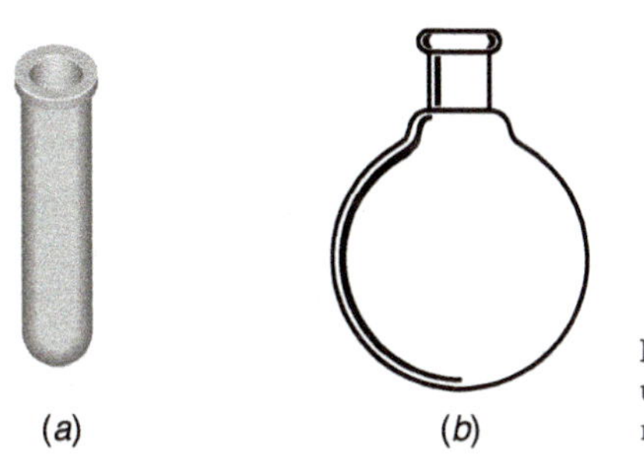


**Figure 3.28 Reactions can be performed using either a sealed tube or an open, round-bottom flask.**

scientists were told to make a series of compounds. One of them used microwave chemistry and the other used conventional methods. Both scientists were given the same preparative route to the molecules to follow. However, after 37 days the chemist using the conventional heating approach concluded that the molecules could not be generated using that route. The microwave chemist on the other hand optimized reaction conditions and produced the final products in two days.

Many reactions utilizing microwave heating have been performed in sealed vessels (Fig. 3.28*a*). These are tubes of varying sizes that can be sealed with a specially designed stopper. Reaction mixtures can then be heated to temperatures well above the boiling point of the solvent inside. This offers a very safe way to perform chemistry at high temperatures and pressures. It is much more convenient than the steel containers used traditionally for this sort of chemistry. Also, it is possible to monitor the temperature and pressure of reaction mixtures very closely, and this means it is possible to report the exact reaction conditions used so that others can use them.

Another option is to use standard laboratory glassware in a microwave. Reactions can be run in round-bottomed flasks equipped with a reflux condenser (Fig. 3.28*b*). The flask sits inside the microwave cavity, and the reflux condenser comes out through the top of the apparatus. Often, just as good results can be obtained using an open vessel arrangement as compared to a sealed tube.

When using a monomode microwave unit, it is possible to perform reactions using sealed tubes of capacity ranging from 0.2–25 mL and open vessels ranging from 10–100 mL. Reactions are performed one at a time. When using sealed tubes, it is possible to automate the unit using robotics so that when one reaction is complete, the tube can be removed from the microwave and the next one put in. This allows for multiple reactions to be performed without the need for the operator to be present. Up to 60 reaction vessels can be lined up and run one after another.

Multimode microwave units can process multiple reaction vessels at the same time. The sealed vessels all sit in a holder (reaction carousel) that fits into the microwave cavity. Working on a scale of up to a few grams, it is possible to process up to 40 reaction vessels at a time. Up to 92 reactions can be run at a time when using microscale quantities of reagents. Another option possible when working in multimode microwave unit is to use one large reaction vessel. This can either be a larger sealed vessel (up to 1 L in volume) or an open round-bottom flask (up to 5 L in volume). This enables chemists to scale up their reactions to make more of their desired compound.

### Equipment Available

There are a number of commercially available scientific microwave units. The four major microwave manufacturers are listed here:

Anton Paar is an Austrian company that manufactures a multimode microwave unit called the Synthos 3000. There are a number of reaction carousels that are available with the unit, allowing for reactions to be performed from the microliter scale up to 100 mL. On the small scale, reactions are run in specially designed silicon carbide plates with either 24 or 48 wells. Using plates made from this inert, highly microwave-absorbing material allows for equal heating of all the wells. Larger reactions are performed in glass or quartz tubes sealed with a specially designed stopper.

Biotage, a company based in Sweden, manufactures a monomode microwave unit called the Initator. Using this instrument, it is possible to run reactions on scales from 0.2–20 mL in sealed tubes. It is possible to automate the unit with a robotic arm, thus allowing up to 60 reactions to be run sequentially.

CEM Corporation, a company based in North Carolina, manufactures a range of microwave units. Its monomode microwave apparatus is called the Discover platform. A number of variants are available. It is possible to run reactions from 0.5–60 mL in sealed tubes using this unit, as well as open round-bottom flasks up to 125 mL in capacity. It is possible to automate the unit, allowing reactions to be run sequentially. In addition, an accessory is available for loading reaction vessels with reactive gases such as hydrogen and carbon monoxide, opening the door for performing a wide range of reactions otherwise not possible using microwave heating. CEM also manufactures a multimode microwave unit called the MARS. There are a number of reaction carousels accommodating sealed tubes that can be used with the unit. In addition, open round-bottom flasks up to 5 L in capacity can be placed into the microwave cavity and standard reflux glassware attached. This allows for scale-up of reactions using batch processing.

Milestone, a company based in Italy, manufactures a number of microwave units. The MultiSYNTH has the capability to act as both a monomode and a multimode microwave unit. This means that conditions can be optimized in a small sealed tube using the monomode functionality and then a series of up to 12 small or 6 larger reactions can run in parallel in multimode. The unit can also accommodate a round-bottom flask of capacity up to 1 L, allowing a reaction to be performed at atmospheric pressure. The MicroSYNTH platform is a multimode microwave unit. There are a number of reaction carousels that can be used, allowing for reactions to be performed in parallel using either glass, Teflon, or quartz tubes. In addition, open round-bottom flasks can be placed into the microwave cavity and standard reflux glassware attached.

All the modern scientific microwave units have the capability to measure temperature during the course of a reaction. This can be done remotely using an infrared sensor located in the wall or the bottom of the microwave cavity. In many cases it is also possible to record the temperature inside a reaction vessel using a fiber-optic probe or thermocouple. Pressure measurement is also possible in many cases. The contents of a reaction mixture can be stirred by means of a magnetic stir plate located beneath the microwave cavity and a Teflon stir bar in the vessel.

When running a reaction, key parameters such as temperature, pressure, and microwave power can be measured throughout the run and data saved to

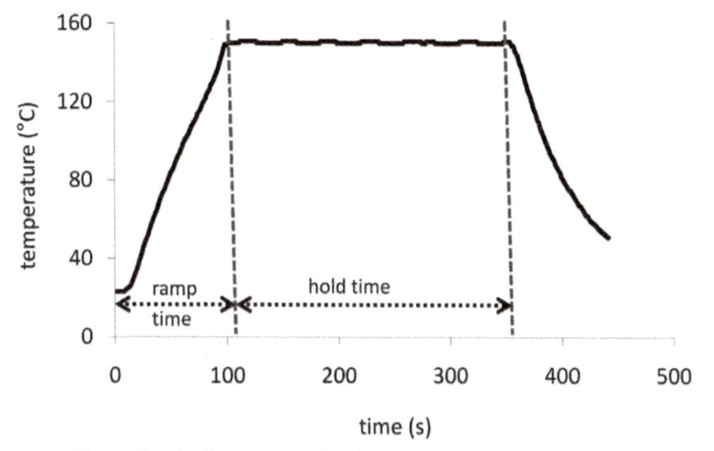


**Figure 3.29 Example of a heating profile for a reaction performed using microwave heating.**

a computer for use in reports and for reproducing the conditions at a later date (Fig. 3.29). Generally, when programming a protocol into a microwave unit, there are two important stages. The first is entering the ramp time. This is the time that the user wants the microwave to take to reach the target temperature. The second is entering the hold time. This is the time that the user wants the reaction mixture to remain at the target temperature before cooling back to room temperature. The microwave unit will use the requisite microwave power to heat the reaction mixture to temperature and then the power will automatically fluctuate to hold the reaction at the set temperature.

### Experimental Protocols

Experimental protocols using microwave heating have been added to Experiments 7, 8, 15, 22, and 30. The experiments in this book can be performed on a range of these commercially available microwave units. The procedures are split into two classes; the first is generally for use with monomode microwave apparatus (Biotage Initiator and CEM Discover) and the second for use with multimode microwave units (Anton Paar Synthos 3000, CEM MARS, and Milestone MicroSYNTH). A modified version of the monomode procedure for use with the Anton Paar Synthos 3000 in conjunction with the silicon carbide plate format for microscale chemistry is added as a footnote in the monomode protocol.

## MICROSCALE LAWS

### Rules of the Trade for Handling Organic Materials at the Microscale Level

Now that we have briefly looked at the equipment we will be using to carry out microscale organic reactions, let us examine the specific techniques that are used to deal with the small quantities of material involved. Microscale synthetic organic reactions, as defined by Cheronis,[9] start with 15–150 mg of the

[9]Cheronis, N. D. *Semimicro Experimental Organic Chemistry*; Hadrion Press: New York, 1958.

limiting reagent. These quantities sound small, and they are. Although 150 mg of a light, powdery material will fill half a 1-mL conical vial, you will have a hard time observing 15 mg of a clear liquid in the same container, even with magnification. This volume of liquid, on the other hand, is reasonably easy to observe if it is in a 0.1-mL conical vial. A vital part of the game of working with small amounts of materials is to become familiar with microscale techniques and to practice them as much as possible in the laboratory.

### Rules for Working With Liquids at the Microscale Level

**1. Liquids are never poured at the microscale level.** Liquid substances are transferred by pipet or syringe. As we are working with small, easy-to-hold glass-ware, the best way to transfer liquids is to hold both containers with the fingers of one hand, with the mouths as close together as possible. The free hand is then used to operate the pipet (syringe) to withdraw the liquid and make the transfer. This approach reduces to a minimum the time that the open tip is not in, or over, one vessel or the other. We use three different pipets and two standard syringes to perform most experiments involving liquids. **This equipment can be a prime source of contamination.** Be very careful to thoroughly clean the pipets and syringes after each use.

- **a. Pasteur pipet (often called a capillary pipet).** A simple glass tube with the end drawn to a fine capillary. These pipets can hold several milliliters of liquid (Fig. 3.30*a*) and are filled using a small rubber bulb or one of the very handy, commercially available pipet pumps. Because many transfers are made with Pasteur pipets, it is suggested that several of them be calibrated for approximate delivery of 0.5, 1.0, 1.5, and 2.0 mL of liquid. This calibration is easily done by drawing the measured amount of a liquid from a graduated cylinder and marking the level of the liquid in the pipet. This mark can be made with transparent tape, or by scratching with a file. Indicate the level with a marking pen before trying to tape or file the pipet.
- **b. Pasteur filter pipet.** A very handy adaptation of the Pasteur pipet is a filter pipet. This pipet is constructed by taking a small cotton ball and placing it in the large open end of the standard Pasteur pipet. Hold the pipet vertically and tap it gently to position the cotton ball in the drawn section of the tube (Fig. 3.30*b*). Now form a plug in the capillary section by pushing the cotton ball down the pipet with a piece of copper wire (Fig. 3.30*c*). Finish by seating the plug flush with the end of the capillary (Fig. 3.30*d*). The optimum-size plug will allow easy movement along the capillary while it is being positioned by the copper wire. Compression of the cotton will build enough pressure against the walls of the capillary (once the plug is in position) to prevent the plug from slipping while the pipet is filled with liquid. If the ball is too big, it will wedge in the capillary before the end is reached, and wall pressure will be so great that liquid flow will be shut off. Even some plugs that are loose enough to be positioned at the end of the capillary will still have developed sufficient lateral pressure to make the filling rate unacceptably slow. If the cotton filter, however, is positioned too loosely, it may be easily dislodged from the pipet by the solvent flow. These

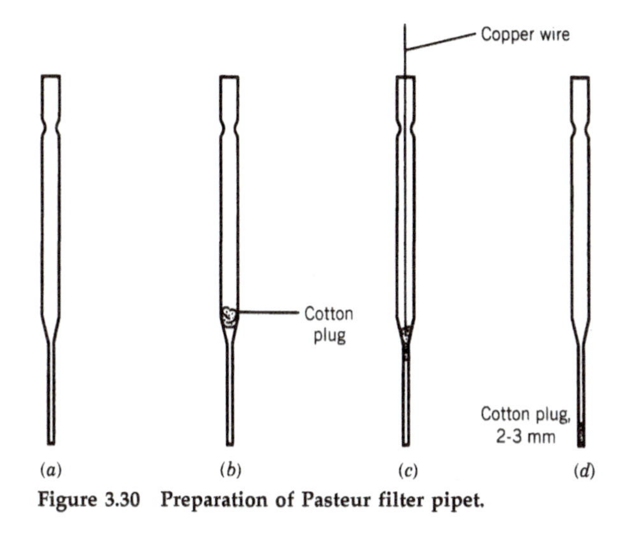


**Figure 3.30 Preparation of Pasteur filter pipet.**

plugs can be quickly and easily inserted with a little practice. Once in place, the plug is rinsed with 1 mL of methanol and 1 mL of hexane, and dried before use.

There are two reasons for placing the cotton plug in the pipet. First, it solves a particular problem with the transfer of volatile liquids via the standard Pasteur pipet: the rapid buildup of back pressure from solvent vapors in the rubber bulb. This pressure quickly tends to force the liquid back out of the pipet and can cause valuable product to drip on the bench top. The cotton plug tends to resist this back pressure and allows much easier control of the solution once it is in the pipet. The time-delay factor becomes particularly important when the Pasteur filter pipet is employed as a microseparatory funnel (see the discussion on extraction techniques in Technique 4, p. 67).

Second, each time a transfer of material is made, the material is automatically filtered. This process effectively removes dust and lint, which are constant problems when working at the microscale level. A second stage of filtration may be obtained by employing a disposable filter tip on the original Pasteur filter pipet as described by Rothchild.[10]

**c. Automatic pipet (considered the Mercedes–Benz of pipets).** Automatic pipets quickly, safely, and reproducibly measure and dispense specific volumes of liquids. These pipets are particularly valuable at the microscale level, because they generate the precise and accurate liquid measurements that are absolutely necessary when handling only microliters of a liquid. The automatic pipet adds considerable insurance for the success of an experiment, since any liquid can be efficiently measured, transferred, and delivered to the reaction flask.

The automatic pipet consists of a calibrated piston pipet with a specially designed disposable plastic tip. It is possible to encounter

[10]Rothchild, R. *J. Chem. Educ.* **1990,** *67*, 425.

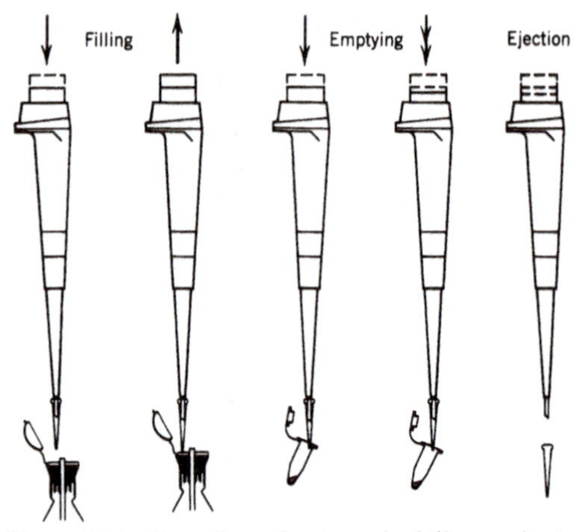


**Figure 3.31 Operation of automatic delivery pipet.**

any one of three pipet styles: single volume, multirange, or continuously adjustable (see Fig. 3.31). The first type is calibrated to deliver only a single volume. The second type is adjustable to two or three predetermined delivery volumes. The third type is the most versatile; it can be set by the user to deliver any volume within the range of the pipet. Obviously, the price of these valuable laboratory tools goes up with increasing features. Automatic pipets are expensive, and usually must be shared in the laboratory. Treat them with respect!

The automatic pipet is designed so that the liquid comes in contact only with the disposable tip.

- Never load the pipet without the tip in place.
- Never immerse the tip completely in the liquid that is being pipetted.
- Always keep the pipet vertical when the tip is attached.
- If an air bubble forms in the tip during uptake, return the liquid, discard the tip, and repeat the sampling process.

If these three rules are followed, most automatic pipets will give many years of reliable service. A few general rules for improving reproducibility with an automatic pipet should also be followed:

- Try to use the same uptake and delivery motion for all samples. Smooth depression and release of the piston will give the most consistent results. Never allow the piston to snap back.
- *Always* depress the piston to the first stop before inserting the tip into the liquid. If the piston is depressed after submersion, formation of an air bubble in the tip becomes likely, which will result in a filling error.
- *Never* insert the tip more than 5 mm into the liquid. It is good practice not to allow the body of the pipet to contact any surface, or bottle neck, that might be wet with a chemical.
- If an air bubble forms in the tip during uptake, return the fluid, discard the tip, and repeat the sampling process.

**d. Syringes.** Syringes are particularly helpful for transferring liquid reagents or solutions to sealed reaction systems from sealed reagent or solvent reservoirs. Syringe needles can be inserted through a septum, which avoids opening the apparatus to the atmosphere. Syringes are also routinely employed in the determination of ultramicro boiling points (10-μL GC syringe). It is critically important to clean the syringe needle after each use. Effective cleaning of a syringe requires as many as a dozen flushes. For many transfers, the microscale laboratory uses a low-cost glass 1-mL insulin syringe in which the rubber plunger seal is replaced with a Teflon seal (ACE Glass). For preparative GC injections, the standard 50- or 100-μL syringes are preferred (see Technique 1).

**2. Liquid volumes may be converted easily to mass measures by the following relationship:**

$$\text{Volume (mL)} = \frac{\text{mass (g)}}{\text{density (g/mL)}}$$

**3. Work with liquids in conical vials,** and work in a vial whose capacity is approximately twice the volume of the material it needs to hold. The trick here is to reduce the surface area of the flask in contact with the sample to an absolute minimum. A conical vial is thus better than the spherical surface of the conventional round-bottom flask.

Liquids may also be weighed directly. A tared container (vial) should be used. After addition of the liquid, the vial should be kept capped throughout the weighing operation. This procedure prevents loss of the liquid by evaporation. If the density of the liquid is known, the approximate volume of the liquid should be transferred to the container using an automatic delivery pipet or a calibrated Pasteur pipet. Use the above expression relating density, mass, and volume to calculate the volume required by the measured mass. Adjustment of the mass to give the desired value can then be made by adding or removing small amounts of liquid from the container by Pasteur pipet.

*NOTE. Before you leave the balance area, be sure to replace all caps on reagent bottles and clean up any spills. A balance is a precision instrument that can easily be damaged by contamination.*

### Rules for Working With Solids at the Microscale Level

**1. General considerations.** Working with a crystalline solid is much easier than working with the equivalent quantity of a liquid. Unless the solid is in solution, a spill on a clean glass working surface usually can be recovered quickly and efficiently. *Be careful, however, when working with a solution. Treat solutions as you would a pure liquid.*

**2. Transfer of solids.** Solids are normally transferred with a microspatula, a technique that is not difficult to develop.

**3. Weighing solids at the milligram level.** Electronic balances can automatically tare an empty vial. Once the vial is tared, the reagent is added in small portions. The weight of each addition is instantly registered; material is added until the desired quantity has been transferred.

Solids are best weighed in glass containers (vials or beakers), in plastic or aluminum weighing trays ("boats"), or on glazed weighing paper. Filter paper or other absorbent materials are not good choices: small quantities of the weighed material will often stick to the fibers of the paper, and vice versa.

## THE LABORATORY NOTEBOOK

Writing is the most important method chemists use to communicate their work. It begins with the record kept in a laboratory notebook. An experiment originally recorded in the laboratory notebook can become the source of information used to prepare scientific papers published in journals or presented at meetings. For the industrial chemist, these written records are critical in obtaining patent coverage for new discoveries.

It is important that you learn to keep a detailed account of your work. A laboratory notebook has several key components. Note how each component is incorporated into the example that follows.

*Key Components of a Laboratory Experiment Write-up*

1. Date experiment was conducted
2. Title of experiment
3. Purpose for running the reaction
4. Reaction scheme
5. Table of reagents and products
6. Details of procedure used
7. Characteristics of the product(s)
8. References to product or procedure (if any)
9. Analytical and spectral data
10. Signature of person performing the experiment and that of a witness, if required

In reference to point 6, it is the obligation of the person doing the work to list the equipment, the amounts of reagents, the experimental conditions, and the method used to isolate the product. Any color or temperature changes should be carefully noted and recorded.

Several additional points can be made about the proper maintenance of a laboratory record.

11. A hardbound, permanent notebook is essential.
12. Each page of the notebook should be numbered in consecutive order. For convenience, an index at the beginning or end of the book is recommended and pages should be left blank for this purpose.
13. If a page is not completely filled, an "X" should be used to show that no further entry was made.
14. Data are always recorded directly into the notebook, *never* on scrap paper! Always record your data in ink. If a mistake is made, draw a neat line through the word or words so that they remain legible. Do not completely obliterate anything; you might learn from your mistakes, if you can read them later.
15. Make the record clear and unambiguous. Pay attention to grammar and spelling.
16. In industrial research laboratories, your signature, as well as that of a witness, is required, because the notebook may be used as a legal document.

**17.** Always write and organize your work so that someone else could come into the laboratory and repeat your directions without confusion or uncertainty. *Completeness* and *legibility* are key factors.

Most of you are newcomers to the organic laboratory, and the reactions you will be performing have probably been worked out and checked in detail. Because of this, your instructor may not require you to keep your notebook in such a meticulous fashion. For example, when you describe the procedure (item 6), it may be acceptable to make a clear reference to the material in the laboratory manual and to note any modifications or deviations from the prescribed procedure. In some cases, it may be more practical to use an outline method. In any event, the following example should be studied carefully. It may be used as a reference when detailed records are important in your work. It is more important to record what you observed and what you actually did, than to record what you were supposed to observe and what you were supposed to do.

*NOTE. Because of its length, the example here is typed. Notebooks are usually handwritten. Many chemists, however, now use computers to record their data.*

*The circled numbers refer to the list on p. 40*

## EXAMPLE OF A LABORATORY NOTEBOOK ENTRY

19 July 2009 ①
PREPARATION OF DIPHENYL SUCCINATE ②

$$\begin{array}{c} CH_2CO_2H \\ | \\ CH_2CO_2H \end{array} + 2\ C_6H_5OH + POCl_3 \rightarrow \begin{array}{c} CH_2CO_2C_6H_5 \\ | \\ CH_2CO_2C_6H_5 \end{array} + HPO_3 + 3\ HCl$$ ④

Diphenyl succinate is being prepared as one of a series of dicarboxylic acid esters that are to be investigated as growth stimulants for selected fungi species. ③

This procedure was adapted from that reported by Daub, G. H., and Johnson, W. S. *Organic Syntheses*, Wiley: New York, 1963; Collect. Vol. IV, p. 390. ⑧

**Physical Properties of Reactants and Products** ⑤

| Compound | MW[a] | Amounts | mmol | mp (°C) | bp (°C) |
|---|---|---|---|---|---|
| Succinic acid | 118.09 | 118 mg | 1.0 | 182 | |
| Phenol | 94.4 | 188 mg | 2.0 | 40–42 | 182 |
| Phosphorous oxychloride | 153.33 | 84 μL | 0.9 | | 105.8 |
| Diphenyl succinate | 270.29 | | | 121 | |

[a]MW = molecular weight.

⑥ In a 3.0-mL conical vial containing a magnetic spin vane and equipped with a reflux condenser protected by a calcium chloride drying tube were placed succinic acid (118 mg, 1.0 mmol), phenol (188 mg, 2.0 mmol), and phosphorous oxychloride (84 μL 0.9 mmol). The reaction mixture was heated with stirring at 115 °C in a sand bath in the **hood** for 1.25 h. It was necessary to conduct the reaction in the **hood,** because hydrogen chloride (HCl) gas evolved during the course of the reaction. The drying tube was removed, toluene (0.5 mL) was added through the top of the condenser using a Pasteur pipet, and the drying tube was replaced. The mixture was then heated for an additional 1 hour at 115 °C.

The hot toluene solution was separated from the red syrupy residue of phosphoric acid using a Pasteur pipet. The toluene extract was filtered by gravity using a fast-grade filter paper and the filtrate was collected in a 10-mL Erlenmeyer flask. The phosphoric acid residue was then extracted with two additional 1.0-mL portions of hot toluene. These extracts were also separated using the Pasteur pipet and filtered, and the filtrate was collected in the same Erlenmeyer flask. The combined toluene solutions were concentrated to a volume of approximately 0.6 mL by warming them in a sand bath under a gentle stream of nitrogen ($N_2$) gas in the **hood.** The pale yellow liquid residue was then allowed to cool to room temperature; the diphenyl succinate precipitated as colorless crystals. The solid was collected by vacuum filtration using a Hirsch funnel, and the filter cake was washed with three 0.5-mL portions of cold diethyl ether. The product was dried in a vacuum oven at 30°C (3 mm Hg) for 30 min.

⑦ ⑧ The 181 mg (67%) of diphenyl succinate had an mp of 120–121 °C (lit. value 121 °C: *CRC Handbook of Chemistry and Physics,* 89th ed.; CRC Press: Boca Raton, FL, 2008–2009; no. 13559, p. 3–220).

⑨ The IR spectrum exhibits the expected peaks for the compound. [*At this point, the data may be listed, or the spectrum attached to a separate page of the notebook.*]

⑩ Marilyn C. Waris

witnessed by
D. Jeanne d'Arc Mailhiot 19 July 2009

## CALCULATING YIELDS

Almost without exception, in each of the experiments presented in this text, you are asked to calculate the percentage yield. For any reaction, it is always important for the chemist to know how much of a product is actually produced (experimental) compared to the theoretical (maximum) amount that could have been formed. The percentage yield is calculated on the basis of the relationship

$$\%\ \text{yield} = \frac{\text{actual yield (experimental)}}{\text{theoretical yield (calculated maximum)}} \times 100$$

The percentage yield is generally calculated on a weight (gram or milligram) or on a mole basis. In the present text, the calculations are made using milligrams.

Several steps are involved in calculating the percentage yield.

**Step 1** Write a *balanced* equation for the reaction. For example, consider the Williamson synthesis of propyl *p*-tolyl ether.

$$CH_3-C_6H_4-OH + CH_3CH_2CH_2-I \xrightarrow[(C_4H_9)_4N^+,\ Br^-]{NaOH} CH_3-C_6H_4-O-(CH_2)_2CH_3 + Na^+, I^-$$

*p*-Cresol    Propyl iodide    Propyl *p*-tolyl ether

**Physical Properties of Reactants**

| Compound | MW (mg/mmol) | Amount | mmol | *d*(mg/μL) |
|---|---|---|---|---|
| *p*-Cresol | 108.15 | 160 μL | 1.56 | 1.5312 |
| 25% (by weight) NaOH soln | 40.0 | 260 μL | ~1.6 | |
| Tetrabutylammonium bromide | 322.38 | 18 mg | 0.056 | |
| Propyl iodide | 169.99 | 150 μL | 1.54 | 1.5058 |

**Step 2** Identify the *limiting* reactant. The ratio of reactants is calculated on a millimole (or mole) basis. In the example, 1.56 mmol of *p*-cresol and ca. 1.6 mmol of sodium hydroxide are used, compared to 1.54 mmol of propyl iodide, which is therefore the limiting reagent. The tetrabutylammonium bromide is not considered because it is used as a catalyst—it is neither incorporated into the product nor consumed in the reaction. The calculation of the theoretical yield is thus based on the amount of propyl iodide, 1.54 mmol.

**Step 3** Calculate the *theoretical* (maximum) amount of the product that could be obtained for the conversion, based on the limiting reactant. Here, one mole of propyl iodide produces one mole of the propyl *p*-tolyl ether. Therefore, the maximum amount of propyl *p*-tolyl ether (molecular weight = 150.2) that can be produced from 1.54 mmol of propyl iodide is 1.54 mmol, or 231 mg.

**Step 4** Determine the *actual* (experimental) yield (milligrams) of product isolated in the reaction. This amount is invariably less than the theoretical quantity, unless the material is impure (one common contaminant is water). For example, student yields for the preparation of propyl *p*-tolyl ether average 140 mg.

**Step 5** Calculate the *percentage yield* using the weights determined in steps 3 and 4. The percentage yield is then

$$\%\ \text{yield} = \frac{140\ \text{mg (actual)}}{231\ \text{mg (theoretical)}} \times 100 = 60.6\%$$

As you carry out each reaction in the laboratory, try to obtain as high a percentage yield of product as possible. The reaction conditions in this book's experiments have been carefully developed; if you master the microscale techniques for transferring reagents and isolating products, your yields will be as high as possible.

## QUESTIONS

**3-1.** Factory A produces the wheels that are used for the frames made in Factory B. Factory C relies exclusively on the materials produced in Factories A and B. Assuming all the necessary parts minus the wheels and frames are housed in Factory C, how many bicycles can be completely assembled when Factories A and B provide 36 wheels and 15 frames, respectively? Explain.

**3-2.** You are provided a vial that contains 180 mg of material. This material represents a 44 percent isolated yield. Calculate the theoretical amount (theoretical yield) that could have been formed.

**3-3.** The density of 2-methyl-2-butanol is 0.806 g/mL. How many mgs represent an aliquot of 430 μL? How many mmols represent an aliquot of 0.650 mL (2-methyl-2-butanol, formula weight is 88.15 g/mol)?

# $C_4H_6$apter 4

# DETERMINATION OF PHYSICAL PROPERTIES

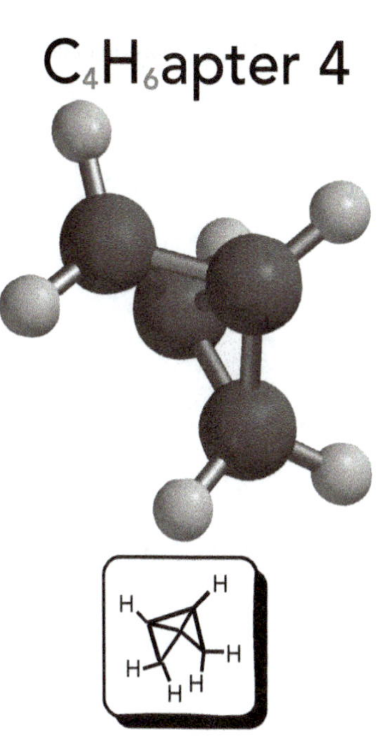

Determination of physical properties is important for substance identification and as an indication of material purity. Historically, the physical constants of prime interest have included boiling point, density, and refractive index in liquids and the melting point in solids. In special cases, optical rotation and molecular weight determinations may be required. Today, with the widespread availability of spectroscopic instrumentation, powerful new techniques may be applied to the direct identification and characterization of materials, including the analysis of individual components of very small quantities of complex mixtures. The sequential measurement of the infrared (IR) and mass spectro-metric (MS) characteristics of a substance resolved "on the fly" by capillary gas chromatography (GC) can be quickly determined and interpreted. This particular combination (GC-IR-MS), which stands out among a number of hyphenated techniques that are now available, is perhaps the most powerful system yet developed for molecular identification. The rapid development of high-field multinuclear magnetic resonance (NMR) spectrometers has added another powerful dimension to identification techniques. NMR sensitivity, however, is still considerably lower than that of either IR or MS. The IR spectrum alone, obtained with one data point per wavenumber can add more than 4000 measurements to the few classically determined properties. *Indeed, even compared to high-resolution MS and pulsed $^1H$ and $^{13}C$ NMR, the infrared spectrum of a material remains a powerful set of physical properties (transmission elements) available to the organic chemist for the identification of an unknown compound.*[1]

Simple physical constants are determined mainly to assist in establishing the purity of *known* materials. Because the boiling point or the melting point of a material can be very sensitive to small quantities of impurities, these data can be particularly helpful in determining whether a starting material needs further purification or whether a product has been isolated in acceptable purity. Gas (GC), high-performance liquid (HPLC), and thin-layer (TLC) chromatography, however, now provide more powerful purity information when such data are required. When a new composition of matter has been formed, an elemental (combustion) analysis is normally reported if sufficient material is available for this destructive analysis. For new substances we are, of course, interested in establishing not only the identity, but also the molecular structure of the materials. In this situation other modern techniques (such as $^1H$ and $^{13}C$ NMR spectroscopy, high-resolution MS, and single-crystal X-ray diffraction) can provide sensitive and powerful structural information.

When comparisons are made between experimental data and values obtained from the literature, it is essential that the latter information be obtained

---

**Chapter 4: $C_4H_6$, Bicyclo[1.1.0]butane**
Lemal, Menger, and Clark (1963).

[1]Griffiths, P. R.; de Haseth, J. A. *Fourier Transform Infrared Spectrometry*, 2nd ed.; Wiley: New York, 2007.

from the most reliable sources available. This is especially true when considering the volume of misinformation found online. Certainly, judgment, which improves with experience, must be exercised in accepting any value as a standard. In addition to the wealth of data found online (e.g., SciFinder Scholar), the known classical properties of a large number of compounds can be obtained from the *CRC Handbook of Chemistry and Physics* and the *Merck Index*. The *Aldrich Catalog Handbook of Fine Chemicals* is also a readily available, inexpensive source. These reference works list physical properties for inorganic, organic, and organometallic compounds. The *Aldrich Catalog* also references IR and NMR data for a large number of substances. New editions of the *CRC Handbook* and the *Aldrich Catalog* are published each year.

## LIQUIDS

### Ultramicro Boiling Point

Upon heating, the vapor pressure of a liquid increases, though in a nonlinear fashion. When the pressure reaches the point where it matches the local pressure, the liquid boils. That is, it spontaneously begins to form vapor bubbles, which rapidly rise to the surface. If heating is continued, both the vapor pressure and the temperature of the liquid will remain constant until the substance has been completely vaporized (Fig. 4.1).

Because microscale preparations generally yield about 30–70 μL of liquid products, using only 5 μL or less of material for boiling point measurements is highly desirable. The modification of the earlier Wiegand ultramicro boiling-point procedure[2] to the ultramicro procedure described here has established that reproducible and reasonably accurate (±1 °C) boiling points can be observed on 3–4 μL of most liquids thermally stable at the required temperatures.

***Procedure.*** Ultramicro boiling points can be conveniently determined in standard (90-mm-length) Pyrex glass capillary melting-point tubes. The melting-point tube replaces the conventional 3- to 4-mm (o.d.) tubing used in the Siwoloboff procedure.[3] The sample (3–4 μL) is loaded into the melting-point capillary via a 10-μL syringe and centrifuged to the bottom if necessary. A small glass bell replaces the conventional melting-point tube as the bubble generator in micro boiling-point determinations. It is formed by heating 3-mm (o.d.) Pyrex tubing with a microburner and drawing it out to a diameter small enough to be readily fit inside the melting-point capillary. A section of the drawn capillary is fused and then cut to yield two small glass bells approximately 5 mm long (Fig. 4.2*a*). It is important that the fused section be reasonably large, because it is more than just a seal. The fused glass must add enough weight to the bell that it will firmly seat itself in the bottom of the melting-point tube.

An alternative technique for preparing the glass bells follows: heat the midsection of an open-ended melting point capillary tube and then draw the glass to form a smaller capillary section. This section is then broken approximately in the middle and each open end is sealed. The appropriate length for the bell is then broken off. Thus, two bells are obtained, one from each section. The sealing

[2]Wiegand, C. *Angew. Chem.* **1955,** *67,* 77. Mayo, D. W.; Pike, R. M.; Butcher, S. S.; Meredith, M. L. *J. Chem. Educ.* **1985,** *62,* 1114.
[3]Siwoloboff, A. *Berichte* **1886,** *19,* 795.

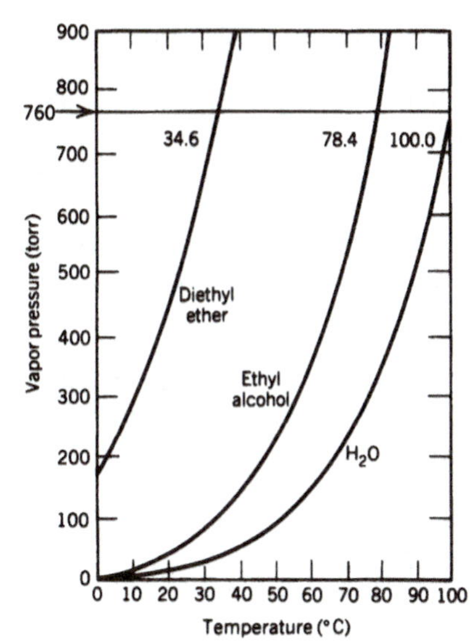


**Figure 4.1 Vapor pressure curves.** (From Brady, J. E.; Humiston, G. E. *General Chemistry,* 3d ed.; Wiley: New York, 1982. (Reprinted by permission of John Wiley & Sons, New York.)

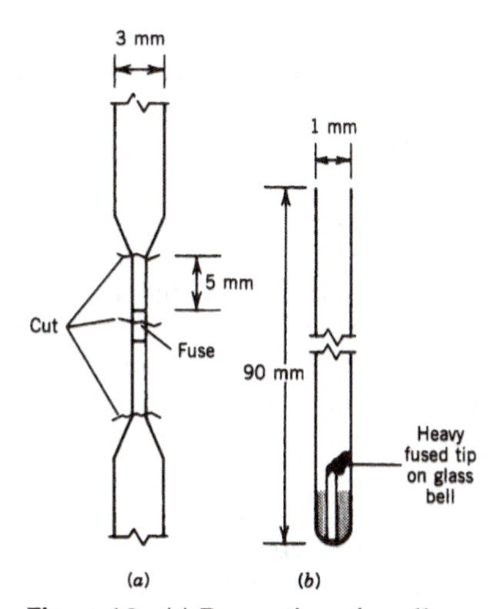


**Figure 4.2 (a) Preparation of small glass bell for ultramicro boiling-point determination. (b) Ultramicro boiling-point assembly.** (From Mayo, D. W.; Pike, R. M.; Butcher, S. S.; Meredith, M. L. *J. Chem. Educ.* ***1985,*** *62,* 1114.)

process (be sure that a significant section of glass is fused during the tube closure to give the bell enough weight) can be repeated on each remaining glass section and thus a series of bells can be prepared in a relatively short period.

A glass bell is now inserted into the loaded melting-point capillary, open end first (down), and allowed to fall (centrifuged if necessary) to the bottom. The assembled system (Fig. 4.2*b*) is then inserted onto the stage of a Thomas-Hoover Uni-Melt Capillary Melting Point Apparatus (Fig. 4.3)[4] or similar system (such as a Mel-Temp).

The temperature is rapidly raised to 15–20 °C below the expected boiling point (the temperature should be monitored carefully in the case of unknown substances), and then adjusted to a maximum rise rate of 2 °C/min and heated until a *fine stream* of bubbles is emitted from the glass bell. The heat control is then adjusted to drop the temperature. The boiling point is recorded at the point where the last escaping bubble collapses (i.e., when the vapor pressure of the substance equals the atmospheric pressure). The heater is then rapidly adjusted to again raise the temperature at 2 °C/min and induce a second stream of bubbles. This procedure may then be repeated several times. *A precise and sensitive temperature control system is essential to the successful application of this cycling technique, but it is not essential for obtaining satisfactory boiling-point data.*

Utilization of the conventional melting-point capillary as the "boiler" tube has the particular advantage that the boiling point of a liquid can readily be determined using a conventional melting-point apparatus. The illumination and magnification available make the observation of rate changes in the bubble stream easily seen. Inexpensive 10-μL GC injection syringes appear to be the most successful instrument to use for transferring the small quantities of liquids involved. The 3-in. needles normally supplied with the 10-μL barrels

**Figure 4.3 Thomas-Hoover melting-point determination device.** *(Courtesy of Thomas Scientific, Swedesboro, NJ.)*

[4]Thomas Scientific, P.O. Box 99, Swedesboro, NJ 08085.

will not reach the bottom of the capillary; liquid samples deposited on the walls of the tube, however, are easily and efficiently moved to the bottom by centrifugation. After the sample is packed in the bottom of the capillary tube, the glass bell is introduced. The glass bell is necessary because a conventional Siwoloboff fused-capillary insert would extend beyond the top of the melting-point tube; thus, capillary action between the "boiler" tube wall and the capillary insert would draw most of the sample from the bottom of the tube up onto the walls. This effect often precludes the formation of the requisite bubble stream.

Little loss of low-boiling liquids occurs (see Table 4.1). Furthermore, if the boiling point is overrun and the sample is suddenly evaporated from the bottom of the "boiler" capillary, it will rapidly condense on the upper (cooler) sections of the tube. These sections extend above the heat-transfer fluid or metal block. The sample can easily be recentrifuged to the bottom of the tube and a new determination of the boiling point begun. Note that if the bell cavity fills completely during the cooling point of a cycle, it is often difficult to reinitiate the bubble stream without first emptying the entire cavity by overrunning the boiling point.

Observed boiling points for a series of compounds, which boil over a wide range of temperatures, are summarized in Table 4.1.

Materials that are thermally stable at their boiling point will give identical values on repeat determinations. Substances that begin to decompose will give values that slowly drift after the first few measurements. The observation of color and/or viscosity changes, together with a variable boiling point, signal the need for caution in making extended repeat measurements.

Comparison of the boiling points obtained experimentally at various atmospheric pressures with reference boiling points at 760 torr is greatly facilitated by the use of pressure–temperature nomographs such as that shown in Figure 4.4. A straight line from the observed boiling point to the observed pressure will pass through the corrected boiling-point value. These values can be of practical importance when carrying out reduced pressure distillations.

***Table 4.1* Observed Boiling Points (°C)**

| Compound | Observed | Literature Value | Reference |
|---|---|---|---|
| Methyl iodide | 42.5 | 42–43 | *a* |
| Isopropyl alcohol | 82.3 | 82.3 | *b* |
| 2,2-Dimethoxypropane | 80.0 | 83.0 | *c* |
| 2-Heptanone | 149–159 | 151.1 | *d* |
| Cumene | 151–153 | 152.4 | *e* |
| Mesitylene | 163 | 164.7 | *f* |
| *p*-Cymene | 175–178 | 177.1 | *g* |
| Benzyl alcohol | 203 | 205.3 | *h* |
| Diphenylmethane | 263–265 | 265 | *i* |

*Note.* (Observed values are uncorrected for changes in atmospheric pressure (corrections all estimated to be less than ±0.5 °C.)

*Source. CRC Handbook of Chemistry and Physics*, 89th ed.; CRC Press: Boca Raton, FL, 2008–2009: [a]no. 6307, p. 3–306; [b]no. 9167, p. 3–442; [c]no. 3883, p. 3–190; [d]no. 5689; p. 3–274; [e]no. 6478, p. 3–314; [f]no. 10509, p. 3–540; [g]no. 6509, p. 3–316; [h]no. 780, p. 3–42; [i]no. 4498, p. 3–218.

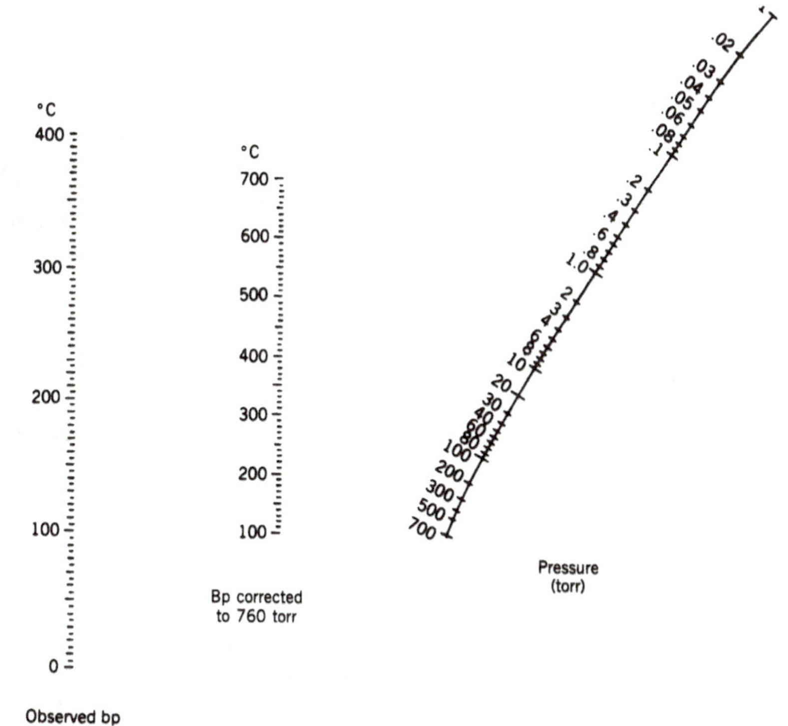


**Figure 4.4 Pressure–temperature nomograph.**

## DENSITY

Density, defined as mass per unit volume, is generally expressed as grams per milli-liter (g/mL) or grams per cubic centimeter (g/cm$^3$) for liquids. Accurate nondestructive procedures have been developed for the measurement of this physical constant at the microscale level. A micropycnometer (density meter), developed by Clemo and McQuillen requires approximately 2 μL (Fig. 4.5).[5]

This very accurate device gives the density to three significant figures. The system is self-filling, and the fine capillary ends do not need to be capped while temperature equilibrium is reached or during weighing (the measured values tend to degrade for substances boiling under 100 °C and when room temperatures rise much above 20 °C). In addition, the apparatus must first be tared, filled, and then reweighed on an *analytical* balance. A technique that results in less precise densities (good to about two significant figures), but which is far easier to use, is simply to substitute a 50- or 100-μL syringe for the pycnometer. The method simply requires weighing the syringe before and after filling it to a measured volume as in the conventional technique. With the volume and the weight of the liquid known, the density can be calculated. A further advantage of the syringe technique is that the pycnometer is not limited to a fixed volume. Although much larger samples are required, it is not inconvenient to utilize the entire sample obtained in the reaction for this measurement, since the material can be efficiently recovered from the syringe for additional

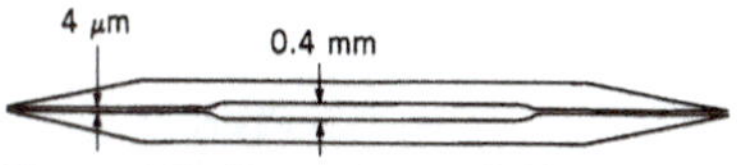

4 μm 0.4 mm

**Figure 4.5 Pycnometer of Clemo and McQuillen.** (From Schneider, F. L. Monographien aus dem Gebiete der qualitativen Mikroanalyse, *Qualitative Organic Microanalysis,* Vol. II; Benedetti-Pichler, A. A., Ed.; Springer: Vienna, 1964.)

[5]Clemo, G. R.; McQuillen, A. *J. Chem. Soc.* **1935,** 1220.

characterization studies. Because density changes with temperature, these measurements should be obtained at a constant temperature.

An alternative to the syringe method is to use *Drummond Disposable Microcaps* as pycnometers. These precision-bore capillary tubes, calibrated to contain the stated volume from end to end (accuracy ±1%), are available from a number of supply houses.[6] These tubes are filled by capillary action or by suction using a vented rubber bulb (provided). The pipets can be obtained in various sizes, but as with the syringe, volumes of 50, 75, or 100 μL are recommended. When using this method, handle the micropipet with forceps and not with your fingers (it's hot). The empty tube is first *tared*, and then filled and weighed again. The difference in these values is the weight of liquid in the pipet. For convenience, the pipet may be placed in a small container (10-mL beaker or Erlenmeyer flask) when the weighing procedure is carried out.

Two inexpensive micropycnometers can also be easily prepared: The first can be made from a Pasteur pipet as reported by Singh et al.[7] The volume of each individual pycnometer can be varied from 20 to 100 μL, or larger if desired. Values to three significant figures are obtained using an *analytical balance*, because evaporation is generally negligible, and if the pycnometer mouth is small.

The second pycnometer, by Pasto and co-workers, is made from a melting-point capillary tube.[8] In both of these techniques, the volume of the pycnometer must be determined. The procedure to determine the density involves the following steps. The empty micropycnometer is tared on an analytical balance, filled with the liquid in question, and reweighed (the difference in weights is the weight of the liquid). The sample is removed and the pycnometer is rinsed with acetone and dried. It is then filled with distilled water and reweighed. From the known[9] density of water at the given temperature the volume of water can be determined and thus the volume of the pycnometer. The volume of the original liquid sample also equals this value. The weight and volume of the sample are used to calculate its density.

## SOLIDS

### Melting Points

In general, the crystalline lattice forces holding organic solids together are distributed over a relatively narrow energy range. The melting points of organic compounds, therefore, are usually relatively sharp, that is, less than 2 °C. The range and maximum temperature of the melting point, however, are very sensitive to impurities. Small amounts of sample contamination by soluble impurities nearly always will result in melting-point depressions.

The drop in melting point is usually accompanied by an expansion of the melting-point range. Thus, in addition to the melting point acting as a useful

---

[6]Drummond Disposable Microcaps are available from Thomas Scientific, P.O. Box 99, Swedesboro, NJ 08085; and Sargent-Welch Scientific Co., a VWR company, P.O. Box 1026, Skokie, IL 60097.

[7]Singh, M. M.; Szafran, Z.; Pike, R. M. *J. Chem. Educ.* **1993,** *70,* A36; see also Ellefson-Kuehn, J., and Wilcox, C. J. *J. Chem. Educ.* **1994,** *71,* A150; and Singh, M. M.; Pike, R. M.; Szafran, Z. *Microscale and Selected Macroscale Experiments for General and Advanced General Chemistry;* Wiley: New York, 1995.

[8]Pasto, D.; Johnson, C.; Miller, M. *Experiments and Techniques in Organic Chemistry;* Prentice-Hall: Englewood Cliffs, NJ, 1992.

[9]Values for the density of water at various temperatures can be found in the *CRC Handbook of Chemistry and Physics.*

guide in identification, it also can be a particularly effective indication of sample purity.

***Procedure.*** In the microscale laboratory, two different types of melting-point determinations are carried out: (1) simple capillary melting points and (2) evacuated melting points.

***Simple Capillary Melting Point.*** Because the microscale laboratory utilizes the Thomas–Hoover Uni-Melt apparatus or a similar system for determining boiling points, melting points are conveniently obtained on the same apparatus. The Uni-Melt system utilizes an electrically heated and stirred silicone oil bath. The temperature readings require no correction in this case because the depth of immersion is held constant. (This assumes, of course, that the thermometer is calibrated to the operational immersion depth.) Melting points are determined in the same capillaries as boiling points. The capillary is loaded by introducing about 1 mg of material into the open end. The sample is then tightly packed (~2 mm) into the closed end by dropping the capillary down a length of glass tubing held vertically to the bench top. The melting-point tube is then ready for mounting on the metal stage, which is immersed in the silicone oil bath of the apparatus. If the melting point of the substance is expected to occur in a certain range, the temperature can be rapidly raised to ~2 °C below the expected value. At that point, the temperature rise should be adjusted to a maximum of 2 °C/min, which is the standard rate of change at which the reference determinations are obtained. The melting-point range is recorded from the temperature at which the first drop of liquid forms (point *e* in Fig. 4.6) to that at which the last crystal melts (point *m* in Fig. 4.6).

***Evacuated Melting Points.*** Many organic compounds begin to decompose at their melting points. This decomposition often begins as the melting point is approached and may adversely affect the values measured. The decomposition can be invariably traced to reaction with oxygen at elevated temperatures. If the melting point is obtained in an evacuated tube, therefore, much more accurate melting points can be obtained. These more reliable values arise not only from increased sample stability, but because several repeat determinations

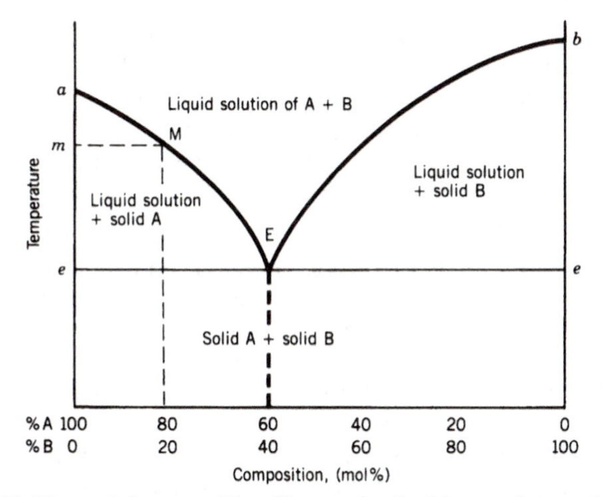


**Figure 4.6 Melting point composition diagram for the binary mixture,** $A + B$. In this diagram, *a* is the melting point of the solid *A*, *b* of solid *B*, *e* of eutectic mixture *E*, and *m* of the 80% *A*:20% *B* mixture, *M*.

can often be made on the same sample. The multiple measurements then may be averaged to provide more accurate data.

Evacuated melting points are quickly and easily obtained with a little practice. The procedure is as follows: Shorten the capillary portion of a Pasteur pipet to approximately the same length as a normal melting-point tube (Fig. 4.7*a*). Seal the capillary end by rotating in a microburner flame. Touch the pipet only to the very edge of the flame, and keep the large end at an angle below the end being sealed (Fig. 4.7*b*). This technique will prevent water from the flame being carried into the tube, where it will condense in the cooler sections. Then load 1–2 mg of sample into the drawn section of the pipet with a microspatula (Fig. 4.7*c*). Tap the pipet gently to seat the solid powder as far down the capillary as it can be worked (Fig. 4.7*d*). Then push the majority of the sample part way down the capillary with the same diameter copper wire that you used to seat the cotton plug in constructing the Pasteur filter pipet (Fig. 4.7*e*). Next, connect the pipet with a piece of vacuum tubing to a mechanical high-vacuum pump. Turn on the vacuum and evacuate the pipet for 30 s (Fig. 4.7*f*). With a microburner, gently warm the surface of the capillary tubing just below the drawn section. On warming, the remaining fragments of the sample (the majority of which has been forced farther down in the tube) will sublime in either direction away from the hot section. Once the traces of sample have been "chased" away, the heating is increased, and the capillary tube is collapsed, fused, and separated from the shank. The shank remains connected to the vacuum system (Fig. 4.7*g*). The vacuum system is then vented and the shank is discarded. The sample is tightly packed into the initially sealed end of the evacuated capillary by dropping it down a section of glass tubing, as in the case of packing open melting-point samples. After the sample is packed (~2 mm in length, see Fig. 4.7*h*), a section of the evacuated capillary about 10–15 mm above the sample is once more gently heated and collapsed by the microburner flame (Fig. 4.7*i*).

This procedure is required to trap the sample below the surface of the heated silicone oil in the melting-point bath, and thus avoid sublimation up the tube to cooler sections during measurement of the melting point. The operation is a little tricky and should be practiced a few times. It is very important that the tubing completely fuse. Now the sample is ready to be placed in the melting-point apparatus. The procedure beyond this point is the same as in the open capillary case, except that after the sample melts, it can be cooled, allowed to crystallize, and remelted several times, and the average value of the range reported. If these values begin to drift downward, the sample can be considered to be decomposing even under evacuated, deoxygenated conditions. In this case the first value observed should be recorded as the melting point and decomposition noted (mp xx dec, where dec = decompose).

***Mixture Melting Point.*** Additional information can often be extracted from the sensitivity of the melting point to the presence of impurities. Where two different substances possess identical melting points (not uncommon), it would be impossible to identify an unknown sample as either material based on the melting point alone. If reference standards of the two compounds are available, however, then mixtures of the unknown and the two standards can be prepared. It is important to prepare several mixtures of varying concentrations for melting-point comparisons, since the point of maximum depression need not occur on the phase diagram at the 50:50 ratio (see Fig. 4.6). In samples that do not exhibit any decomposition at the melting point, the prepared

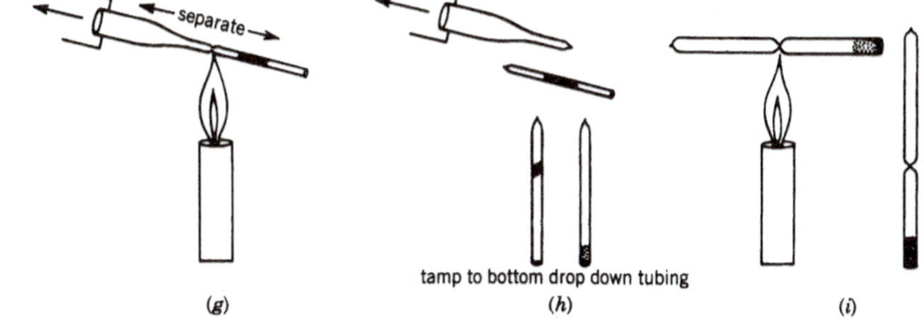


**Figure 4.7 Procedure for obtaining evacuated melting-point capillaries.**

mixtures should be first heated until a homogeneous melt is obtained. Each is then cooled and ground to a fine powder, and the definitive melting point is obtained on the ground sample. The melting points of the unknown and the mixed samples should be obtained simultaneously (the Uni-Melt stage will accept up to seven capillaries at one time). This is desirable because all the samples will then be heated at the same rate. The unknown sample and the mixture of the unknown with the correct reference will have identical values, but the mixture of the reference with a different substance will give a depressed melting point. This procedure is the classical step to positive identification of crystalline solids.

Mixtures of two different compounds only rarely fail to exhibit mixture melting-point depression, but it can happen. Some mixtures may not show a depression or show only a very small one, due to eutectic or compound formation. Elevation of the melting point has also been observed. Therefore, if mixture melting-point data are used for identification purposes, comparison of other physical constants or spectroscopic data is advocated to establish identity beyond any reasonable doubt.

## QUESTIONS

**4-1.** Room temperature is recorded when a density determination for a given substance is performed in the laboratory. Why?

**4-2.** Describe how you would determine the melting point of a substance that sublimes before it melts.

**4-3.** You are presented with four vials, each containing a white crystalline solid. Two are unlabeled vials containing pure samples of *trans*-cinnamic acid and urea, respectively. The other two are labeled reference standards for each sample. Devise a method for the proper identification of the unlabeled vials, knowing that the literature melting point for both *trans*-cinnamic acid and urea is 132.5–133 °C.

**4-4.** In the microscale method of determining boiling points, one heats the liquid until a steady stream of bubbles is observed coming out of the bell. The temperature is then lowered and the boiling point is read just as the bubbles stop. Why is this technique preferable to measuring the boiling point when the bubbles first start to appear?

**4-5.** What would you expect the observed boiling point to be at 10 torr of a liquid which has a boiling point of 300 °C at 760 torr?

$C_5H_6$apter 5

# MICROSCALE LABORATORY TECHNIQUES

H H H H H H

This chapter introduces the microscale organic laboratory techniques used throughout the experimental sections of the textbook. These must be mastered to be successful when working at this scale. Detailed discussions are given for each individual experimental technique. At the end of each discussion there is a list of the experiments in Chapters 6, 7 and 10W that use the technique. These lists should prove useful to instructors compiling experiments to be covered in the laboratory. The lists will also be handy for students who wish to examine the application of a particular technique to other experiments not covered in their laboratory sequence.

*As was the case with the fourth edition, a continued effort has been made to streamline the basic reference material from the text using our accompanying website (www.wiley. com/college/MOL5). The icon at the right is used throughout the text to indicate website material that will be of interest to the user. We hope this treatment of the laboratory will make the more important aspects of the basic text easier to access and will speed your laboratory work along.*

www

One of the principal hurdles in dealing with experimental chemistry is the isolation of pure materials. Characterization (identification) of a substance almost always requires a pure sample of the material. This is a particularly difficult demand of organic chemistry because most organic reactions generate several products. We are generally satisfied if the desired product is the major component of the mixture obtained. This chapter places a heavy emphasis on separation techniques.

## *Gas Chromatography*

TECHNIQUE 1

Technique 1 begins the discussion of the resolution (separation) of microliter quantities of liquid mixtures via preparative gas chromatography. Techniques 2 and 3 deal with semimicro adaptations of classical distillation routines that focus on the separation of liquid mixtures involving one to several milliliters of material.

Chromatography methods revolutionized experimental organic chemistry. These methods are by far the most powerful of the techniques for separating mixtures and isolating pure substances, either solids or liquids. Chromatography is the resolution (separation) of a multicomponent mixture (several hundred components in some cases) by distribution between two phases, one stationary and one mobile. The various methods of chromatography are categorized by the phases involved: column, thin-layer, and paper (all solid–liquid chromatography); partition (liquid–liquid chromatography); and vapor phase

**Chapter 5: $C_5H_6$, Propellane**
Wiberg and Walker (1982).

(gas–liquid chromatography, or simply gas chromatography). The principal mechanism these separations depend on is differential solubility, or adsorbtivity, of the mixture components in the two phases involved. That is, the components must exhibit different partition coefficients (see also Technique 4 for a detailed discussion of partition coefficients).

Gas chromatography (GC, sometimes called vapor-phase chromatography) is an extraordinarily powerful technique for separating mixtures of organic compounds. The stationary phase in GC is a high-boiling liquid and the mobile phase is a gas (the carrier gas). Gas chromatography can separate mixtures far better than distillation techniques can (see Technique 2 discussion).

**Preparative GC** separations, which involve perhaps 5–100 μL of material, require relatively simple instrumentation but sacrifice resolution for the ability to separate larger amounts of material.

**Analytical GC** separations require tiny amounts of material (often 0.1 μL of a very dilute solution), and can separate incredibly complex mixtures. The ability to work with small quantities of materials in analytical GC separations is an advantage at the microscale level. This analytical tool is used to analyze distillation fractions.

### GC Instrumentation

GC instrumentation can range from straightforward and relatively simple systems to systems with complex, highly automated, and relatively expensive components. A diagram of a common and simple GC typically used in an instructional laboratory is shown in Figure 5.1.

***Injection Port*** The analysis begins in a heated injection port. The sample mixture is introduced by syringe through a septum into the high-temperature chamber (injection port) through which the inert carrier gas (the mobile phase) is flowing. Helium and nitrogen are common carrier gases. The solubility of the sample in the carrier gas depends mostly on the vapor pressure of the substances in the sample. Heating the injection port helps to ensure the vaporization of less volatile samples. There are two major constraints on GC:

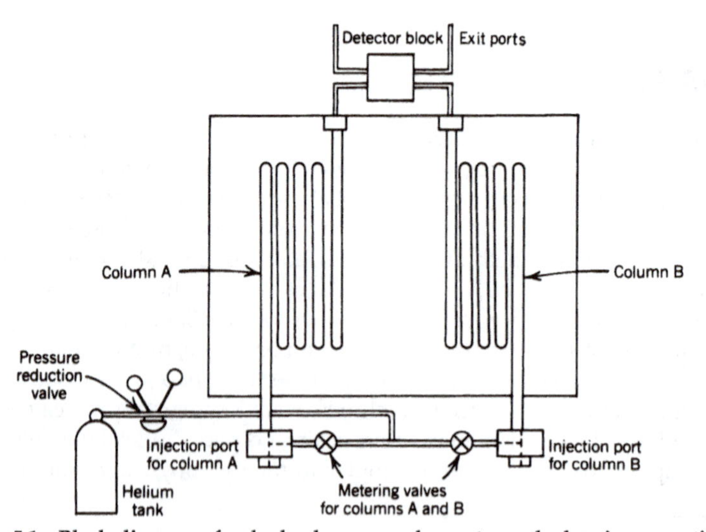


**Figure 5.1 Block diagram of a dual-column gas chromatograph showing essential parts.** *(Courtesy of GOW-MAC Instrument Co., 277 Brodhead Rd., Bethlehem, PA 18017.)*

The sample must be stable at the temperature required to cause vaporization, and the sample must have sufficient vapor pressure to be completely soluble in the carrier gas at the column operating temperatures.

*NOTE. When injecting a sample, always position your thumb or finger over the syringe plunger. This prevents a blow-back of the sample by the carrier gas pressure in the injection port.*

***Column*** The vaporized mixture is swept by the carrier gas from the injection port onto the column. Bringing the sample mixture into intimate contact with the column begins the separation process. The stationary liquid phase in which the sample will partially dissolve is physically and/or chemically bonded to inert packing material (often called the support) in the column. Gas-chromatographic columns are available from manufacturers in a variety of sizes and shapes. In the diagram of the GOW-MAC instrument (Fig. 5.1), two parallel coiled columns are mounted in an oven. Considerable oven space can be saved and better temperature regulation achieved if the columns are coiled. Temperature regulation is particularly important, because column resolution degrades rapidly if the entire column is not at the same temperature. Most liquid mixtures need a column heated above ambient temperatures to achieve the vapor pressure the separation requires.

The mixture is separated as the carrier gas sweeps the sample through the column. Columns are usually made from stainless steel, glass, or fused silica. The diameter and length of the column are critical factors in separating the sample mixture.

***Packed Columns.*** In packed columns the liquid (stationary) phase in contact with the sample contained in the mobile gas phase is maximized by coating a finely divided inert support with the nonvolatile liquid. The coated support is carefully packed into the column so as not to develop empty spaces. Packed columns are usually $\frac{1}{4}$ or $\frac{1}{8}$ inch in diameter and range from 4 to 12 feet in length. These columns are particularly useful in the microscale laboratory, since they can be used for both analytical and preparative GC. Simple mixtures of 20–80 μL of material can often be separated into their pure components and collected at the exit port of the detector. Smaller samples (0.2–2.0 μL range) will exhibit better separation.

***Capillary Columns.*** Capillary columns have no packing; the liquid phase is simply applied directly to the walls of the column. These columns are referred to as wall-coated, open-tubular (WCOT) columns. The reduction in surface area (compared to packed columns) is compensated for by tiny column diameters (perhaps 0.1 mm) and impressive lengths (100 m is not uncommon). Capillary columns are the most powerful columns used for analytical separations. Mixtures of several hundred compounds can be completely resolved on a capillary GC column. These columns require a more sophisticated and expensive chromatography instrument. Capillary columns, because of their tiny diameters, can accommodate only very small samples, perhaps 0.1 μL or less of a dilute solution. Capillary columns cannot be used for preparative separations.

***Liquid Phase*** Once the sample is introduced on the column (in the carrier gas), it will undergo partition into the liquid phase. The choice of the liquid phase is particularly important because it directly affects the relative distribution coefficients.

In general, the stationary liquid phase controls the partitioning of the sample by two criteria. First, if little or no interaction occurs between the sample components and the stationary phase, the boiling point of the materials will determine the order of elution. Under these conditions, the highest boiling species will be the last to elute. Second, the functional groups of the components may interact directly with the stationary phase to establish different partition coefficients. Elution then depends on the particular binding properties of the sample components.

Some typical materials used as stationary phases are shown below.

| Name | Stationary Phase | Maximum Temperature (°C) | Mechanism of Interaction |
|---|---|---|---|
| Silicone oil DC710, etc. | $R_3Si[OSiR_2]_nOSiR_3$ | 250 | According to boiling point |
| Polyethylene glycol (Carbowax) | $HO[CH_2CH_2O]_nCH_2CH_2OH$ | 150 | Relatively selective toward polar components |
| Diisodecyl phthalate | $o$-$C_6H_4[CO_2$-isodecyl$]_2$ | 175 | According to boiling point |

***Oven Temperature*** The temperature of the column will also affect the separation. In general, the elution time of a sample will decrease as the temperature is increased. That is, retention times are shorter at higher temperatures. Higher boiling components tend to undergo diffusion broadening at low column temperatures because of the increase in retention times. If the oven temperature is too high, however, equilibrium partitioning of the sample with the stationary phase will not be established. Then the components of the mixture may elute together or be incompletely separated. Programmed oven temperature increases can speed up elution of the higher boiling components, but suppress peak broadening and therefore increase resolution. Temperature-programming capabilities require more sophisticated ovens and controllers.

***Flow Rate*** The flow rate of the carrier gas is another important parameter. The rate must be slow enough to allow equilibration between the phases, but fast enough to ensure that diffusion will not defeat the separation of the components.

***Column Length*** As noted, column length is an important factor in separation performance. As in distillations, column efficiency is directly proportional to column height, which determines the number of evaporation–condensation cycles. In a similar manner, increasing the length of a GC column allows more partition cycles to occur. Difficult-to-separate mixtures, such as the xylenes (very similar boiling points: *o*-xylene, 144.4 °C; *m*-xylene, 139.1 °C; and *p*-xylene, 138.3 °C), have a better chance of being separated on longer columns. In fact, both GC and distillation resolution data are described using the same term, *theoretical plates* (see Technique 2 and Experiments [3C] and [3D]).

***Detector and Exit Port*** A successfully separated mixture will elute as its individual components at the instrument's exit port (also temperature controlled). To monitor the exiting vapors, a detector is placed in the gas stream just before the exit port (Fig. 5.1). After passing through the detector, the carrier gas and the separated sample components are then vented.

One widely used detector is the nondestructive, thermal conductivity detector, sometimes called a hot-wire detector. A heated wire in the gas stream changes its electrical resistance when a substance dilutes the carrier gas and thus

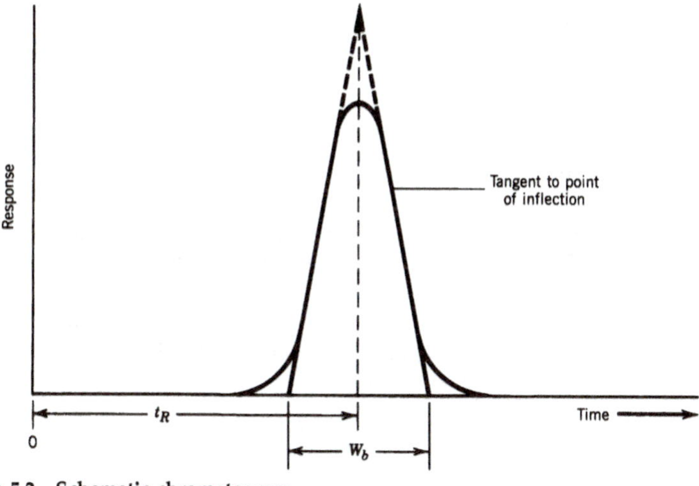


**Figure 5.2 Schematic chromatogram.**

changes its thermal conductivity. Helium has a higher thermal conductivity than most organic substances. When substances other than helium are present, the conductivity of the gas stream changes, which changes the resistance of the heated wire. The change in resistance is measured by comparing it to a reference detector mounted in a second (parallel) gas stream (Wheatstone bridge). The resulting electrical signal is plotted on a chart recorder, where the horizontal axis is time and the vertical axis is the magnitude of the resistance difference. The plot of resistance difference versus time is referred to as the gas *chromatogram.* The retention time ($t_R$) is defined as the time from sample injection to the time of maximum peak intensity. The baseline width ($W_b$) of a peak is defined as the distance between two points where tangents to the points of inflection cross the baseline (Fig. 5.2).

Capillary GC systems, and other GC systems used only for analytical separations, often use a flame-ionization detector (FID). In a flame-ionization detector, the gas eluting from the GC column is mixed with air (or oxygen) and hydrogen, and burned. The conductivity of the resulting flame is measured; it changes with the ionic content of the flame, which is proportional to the amount of carbon (from organic material) in the flame. The advantage of an FID is its high sensitivity; amounts of less than a microgram are easily detected. Its disadvantage is that it destroys (burns) the material it detects.

***Theoretical Plates*** It is possible to estimate the number of theoretical plates (directly related to the number of distribution cycles) present in a column for a particular substance. The parameters are given in the relationship[1]

$$n = 16\left(\frac{t_R}{W_b}\right)^2$$

where the units of retention time ($t_R$) and baseline width ($W_b$) are identical (minutes, seconds, or centimeters). As in distillation columns, the larger the number of theoretical plates, *n,* the higher the resolution of the column and the better the separation.

[1]Berg, E. W. *Physical and Chemical Methods of Separation;* McGraw-Hill: New York, 1963, p. 111.

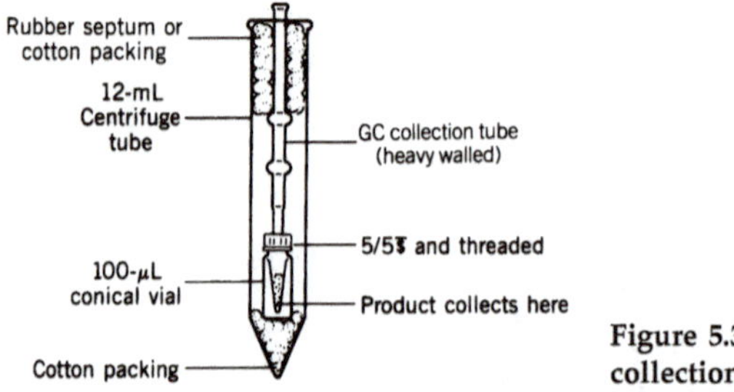


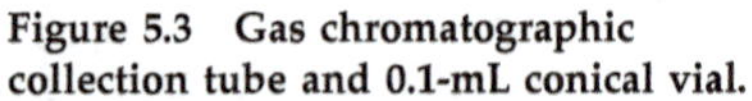
**Figure 5.3 Gas chromatographic collection tube and 0.1-mL conical vial.**

The efficiency of a system may be expressed as the *height-equivalent theoretical plate* (HETP) in centimeters (or inches) per plate. The HETP is related to the number of theoretical plates $n$ by

$$\text{HEPT} = \frac{L}{n}$$

where $L$ is the length of the column, usually reported in centimeters. The smaller the HETP, the more efficient the column.

The number of theoretical plates available in fractional distillation columns is limited by column holdup (see Techniques 2, 3, and website discussion of distillation theory). Thus, distillations of less than 500 μL are generally not practical. Gas-chromatographic columns, on the other hand, operate most efficiently at the microscale or submicroscale levels, where 500 μL would be an order of magnitude (even 3–8 orders of magnitude in the case of capillary columns) too large.

WWW

***Fraction Collection*** Sequential collection of separated materials can be made by attaching suitable sample condensing tubes to the exit port (see Fig. 3.6).

***Procedure for Preparative Collection.*** The collection tube (oven dried until 5 min before use) is attached to the heated exit port by the metal 5/5 ₮ joint. Sample collection is begun 30 s before detection of the expected peak on the recorder (based on previously determined retention values; refer to your local laboratory instructions) and continued until 30 s after the recorder's return to baseline. After the collection tube is detached, the sample can be analyzed directly when collected into a GC NMR collection tube[2] or transferred to the 0.1-mL conical GC collection vial. After the collection tube is joined to the vial (preweighed with cap) by the 5/5 ₮ joints, the system is centrifuged to force the sample down into the vial (Fig. 5.3). The collection tube is then removed, and the vial is capped and reweighed.

The efficiency of collection can exceed 90% with most materials, even with relatively low-boiling substances. In the latter case, the collection tube, after attachment to the instrument, is wrapped with a paper tissue. As the (oven-dried) tube is being wrapped, it is also being flushed by the carrier gas, which removes any traces of water condensation. The wrapping is then saturated with liquid nitrogen to cool the collection tube.

Preparative GC in the microscale laboratory often replaces the macroscale purification technique of fractional distillation. Distillation is impractical with less than 500 μL of liquid.

[2]Bressette, A. R. *J. Chem. Educ.* **2001**, *78*, 366.

Refer to Experiment 3105-7 for specific experimental details on preparative GC applied to the separation of a number of binary (two-component) mixtures. These are designed as practice examples to give you experience with sample collection.

## QUESTIONS

**5-1.** What is the main barrier to separating liquid mixtures of less than 500 μL by distillation?

**5-2.** A sample mixture of ethyl benzoate (bp 212 °C) and dodecane (bp 216.2 °C) is injected on two GC columns. Column A has DC710 silicone oil as the stationary phase, and column B uses polyethylene glycol as the stationary phase. Which substance would be certain to elute first from column A and would the same material be expected to elute first from column B? Which column, A or B, would be expected to give the better separation of these two substances?

**5-3.** Question 5-2 refers to separating a mixture of two high boiling liquids by gas chromatography. These materials have similar boiling points. List several GC variables and conditions that would make it easier to separate these substances by gas chromatography.

**5-4.** Capillary GC columns have better resolution than packed columns even though the enormous surface area provided by the packing material is absent in capillary columns. Why?

**5-5.** Preparative GC requires packed columns. Why is this technique limited to these lower resolution columns?

*NOTE. Gas chromatographic purification of reaction products is suggested in the following experiments: Experiments [2], [3C], [3D], [5A], [5B], [8C], [9], [10], [13], [17], and [32].*

# Simple Distillation

TECHNIQUE 2

Distillation is the process of heating a liquid to the boiling point, condensing the heated vapor by cooling, and returning either a portion of, or none of, the condensed vapors to the distillation vessel. Distillation differs from reflux (see p. 23) only in that at least some of the condensate is removed from the boiling system. Distillations in which a fraction of the condensed vapors are returned to the boiling system are often referred to as being under "partial reflux." Two types of distillations will be described. Students are encouraged to refer to and study the more detailed discussion of distillation theory. The website also has a detailed discussion of the theory of steam distillation, which is used in Experiments [11C] and [32]. There are times when ordinary distillation may not be feasible for the separation of a liquid from dissolved impurities. The compound of interest may boil at a high temperature that is difficult to control with a simple apparatus, or it may tend to decompose or oxidize at high temperatures. If the compound is only sparingly soluble in water and any small amount of water can be removed with a drying agent, steam distillation may be the technique of choice. Compounds that are immiscible in water have very large positive deviations from Raoult's law. Therefore, the boiling temperature is generally lower than that of water and the compound.

www

Simple distillation involves the use of the distillation process to separate a liquid from minor components that are nonvolatile, or that have boiling points at least 30–40 °C above that of the major component. A typical setup for a macroscale distillation of this type is shown in Figure 5.4. At the microscale

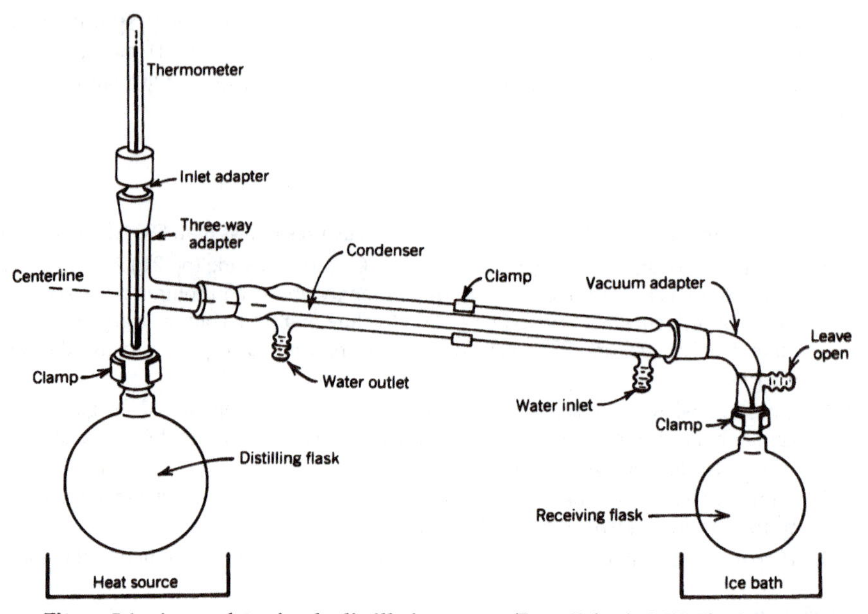


**Figure 5.4 A complete simple distillation setup.** (From Zubrick, J. W. *The Organic Chem Lab Survival Manual,* 7th ed.; Wiley: New York, 2008. Reprinted by permission of John Wiley & Sons, Inc., New York.)

level, working with volumes smaller than 500 μL, GC techniques (see Technique 1) have replaced conventional microdistillation processes.[3] Semimicroscale simple distillation is an effective separation technique for volumes in the range of 0.5–2 mL. Apparatus that achieve effective separation of mixture samples in this range have been developed. One of the most significant of these designs is the classic Hickman still, shown in Figure 5.5. This still is used in several ways in the experiments described in Chapters 6, 7, and 10W for purifying solvents, carrying out reactions, and concentrating solutions. Experiment 3105-1 introduces the use of the Hickman still.

www

In a distillation where liquid is to be separated from a nonvolatile solute, the vapor pressure of the liquid is lowered by the presence of the solute, but the vapor phase consists of only one component. Thus, except for the incidental transfer of non-volatile material by splashing, the material condensed should consist only of the volatile component (see Experiment 3105-1).

We can understand what is going on in a simple distillation of two volatile components by referring to the phase diagrams shown in Figures 5.6 and 5.7. Figure 5.6 is the phase diagram for hexane and toluene. The boiling points of these liquids are separated by 42 °C. Figure 5.7 is the phase diagram for methylcyclohexane and toluene. Here the boiling points are separated by only 9.7 °C.

Imagine a simple distillation of the hexane–toluene pair in which the liquid in the pot is 50% hexane. In Figure 5.6, when the liquid reaches 80.8 °C it will be in equilibrium with vapor having a composition of 77% hexane. This result is indicated by the line *A–B*. If this vapor is condensed to a liquid of the same composition, as shown by line *B–C*, we will have achieved a significant enrichment of the condensate with respect to hexane. This change in composition is referred to

[3]Schneider, F. L. In *Monographien aus dem Gebiete der qualitativen Mikroanalyse,* Vol. II: *Qualitative Organic Microanalysis;* Benedetti-Pichler, A. A., Ed.; Springer-Verlag: Vienna, 1964; p. 31.

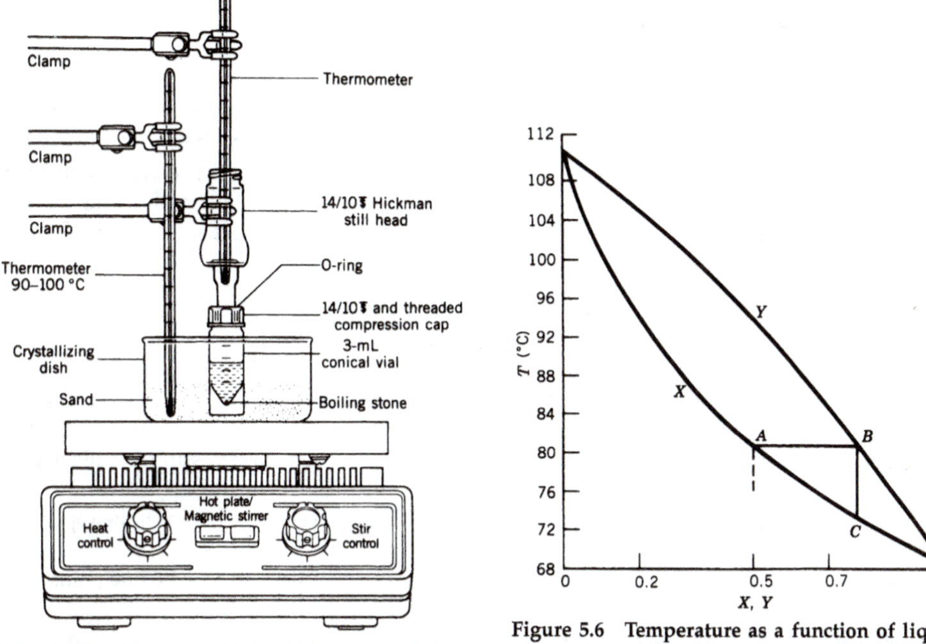


**Figure 5.5 Hickman still (14/10 ₸ with conical vial [3 mL]).**

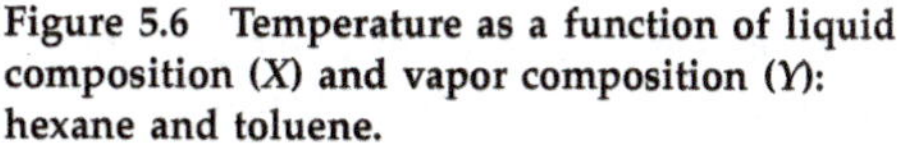

**Figure 5.6 Temperature as a function of liquid composition (X) and vapor composition (Y): hexane and toluene.**

as a simple distillation. The process of evaporation and condensation is achieved by the theoretical construct known as a *theoretical plate*. When this distillation is actually done with a Hickman still, some of the mixture will go through one evaporation and condensation cycle, some will go through two of these cycles, and some may be splashed more directly into the collar. A resolution (separation) of between one and two theoretical plates is generally obtained.

Referring to Figure 5.7, if we consider the same process for a 50% mixture of methylcyclohexane and toluene, the methylcyclohexane composition will increase to 58% for a distillation with one theoretical plate. Simple distillation may thus provide adequate enrichment of the MVC (more volatile constituent) if the boiling points of the two liquids are reasonably well separated, as they are for hexane and toluene. If the boiling points are close together, as they are for methylcyclohexane and toluene, the simple distillation will not provide much enrichment.

As we continue the distillation process and remove some of the MVC by condensing it, the residue in the heated flask becomes less rich in the MVC. This means that the next few drops of condensate will be less rich in the MVC. As the distillation is continued, the condensate becomes less and less rich in the MVC.

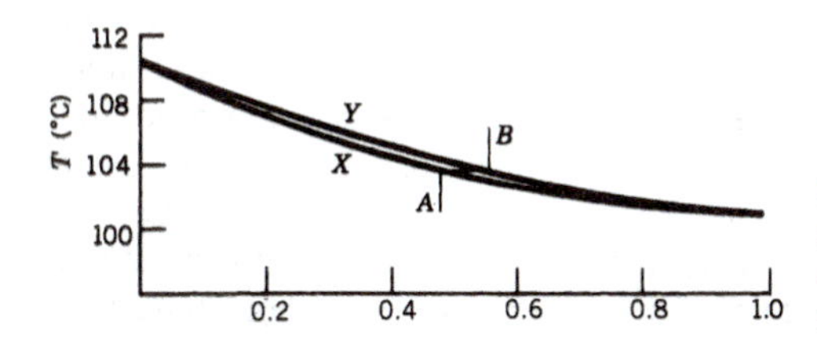


**Figure 5.7 Temperature as a function of liquid composition (X) and vapor composition (Y): methylcyclohexane and toluene.**

We can improve on simple distillation by repeating the process. For example, we could collect the condensate until about one-third is obtained. Then we could collect a second one-third aliquot in a separate container. Our original mixture would then be separated into three fractions. The first third would be richest in the MVC and the final third (the fraction remaining in the distillation pot) would be the richest in the least volatile component. If the MVC were the compound of interest, we could re-distill the first fraction collected (from a clean flask!) and collect the first third of the material condensing in that process. This simplest of all fractional distillation strategies is used in Experiment 3105-1.

## QUESTIONS

**5-6.** What is the major drawback of trying to distill a 500-μL mixture of liquids, all with boiling points below 200 °C?

**5-7.** How might you separate the mixture discussed in question 5-6 if distillation were unsuccessful? Explain your choice.

**5-8.** If starting with an equal mixture of hexane and toluene, approximate the composition of hexane if the vapor at 94 °C is condensed to a liquid using the data presented in Figure 5.6.

**5-9.** Why do simple distillations require that the components of the mixture to be separated have boiling points that are separated by 40 °C or more?

**5-10.** Which constituent of an equimolar mixture makes the larger contribution to the vapor pressure of the mixture, the higher or lower boiling component? Explain.

*NOTE. The following experiments use Technique 2: Experiments [3A], [3B], [11C], [29], and [32].*

www → *[$3A_{adv}$].*

## TECHNIQUE 3

## *Fractional Distillation*

Fractional distillation can occur in a distillation system containing more than one theoretical plate. This process must be used when the boiling points of the components differ by less than 30–40 °C and fairly complete separation is desired. A fractionating column is needed to accomplish this separation. As discussed previously, a liquid– vapor composition curve (Fig. 5.8) shows that the lower boiling component of a binary mixture makes a larger contribution to the vapor composition than does the higher boiling component. On condensation, the liquid formed will be richer in the lower boiling component. This condensate will not be pure, however, and, in the case of components with close boiling points, it may be only slightly enriched. If the condensate is vaporized a second time, the vapor in equilibrium with this liquid will show a further enrichment in the lower boiling component. The trick to separating liquids with similar boiling points is to repeat the vaporization–condensation cycle many times. Each cycle is one *theoretical plate.* Several column designs, which achieve varying numbers of theoretical plates, are available for use at the macro level (Fig. 5.9).

Most distillation columns are designed so that fractionation efficiency is achieved by the very large surface area in contact with the vapor phase (and

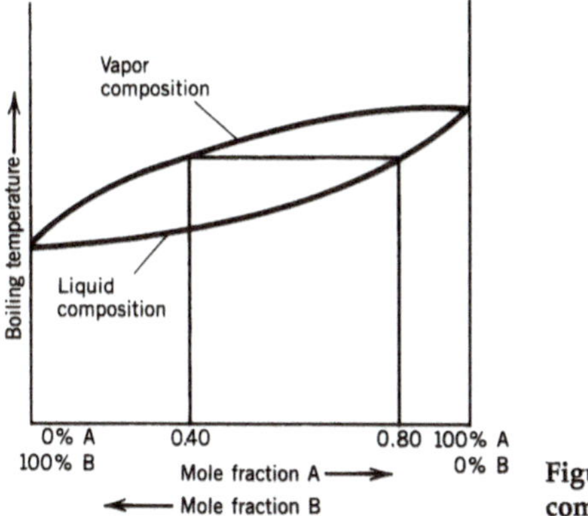


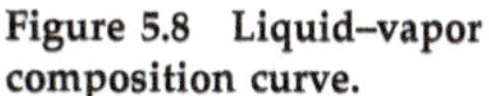

**Figure 5.8 Liquid–vapor composition curve.**

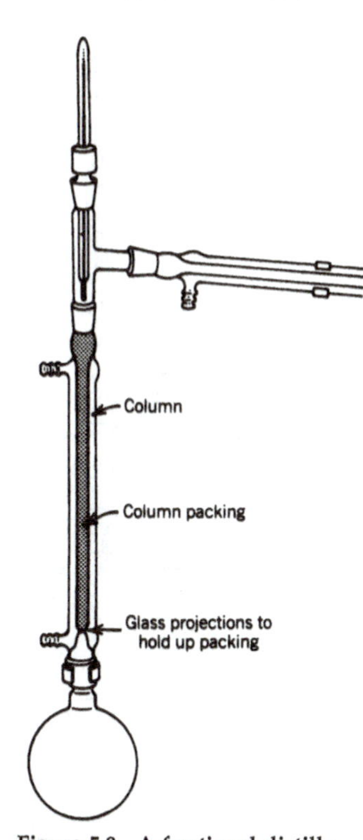


**Figure 5.9 A fractional distillation setup.** (From Zubrick, J.W. *The Organic Chem Lab Survival Manual,* 7th ed.; Wiley: New York, 2008. Reprinted with permission of John Wiley & Sons, Inc., New York.)

very similar to the way increased resolution is obtained on a GC column (see Technique 1). This increased surface area can be accomplished by packing the fractionating column with wire gauze or glass beads. Unfortunately, a large volume of liquid must be distributed over the column surface in equilibrium with the vapor. Furthermore, the longer the column the more efficient it becomes (see Technique 1), but longer columns also require additional liquid phase. The amount of liquid phase required to fill the column with a liquid—vapor equilibrium is called *column holdup*. Column holdup is essentially lost from the distillation because this volume can never go past the top of the distillation column; it can only return to the distillation pot upon cooling. Column holdup can be large compared to the total volume of material available for the distillation. With mixtures of less than 2 mL, column holdup precludes the use of the most common fractionation columns. Columns with rapidly spinning bands of metal gauze or Teflon have very low column holdup and have a large number of theoretical plates relative to their height (Fig. 5.10).

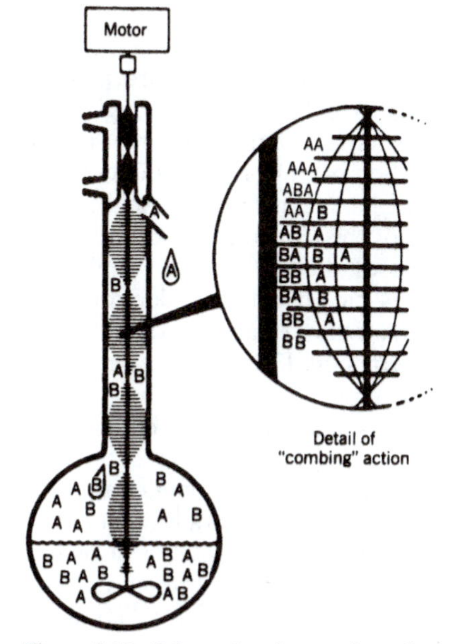


**Figure 5.10 Schematic of a metal-mesh, spinning-band still.**

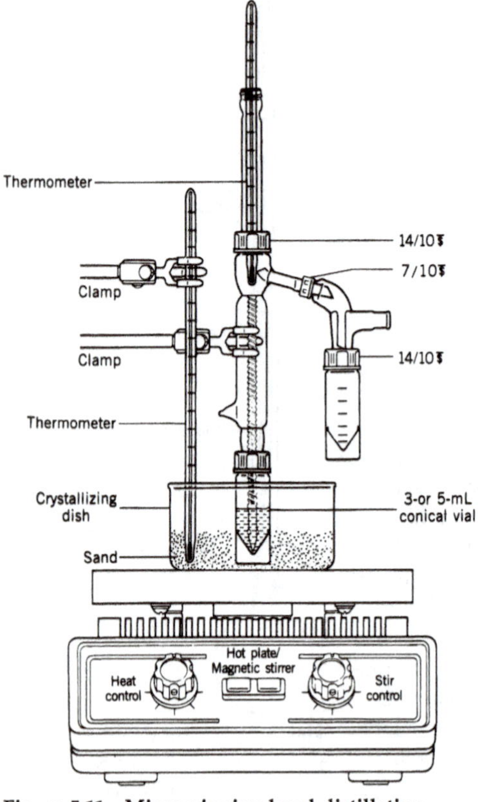


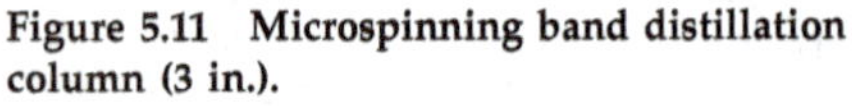

**Figure 5.11 Microspinning band distillation column (3 in.).**

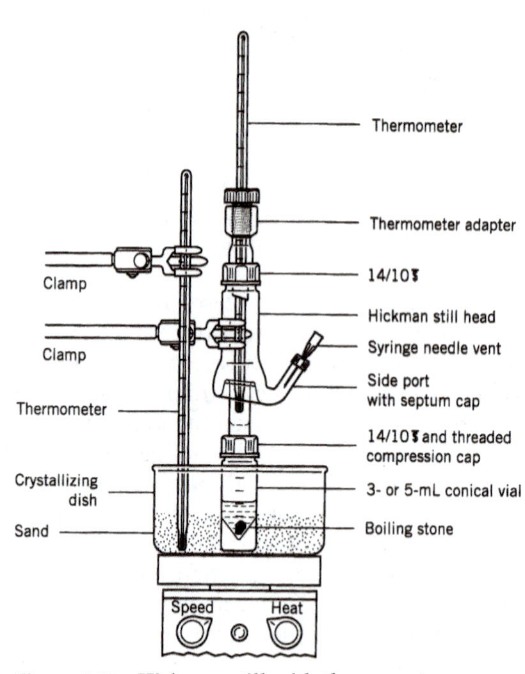


**Figure 5.12 Hickman still with thermometer adapter.**

Microscale spinning-band distillation apparatus (Fig. 5.11) can achieve nearly 12 theoretical plates and are simple enough to be used in the instructional laboratory. This still has a Teflon band that fits closely inside an insulated glass tube. The Teflon band has spiral grooves which, when the band is spun (1000–1500 rpm), rapidly return condensed vapor to the distillation pot. A powerful extension of this apparatus uses a short spinning band inside a modified Hickman still head (see Fig. 3.15). These stills are called Hickman–Hinkle stills; 4-cm Hickman–Hinkle columns can have more than 10 theoretical plates. The commercially available 2.5-cm version is rated at 6 theoretical plates. Experiments [3C] and [3D] involve fractional distillation with spinning-band columns.

The thermometer is positioned directly down the center of the distillation column, with the bulb just at the bottom of the well. It is very important to position both the still and the thermometer as vertically as possible; the thermometer must not touch the glass walls of the column (Fig. 5.12). Experiment 3105-1 uses the Hickman still for fractional distillation. A two-theoretical-plate distillation is obtained with this system on a two-component mixture by carrying out two sequential fractional distillations.

WWW → For a more detailed discussion of how spinning bands work, see the discussion of distillation.

## QUESTIONS

**5-11.** Why is it very important that the hot vapor in microscale distillations climb the column very slowly?

**5-12.** Why might Teflon be the material of choice for constructing microscale spinning bands?

**5-13.** The spinning band overcomes two major problems of microscale distillations by wiping the liquid condensate rapidly from the column walls. What are these problems?

**5-14.** Why are spinning bands so effective at increasing the number of theoretical plates in distillation columns?

**5-15.** Why is steam distillation often used to isolate and purify naturally occurring plant substances?

*NOTE. The following experiments use Technique 3: Experiments [3C] and [3D].*

# Solvent Extraction

Solvent extraction is frequently used in the organic laboratory to separate or isolate a desired compound from a mixture or from impurities. Solvent extraction methods are readily adapted to microscale work because small quantities are easily manipulated in solution. Solvent extraction methods are based on the solubility characteristics of organic substances in the solvents used in a particular separation procedure. Liquid–liquid and solid–liquid extractions are the two major types of extractions used in the organic laboratory.

### Intermolecular Properties: Solubility

Substances vary greatly in their solubility in various solvents, but a useful and generally true principle is that a substance tends to dissolve in a solvent that is chemically similar to itself. In other words, *like dissolves like.* The significant exceptions to this general statement are seen when solubilities are determined by acid–base properties.

Thus, to be soluble in water a compound needs to have some of the molecular characteristics of water. Alcohols, for example, have a hydroxyl group (—OH) bonded to a hydrocarbon chain or framework (R—OH). The hydroxyl group can be thought of as effectively half a water ($H_2O$) molecule; its polarity is similar to that of water. This polarity is due to the charge separation arising from the different electronegativities of the hydrogen and oxygen atoms. The O—H bond, therefore, is considered to have partial ionic character.

$$-\overset{\delta-}{\ddot{\underset{..}{O}}}-\overset{\delta+}{H}$$

Partial ionic character of the hydroxyl group

This polar, or partial ionic, character leads to relatively strong hydrogen bond formation between molecules with hydroxyl groups. Strong hydrogen bonding (shown here for the ethanol–water system) occurs in molecules that have a hydrogen atom attached to an oxygen, nitrogen, or fluorine atom—all three are quite electronegative atoms.

$$CH_3-CH_2-\overset{\delta-}{\ddot{O}}:-\overset{\delta+}{H}$$

Ethanol

$$CH_3-CH_2-\overset{\delta-}{O}(H)\cdots\overset{\delta+}{H}-O(H)\cdots\overset{\delta+}{H}\cdots\overset{\delta-}{O}-CH_2-CH_3$$

Hydrogen bond formation

**Table 5.1 Comparison of Boiling Point Data**

| Name | Formula | MW | bp (°C) |
|---|---|---|---|
| Ethanol | $CH_3CH_2OH$ | 46 | 78.3 |
| Propane | $CH_3CH_2CH_3$ | 44 | −42.2 |
| Methyl acetate | $CH_3CO_2CH_3$ | 74 | 54 |
| Diethyl ether | $(CH_3CH_2)_2O$ | 74 | 34.6 |
| Ethene | $CH_2{=}CH_2$ | 28 | −102 |
| Methylamine | $CH_3NH_2$ | 31 | −6 |

The hydroxyl end of the ethanol molecule is very similar to water. When ethanol is added to water, therefore, they are miscible in all proportions. That is, ethanol is completely soluble in water and water is completely soluble in ethanol. This degree of solubility occurs because the attractive forces between the two molecules are nearly as strong as those between two water molecules; however, the attraction in the first case is somewhat weakened by the presence of the nonpolar ethyl group, $CH_3CH_2—$. Hydrocarbon groups attract each other only weakly, as demonstrated by their low melting and boiling points. Three examples of the contrast in boiling points between compounds of different structure, but similar molecular weight, are summarized in Table 5.1. Molecules that attract each other weakly (lower intermolecular forces) have lower boiling points.

Ethanol is completely miscible with water, but the solubility of octanol in water is less than 1%. Why the difference in solubilities between these two alcohols? The dominant structural feature of ethanol is its polar hydroxyl group; the dominant structural feature of octanol is its nonpolar alkyl group:

$$CH_3—CH_2—CH_2—CH_2—CH_2—CH_2—CH_2—CH_2—\overset{\delta-}{\ddot{O}}\!:—\overset{\delta+}{H}$$

Octanol

$$CH_3—CH_2—\ddot{O}—CH_2—CH_3$$

Diethyl ether

As the size of the hydrocarbon section of the alcohol molecule increases, the intermolecular attraction between the polar hydroxyl groups of the alcohol and the water molecules is no longer strong enough to overcome the hydrophobic (lacking attraction to $H_2O$) nature of the nonpolar hydrocarbon section of the alcohol. On the other hand, octanol has a large nonpolar hydrocarbon group as its dominant structural feature. We might, therefore, expect octanol to be more soluble in less polar solvents, and, in fact, octanol is completely miscible with diethyl ether. Ethers are weakly polar solvents because a C—O bond is much less polar than an O—H bond (carbon is less electronegative than oxygen). Because both octanol and diethyl ether are rather nonpolar, each is completely soluble in the other. For compounds with both polar and nonpolar groups, in general, those compounds with five or more carbon atoms in the hydrocarbon portion of the molecule will be more soluble in nonpolar solvents, such as pentane, diethyl ether, or methylene chloride. Figure 5.13 summarizes the solubilities of a number of straight-chain alcohols, carboxylic acids, and hydrocarbons in water. As expected, most monofunctional compounds with more than five carbon atoms have solubilities similar to the hydrocarbons.

Several additional relationships between solubility and structure have been observed.

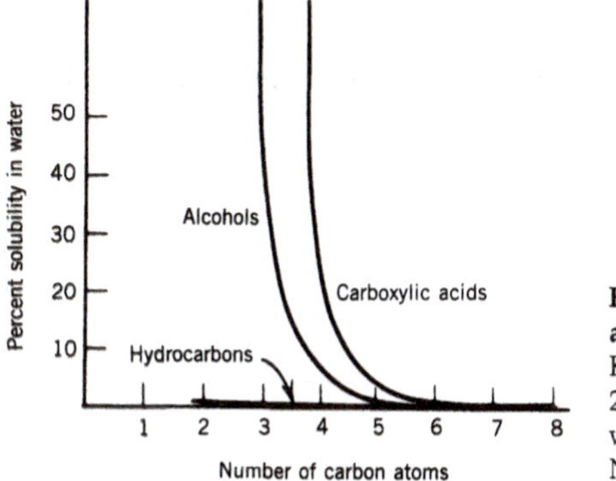


**Figure 5.13 Solubility curve of acids, alcohols, and hydrocarbons.** (From Kamm, O. *Qualitative Organic Analysis*, 2nd ed.; Wiley: New York, 1932. Reprinted with permission of John Wiley & Sons, New York.)

**1.** Branched-chain compounds have greater water solubility than their straight-chain counterparts, as illustrated in Table 5.2 with a series of alcohols.

**2.** The presence of more than one polar group in a compound will increase that compound's solubility in water and decrease its solubility in nonpolar solvents. For example, sugars, such as cellobiose, contain multiple hydroxyl and/or acetal groups and are water soluble and ether insoluble. Cholesterol, which has only a single hydroxyl group on its 27 carbon atoms, is insoluble in water and quite soluble in ether.

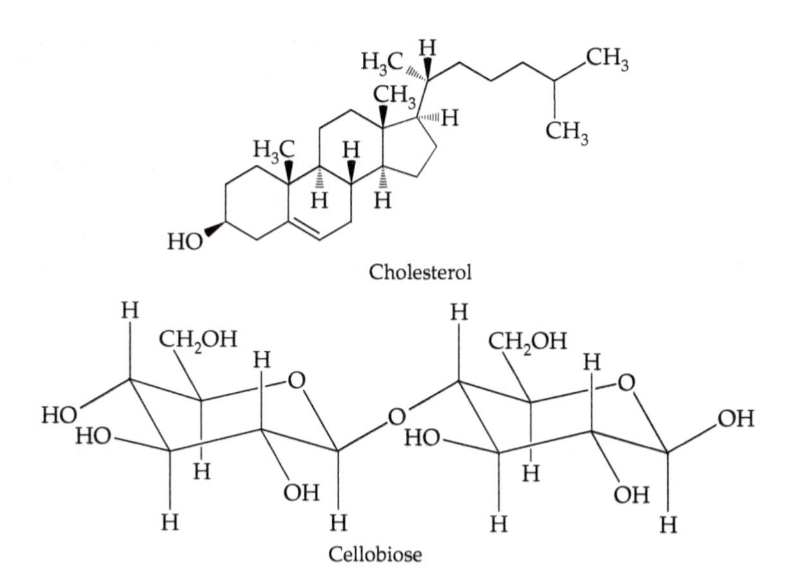


***Table 5.2*** **Water Solubility of Alcohols**

| Name | Structural Formula | Solubility (g/100 g $H_2O$ at 20 °C) |
|---|---|---|
| Hexanol | $CH_3(CH_2)_4CH_2OH$ | 0.6 |
| Pentanol | $CH_3(CH_2)_3CH_2OH$ | 2.2 |
| 2-Pentanol | $CH_3(CH_2)_2CH(OH)CH_3$ | 4.3 |
| 2-Methyl-2-butanol | $(CH_3)_2C(OH)CH_2CH_3$ | 11.0 |

**Table 5.3 Water Solubility of Amines**

| Name | Structural Formula | Solubility (g/100 g $H_2O$ at 25 °C) |
|---|---|---|
| Ethylamine | $CH_3CH_2NH_2$ | ∞ |
| Diethylamine | $(CH_3CH_2)_2NH$ | ∞ |
| Trimethylamine | $(CH_3)_3N$ | 91 |
| Triethylamine | $(CH_3CH_2)_3N$ | 14 |
| Aniline | $C_6H_5$—$NH_2$ | 3.7 |
| 1,4-Diaminobenzene | $H_2N$—$C_6H_4$—$NH_2$ | 3.8 |

**3.** The presence of a chlorine atom, even though it lends some partial ionic character to the mostly covalent C—Cl bond, does not normally impart water solubility to a compound. In fact, compounds such as methylene chloride ($CH_2Cl_2$), chloroform ($CHCl_3$), and carbon tetrachloride ($CCl_4$) have long been used as solvents for extracting aqueous solutions. The latter two solvents are not often used nowadays, unless strict safety precautions are exercised, because they are potentially carcinogenic.

**4.** Most functional groups capable of forming a hydrogen bond with water increased the water solubility of a substance. For example, smaller alkyl amines have significant water solubility; the water-solubility data for a series of amines are summarized in Table 5.3.

The solubility characteristics of any given compound govern its distribution (*partition*) between the phases of two immiscible solvents (in which the material has been dissolved) when these phases are intimately mixed.

## PARTITION (OR DISTRIBUTION) COEFFICIENT

A given substance $X$ is partially soluble in each of two immiscible solvents. If $X$ is placed in a mixture of these two solvents and shaken, an equilibrium will be established between the two phases. That is, substance $X$ will partition (distribute) itself in a manner that is a function of its relative solubility in the two solvents:

$$X_{\text{solvent 1}} \rightleftharpoons X_{\text{solvent 2}}$$

The equilibrium constant, $K_p$, for this equilibrium expression is known as the *partition* or *distribution coefficient*:

$$K_p = \frac{[X_{\text{solvent 2}}]}{[X_{\text{solvent 1}}]}$$

The equilibrium constant is thus the ratio of the concentrations of the species, $X$, in each solvent for a given system at a given temperature. The partition coefficient can be conveniently estimated as the ratio of the solubility of $X$ in solvent 1 vs. solvent 2:

$$K_p = \frac{\text{solubility of } X \text{ in solvent 2}}{\text{solubility of } X \text{ in solvent 1}}$$

When solvent 1 is water and solvent 2 is an organic solvent such as diethyl ether, the basic equation used to express the coefficient $K_p$ is

$$K_p = \frac{(g/100\ mL)_{organic\ layer}}{(g/100\ mL)_{water\ layer}}$$

This expression uses grams per 100 mL for the concentration units. Note that the partition coefficient is dimensionless, so any concentration units may be used if the units are the same for both phases. For example, grams per liter (g/L), parts per million (ppm), and molarity (M) can all be used. If equal volumes of both solvents are used, the equation reduces to the ratio of the weights ($g_{organic}/g_{water}$) of the given species in the two solvents:

$$K_p = \frac{g_{organic\ layer}}{g_{water\ layer}}$$

Determination of the partition coefficient for a particular compound in various immiscible-solvent combinations often can give valuable information for isolating and purifying the compound by using extraction techniques. Thus, liquid–liquid extraction is a common separation technique used in organic as well as analytical laboratories.

Table 5.4 provides some examples of $K_p$ values determined at room temperature for a number of compounds in the water–methylene chloride system.

Let us now look at a typical calculation for the extraction of an organic compound P from an aqueous solution using diethyl ether. We will assume that the $K_{p\ ether/water}$ value (partition coefficient of P between diethyl ether and water) is 3.5 at 20 °C. If a solution of 100 mg of P in 300 μL of water is extracted at 20 °C with 300 μL of diethyl ether, the following expression holds:

$$K_{p\ ether/water} = \frac{C_e}{C_w} = \frac{W_e/300\ \mu L}{W_w/300\ \mu L}$$

*where*

$W_e$ = weight of P in the ether layer
$W_w$ = weight of P in the water layer
$C_e$ = concentration of P in the ether layer
$C_w$ = concentration of P in the water layers

Since $W_w = 100 - W_e$, the preceding relationship can be written as

$$K_{p\ ether/water} = \frac{W_e/300\ \mu L}{(100 - W_e)/300\ \mu L} = 3.5$$

If we solve for the value of $W_e$, we obtain 77.8 mg; the value for $W_w$ is 22.2 mg. Thus, we see that after one extraction with 300 μL of ether, 77.8 mg of P (77.8% of the total) is removed by the ether and 22.2 mg (22.2% of the total) remains in the water layer. Is it preferable to make a single extraction with the total quantity of solvent available, or to make multiple extractions with portions of the solvent? The second method is usually more efficient.

**Table 5.4 Representative $K_p$ Values in $CH_2Cl_2$—$H_2O$**

| Compound | $K_p$ Value |
|---|---|
| Nitrobenzene | 51.5 |
| Aniline | 3.3 |
| 1,2-Dihydroxybenzene | 0.2 |

To illustrate, consider extracting the 100 mg of P in 300 μL of water with *two 150*-μL portions of diethyl ether instead of one 300-μL portion.

For the first 150-μL extraction,

$$\frac{W_e/150\ \mu L}{W_w/300\ \mu L} = \frac{W_e/150\ \mu L}{(100 - W_e)/300\ \mu L}$$

Solving for $W_e$, we obtain 63.6 mg. The amount of P remaining in the water layer ($W_w$) is then 36.4 mg. The aqueous solution is now extracted with the second portion of ether (150 μL). We then have

$$\frac{W_e/150\ \mu L}{(36.4 - W_e)/300\ \mu L} = 3.5$$

As before, by solving for $W_e$, we obtain 23.2 mg for the amount of P in the ether layer; $W_w = 13.2$ mg in the water layer.

The two extractions, each with 150 μL of ether, removed a total of 63.6 mg + 23.2 mg = 86.8 mg of P (86.8% of the total). The P left in the water layer is then 100 − 86.8, or 13.2 mg (13.2% of the total).

It can be seen from these calculations that the multiple-extraction technique is more efficient. The single extraction removed 77.8% of P; the double extraction (with the same total volume of ether) increased this to 86.8%. To extend this relationship, three extractions with one-third the total quantity of ether in each portion would be even more efficient. You might wish to calculate this to prove the point. Of course, there is a practical limit to the number of extractions that can be performed.

The multiple-extraction example shown here illustrates that several extractions with small volumes is more efficient than a single extraction procedure. This is always true *provided* the partition coefficient is neither very large nor very small. If the partition coefficient $K_p$ for a substance between two solvents is very large ($K_p > 100$) or very small ($K_p < 0.01$), multiple extractions (using the same total amount of solvent) do not significantly increase the efficiency of the extraction process.[4]

### Extraction

***Liquid–Liquid Extraction.*** The more common type of extraction, liquid–liquid extraction, is used extensively. It is a very powerful method for separating and isolating materials at the microscale level. It is operationally not a simple process, so attention to detail is critical.

There are several important criteria to consider when choosing a solvent for the extraction and isolation of a component from a solution:

- The chosen extraction solvent must be immiscible with the solution solvent.
- The chosen extraction solvent must be favored by the distribution coefficient for the component being extracted.
- The chosen extraction solvent should be readily separated from the desired component after extraction. This usually means that it should have a low boiling point.
- The chosen organic extraction solvent must not react chemically with any component in the aqueous mixture being extracted.

[4]Palleros, D. R. *J. Chem. Educ.* **1995,** *72,* 319.

*NOTE. The aqueous phase may be modified, as in acid–base extractions, but the organic solvent does not react with the components in the aqueous mixture. See Experiments [4B, 4C], pp. 146–150.*

***Microscale Extraction.*** A capped conical vial or a stoppered centrifuge tube is the best container for most microscale extractions, but a small test tube may be used. Note that a conical vial and a centrifuge tube have the same inner shape. This shape has the advantage that as the lower phase (layer) is withdrawn by pipet, the interface (boundary) between the two liquid phases becomes narrower and narrower, and thus easier to see, at the bottom of a conical container. This is not the case for a test tube. The centrifuge tube has the added advantage that if a solid precipitate must be separated or an emulsion broken up, it can easily be done using a centrifuge.

A good rule of thumb is that the container to be used for the extraction should be *at least three times the volume of liquid* you wish to extract.

Regardless of the container used, in any liquid–liquid extraction, the two immiscible solvents must be completely mixed to maximize the surface area of the interface between the two and allow partitioning of the solute. This can be accomplished by shaking (carefully to avoid leakage around the cap), using a Vortex mixer, or by adding a magnetic spin vane and then stirring with a magnetic stirrer.

Another important rule in the extraction process is that you should *never discard any layer until the isolation is complete.*

Let us consider a practical example. Benzanilide can be prepared by the in situ rearrangement of benzophenone oxime in acid solution:

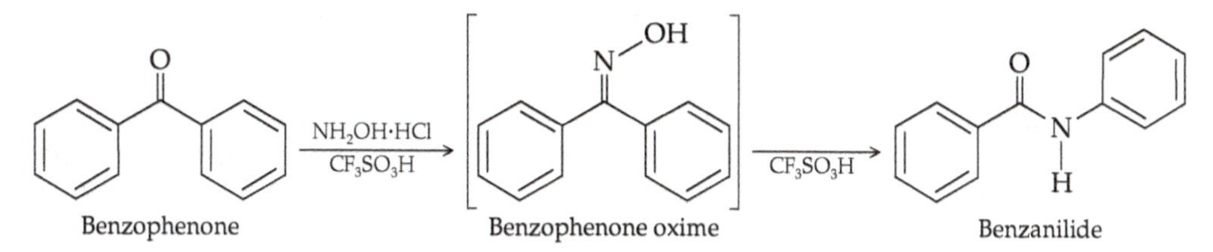


The benzanilide is separated from the reaction mixture by extraction with three 1.0-mL portions of methylene chloride solvent.

*NOTE. Saying, for example, "extracted with three 1.0-mL portions of methylene chloride" means that three extractions are performed (one after the other, each using 1.0 mL of methylene chloride) and the three methylene chloride extracts are combined.*

A microscale extraction process consists of two parts: (1) mixing the two immiscible solutions, and (2) separating the two layers after the mixing process.

**1. Mixing.** In the experimental procedure for the isolation of the benzanilide product, methylene chloride (1.0 mL) is added to the aqueous reaction mixture contained in a 5.0-mL conical vial (or centrifuge tube). The extraction procedure is outlined in the following steps:

**Step 1.** Cap the vial.

**Step 2.** Shake the vial gently to thoroughly mix the two phases (careful!)

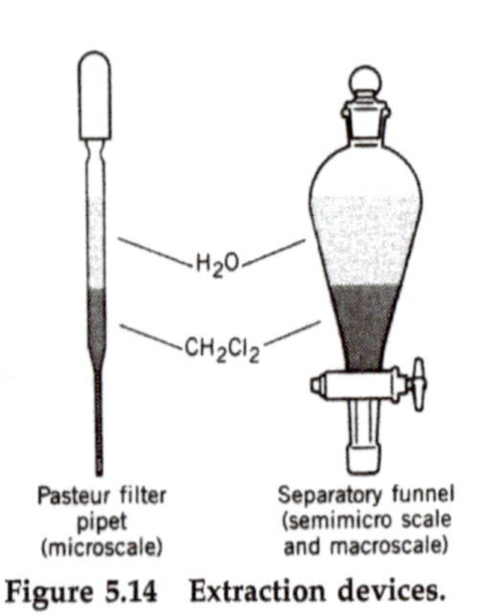


**Figure 5.14 Extraction devices.**

*NOTE. The mixing may be carried out using a Vortex mixer or magnetic stirrer—see previous discussion.*

**Step 3.** Carefully vent the vial by loosening the cap to release any pressure that may have developed.

**Step 4.** Allow the vial to stand on a level surface to permit the two phases to separate. A sharp phase interface should appear.

*NOTE. For safety reasons it is advisable to place the vial in a small beaker to prevent tipping. If a volatile solvent such as ether is used, it is advisable to place the vial or centrifuge tube in a beaker of ice water to prevent loss of solvent during the transfers.*

**2. Separation.** At the microscale level, the two phases are separated with a Pasteur filter pipet (a simple Pasteur pipet can be used in some situations), which acts as a miniature separatory funnel. The separation of the phases is shown in Figure 5.14.

A major difference between macro and micro techniques is that when microscale volumes are used, as just discussed, the mixing and separation are done in two parts. When macroscale volumes are used in a separatory funnel, mixing and separation are both done in the funnel in one step. The separatory funnel is an effective device for extractions with larger volumes, but it is not practical for microscale extractions because of the large surface areas involved.

Benzanilide is more soluble in methylene chloride than in water. Multiple extractions are performed to ensure complete removal of the benzanilide from the aqueous phase. The methylene chloride solution is the lower layer because it is more dense than water. The following list outlines the general method for an organic solvent more dense than water.

*NOTE. (a) One technique is to hold the pipet across the palm of the hand and squeeze the bulb with the thumb and index finger. (b) Remember to have an empty tared vial available in which to place the separated phase. (c) A pipet pump (Figure 5.15) may be used to replace the bulb. One advantage with using a pipet pump is the dispensing of liquids in a more controlled fashion.*

Pipet pump with pipet

**Figure 5.15 Pipet pump with pipet.** (Reprinted with permission of John Wiley and Sons, Inc. from Szafran, Z.; Pike, R. M.; Foster, J. C. Microscale *General Chemistry Laboratory*, 2nd ed., p. 36. 2003.)

The recommended procedures are shown in Figures 5.16 and 5.17.

**Step 1.** Squeeze the pipet bulb to force air from the pipet.

**Step 2.** Insert the pipet into the vial until it is close to the bottom. Be sure to hold the pipet vertically.

**Step 3.** Carefully allow the bulb to expand, drawing only the lower methylene chloride layer into the pipet. This should be done in a smooth, steady manner so as not to disturb the interface between the layers. With practice, you can judge the amount that the bulb must be squeezed to just separate the layers. Keep the pipet vertical. *Do not tip the pipet back and allow liquid to enter the bulb! Do not suck liquid into the bulb!*

**Step 4.** (Step 4 is not shown in the figure.) Holding the pipet vertical, place it over and into the neck of an empty vial (as shown in Fig. 5.16, Step 2), and gently squeeze the bulb to transfer the methylene chloride solution into the vial. A second extraction can now be performed after adding

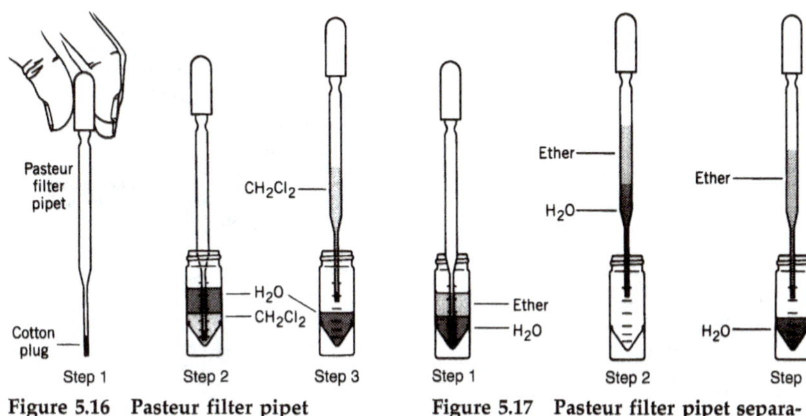


**Figure 5.16 Pasteur filter pipet separation of two immiscible liquid phases; the more dense layer contains the product.**

**Figure 5.17 Pasteur filter pipet separation of two immiscible liquid phases; the less dense layer contains the product.**

another portion of methylene chloride to the original vial. The procedure is repeated. Multiple extractions can be performed in this manner. Each methylene chloride extract is transferred to the same vial—that is, the extracts are combined. The reaction product has now been transferred from the aqueous layer (aqueous phase) to the methylene chloride layer (organic phase), and the phases have been separated.

In a diethyl ether–water extraction, the ether layer is less dense and thus is the upper layer (phase). An organic reaction product generally dissolves in the ether layer and is thus separated from water-soluble byproducts and other impurities. The procedure followed to separate the water–ether phases is identical to that described above for methylene chloride–water systems, except that here the top layer (organic layer) is transferred to the new container. The following list outlines the general method for an organic solvent less dense than water (refer to Fig. 5.17).

**Step 1.** Squeeze the pipet bulb to force air from the pipet and insert the pipet into the vial until it is close to the bottom. Then, draw **both** phases slowly into the pipet. Keep the pipet vertical. *Do not tip the pipet back and allow liquid to enter the bulb! Do not suck liquid into the bulb!* Try not to allow air to be sucked into the pipet, as this tends to mix the phases in the pipet. If mixing does occur, allow time for the interface to re-form.

**Step 2.** Return the aqueous layer (bottom layer) to the **original** container by gently squeezing the pipet bulb.

**Step 3.** Transfer the separated ether layer (top layer) to a new tared vial.

***Separatory Funnel—Semimicroscale and Macroscale Extractions.*** A separatory funnel (Fig. 5.14) is effective for extractions carried out at the semimicroscale and macroscale levels. The mixing and separation are done in the funnel itself in one step. Many of you may be familiar with this device from the general chemistry laboratory. The same precautions as outlined above for microscale extraction should be observed here.

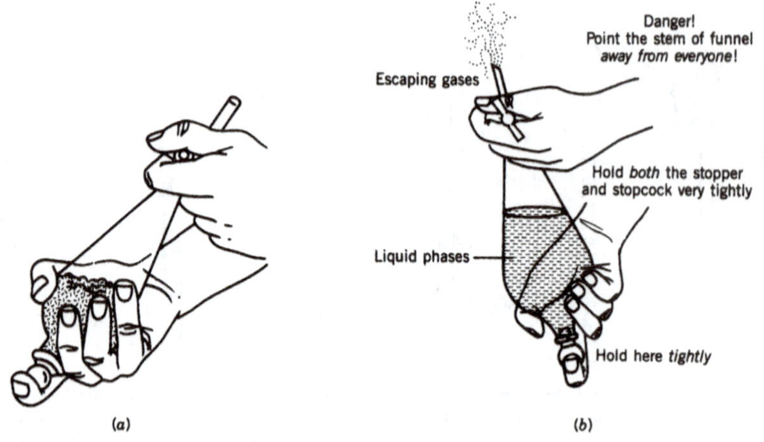


**Figure 5.18 (a) Correct position for holding a separatory funnel while shaking. (b) Correct method for venting a separatory funnel.**

*NOTE. The funnel size should be such that the total volume of solution is less than half the total volume of the funnel. If the funnel has a ground-glass stopcock and/or stopper, the ground-glass surfaces must be lightly greased to prevent sticking, leaking, or freezing. If Teflon stoppers and stopcocks are used, grease is not necessary because these are self-lubricating.*

**Step 1.** Close the stopcock of the separatory funnel.

**Step 2.** Add the solution to be extracted, after first making sure that the *stopcock is closed.* The funnel should be supported in an iron ring attached to a ring stand or rack on the lab bench.

**Step 3.** Add the proper amount of extraction solvent (about one-third of the volume of the solution to be extracted is a good rule of thumb) and place the stopper on the funnel.

**Step 4.** Remove the funnel from the ring stand, keeping the stopper in place with the index finger of one hand, and holding the funnel in the other hand with your fingers positioned so they can operate the stopcock (Fig. 5.18*a*).

**Step 5.** Carefully invert the funnel (make sure its stem is pointing up, and not pointing at you or anyone else). Slowly open the stopcock to release any built-up pressure (Fig. 5.18*b*). Close the stopcock and then shake the funnel for several seconds. Position the funnel for venting (make sure the stem is pointing up, and not pointing at you or anyone else). Open the stopcock to release built-up pressure. Repeat this process 2–4 times. Then, close the stopcock and return the funnel upright to the iron ring.

**Step 6.** Allow the layers to separate and then remove the stopper.

**Step 7.** Place a suitable clean container just below the tip of the funnel. Gradually open the stopcock and drain the bottom layer into the clean container.

**Step 8.** Remove the upper layer by pouring it from the top of the funnel. This way it will not become contaminated with traces of the lower layer found in the stem of the funnel.

When aqueous solutions are extracted with a *less dense solvent,* such as ether, the bottom, aqueous layer can be drained *into its original container.* Once the top (organic) layer is removed from the funnel, the aqueous layer can then be returned for further extraction. Losses can be minimized by rinsing the original container with a small portion of the extraction solvent, which is then added to the funnel. When the extraction solvent is denser than the aqueous phase (e.g., methylene chloride), the aqueous phase is the top layer, and therefore is kept in the funnel for subsequent extractions.

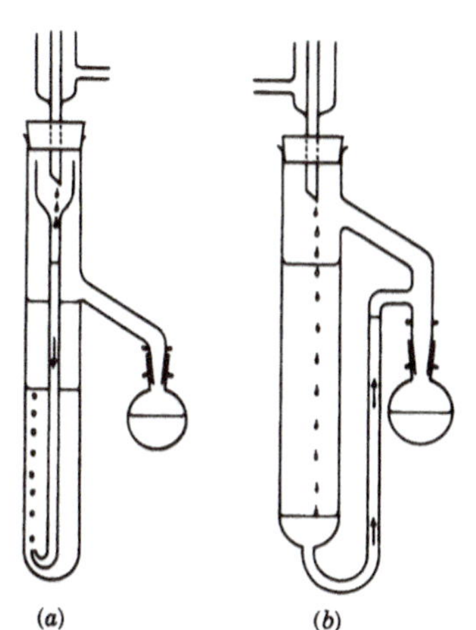


**Figure 5.19 Early designs for single-stage extractors: (a) Kutscher–Steudel extractor; (b) Wehrli extractor.**

***Continuous Liquid–Liquid Extraction.*** Continuous extraction of liquid–liquid systems is also possible and particularly valuable when the component to be separated is only slightly soluble in the extraction solvent. The advantage of using continuous extraction is that it can be carried out with a limited amount of solvent. In batchwise extractions a prohibitive number of individual extractions might have to be performed to accomplish the same overall extraction. Specialized apparatus, however, is required for continuous liquid–liquid extraction.

Two types of continuous extraction apparatus are often used to isolate various species from aqueous solutions using less dense and more dense immiscible solvents (e.g., diethyl ether and methylene chloride) (Fig. 5.19).

The extraction is carried out by allowing the condensate of the extraction solvent, as it forms on the condenser on continuous distillation, to drop through an inner tube (see Fig. 5.19*a* in the case of the less dense solvent) and to percolate up through the solution containing the material to be extracted. This inner tube usually has a sintered glass plug on its end, which generates smaller droplets of the solvent and thus increases the efficiency of the procedure. The extraction solution is then returned to the original distilling flask. Eventually, in this manner, the desired material, extracted in small increments, is collected in the boiling flask and can then be isolated by concentrating the collected solution. This method works on the premise that fresh portions of the less-dense phase are continuously introduced into the system, and it is often used in those instances where the organic material to be isolated has an appreciable solubility in water. In the case of a more dense extraction solvent (see Fig. 5.19*b*) the system functions in much the same fashion, but in this case the inner tube is removed and the condensed vapors percolate directly through the lighter phase (the phase to be extracted) to form the lower layer. This layer can cycle back to the distillation flask through a small-bore tubing connection from the bottom of the receiver flask to the distillation flask. Continuous liquid–liquid extraction is useful for removing extractable components from those having partition ratios that approach zero. Note that this method requires a very long period of time.

***Separation of Acids and Bases.*** The separation of organic acids and bases is another important and extensive use of the extraction method. The distribution coefficients of organic acids and bases are affected by pH when one of the solvents is water. An organic acid that is insoluble in neutral water (pH 7) becomes soluble when the water is made basic with an aqueous sodium hydroxide solution. The acid and the sodium hydroxide quickly react to form a sodium carboxylate salt, $RCO_2^{-}$, $Na^{+}$. The salt is, of course, ionic and therefore it readily dissolves in the water. Thus, the acid–base reaction reverses the solubility characteristics of a water-insoluble organic acid.

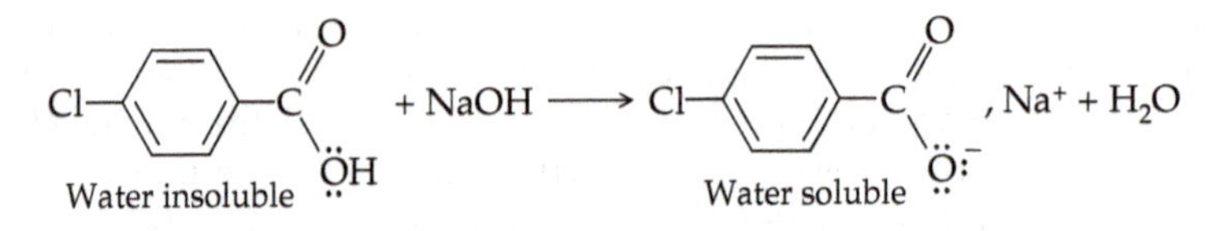


The water phase may then be extracted with an immiscible organic solvent to remove any impurities, leaving the acid salt in the water phase. Neutralizing the water layer with hydrochloric acid (to pH ≤ 7) reprotonates the carboxylate salt to reform the carboxylic acid, and causes the purified water-insoluble organic acid to precipitate (if it's a solid). In a similar fashion, water-insoluble organic bases, such as amines ($RNH_2$), can be rendered water soluble by treatment with dilute hydrochloric acid to form water-soluble hydrochloride salts.

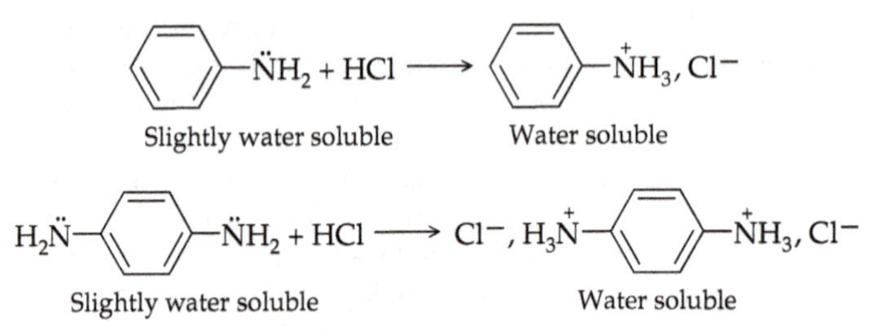


Extraction procedures can be used to separate mixtures of solids. For example, the flow chart below diagrams a sequence used to separate a mixture made up of an aromatic carboxylic acid ($ArCO_2H$), an aromatic base ($ArNH_2$), and a neutral aromatic compound (ArH). Aromatic compounds are discussed here simply because they are likely to be crystalline solids.

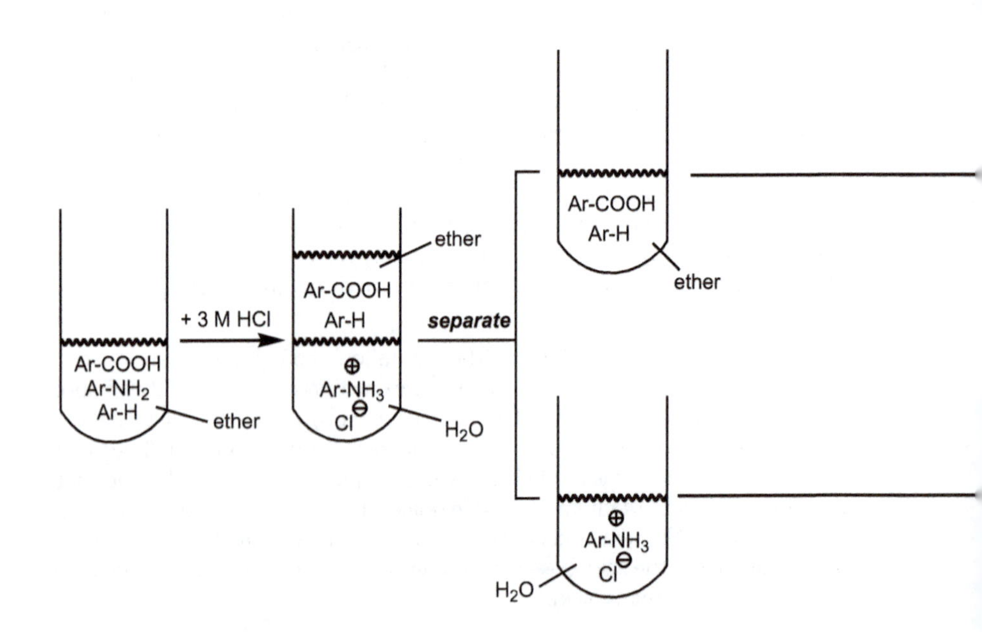

In this example, we assume that the organic acid and base are solids. If either or both were liquids, an additional extraction of the final acidic aqueous or alkaline solution with ether, followed by drying and concentration, would be required to isolate the acidic or basic component.

***Salting Out.*** Most extractions in the organic laboratory involve water and an organic solvent. Many organic compounds have partial solubility in both solvents. To extract them from water, the partition coefficient (between the organic solvent and water) can be shifted in favor of the organic layer by saturating the water layer with an inorganic salt, such as sodium chloride. Water molecules prefer to solvate the polar ions (in this case sodium and chloride ions), and thus free the neutral organic molecules to migrate into the organic phase. Another way to think of this is to realize that the ionic solution is more polar than pure water, so the less polar organic molecules are less soluble than in pure water. Forcing an organic material out of a water solution by adding an inorganic salt is called *salting out.*

Salting out can also be effectively used for the *preliminary drying* of the wet organic layer that results from an extraction process. (Diethyl ether, in particular, can dissolve a fair amount of water.) Washing this organic layer with a saturated salt solution removes most of the dissolved water into the aqueous phase. This makes further drying of the organic phase with solid drying agents easier and much more effective (see Drying Agents below).

### Solid–Liquid Extraction

The simplest form of solid–liquid extraction involves treating a solid with a solvent and then decanting or filtering the solvent extract away from the solid. An example of this technique is extracting usnic acid from

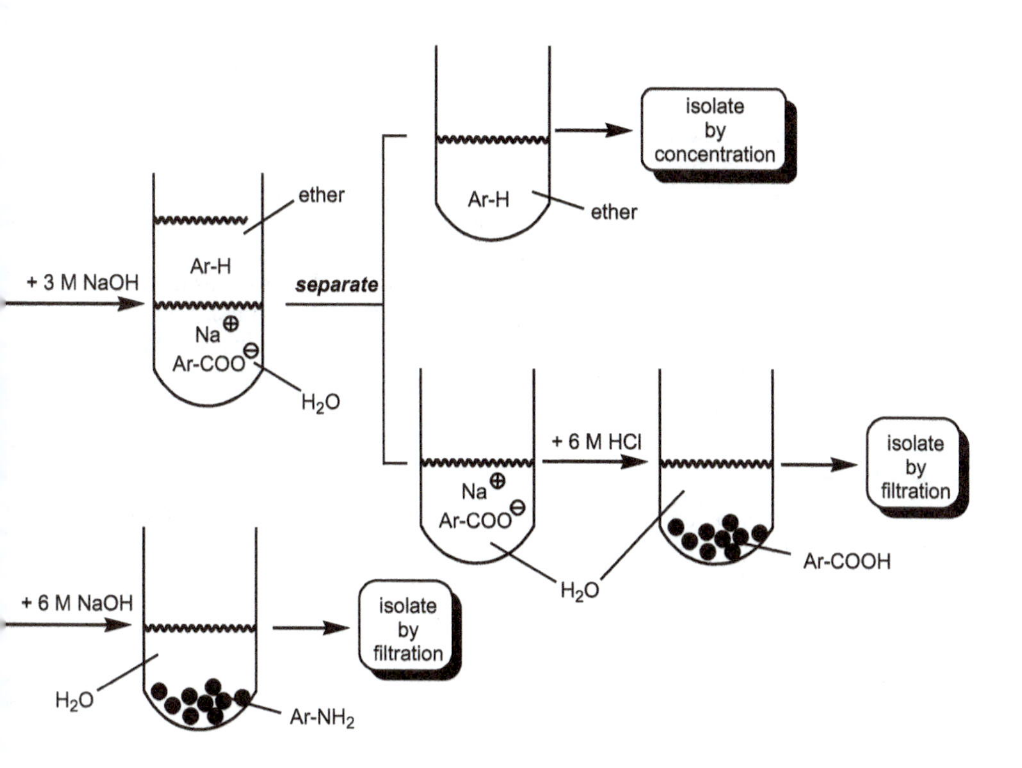

**Figure 5.20 A solid–liquid continuous extraction apparatus.**

its native lichen with acetone. This type of extraction is most useful when only one main component of the solid phase has appreciable solubility in the solvent. The extraction of caffeine from tea (see Experiment 3105-2) is another example of this method; it is accomplished by heating the tea in an aqueous solution of sodium carbonate. This approach works well because the water swells the tea leaves and allows the caffeine to be extracted more readily.

Microscale extractions of trimyristin from nutmeg, and cholesterol from gallstones, have been described.[5] Diethyl ether was used as the solvent in both cases. A packed Pasteur pipet column was used for filtering, drying (nutmeg experiment), and decolorizing (gallstone experiment).

Herrera and Almy described a simple continuous extraction apparatus (Fig. 5.20).[6] The apparatus is constructed from a 50-mL beaker and a paper cone prepared from a 9-cm disk of filter paper (nonfluted), which rests on the lip of the beaker. A small notch is cut in the cone to allow solvent vapor to pass around it. The extraction solvent is placed in the beaker; the solid material to be extracted is placed in the cone. A watch glass containing 2–3 g of ice is placed on top of the assembly to act as the condenser and to hold the paper cone in place. As the ice melts, the water is removed and replaced with fresh ice. The beaker is heated on a hot plate in the hood (some solvent evaporates during the extraction process and may need to be replaced). The concentrated solution collected in the beaker is then cooled and the solid product is isolated by filtration or is recrystallized. This system needs to be attended at all times, but works reasonably well for brief extractions.

Various apparatus have been developed for use when longer extraction periods are required. They all use what is called a *countercurrent process*. The best-known apparatus is the Soxhlet extractor, first described in 1879 (Fig. 5.21).[7] The solid sample is placed in a porous thimble. The extraction-solvent vapor, generated by refluxing the extraction solvent contained in the distilling pot, passes up through the vertical side tube into the condenser. The liquid condensate then drips onto the solid, which is extracted. The extraction solution passes through the pores of the thimble, eventually filling the center section of the Soxhlet. The siphon tube also fills with this extraction solution and when the liquid level reaches the top of the tube, siphoning action returns the thimbleful of extract to the distillation pot. The cycle is automatically repeated many times, concentrating the extract in the distillation pot. The advantage of this arrangement is that the process may be continued automatically and unattended for as long as necessary. The solvent is then removed from the extraction solution collected in the pot, providing the extracted compound(s). Soxhlet extractors are available from many supply houses and can be purchased in various sizes. Of particular interest to us is the microscale variety, which is effective for small amounts of material and is now commercially available.[8]

## Drying Agents

Organic extracts separated from aqueous phases usually contain traces of water. Even washing with saturated salt solution (see Salting Out above) cannot

[5]Vestling, M. M. *J. Chem. Educ.* **1990,** *67,* 274.
[6]Herrera, A.; Almy, J. *J. Chem. Educ.* **1998,** *75,* 83.
[7]Soxhlet, F. *Dinglers Polytech. J.* **1879,** *232,* 461.
[8]Microscale Soxhlet equipment is available from ACE Glass, Inc., 1430 Northwest Boulevard, Vineland, NJ 08360.

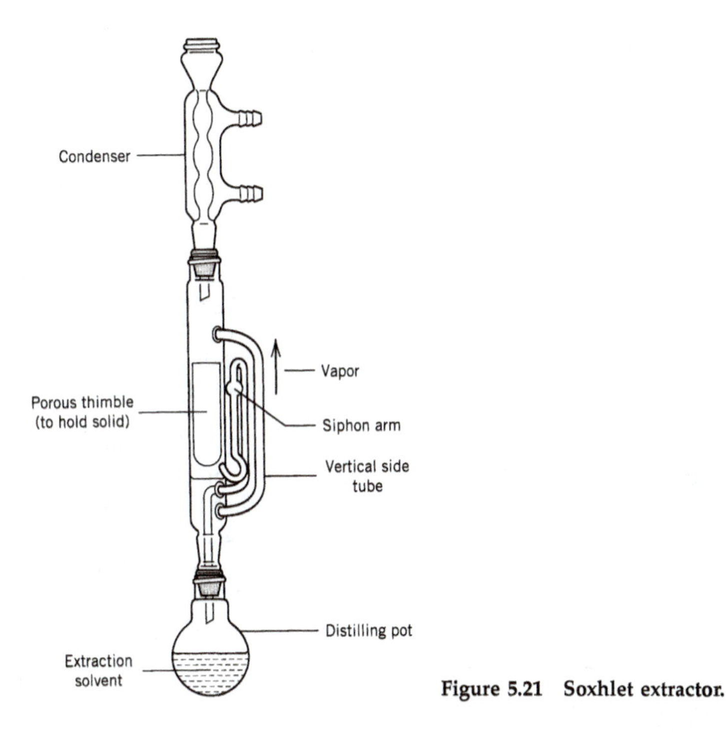


**Figure 5.21 Soxhlet extractor.**

remove all of the water. Organic extracts must therefore be dried to remove any residual water before the solvent is evaporated or further purification is performed. Organic extracts can be conveniently dried with an anhydrous inorganic salt, such as magnesium sulfate, sodium sulfate, or calcium sulfate. These salts readily absorb water and form insoluble hydrates, thus removing the water from the wet organic phase. The hydrated solid can then be removed from the dried solution by filtration or by decanting (pouring) the solution away from the solid. Although many drying agents are known, not every drying agent can be used in every case. The ideal drying agent should dry the solution quickly, have a high capacity for water, cost little, and not react with the material being dried.[9]

Table 5.5 summarizes the properties of some of the more common drying agents used in the laboratory.

Make sure that the solid drying agent is in its *anhydrous* form. Sodium sulfate is a good general-purpose drying agent and is usually the drying agent of choice at room temperature. Use the granular form, if at all possible.

Magnesium sulfate is supplied as a fine powder (high surface area). It has a high water capacity and is inexpensive; it dries solutions more quickly than does sodium sulfate. The disadvantage of magnesium sulfate is that the desired product (or water molecules) can become trapped on the surface of the fine particles. If it is not thoroughly washed after separation, precious product may

[9]Quantitative studies on the efficiency of drying agents for a wide variety of solvents have been reported. See Burfield, D. R.; Smithers, R. H. *J. Org. Chem.* **1983,** *48,* 2420, and references therein. Other useful information can be found in Armarego, W. L. F.; Chai, C. L. L. *Purification of Laboratory Chemicals,* 5th ed.; Elsevier: New York, 2003, and in Ridduck, J. A.; Bunger, W. B.; Sakano, T. K. *Organic Solvents, Physical Properties and Methods of Purification,* 4th ed.; Wiley: New York, 1986.

**Table 5.5 Properties of Common Drying Agents**

| Drying Agent | Formula of Hydrate | Comments |
|---|---|---|
| Sodium sulfate | $Na_2SO_4 \cdot 10H_2O$ | Slow to absorb water and inefficient, but inexpensive and has a high capacity. Loses water above 32 °C Granular form available. |
| Magnesium sulfate | $MgSO_4 \cdot 7H_2O$ | One of the best. Can be used with nearly all organic solvents. Usually in powder form. |
| Calcium chloride | $CaCl_2 \cdot 6H_2o$ | Relatively fast drying, but reacts with many oxygen- and nitrogen-containing compounds. Usually in granular form. |
| Calcium sulfate | $CaSO_4 \cdot \frac{1}{2} H_2O$ | Very fast and efficient, but has a low dehydration capacity. |
| Silica gel | $(SiO_2)_m \cdot nH_2O$ | High capacity and efficient. Commercially available t.h.e. $SiO_2$ drying agent is excellent.[a] |
| Molecular sieves | $[Na_{12}(Al_{12}Si_{12}O_{48})] \cdot 27H_2O$ | High capacity and efficient. Use the 4-Å size.[b] |

[a]Available from EMD Chemicals, 10394 Pacific Center Court, San Diego, CA 92121.
[b]Available from Sigma-Aldrich, Inc., 940 West Saint Paul Ave., Milwaukee, WI 53233.

be lost. Furthermore, it is usually more difficult to remove a finely powdered solid agent, which may pass through the filter paper (if used) or clog the pores of a fine porous filter. A smaller surface area translates into less adsorption of product on the surface and easier separation from the dried solution.

Molecular sieves have pores or channels in their structures. A small molecule such as water can diffuse into these channels and become trapped. The sieves are excellent drying agents, have a high capacity, and dry liquids completely. The disadvantages are that they dry slowly and are more expensive than the more common drying agents.

Calcium chloride is very inexpensive and has a high capacity. Use the granular form. Do not use it to dry solutions of alcohols, amines, or carboxylic acids because it can react with these substances.

Calcium sulfate is often sold under the trade name of Drierite. It is a somewhat expensive drying agent. Do not use the blue Drierite (commonly used to dry gases) because the cobalt indicator (blue when dry, pink when wet) may leach into the solvent.

The amount of drying agent needed depends on the amount of water present, on the capacity of the solid desiccant to absorb water, and on its particle size (actually, its surface area). If the solution is wet, the first amount of drying agent will clump (molecular sieves and t.h.e. $SiO_2$ are exceptions). Add more drying agent until it appears mobile when you swirl the liquid. A solution that is no longer cloudy is a further indication that the solution is dry. Swirling the contents of the container increases the rate of drying; it helps establish the equilibrium for hydration:

$$\underset{\text{Anhydrous solid}}{\text{Drying agent} + nH_2O} \rightleftharpoons \underset{\text{Solid hydrate}}{\text{Drying agent} \cdot nH_2O}$$

Most drying agents achieve approximately 80% of their drying capacity within 15 min; longer drying times are generally unnecessary. The drying agent may be added directly to the container of the organic extract, or the extract may be passed through a Pasteur filter pipet packed with the drying agent. A funnel fitted with a cotton, glass wool, or polyester plug to hold the drying agent may also be used.

As for the most common question asked with this technique—Is this "dry"?—you should be encouraged to have in your lab a series of flasks which contain a set quantity of solvent and drying agent. The difference with each flask within the series is the percentage of water which allows for a visual comparison of what is and what is not "dry."

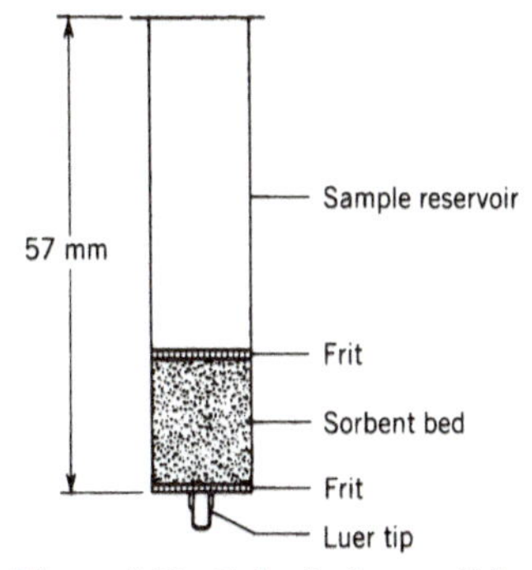


**Figure 5.22 Polyethylene solid-phase extraction column.**

### Solid-Phase Extraction

In the modern research laboratory, the traditional liquid–liquid extraction technique may be replaced by the solid-phase extraction method.[10] The advantages of this newer approach are that it is rapid, it uses only small volumes of solvent, it does not form emulsions, isolated solvent extracts do not require a further drying stage, and it is ideal for working at the microscale level. This technique is finding wide acceptance in the food industry and in the environmental and clinical area, and it is becoming the accepted procedure for the rapid isolation of drugs of abuse and their metabolites from urine. Solid-phase extraction is accomplished using prepackaged, disposable, extraction columns. A typical column is shown in Figure 5.22. The columns are available from several commercial sources.[11]

The polypropylene columns can be obtained packed with 100–1000 mg of 40-μm sorbent sandwiched between two 20-μm polyethylene frits. The columns are typically 5–6 cm long. Sample volumes are generally 1–6 mL.

The adsorbent (stationary phase) used in these columns is a nonpolar adsorbent chemically bonded to silica gel. In fact, they are the same nonpolar adsorbents used in the reversed-phase high-performance liquid chromatography (HPLC). More specifically, the adsorbents are derivatized silica gel where the —OH groups of the silica gel have been replaced with siloxane groups by treating silica gel with the appropriate organochlorosilanes.

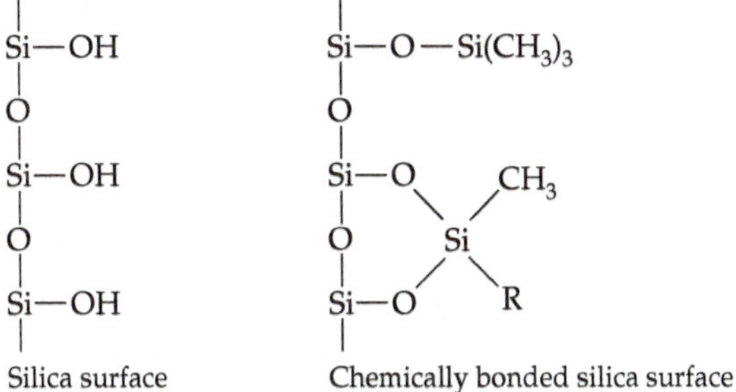


[10]For a description of this method see Zief, M.; Kiser, R. *Am. Lab.* **1990,** *22 70;* Zief, M. *NEACT J.* **1990,** *8,* 38; Hagen, D. F.; Markell, C. G.; Schmitt, G. A.; Blevins, D. D. *Anal. Chim. Acta* **1990,** *236,* 157; Arthur, C. L.; Pawliszyn, J. *Anal. Chem.* **1990,** *62,* 2145; Dorsey, J.; Dill, K. A. *Chem. Rev.* **1989,** *89,* 331; Zubrick, J. W. *The Organic Chem Lab Survival Manual,* 7th ed., Wiley: New York, 2008; Simpson, N. J. K, Ed. *Solid-Phase Extraction: Principles, Techniques and Applications,* Marcel Dekker: New York, 2000; "Solid Phase Extraction." Retrieved March 19, 2009 from www.sigmaaldrich.com/analytical-chromatography/sample-preparation/spe.html.

[11]These columns are available from Analytichem International, J. T. Baker, Inc., Supelco, Inc., Aldrich Chemical, Waters Associates and Biotage (a Division of Dyax Corp).

Two of the most popular nonpolar packings are those containing R groups consisting of an octadecyl ($C_{18}H_{37}$—)or phenyl ($C_6H_5$—) group. These packing materials (stationary phases) can adsorb nonpolar (like attracts like) organic material from aqueous solutions. The adsorbed material is then eluted from the column using a solvent strong (nonpolar) enough to displace it, such as methanol, methylene chloride, or hexane. The analyte capacity of bonded silica gels is about 10–20 mg of analyte per gram of packing.

An example of a typical solid-phase extraction is the determination of the amount of caffeine in coffee using a 1-mL column containing 100 mg of octadecyl-bonded silica. This efficient method isolates about 95% of the available caffeine. The column is conditioned by flushing 2 mL of methanol followed by 2 mL of water through the column. One milliliter of a coffee solution (~0.75 mg of caffeine/mL) is then drawn through the column at a flow rate of 1 mL/min. The column is washed with 1 mL of water and air dried (vacuum) for 10 min. The adsorbed caffeine is then eluted with two 500-μL portions of chloroform.

## QUESTIONS

**5-16.** You are presented a two-phase system. The two liquids are immiscible. The top phase is blue and the bottom, orange. One phase is water. Please devise an experiment to **definitively** differentiate which phase is water.

**5-17.** Which layer (upper or lower) will each of the following organic solvents usually form when being used to extract an aqueous solution?

toluene methylene chloride diethyl ether
hexane acetone

**5-18.** Construct a flow chart to demonstrate how you could separate a mixture of 1,4-dichlorobenzene, 4-chlorobenzoic acid, and
4-chloroaniline using an extraction procedure.

**5-19.** A slightly polar organic compound partitions itself between ether and water phases. The $K_p$ (partition coefficient) value is 2.5 in favor of the ether solvent. What simple procedure could you use to increase this $K_p$ value?

**5-20.** You weight out exactly 1.00 mg of benzoic acid and dissolve it in a mixture of 2.0 mL of diethyl ether and 2.0 mL of water. After mixing and allowing the layers to separate, the ether layer is removed, dried, and concentrated to yield 0.68 mg of benzoic acid. What is the $K_p$ value (ether/water) for this system?

**5-21.** If asked to separate an equal mixture of benzoic acid [$pK_a$ = 4.2] and 2-naphthol [$pK_a$ = 9.5] using a liquid–liquid extraction technique, explain why an aqueous solution of $NaHCO_3$[$pK_a$ = 6.4] would be far more effective than the stronger aqueous solution of NaOH[$pK_a$ = 15.7].

## BIBLIOGRAPHY

**For overviews on extraction methods see the following general references:**

Dean, J. R. *Extraction Methods for Environmental Analysis;* Wiley: New York, 1998.

Handley, A. J. *Extraction Methods for Organic Analysis;* CRC Press LLC: Boca Raton, FL, 1999.

*Kirk–Othmer Encyclopedia of Chemical Technology,* 4th ed.; Wiley: New York, 1993; Vol. 10, p. 125.

Schneider, Frank L. *Qualitative Organic Microanalysis;* Vol. II of *Monographien aus dem Gebiete der qualitätiven Mikroanalysis,* A. A. Benedetti-Pichler, Ed.; Springer-Verlag: Vienna, 1964, p. 61.

Shugar, G. J. *Chemical Technicians' Ready Reference Handbook,* 3rd ed.; McGraw-Hill: New York, 1990.

Zubrick J. W. *The Organic Chem Lab Survival Manual: A Student's Guide to Techniques,* 7th ed.; Wiley: New York, 2008.

Zubrick, J. W. *The Organic Chem Lab Survival Manual,* 7th ed. Wiley: New York, 2008.

*NOTE The following experiments use Technique 4: Experiments [4A], [4B], [5A], [5B], [7], [8A], [11A], [11B], [11C], [12], [13], [16], [17], [19A], [19B], [19C], [22A], [22B], [23], [27],[30], [32], [34A], [34B], [A1$_b$], [D3], [E3], [F1], [F2], [F3], and [F4].*

www → *[3A$_{adv}$], [4$_{adv}$], and [6$_{adv}$].*

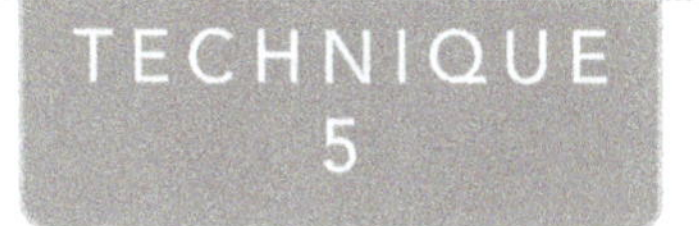


# Crystallization

This discussion introduces the basic technique of purifying solid organic substances by crystallization. The technique of crystallizing an organic compound is fundamental; it must be mastered if you are going to purify solids. *It is not an easy art to acquire.* Organic solids tend not to crystallize as easily as inorganic substances.

Legend has it that an organic chemist resisted an invitation to leave a well-worn laboratory for new quarters because he suspected that the older laboratory (in which many crystallizations had been carried out) harbored seed crystals for a large variety of substances the chemist needed. Carried by dust from the earlier work, these traces of material presumably aided the successful initiation of crystallization of reluctant materials. Further support for this legend comes from the often quoted (but never substantiated) belief that after a material was first crystallized in a particular laboratory, subsequent crystallizations of the material, regardless of its purity or origin, were always easier.

In several areas of chemistry, the success or failure of an investigation can depend on the ability of a chemist to isolate tiny quantities of crystalline substances. Often the compounds of interest must be extracted from enormous amounts of extraneous material. In one of the more spectacular examples, Reed et al. isolated 30 mg of the crystalline coenzyme lipoic acid from 10 tons of beef liver residue.[12]

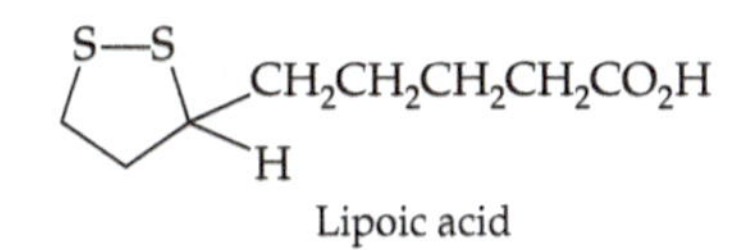


Lipoic acid

### General Crystallization Procedure

The following steps are the essentials of crystallization:

**Step 1.** Select a suitable solvent.

**Step 2.** Dissolve the material to be purified in the minimum amount of warm solvent. *Remember that most organic solvents are extremely flammable and that many produce very toxic vapor.*

**Step 3.** Once the solid mixture has fully dissolved, filter the heated solution, and then bring it to the point of saturation by evaporating a portion of the solvent.

**Step 4.** Cool the warm saturated solution to reduce the solubility of the solute; this usually causes the solid material to precipitate. If the material has a low melting point or is very impure it may come out of solution sometimes as an oil. If so, reheat the solution and allow it to recool slowly.

**Step 5.** Isolate the solid by filtration, and then remove the last traces of solvent.

The crystallization is successful if the solid is recovered in good yield and is purer than it was before the crystallization. This cycle, from solid state to solution and back to solid state is called *recrystallization* when both the initial and final materials are crystalline.

---

[12]Reed, L. J.; Gunsalus, I. C.; Schnakenberg, G. H. F.; Soper, Q. F.; Boaz, H. E.; Kem, S. F.; Parke, T. V. *J. Am. Chem. Soc.* **1953,** *75,* 1267.

Although the technique sounds fairly simple, in reality it is demanding. Successful purification of microscale quantities of solids will require your utmost attention. Choosing a solvent system is critical to a successful crystallization. To achieve high recoveries, the compound to be crystallized should ideally be very soluble in the hot solvent, but nearly insoluble in the cold solvent. To increase the purity of the compound, the impurities should be either *very soluble* in the solvent at all temperatures or *not soluble* at any temperature. The solvent should have as low a boiling point as possible so that traces of solvent can be easily removed (evaporated) from the crystals after filtration. It is best to use a solvent that has a boiling point at least 10 °C lower than the melting point of the compound to be crystallized to prevent the solute from "oiling out" of solution. Thus, the choice of solvent is critical to a good crystallization. Table 5.6 lists common solvents used in the purification of organic solids.

When information about a suitable solvent is not available, the choice of solvent is made on the basis of solubility tests. Craig's rapid and efficient procedure for microscale solubility testing works nicely; it requires only milligrams of material and a nine-well, Pyrex spot plate.[13]

Place 1–2 mg of the solid in each well and pulverize each sample with a stirring rod. Add 3–4 drops of a given solvent to the first well and observe whether the material dissolves at ambient temperature. If not, stir the mixture for 1.5–2 min and observe and record the results. Repeat this process with the chosen set of solvents, using a separate well for each solubility test. *Keep track of which well contains which solvent.* Place your test plate (containing the samples) on a hot plate (set at its lowest setting) in the **hood;** add additional solvent if necessary. Record the solubility characteristics of the sample in each hot solvent. Cool the plate and see if crystallization occurs in any of the wells. On the basis of your observations, choose an appropriate solvent or a solvent pair (see the following paragraph) to recrystallize your material.

Solubility relationships are seldom ideal for crystallization; most often a compromise is made. If there is no suitable single solvent available, it is possible to use a mixture of two solvents, called a *solvent pair*. A solvent is chosen that will

**Table 5.6 Common Solvents**

| Solvent | bp (°C) | Polarity |
|---|---|---|
| Acetone | 56 | Polar |
| Cyclohexane | 81 | Nonpolar |
| Diethyl ether | 35 | Intermediate polarity |
| Ethanol, 95% | 78 | Polar |
| Ethyl acetate | 77 | Intermediate polarity |
| Hexane | 68 | Nonpolar |
| Ligroin | 60–90 | Nonpolar |
| Methanol | 65 | Polar |
| Methylene chloride | 40 | Intermediate polarity |
| Methyl ethyl ketone | 80 | Intermediate polarity |
| Petroleum ether | 30–60 | Nonpolar |
| Toluene | 111 | Nonpolar |
| Water | 100 | Polar |

[13]Craig, R. E. R. *J. Chem. Educ.* **1989,** *66,* 88.

readily dissolve the solid. Once the solid is dissolved in the minimum amount of hot solvent, the solution is filtered. A second solvent, miscible with the first, in which the solute has much lower solubility, is then added dropwise to the hot solution to achieve saturation. In general, polar organic molecules have higher solubilities in polar solvents, and nonpolar materials are more soluble in nonpolar solvents (like dissolves like). Table 5.7 lists some common solvent pairs.

It can take a long time to work out an appropriate solvent system for a particular reaction product. In most instances, with known compounds, the optimum solvent system has been established. Most crystallizations are not very efficient because many impurities have solubilities similar to those of the compounds of interest. Recoveries of 50–70% are not uncommon.

Several microscale crystallization techniques are available.

*Table 5.7* **Common Solvent Pairs**

| Solvent 1 (more polar) | Solvent 2 (less polar) |
|---|---|
| Acetone | Diethyl ether |
| Diethyl ether | Hexane |
| Ethanol | Acetone |
| Ethyl acetate | Cyclohexane |
| Methanol | Methylene chloride |
| Acetone | Water |
| Water | Ethanol |
| Toluene | Ligroin |

### Simple Crystallization

Simple crystallization works well with large quantities of material (100 mg and up), and it is essentially identical to that of the macroscale technique.

**Step 1.** Place the solid in a small Erlenmeyer flask or test tube. A beaker is not recommended because the rapid and dangerous loss of flammable vapors of hot solvent occurs much more easily from the wide mouth of a beaker than from an Erlenmeyer flask. Furthermore, solid precipitate can rapidly collect on the walls of the beaker as the solution becomes saturated because the atmosphere above the solution is less likely to be saturated with solvent vapor in a beaker than in an Erlenmeyer flask.

**Step 2.** Add a minimal amount of solvent and heat the mixture to the solvent's boiling point in a sand bath. Stir the mixture by twirling a spatula between the thumb and index finger. A magnetic stir bar may be used if a magnetic stirring hot plate is used.

**Step 3.** Continue stirring and heating while adding solvent dropwise until all of the material has dissolved.

**Step 4.** Add a decolorizing agent (powdered charcoal, ~2% by weight; or better, activated-carbon Norit pellets,[14] ~0.1% by weight), to remove colored minor impurities and other resinous byproducts.

**Step 5.** Filter (by gravity) the hot solution into a second Erlenmeyer flask (pre-heat the funnel with hot solvent). This removes the decolorizing agent and any insoluble material initially present in the sample.

**Step 6.** Evaporate enough solvent to reach saturation.

**Step 7.** Cool to allow crystallization (crystal formation will be better if this step takes place slowly). After the system reaches room temperature, cooling it in an ice bath may improve the yield.

**Step 8.** Collect the crystals by filtration on a Büchner or Hirsch funnel. Save the mother liquor (this is the term used to describe the solution that was separated from the original crystals) until the identity of the product has been established. In some cases, it is possible to recover more product by concentrating and further cooling the mother liquor. The second crop of crystals, however, is usually not as pure as the first.

**Step 9.** Wash (rinse) the crystals carefully.

**Step 10.** Dry the crystals.

[14]Available from Sigma-Aldrich Chemical Co., 940 West St. Paul Ave., Milwaukee, WI 53233.

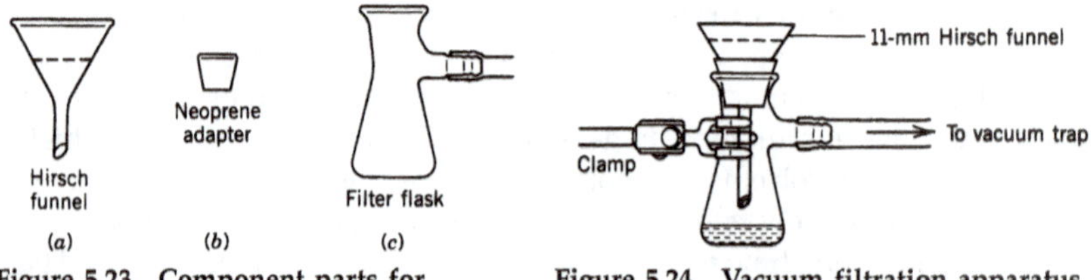


**Figure 5.23 Component parts for vacuum filtration.**

**Figure 5.24 Vacuum filtration apparatus.**

## Filtration Techniques

***Use of the Hirsch Funnel.*** The standard filtration system for collecting products purified by recrystallization in the microscale laboratory is vacuum (suction) filtration with an 11-mm Hirsch funnel. Many reaction products that do not require recrystallization can also be collected directly by vacuum filtration. The Hirsch funnel (Fig.5.23*a*) is composed of a ceramic cone with a circular flat bed perforated with small holes. The diameter of the bed is covered by a flat piece of filter paper of the same diameter. The funnel is sealed into a filter flask with a Neoprene adapter (Fig. 5.23*b*). Plastic and glass varieties of this funnel that have a polyethylene or glass frit are now available. It is still advisable to use the filter paper disk with these funnels to prevent the frit from clogging or becoming discolored. Regardless of the type of filter used, *always wet the filter paper disk with the solvent being used in the crystallization and then apply the vacuum.* This ensures that the filter paper disk is firmly seated on the bed of the filter.

Filter flasks have thick walls, and a side arm to attach a vacuum hose, and are designed to operate under vacuum (see Fig. 5.23*c*). The side arm is connected with *heavy-walled* rubber vacuum tubing to a water aspirator (water pump). The water pump uses a very simple aspirator based on the Venturi effect. Water is forced through a constricted throat in the pump. (See the detailed discussion of the Venturi effect and water pumps in the section on reduced pressure [vacuum] distillations.) When water flows through the aspirator, the resulting partial vacuum sucks air down the vacuum tubing from the filter flask. *Always turn the water on full force.* With the rubber adapter in place, air is pulled through the filter paper, which is held flat on the bed of the Hirsch (or Büchner) funnel by the suction. The mother liquors are rapidly forced into the filter flask, where the pressure is lower, by atmospheric pressure. The crystals retained by the filter are dried by the stream of air passing through them (Fig. 5.24).

www

When you use a water pump, *it is very important to have a safety trap mounted in the vacuum line leading from the filter flask.* Any drop in water pressure at the pump (easily created by other students on the same water line turning on other aspirators at the same time) can result in a backup of water into the flask. As the flow through the aspirator decreases, the pressure at that point rapidly increases and water is forced up the vacuum tubing toward the filter flask (Fig. 5.24). It is also important to vent the vacuum by opening the vent stopcock or disconnecting the rubber tubing from the filter flask, *before* the water is turned off (see Fig. 5.25).

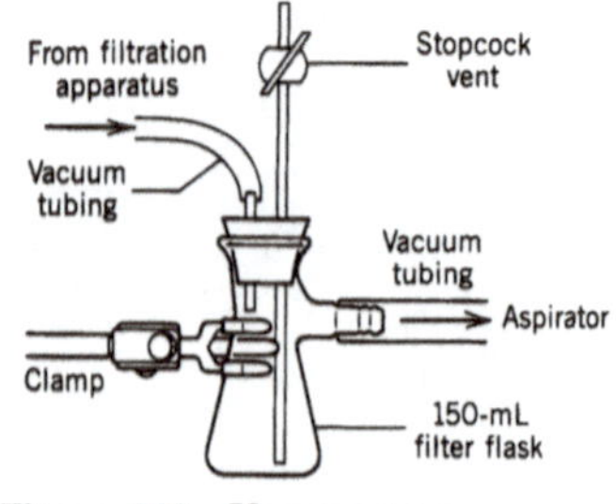


**Figure 5.25 Vacuum trap.**

In some cases, the precipitate collected on the Hirsch funnel is not highly crystalline. The filter cake may be too thick or pasty to dry simply by pulling air through it. A thin, flexible rubber sheet or a piece of plastic food wrap placed over the mouth of the funnel, such that the suction generated from the vacuum pulls the sheeting down onto the filter cake (collected crystals), will place atmospheric pressure on the solid cake. This pressure can force much of the remaining solvent from the collected material, and thus further dry it. Use a piece of sheeting large enough to cover the entire filter cake. Otherwise, a vacuum may not be created and adequate drying may not occur.

In some instances, substances may retain water or other solvents with great tenacity. To dry these materials, a *desiccator* is often used. This is generally a glass or plastic container containing a *desiccant* (a material capable of absorbing water). The substance to be dried, held in a suitable container, is then placed on a support above the desiccant. This technique is often used in quantitative analysis to dry collected precipitates. Vacuum desiccators are available (Fig. 5.26). If this method of drying is still insufficient, a *drying pistol* can be used (Fig. 5.27). The sample, in an open container (vial) is placed in the apparatus, which is then evacuated. The pistol has a pocket in which a strong adsorbing agent, such as $P_4O_{10}$ (for water), NaOH or KOH (for acidic gases), or paraffin wax (for organic solvents), is placed. The pistol is heated by refluxing vapors that surround the barrel. A simple alternative to this method is the use of a side-armed test tube (Fig. 5.28).

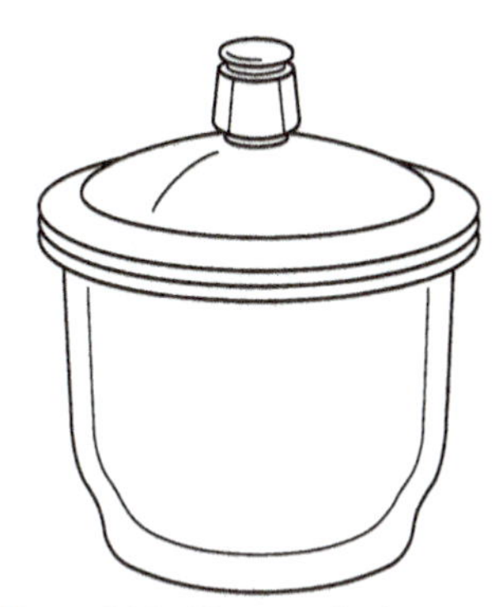

**Figure 5.26 Vacuum desiccator.**

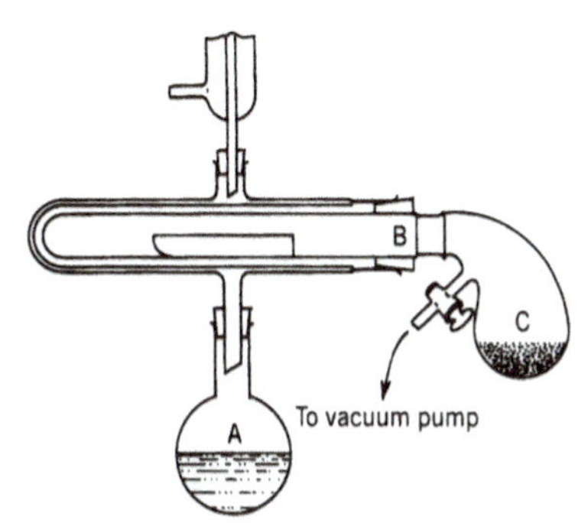


**Figure 5.27 Abderhalden vacuum drying apparatus.** A, refluxing heating liquid; B, vacuum drying chamber; C, desiccant.

***A Hirsch-Funnel Alternative—A Nail-Filter Funnel.*** A nail-filter funnel is a low-cost substitute for a Hirsch funnel (Fig. 5.29).[15] This apparatus is easily assembled from common laboratory glassware. Obtain a soft-glass rod that fits in the stem of a small glass funnel. Cut the rod to a suitable size and heat the tip of one end over a burner flame. When the tip becomes soft, hold the rod vertically and press the hot tip against a cold metal surface to flatten it to form a flat nail-like head. *The nail head should not be perfectly round or it will block the flow of liquid.* Cut the cooled rod to a suitable length and place the "nail" inside the stem of the funnel so that the flattened head of the nail rests on the top opening of the funnel. Cut a piece of filter paper just slightly larger than the nail head, place it on the nail head, and then place the funnel in a filter flask with a neoprene adapter. Be sure to wet the filter paper before filtering.

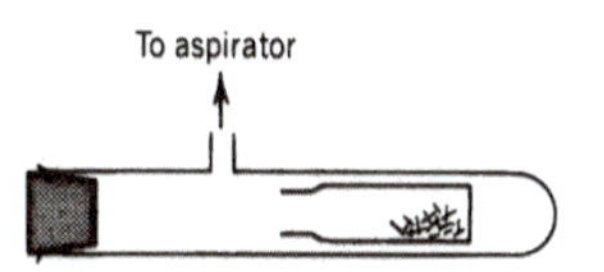


**Figure 5.28 Side-arm test tube as a vacuum dryhing apparatus.**

***Craig Tube Crystallizations.*** The Craig tube[16] is commonly used for microscale crystallizations in the range of 10–100 mg of material (Fig. 5.30). The process consists of the following steps.

**Step 1.** Place the sample in a small test tube (10 × 75 mm)

**Step 2.** Add the solvent (0.5–2 mL), and dissolve the sample by heating in the sand bath; add drops of solvent as needed. Rapid stirring with a microspatula (roll the spatula rod between your thumb and index finger) helps dissolve the material and protects against boilover. Add several drops of solvent by Pasteur pipet after the sample has completely dissolved. It will be easy to remove this excess at a later stage, since the volumes involved are very small. The additional solvent ensures that the solute will stay in solution during the hot transfer. Norit charcoal pellets may be added at this stage, if needed to remove colored impurities.

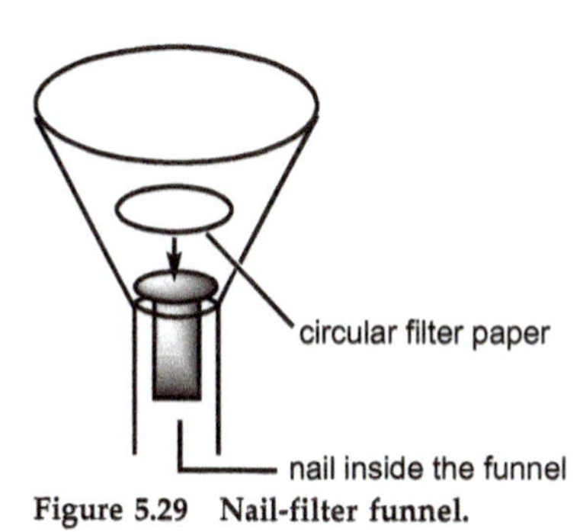


**Figure 5.29 Nail-filter funnel.**

[15]Singh, M. M.; Pike, R. M.; Szafran, Z. *Microscale & Selected Macroscale Experiments for General & Advanced General Chemistry;* Wiley: New York, NY, 1995, pp. 47, 63; Claret, P. A. *Small Scale Organic Preparations;* Pitman: London, 1961, p. 15.
[16]Craig, L. C.; Post, O. W. *Ind. Eng. Chem., Anal. Ed.* **1944,** *16,* 413.

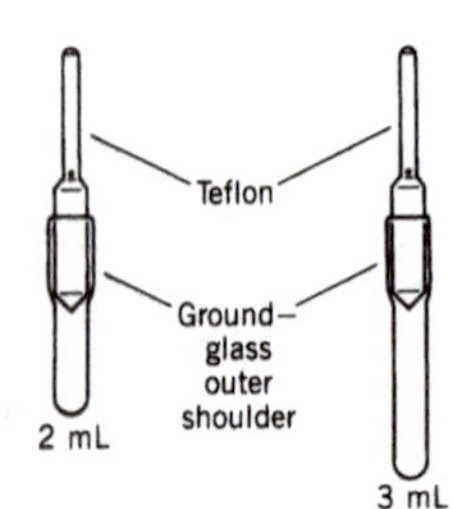


**Figure 5.30 Craig tubes.**

**Step 3.** Transfer the heated solution to the Craig tube by Pasteur filter pipet (see Fig. 3.30) that has been preheated with hot solvent. This transfer automatically filters the solution. A second filtration is often necessary if powdered charcoal has been used to decolorize the solution.

**Step 4.** The hot, filtered solution is then concentrated to saturation by gentle boiling in the sand bath. Constant agitation of the solution with a microspatula during this short period can avoid the use of a boiling stone and prevent boilover. Ready crystallization on the microspatula just above the solvent surface is a good indication that saturation is close at hand.

**Step 5.** The upper section of the Craig tube (the "head" or stopper) is set in place and the saturated solution is allowed to cool in a safe place. As cooling commences, seed crystals, if necessary, may be added by crushing them against the side of the Craig tube with a microspatula just above the solvent line. A good routine, if time is available, is to place the assembly in a small Erlenmeyer, then place the Erlenmeyer in a beaker, and finally cover the first beaker with a second inverted beaker. This procedure will ensure slow cooling, which will enhance good crystal growth (Fig. 5.31). A Dewar flask may be used when very slow cooling and large crystal growth are required.

**Step 6.** After the system reaches room temperature, cooling in an ice bath may improve the yield.

**Step 7.** Remove the solvent by inverting the Craig tube assembly into a centrifuge tube and spinning the mother liquors away from the crystals (Fig. 5.32). *This operation should be carried out with care.* First, fit the head with a thin copper wire (Fig. 5.32), held in place by a loop at the end of the wire that is placed around the narrow part of the neck. Some Teflon heads have a hole in the neck to anchor the wire. The copper wire should not be much longer than the centrifuge tube.

**Step 8.** Now insert the Craig tube into a centrifuge tube. To do this, hold the Craig tube upright (with the head portion up) and place the centrifuge tube *down over* the Craig tube. Push the Craig tube up with your finger so that the head is against the inverted bottom of the centrifuge tube, and then invert the whole assembly (Fig 5.32).

**Step 9.** Place the assembly into a centrifuge tube, *balance the centrifuge,* and spin the mother liquors away from the crystals (Fig. 5.32). This

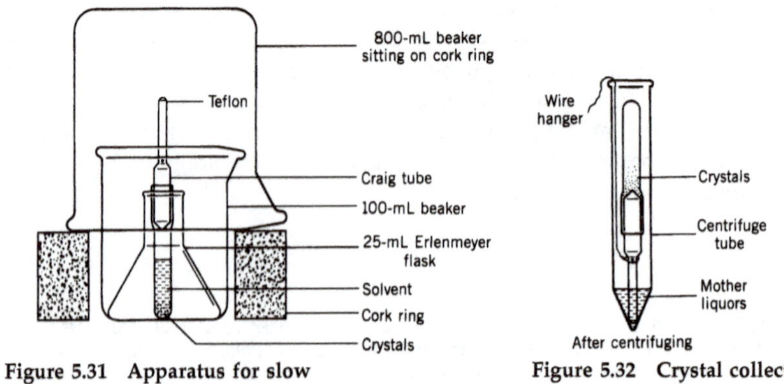


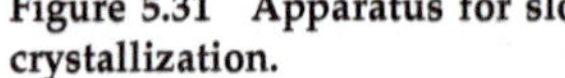

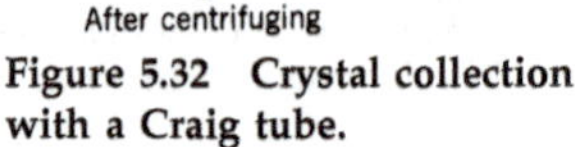

**Figure 5.31 Apparatus for slow crystallization.**

**Figure 5.32 Crystal collection with a Craig tube.**

replaces the usual filtration step in simple crystallizations. It avoids another transfer of material and also avoids product contact with filter paper.

**Step 10.** Remove the apparatus from the centrifuge, and then carefully remove the Craig tube from the centrifuge tube. Gently pull upward on the copper wire while at the same time applying downward pressure with your fingers to the bottom of the inverted Craig tube (this will keep the Craig tube assembly together and not let any of the crystalline product fall back into the centrifuge tube and into the mother liquors). Once the Craig tube assembly is removed from the centrifuge tube, turn it so that the neck of the tube is up, and then disassemble it. At this point scrape any crystalline product clinging to the head into the lower section. If the lower section is tared, it can be left to air dry to constant weight or placed in a warm vacuum oven (use a rubber band or thin wire to wrap a piece of filter paper over the open end to prevent dust from collecting on the product while drying).

The cardinal rule in carrying out the purification of small quantities of solids is *Keep the number of transfers to an absolute minimum!* The Craig tube is very helpful in this regard.

For other recrystallization and filtration methods, see reference material online; Chapter 5W, Crystallization. WWW

## QUESTIONS

**5-22.** What is the purpose of using activated carbon in a recrystallization procedure?

**5-23.** List several advantages and disadvantages of using a Craig tube for recrystallization.

**5-24.** Why is it advisable to use a stemless or a short-stemmed funnel when carrying out a gravity filtration?

**5-25.** Which of the following solvent pairs could be used in a recrystallization? Why or why not?

**(a)** Acetone and ethanol

**(b)** Hexane and water

**(c)** Hexane and diethyl ether

**5-26.** You perform a recrystallization on 60 mg of a solid material and isolate 45 mg of purified material. What is the percent recovery? Further concentration of the mother liquor provides an additional 8 mg of material. What is the total percent recovery?

**5-27.** Describe two techniques that can be used to induce crystallization.

**5-28.** When would you advise someone to use a solvent pair to carry out a recrystallization?

**5-29.** You are provided a solid which you suspect is not pure and the accompanying data sheet has very limited information. Two items which are recognizable are that an ethereal solvent has worked well when performing recrystallizations and the literature melting point is 61 °C. When looking at what ethereal solvents you have to choose from, you see two: *t*-butyl methyl ether and diethyl ether. Why would the latter be the far better choice knowing that aside from relative boiling points, the two ethers are very similar when considering their physical properties.

*NOTE. The following experiments use Technique 5: Experiments [6], [7], [15], [16], [18], [19A], [19B], [19C], [19D], [20], [23A], [23B], [24A], [24B], [25A], [25B], [26], [28], [29A], [29B], [29C], [29D], [30], [31], [33A], [33B], [34A], [34B], [A1$_a$], [A2$_a$], [A3$_a$], [A1$_b$], [A2$_b$], [A3$_b$], [A4$_{ab}$], [B1], [C2], [D1], [D2], [E1], [E2], [F1], [F2], and [F4].*

*[2$_{adv}$], [3A$_{adv}$], [3B$_{adv}$], [5$_{adv}$], [6$_{adv}$], and [7$_{adv}$].* WWW

Technique 6A

# Chromatography

## Column, Flash, High-Performance Liquid, and Thin-Layer Chromatography

The basic theory of chromatography is introduced in Technique 1 in the discussion of gas-phase separations. The word *chromatography* is derived from the Greek word for color, *chromatos*. Tswett discovered the technique in 1903 while studying ways to separate mixtures of natural plant pigments.[17] The chromatographic zones were detected simply by observing the visual absorption bands. Thus, as originally applied, the name was not an inconsistent use of terminology. Today, however, most mixtures are colorless. The separated zones in these cases are detected by other methods.

Two chromatographic techniques are discussed in this section. Both depend on adsorption and distribution between a stationary solid phase and a moving liquid phase. The first is column chromatography, which is used extensively throughout organic chemistry. It is one of the oldest of the modern chromatographic methods. The second technique, thin-layer chromatography (TLC), is particularly effective in rapid assays of sample purity. It can also be used as a preparative technique for obtaining tiny amounts of high-purity material for analysis.

### Column Chromatography

Column chromatography, as its name implies, uses a column packed with a solid stationary phase. A mobile liquid phase flows by gravity (or applied pressure) through the column. Column chromatography uses polarity differences to separate materials. A sample on a chromatographic column is subjected to two opposing forces: (1) the solubility of the sample in the elution solvent system, and (2) the adsorption forces binding the sample to the solid phase. These interactions comprise an equilibrium. Some sample constituents are adsorbed more tightly; other components of the sample dissolve more readily in the liquid phase and are eluted more rapidly. The more rapidly eluting materials, thus, are carried further down the column before becoming readsorbed, and thus exit the column before more tightly bound components. The longer the column, the larger the number of adsorption–dissolution cycles (much like the vaporization–condensation cycles in a distillation column), and the greater the separation of sample components as they elute down the column. A molecule that is strongly adsorbed on the stationary phase will move slowly down the column; a molecule that is weakly adsorbed will move at a faster rate. Thus, a complex mixture can be resolved into separate bands of pure materials. These bands of purified material eventually elute from the column and can be collected.

Many materials have been used as the stationary phase in column chromatography. Finely ground alumina (aluminum oxide, $Al_2O_3$) and silicic acid (silica gel, $SiO_2$) are by far the most common adsorbents (stationary phases). Many common organic solvents are used as the liquids (sometimes called *eluents*) that act as the mobile phase and elute (wash) materials through the column. Table 5.8 lists the better known column packing and elution solvents.

[17]Tswett, M. *Ber. Deut. Botan. Ges.* **1906,** *24,* 235.

| *Table 5.8* Column Chromatography Materials | | | |
|---|---|---|---|
| Stationary Phase | | Moving Phase | |
| Alumina<br>Silicic acid<br>Magnesium sulfate<br>Cellulose paper | ↑ *Increasing adsorption of polar materials* | Water<br>Methanol<br>Ethanol<br>Acetone<br>Ethyl acetate<br>Diethyl ether<br>Methylene chloride<br>Cyclohexane<br>Pentane | ↑ *Increasing solvation of polar materials* |

Silica gel impregnated with silver nitrate (usually 5–10%) is also a useful adsorbent for some functional groups. The silver cation selectively binds to unsaturated sites via a silver-ion $\pi$ complex. Traces of alkenes are easily removed from saturated reaction products by chromatography with this system. This adsorbent, however, must be protected from light until used, or it will quickly darken and become ineffective.

Column chromatography is usually carried out according to the procedures discussed in the following five sections.

***Packing the Column.*** The quantity of stationary phase required is determined by the sample size. A common rule of thumb is to use a weight of packing material 30– 100 times the weight of the sample to be separated. Columns are usually built with roughly a 10:1 ratio of height to diameter. In the microscale laboratory, two standard chromatographic columns are used:

1. A Pasteur pipet, modified by shortening the capillary tip, is used to separate smaller mixtures (10–100 mg). Approximately 0.5–2.0 g of packing is used in the pipet column (Fig. 5.33*a*).
2. A 50-mL titration buret (modified by shortening the column to 10 cm above the stopcock) is used for larger (50–200 mg) or difficult-to-separate sample mixtures. A buret column uses approximately 5–20 g of packing (Fig. 5.33*b*).

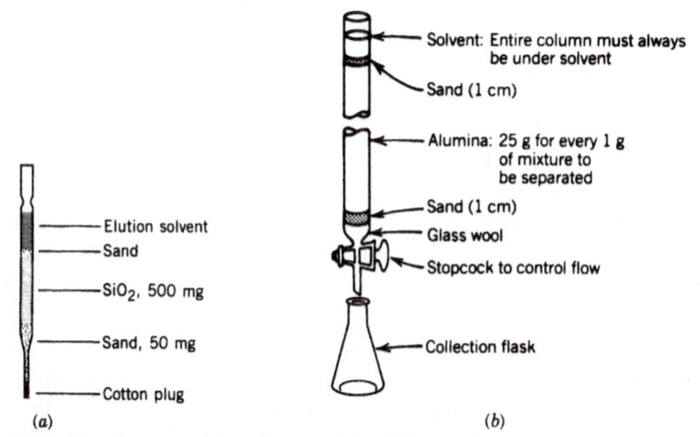


**Figure 5.33 Chromatographic columns: (a) a Pasteur pipet column; (b) a buret column.** (From Zubrick, James W. *The Organic Chem Lab Survival Manual,* 7th ed.; Wiley: New York, 2008. Reprinted by permission of John Wiley & Sons, Inc., New York.)

Both columns are prepared by first clamping the empty column in a vertical position and then seating a small cotton or glass wool plug at the bottom. For a buret column, cover the cotton with a thin layer of sand. The Pasteur pipets are loaded by adding the dry adsorbent with gentle tapping,"dry packing."The pipet column (*dry column*) is then premoistened just prior to use.

The burets (*wet columns*) are packed by a slurry technique. In this procedure the column is filled part way with solvent; then the stopcock is opened slightly, and as the solvent slowly drains from the column a slurry of the adsorbent–solvent is poured into the top of the column. The column should be gently tapped while the slurry is added. The solvent is then drained to the top of the adsorbent level and held at that level until used. Alternatively, the wet-packed column can be loaded by sedimentation techniques rather than using a slurry. One such routine is to initially fill the column with the least-polar solvent to be used in the intended chromatographic separation. Then the solid phase is slowly added with gentle tapping, which helps to avoid subsequent channeling. As the solid phase is added, the solvent is slowly drained from the buret at the same time. After the adsorbent has been fully loaded, the solvent level is then lowered to the top of the packing as in the slurry technique.

***Sample Application.*** Using a Pasteur pipet, apply the sample in a minimum amount of solvent (usually the least polar solvent in which the material is readily soluble) to the top of the column. *Do not disturb the sand layer!* Rinse the pipet, and add the rinses to the column just as the sample solution drains to the top of the adsorbent layer.

***Elution of the Column.*** The critical step in resolving the sample mixture is eluting the column. Once the sample has been applied to the top of the column, the elution begins (a small layer of sand can be added to the top of the buret column after addition of the first portion of elution solvent).

*NOTE: Do not let the column run dry: This can cause channels to form in the column.*

In a buret column, the flow is controlled by the stopcock. The flow rate should be set to allow time for equilibrium to be established between the two phases; this will depend on the nature and amount of the sample, the solvent, and how difficult the separation will be. The Pasteur pipet column is *free flowing* (the flow rate is controlled by the size of the capillary tip and its plug); once the sample is on the column, the chromatography will require constant attention.

If necessary, it is possible to ease this restriction somewhat by modifying the pipet. Place a Tygon connector (short sleeve) at the top of the pipet column. Once the sample is on the column, insert a second pipet into this connector with its tip just below the liquid level on the top of the column. Add additional solvent through the second pipet (use a bulb, if necessary, but remove it before the elution begins), which acts as a solvent reservoir. As the solvent level in the column pipet drops below the tip of the top pipet, air is admitted, and additional solvent is automatically delivered to the chromatographic column. Thus, the solvent head on the column is maintained at a constant volume. The top pipet need be filled only at necessary intervals; larger volumes of solvent can thus be added to this reservoir. This arrangement also prevents dislodging of the absorbent as new solvent is added.

*NOTE It is exceedingly important that solvents do not come in contact with the Tygon sleeve holding the second pipet. These sleeves contain plasticizers that will readily dissolve and contaminate the sample.*

The choice of solvent is dictated by a number of factors. A balance between the adsorption power of the stationary phase and the solvation power of the elution solvent governs the rate of travel of the material down the column. If the material travels rapidly down the column, then too few adsorption–elution cycles will occur and the materials will not separate. If the sample travels too slowly, diffusion broadening takes over and separation is degraded. Solvent choices and elution rates can strike a balance between these factors and maximize the separation. It can take considerable time to develop a solvent or mixture of solvents that produces a satisfactory separation of a particular mixture.

***Fraction Collection.*** As the solvent elutes from the column, it is collected in a series of "fractions" by using small Erlenmeyer flasks or vials. Under ideal conditions, as the mixture of material travels down the column, it will separate into several individual bands (zones) of pure substances. By careful collection of the fractions, these bands can be separated as they sequentially elute from the column (similar to the collection of GC fractions in the example described in Technique 1). The bands of eluted material can be detected by a number of techniques (weighing fraction residues, colored materials, TLC, etc.).

Column chromatography is a powerful technique for the purification of organic materials. In general, it is significantly more efficient than crystallization procedures. Recrystallization is often best avoided until the last stages of purification, where it will be most efficient. Rely instead on chromatography to do most of the separation.

Column chromatography of a few milligrams of product usually takes no more than 30 min, but chromatographing 10 g of product might easily take several hours, or even all day. Large-scale column chromatography (50–100 g) of a complex mixture could take several days to complete using this type of equipment.

### Flash Chromatography

Flash chromatography, first described by Still and co-workers in 1978, is a common method for separating and purifying nonvolatile mixtures of organic compounds.[18] The technique is rapid, easy to perform, relatively inexpensive, and gives good separations. Many laboratories routinely use flash chromatography to separate mixtures ranging from 10 mg to 10 g.

This moderate-resolution, preparative chromatography technique was originally developed using silica gel (40–63 μm). Bonded-phase silica gel of a larger particle size can be used for reversed-phase flash chromatography. Flash chromatography columns are generally packed dry to a height of approximately 6 in. Thin-layer chromatography is a quick way to choose solvents for flash chromatography. A solvent that gives differential retardation factor ($DR_f$) values (of the two substances requiring seperation) $\geq 0.15$ on TLC usually gives effective separation with flash chromatography. Table 5.9 lists typical experimental parameters for various sample sizes, as a guide to separations using flash chromatography. In general, a mixture of organic compounds separable by TLC can be separated preparatively using flash chromatography.

[18]Still, W. C.; Kahn, M.; Mitra, A. *J. Org. Chem.* **1978,** *43,* 2923.

*Table 5.9* **Typical Experimental Parameters**

| Column Diameter (mm) | Total Volume of Eluent (mL)[a] | Typical Sample Loading (mg) $DR_{f>0.2}$ | Typical Sample Loading (mg) $DR_{f>0.1}$ | Typical Fraction (mL) |
|---|---|---|---|---|
| 10 | 100–150 | 100 | 40 | 5 |
| 20 | 200–250 | 400 | 160 | 10 |
| 30 | 400–450 | 900 | 360 | 20 |
| 40 | 500–550 | 1600 | 600 | 30 |
| 50 | 1000–1200 | 2500 | 1000 | 50 |

*Source.* Data from Majors, R. E., and Enzweiler, T. *LC, GC* **1998,** *6,* 1046.
[a]Required for both packing and elution.

Flash chromatography apparatus generally consists of a glass column equipped to accept a positive pressure of compressed air or nitrogen applied to the top of the column. A typical commercially available arrangement is shown in Figure 5.34.[19]

Generally, a 20–25% solution of the sample in the elution solvent is recommended, as is a flow rate of about 2 in./min. The column must be conditioned, before the sample is applied, by flushing the column with the elution solvent to drive out air trapped in the stationary phase and to equilibrate the stationary phase and the solvent.

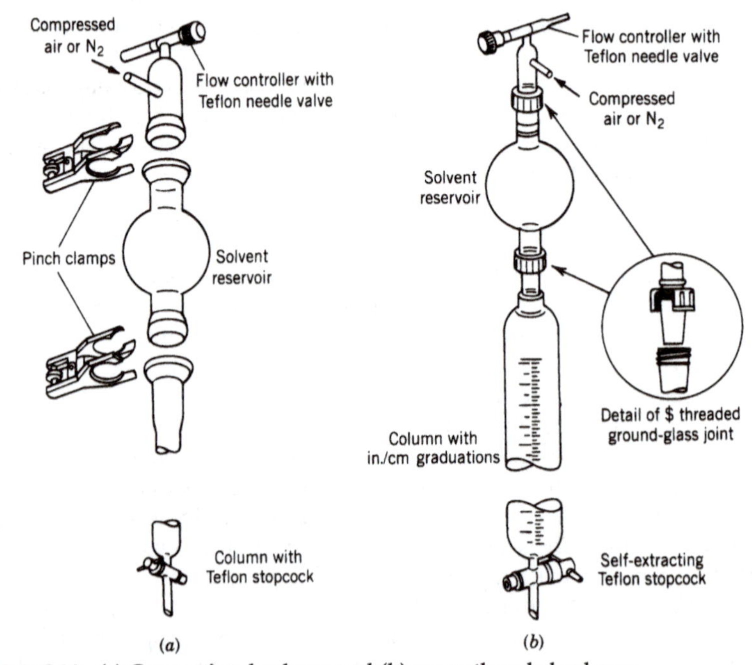


**Figure 5.34 (a) Conventional column and (b) screw-threaded column.**

[19]A complete line of glass columns, reservoirs, clamps, and packing materials for flash chromatography is offered by Sigma-Adrich Chemical Co., P.O. Box 355, Milwaukee, WI 53201. Silica gels for use in this technique are also available from Amicon, Danvers, MA; J. T. Baker, Phillipsburg, NJ; EM Science/Merck, Gibbstown, NJ.; ICN Biomedicals, Inc., Cleveland, OH.

Several modifications of the basic arrangement have been reported, especially in regard to the adaptation of the technique to the instructional laboratory. These involve inexpensive pressure control valves, use of an aquarium "vibrator" air pump, and adapting a balloon reservoir to supply the pressurized gas.

At the microscale level, a pipet bulb or pump on the pipet column can be used to supply pressure to the column. If a bulb is used, squeeze it to apply pressure, and *remove the bulb from the pipet before releasing it!* Otherwise, material may be sucked up into the bulb and most likely disturb the column. Reapply the bulb to re-create pressure. If a pump is used, do not back off the pressure once it has been applied.

An improved method, utilizing a capillary Pasteur pipet for introducing the sample onto the chromatographic column approximately doubles the effectiveness (theoretical plates) of the column.[20] Dry-column flash chromatography[21] has been adapted for use in the instructional laboratory.[22] The "column" consists of a dry bed of silica gel in a sintered glass funnel placed in a standard vacuum filtration flask; the solvent is eluted by suction. Small (16 × 150-mm) test tubes inserted into the flask below the stem of the funnel are used to collect the fractions. This technique has been used successfully to separate mixtures ranging from 150 to 1000 mg.

### Thin-Layer Chromatography

Thin-layer chromatography (TLC) is closely related to column chromatography, in that the phases used in both techniques are essentially identical. Alumina and silica gel are typical stationary phases, and the usual solvents are the mobile phases. There are, however, some distinct differences between TLC and column chromatography. The mobile (liquid) phase *descends* in column chromatography; the mobile phase *ascends* in TLC. The column of stationary-phase material used in column chromatography is replaced by a thin layer (100 μm) of stationary phase spread over a flat surface. A piece of window glass, a microscope slide, or a sheet of plastic can be used as the support for the thin layer of stationary phase. It is possible to prepare your own glass plates, but plastic-backed thin-layer plates are only commercially available. Plastic-backed plates are particularly attractive because they can easily be cut with scissors into strips of any size. Typical strips measure about 1 × 3 in., but even smaller strips can be satisfactory.

Thin-layer chromatography has some distinct advantages: it needs little time (2– 5 min) and it needs *very* small quantities of material (2–20 μg). The chief disadvantage of this type of chromatography is that it is not very amenable to preparative scale work. Even when large surfaces and thicker layers are used, separations are most often restricted to a few milligrams of material.

*NOTE. Do not touch the active surface of the plates with your fingers. Handle them only by the edge.*

TLC is performed as follows:

**Step 1.** Draw a light pencil line parallel to the short side of the plate, 5–10 mm from the edge. Mark one or two points, evenly spaced, on the line.

---

[20]Pivnitsky, K. K. *Aldrichimica Acta* **1989,** *22,* 30.
[21]Harwood, L. M.; *Aldrichimica Acta* **1985,** *18,* 25; Sharp, J. T.; Gosney I.; Rowley A. G. *Practical Organic Chemistry;* Chapman & Hall: New York, 1989.
[22]Shusterman, A. J.; McDougal, P. G.; Glasfeld, A. *J. Chem. Educ.* **1997,** *74,* 1222.

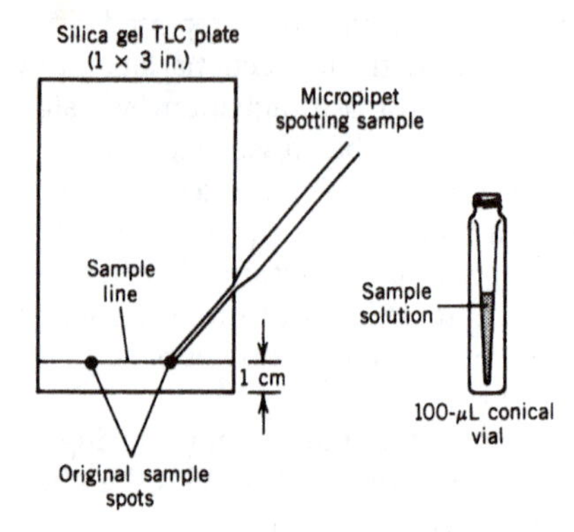


**Figure 5.35 Applying a sample to a TLC plate.**

Place the sample to be analyzed (1 mg or less) in a 100-μL conical vial and add a few drops of a solvent to dissolve the sample. Use a capillary micropipet to apply a small fraction of the solution from the vial to the plate (Fig. 5.35). (These pipets are prepared by the same technique used for constructing the capillary insert for ultramicro boiling-point determinations, see Chapter 4.) Apply the sample to the adsorbent side of the TLC plate by gently touching the tip of the filled capillary to the plate. Remove the tip from the plate before the dot of solvent grows to much more than a few millimeters in diameter. If it turns out that you need to apply more sample, let the dot of solvent evaporate and then reapply more sample to exactly the same spot.

**Step 2.** Place the spotted thin-layer plate in a screw-capped, wide-mouth jar, or a beaker with a watch glass cover, containing a small amount of elution solvent (Fig. 5.36). It helps if one side of the jar's (beaker's) interior is covered with a piece of filter paper that wicks the solvent up to increase the surface area of the 100 solvent. The TLC plate must be positioned so that the spot of your sample is *above* the solvent. Cap the jar, or replace the watch glass on the beaker, to maintain an atmosphere saturated with the elution solvent. The elution solvent climbs the plate by capillary action, eluting the sample up the plate. *Do not move the developing chamber after the action has started.* Separation of mixtures into individual spots occurs by exactly the same mechanism as in column chromatography. Stop the elution by removing the plate from the jar or beaker when the solvent front nears the top of the TLC plate. Quickly (before the solvent evaporates) mark the position of the solvent front on the plate.

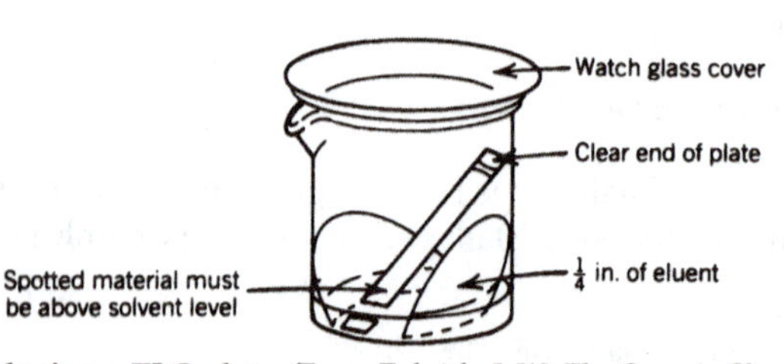


**Figure 5.36 Developing a TLC plate.** (From Zubrick, J. W. *The Organic Chem Lab Survival Manual,* 7th ed.; Wiley: New York. 2008. Reprinted by permission of John Wiley & Sons, Inc., New York.)

**Step 3.** Colorless, separated components of a mixture can often be observed in a developed TLC plate by placing the plate in an iodine-vapor chamber (a sealed jar containing solid $I_2$) for a minute or two. Iodine forms a reversible complex with most organic substances and dark spots will thus appear in those areas containing sample material. Mark the spots with a pencil soon after removing the TLC plate from the iodine chamber because the spots may fade. Samples that contain a UV-active chromophore (see Chapter 8) can be observed without using iodine. TLC plates are commonly prepared with an UV-activated fluorescent indicator mixed in with the silica gel. Sample spots can be detected with a hand-held UV lamp; the sample quenches the fluorescence induced by the lamp and appears as a dark spot against the fluorescent blue-green background.

**Step 4.** The TLC properties of a compound are reported as $R_f$ values (retention factors). The $R_f$ value is the distance traveled by the substance divided by the distance traveled by the solvent front (this is why the position of the solvent front should be quickly marked on the plate when the chromatogram is terminated; see Fig. 5.37). TLC $R_f$ values vary with the moisture content of the adsorbent. Thus, the actual $R_f$ of a compound in a given solvent can vary from day to day and from laboratory to laboratory. The best way to determine if two samples have identical $R_f$ values is to elute them together on the same plate.

Thin-layer chromatography is used in a number of applications. The speed of the technique makes it quite useful for monitoring large-scale column chromatography. Analysis of fractions can guide decisions on the solvent–elution sequence. TLC analysis of column-derived fractions can also determine how best to combine collected fractions. Following the progress of a reaction by periodically removing small aliquots for TLC analysis is an extremely useful application of thin-layer chromatography.

## Paper Chromatography

The use of cellulose-paper as an adsorbent is referred to as *paper chromatography*. This technique has many of the characteristics of TLC in that sheets or

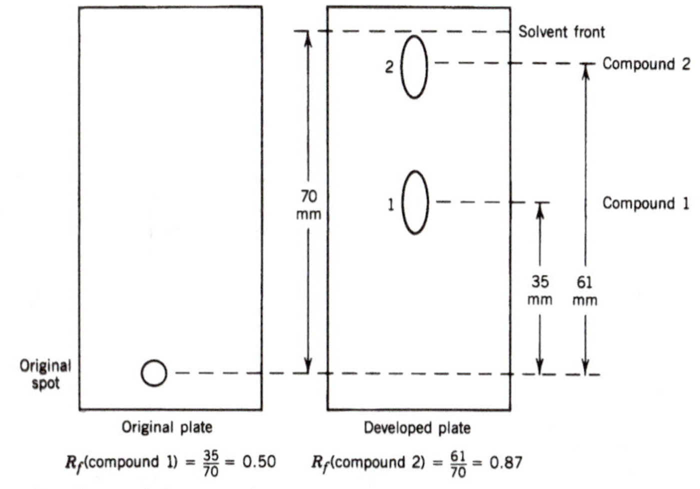


**Figure 5.37 Determining $R_f$ values.**

strips of filter paper are used as the stationary phase. In this case, however, the paper is usually positioned to hang down from trays holding the paper and the elution solvent. The solvent front, therefore, descends downward rather than upward as in TLC. Paper chromatography has a distinct advantage: It is very amenable to the use of aqueous mobile phases and very small sample sizes. It is primarily used for the separation of highly polar or polyfunctional species such as sugars and amino acids. It has one major disadvantage: It is very slow. Paper chromatograms can easily take three to four hours or more to elute.

### High-Performance Liquid Chromatography

Although gas chromatography is a powerful chromatographic method, it is limited to compounds that have a significant vapor pressure at temperatures up to about 200 °C. Thus, compounds of high molecular weight and/or high polarity cannot be separated by GC. High-performance liquid chromatography (HPLC) does not present this limitation.

GC and HPLC are somewhat similar, in the instrumental sense, in that the analyte is partitioned between a stationary phase and a mobile phase. Whereas the mobile phase in GC is a gas, the mobile phase in HPLC is a liquid. As shown schematically in Figure 5.38, the mobile phase (solvent) is delivered to the system by a pump capable of pressures up to about 6000 psi. The sample is introduced by the injection of a solution into an injection loop. The injection loop is brought in line between the pump and the column (stainless steel) by turning a valve; the sample then flows down the column, is partitioned, and flows out through a detector.

The solid phase in HPLC columns used for organic monomers is usually some form of silica gel. "Normal" HPLC refers to chromatography using a solid phase (usually silica gel) that is more polar than the liquid phase, or solvent, so that the less polar compounds elute more rapidly. Typical solvents include ethyl acetate, hexane, acetone, low molecular weight alcohols, chloroform, and acetonitrile. For extremely polar compounds, such as amino

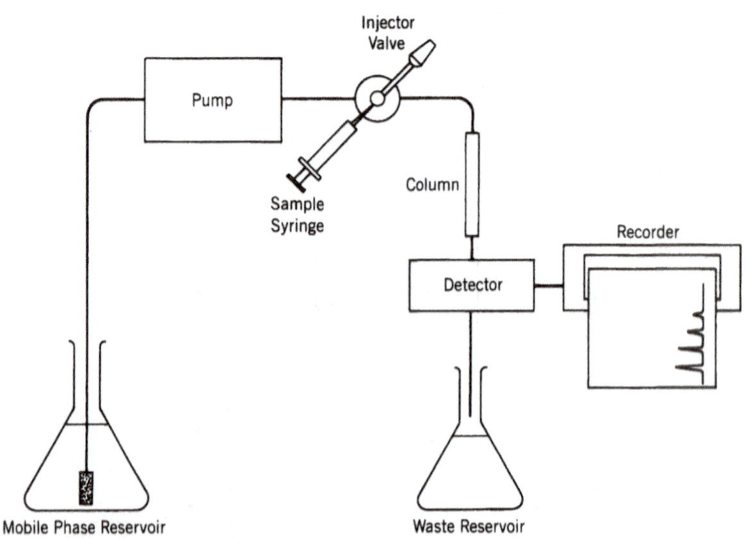


**Figure 5.38 High-performance liquid chromatography system block diagram.** *(Courtesy of the Perkin-Elmer Corp., Norwalk, CT.)*

acids, "reversed-phase" HPLC is used. Here, the liquid phase is more polar than the stationary phase, and the more polar compounds elute more rapidly. The mobile phase is usually a mixture of water and a water-miscible organic solvent such as acetonitrile, dioxane, methanol, isopropanol, or acetone. The stationary phase is usually a modified silica gel where the —OH groups of the silica gel have been replaced by —OSiR groups; R is typically a linear $C_{18}$ alkyl chain. These so-called "bonded-phase" columns are not capable of handling as much analyte as normal silica gel columns, and are thus easily overloaded and are less useful for preparative work. (For further discussion see Solid-Phase Extraction, Technique 4, page 83.)

A wide variety of detection systems are available for HPLC. UV detection is common, inexpensive, and sensitive. The solvent flowing off the column is sent through a small cell where the UV absorbance is recorded over time. Many detectors are capable of variable wavelength operation so the detector can be set to the wavelength most suitable to the compound or compounds being analyzed. Photodiode array detectors are available; these can obtain a full UV spectrum in a fraction of a second, so that more information can be obtained on each component of a mixture. For compounds that absorb light in the visible (vis) spectrum, many detectors can be set to visible wavelengths. The principal shortcoming of UV-vis detection is that to be detected, compounds being studied must have a UV chromophore, such as an aromatic ring or other conjugated $\pi$ system (see Chapter 8).

For compounds that lack a UV-vis chromophore, refractive index (RI) detection is a common substitute. An RI detector measures the difference in refractive index between the eluant and a reference cell filled with the elution solvent. Refractive index detection is significantly less sensitive than UV-vis detection, and the detector is quite sensitive to temperature changes during the chromatographic run.

More sophisticated HPLC instruments offer the ability to mix two or three different solvents and to use solvent gradients by changing the solvent composition as the chromatographic run progresses. This allows the simultaneous analysis of compounds that differ greatly in their polarity. For example, a silica gel column might begin elution with a very nonpolar solvent, such as hexane. The solvent polarity is then continuously increased by blending in more and more ethyl acetate until the elution solvent is pure ethyl acetate. This effect is directly analogous to temperature programming in GC.

For analytical work, typical HPLC columns are about 5 mm in diameter and about 25 cm in length. The maximum amount of analyte for such columns is generally less than 1 mg, and the minimum amount is determined by the detection system. High-performance liquid chromatography can thus be used to obtain small amounts of purified compounds for infrared (IR), nuclear magnetic resonance (NMR) or mass spec-trometric (MS) analysis. Larger "semipreparative" columns that can handle up to about 20 mg without significant overloading are useful for obtaining material for $^{13}C$ NMR spectroscopy or further synthetic work.

HPLC has the advantage that it is rapid, it uses relatively small amounts of solvent, and it can accomplish very difficult separations.

## Concentration of Solutions

Technique 6B

The solvent can be removed from chromatographic fractions (extraction solutions, or solutions in general) by a number of different methods.

### Distillation

Concentration of solvent by distillation is straightforward, and the standard routine is described in Technique 2 (page 61). This approach allows for high recovery of volatile solvents and often can be done outside a hood. The Hickman still head and the 5- or 10-mL round-bottom flask are useful for this purpose. Distillation should be used primarily for concentration of the chromatographic fraction, followed by transfer of the concentrate with a Pasteur filter pipet to a vial for final isolation.

### Evaporation with Nitrogen Gas

A very convenient method for removal of final solvent traces is the concentration of the last 0.5 mL of a solution by evaporation with a gentle stream of nitrogen gas while the sample is warmed in a sand bath. This process is usually done at a hood station where several Pasteur pipets can be attached to a manifold leading to a source of dry nitrogen gas. Gas flow to the individual pipets is controlled by needle valves. *Always test the gas flow with a blank vial of solvent.*

Ruekberg described an alternative way to remove solvent from solutions of compounds that are not readily oxidized.[23] The setup includes an aquarium air pump, a pressure safety valve and ballast container, a drying tube, and a manifold. Blunted hypodermic needles are used in place of Pasteur pipets.

The sample vial will cool as the solvent evaporates, and gentle warming and agitation of the vial will thus help remove the last traces of the solvent. This avoids possible moisture condensation on the sample residue, as long as the gas itself is dry. *Do not leave the heated vial in the gas flow after the solvent is removed!* This precaution is particularly important in the isolation of liquids. Tare the vial before filling it with the solution to be concentrated; constant weight over time is the best indication that all solvent has been removed.

### Removal of Solvent Under Reduced Pressure

Concentration of solvent under reduced pressure is very efficient. It reduces the time for solvent removal in microscale experiments to a few minutes. In contrast, distillation or evaporation procedures require many minutes for even relatively small volumes. Several methods are available.

***Filter Flask Method.*** This vacuum-concentration technique can be tricky and should be practiced prior to committing hard-won reaction product to this test. The procedure is most useful with fairly large chromatographic fractions (5–10 mL). The sequence of operations is as follows (see also Fig. 5.39):

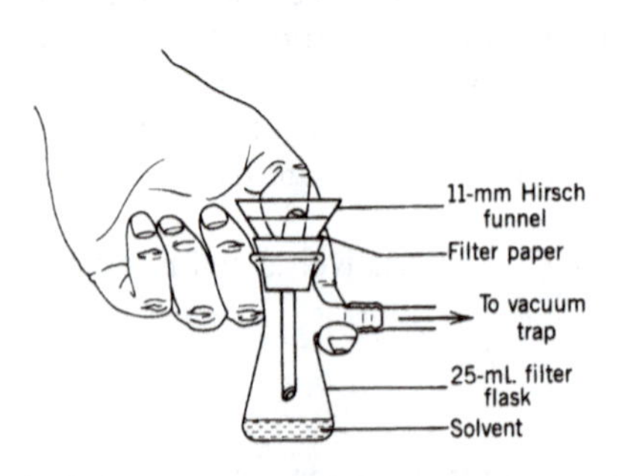


**Figure 5.39 Removal of solvent under reduced pressure.**

[23]Ruekberg, B. *J. Chem. Educ.* **1995,** *72,* A200.

**Step 1.** Transfer the chromatographic fraction to the 25-mL filter flask.

**Step 2.** Insert the 11-mm Hirsch funnel and rubber adapter into the flask.

**Step 3.** Turn on the water pump (with trap) and connect the vacuum tubing to the pressure flask side arm while holding the flask in one hand.

**Step 4.** Place the thumb of the hand holding the filter flask over the Hirsch funnel filter bed to shut off the air flow through the system (see Fig. 5.39). This will result in an immediate drop in pressure. The volatile solvent will rapidly come to a boil at room temperature. Thumb pressure adjusts air leakage through the Hirsch funnel and thereby controls the pressure in the system. It is also good practice to learn to manipulate the pressure so that the liquid does not foam up into the side arm of the filter flask.

The filter flask must be warmed by the sand bath during this operation; rapid evaporation of the solvent will quickly cool the solution. The air leak used to control the pressure results in a stream of moist laboratory air being rapidly drawn over the surface of the solution. If the evaporating liquid becomes cold, water will condense over the interior of the filter flask and contaminate the isolated residue. Warming the flask while evaporating the solvent will avoid this problem and help speed solvent removal. The temperature of the flask should be checked from time to time by touching it with the palm of the free hand. The flask is kept slightly above room temperature by adjusting the heating and evaporation rates. It is best to practice this operation a few times with pure solvent (blanks) to see whether you can avoid boilovers and accumulating water residue in the flask.

***Rotary Evaporator Method.*** In most research laboratories, the most efficient way to concentrate a solution under reduced pressure is to use a **rotary evaporator.** A commercial micro-rotary evaporator is shown in Figure 5.40.

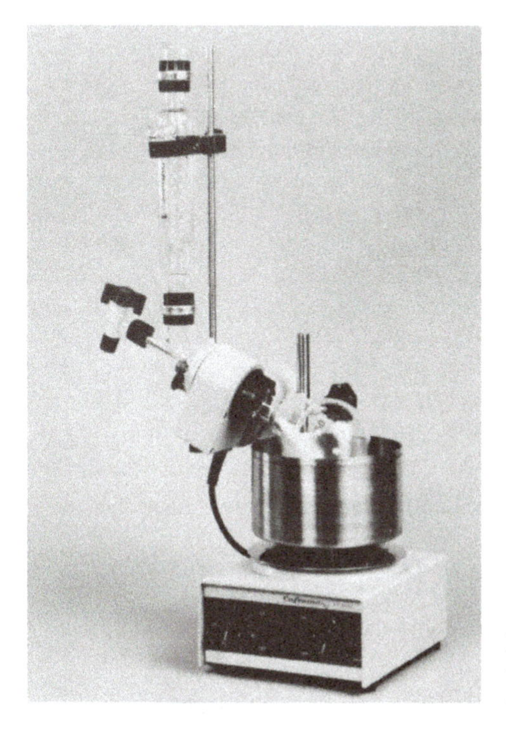

**Figure 5.40 Heidolph micro-rotary evaporator.** *(Courtesy of Caframo, Ltd., Wiarton, Ontario, Canada.)*

This equipment makes it possible to recover the solvent removed during the operation.

The rotary evaporator is a motor-driven device that rotates the flask containing the solution to be concentrated under reduced pressure. The rotation continuously exposes a thin film of the solution for evaporation. This process is very rapid, even well below the boiling point of the solvent being removed. Since the walls of the rotating flask are constantly rewetted by the solution, bumping and superheating are minimized. The rotating flask may be warmed in a water bath or other suitable device that controls the rate of evaporation. A suitable adapter (a "bump bulb") should be used on the rotary evaporator to guard against splashing and sudden boiling, which may lead to lost or contaminated products.

In microscale work, never pour a recovered liquid product from the rotary flask. Always use a Pasteur pipet.

***Hickman Still–Rotary Evaporation Apparatus.*** A simple microscale rotary evaporator for use in the instructional laboratory consists of a 10-mL round-bottom flask connected to a capped Hickman still (side-arm type), which in turn is attached to a water aspirator (with trap).[24] The procedure involves transferring the solution to be concentrated to the preweighed 10-mL flask. The flask is then attached to a Hickman still with its top joint sealed with a rubber septum and threaded compression cap. The apparatus is connected by the still side arm to the trap–vacuum source with a vacuum hose. With the aspirator on, one shakes the apparatus while warming the flask in the palm of the hand. In this manner, bumping is avoided and evaporation is expedited. The still acts as a splash guard. Heat transfer is very effective, and once the flask reaches ambient temperature, the vacuum is released by venting through the trap stopcock.

## QUESTIONS

**5-30.** When marking the sample line on a TLC plate, why is it inadvisable to use a ball-point pen?

**5-31.** A series of dyes is separated by TLC. The data are given below. Calculate the $R_f$ value for each dye.

| Material | Distance moved (cm) |
|---|---|
| Solvent | 6.6 |
| Bismarck brown | 1.6 |
| Lanacyl violet BF | 3.8 |
| Palisade yellow 3G | 5.6 |
| Alizarine emerald G | 0.2 |

**5-32.** Why is it important not to let the level of the elution solvent in a packed chromatographic column drop below the top of the solid-phase adsorbent?

**5-33.** What are some advantages of using column chromatography to purify reaction products in the microscale laboratory?

**5-34.** Discuss the similarities and dissimilarities of TLC, paper, and column chromatography.

[24]Maynard, D. F. *J. Chem. Educ.* **1994,** *71,* A272.

**5-35.** Discuss the similarities and dissimilarities of HPLC and gas chromatography.

**5-36. (a)** What are the main advantages of using flash chromatography?

**(b)** How can TLC be used in connection with flash chromatography?

**5-37.** Using the information presented on the right, please identify and explain which spot has an $R_f$ value of 0.5.

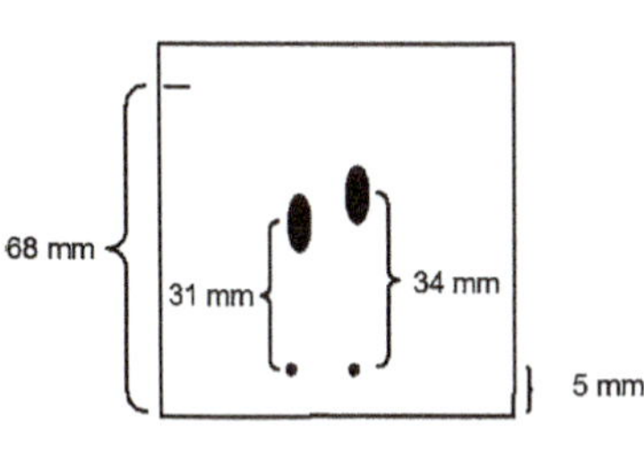


*NOTE. The following experiments use Technique 6: Experiments [8A], [8B], [8C], [11C], [12], [13], [16], [17], [19A], [19B], [19C], [19D], [22A], [22B], [27], [29A], [29B], [29C], [29D], [30], [33A], [33B], [35], [A2$_a$], [A1$_b$], [E1], and [E3]. [1A$_{adv}$], [1B$_{adv}$], [4$_{adv}$], and [7$_{adv}$].*

www

# Collection or Control of Gaseous Products

TECHNIQUE 7

## Water-Insoluble Gases

Numerous organic reactions lead to the formation of gaseous products. If the gas is insoluble in water, collection is easily accomplished by displacing water from a collection tube. A typical experimental setup for the collection of gases is shown in Figure 5.41.

As illustrated, the glass capillary efficiently transfers the evolved gas to the collection tube. The delivery system need not be glass; small polyethylene or polypropylene tubing may also serve this purpose. In this arrangement, a syringe needle is inserted through a septum to accommodate the plastic tubing as shown in Figure 5.42. An alternative to this connector is a shortened Pasteur pipet inserted through a thermometer adapter (Fig. 5.42). Another alternative to the syringe needle or glass pipet tip is suggested by Jacob.[25] The lower half of the tapered tip of a plastic automatic delivery pipet

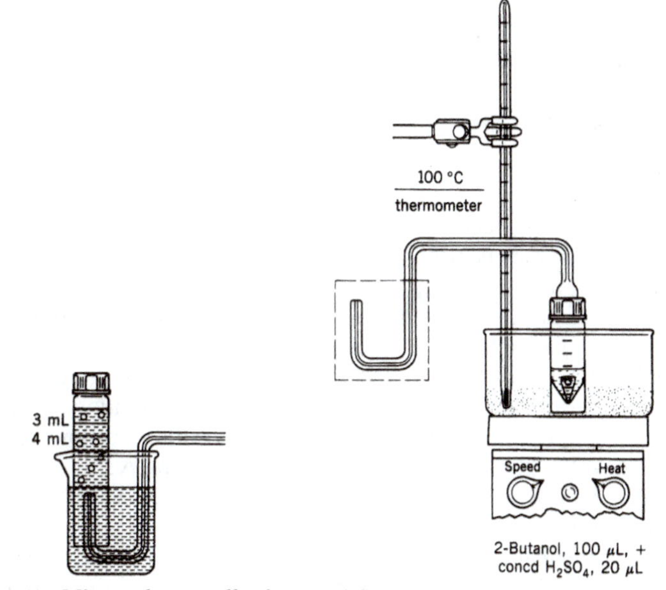


**Figure 5.41 Microscale gas collection apparatus.**

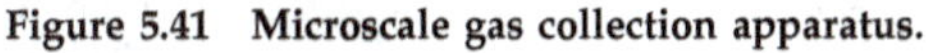

[25]Jacob, L. A. *J. Chem. Educ.* **1992,** *69,* A313.

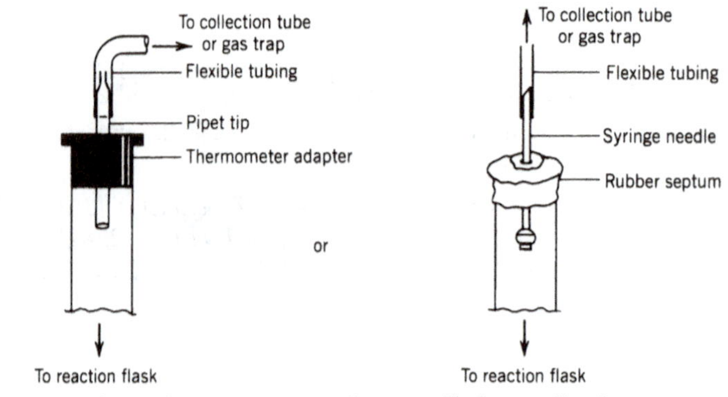


**Figure 5.42 Alternative arrangements for controlled gas collection.**

tip is cut off to prevent buildup of excess pressure in the reaction vessel. The pipet tip is then inserted through a previously pierced rubber septum or into a thermometer adapter. The narrow end of the tip is then inserted into the plastic tubing.

An example of a reaction leading to gaseous products that can use this collection technique is the acid-catalyzed dehydration of 2-butanol. The products of this reaction are a mixture of alkenes: 1-butene, *trans*-2-butene, and *cis*-2-butene, which boil at −6.3, 0.9, and 3.7 °C respectively and *sec*-butyl ether (2,2′-oxybisbutane) which boils at 123 °C. While all four compounds are formed in the reaction mixture, the setup is designed to collect the gases and thus the three alkenes.

In Figure 5.41 the gas collection tube is capped with a rubber septum. This arrangement allows for convenient removal of the collected gaseous butenes using a gas-tight syringe, as shown in Figure 5.43. In this particular reaction, the mixture of gaseous products is conveniently analyzed at ambient temperature by GC (see Technique 1).

## Trapping Byproduct Gases

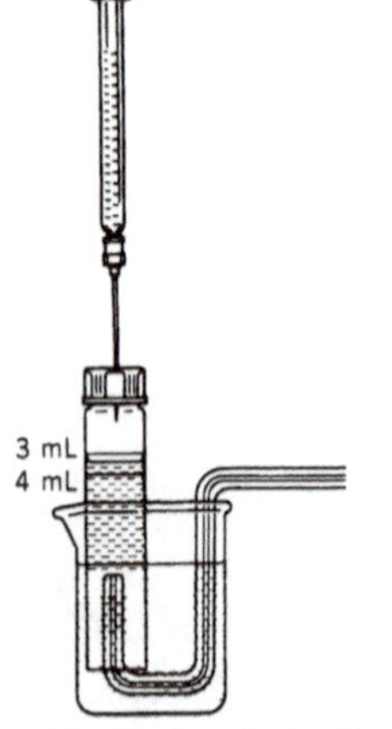


**Figure 5.43 Removal of collected gases.**

Some organic reactions release poisonous or irritating gases as byproducts. For example, hydrogen chloride, ammonia, and sulfur dioxide are typical byproducts in organic reactions. In these cases, the reaction is generally run in a **hood.** A gas trap may be used to prevent the gases from being released into the laboratory atmosphere. If the evolved gas is water soluble, the trap technique works well at the microscale level. The evolved gas is directed from the reaction vessel to a container of water or other aqueous solution, wherein it dissolves (reacts). For example, a dilute solution of sodium or ammonium hydroxide is suitable for acidic gases (such as HCl); a dilute solution of sulfuric or hydrochloric acid is suitable for basic gases (such as $NH_3$ or low molecular weight amines). Various designs are available for gas traps. A simple, easily assembled one for a gas that is very soluble in water is shown in Figure 5.44. Note that the funnel is not immersed in the water. If the funnel is held below the surface of the water and a large quantity of gas is absorbed or dissolved, the water easily could be drawn back into the reaction assembly. If the gas to be collected is not very soluble, the funnel may be immersed just below (1–2 mm) the surface of the water.

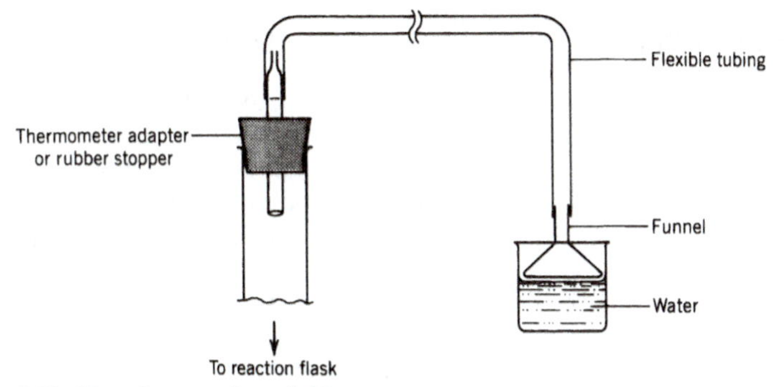


**Figure 5.44 Trapping a water-soluble gas.**

At the microscale level, small volumes of gases are evolved that may not require the funnel. Three alternatives are available:

**a.** Fill the beaker (100 mL) in Figure 5.41 with moistened fine glass wool and lead the gas delivery tube directly into the wool.

**b.** Place moistened glass wool in a drying tube and attach the tube to the reaction apparatus (see Chapter 3W, Fig. 3.10W). However, be careful not to let the added moisture drip into the reaction vessel; place a small section of **dry** glass wool in the tube before the moist section is added.

www

**c.** Use a water aspirator. An inverted funnel can be placed over the apparatus opening where the evolved gas is escaping (usually the top of a condenser) and connected with flexible tubing (through a water trap) to the aspirator. A second arrangement is to use a glass or plastic T-tube (open on one end) inserted in the top of the condenser, by use of a rubber stopper, in place of the funnel.[26] If the reaction must be run under anhydrous conditions, a drying tube is inserted between the condenser and T-tube. This arrangement is very efficient, easy to assemble, and inexpensive. The simplest method is to clamp a Pasteur pipet so that its tip is inserted well into the condenser, and connect it (through a water trap) to the aspirator.

## QUESTIONS

**5-38.** In Figure 5.41 why is a septum, not just a plain cap, used on the top of the gas collection tube?

**5-39.** An evolved gas is directed from the reaction vessel to a container of water or other aqueous solution, wherein it dissolves (reacts). For example, a dilute solution of sodium or ammonium hydroxide is suitable for acidic gases (such as HCl). What solution would be appropriate to trap thiols and sulfides? (*Hint:* Consult a qualitative analysis text.)

**5-40.** One way to eliminate emissions is to place moistened glass wool in a drying tube, which is then attached to the reaction apparatus (Fig. 3.10W). What precautions must be taken when using this method?

www

**5-41.** In the collection of water-insoluble gases with the apparatus shown in Figure 5.41, describe how one might measure the rate at which a gas is evolved from a reaction mixture.

*NOTE. The following experiments use Technique 7: Experiments [9], [10], [14], [A2$_a$], and [B2].*

[26]Horodniak, J. W.; Indicator, N. *J. Chem. Educ.* **1970,** 47, 568.

# Measurement of Specific Rotation

Solutions of optically active substances, when placed in the path of a beam of polarized light, may rotate the plane of the polarized light. Enantiomers (two molecules that are nonidentical mirror images) have identical physical properties (melting points, boiling points, infrared and nuclear magnetic resonance spectra, etc.) except for their interaction with plane polarized light, their *optical activity*. Optical rotation data can provide important information concerning the absolute configuration and the enantiomeric purity of a sample.

Optical rotation is measured using a *polarimeter*. This technique is applicable to a wide range of analytical problems, from purity control to the analysis of natural and synthetic compounds. The results obtained from measuring the observed angle of rotation $\alpha$ are generally expressed as the *specific rotation* $[\alpha]$.

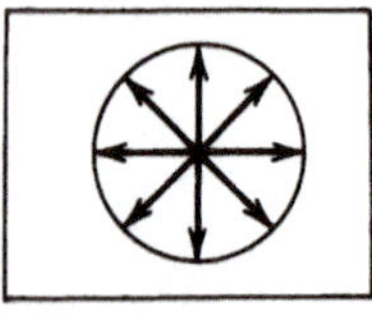

**Figure 5.45 Oscillation of the electric field of ordinary light occurs in all possible planes perpendicular to the direction of propagation.** (From Solomons, T. W. G. *Organic Chemistry*, 9th ed.; Wiley: New York, 2008. Reprinted by permission of John Wiley & Sons, Inc., New York.)

## Theory

Ordinary light behaves as though it were composed of electromagnetic waves in which the oscillating electric field vectors are distributed among the infinite number of possible orientations around the direction of propagation (see Fig. 5.45).

*NOTE. A beam of light behaves as though it is composed of two, mutually perpendicular, oscillating fields: an electric field and a magnetic field. The oscillating magnetic field is not considered in the following discussion.*

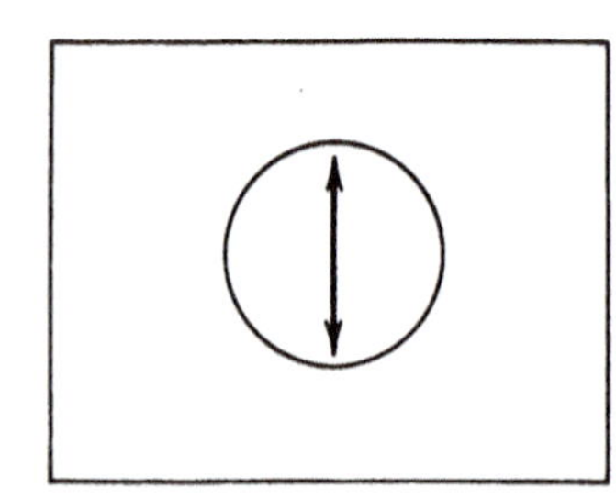

**Figure 5.46 The plane of oscillation of the electric field of plane-polarized light. In this example the plane of polarization is vertical.** (From Solomons, T. W. G. *Organic Chemistry*, 9th ed.; Wiley: New York, 2008. Reprinted by permission of John Wiley & Sons, Inc., New York.)

The planes in which the electrical fields oscillate are perpendicular to the direction of propagation of the light beam. If one separates one particular plane of oscillation from all other planes by passing the beam of light through a polarizer, the resulting radiation is plane-polarized (Fig. 5.46). In the interaction of light with matter, this plane-polarized radiation is represented as the vector sum of two circularly polarized waves. The electric vector of one of the waves moves in a clockwise direction; the other moves in a counterclockwise direction. Both waves have the same amplitude (Fig. 5.47). These two components add vectorially to produce plane-polarized light.

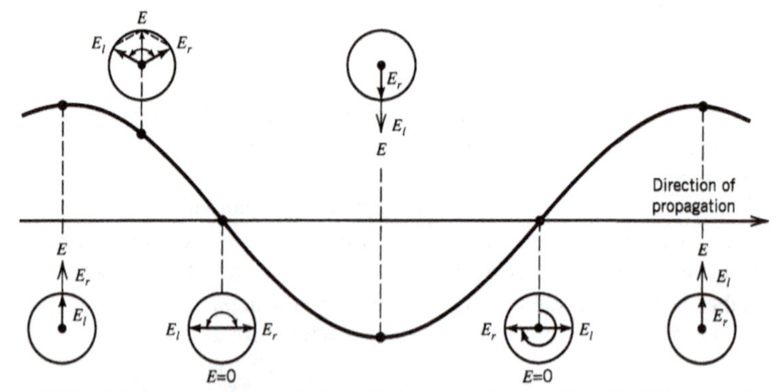


**Figure 5.47 A beam of plane-polarized light viewed from the side (sine wave) and along the direction of propagation at specific times (circles) where the resultant vector $E$ and the circularly polarized components $E_l$ and $E_r$ are shown.** (From Douglas, B., McDaniel, D. H., and Alexander, J. J. *Concepts and Models of Inorganic Chemistry*, 3rd ed. Wiley, New York, 1994. Reprinted by permission of John Wiley & Sons, Inc., New York.)

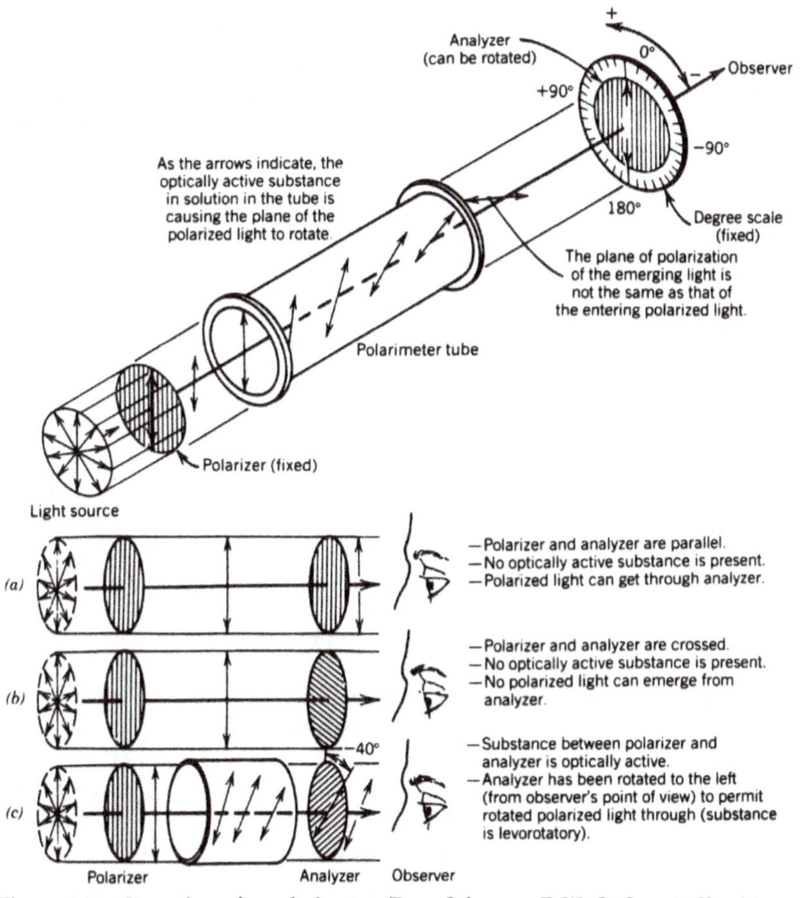


**Figure 5.48 Operation of a polarimeter.** (From Solomons, T. W. G. *Organic Chemistry*, 9th ed. Wiley, New York, 2008. Reprinted by permission of John Wiley & Sons, Inc. New York.)

If the passage of plane-polarized light through a material reduces the velocity of one of the circularly polarized components more than the other by interaction with bonding and nonbonding electrons, the transmitted beam of radiation has its plane of polarization rotated from its *original* position (Figs. 5.48 and 5.49). A **polarimeter** is used to measure this angle of rotation.

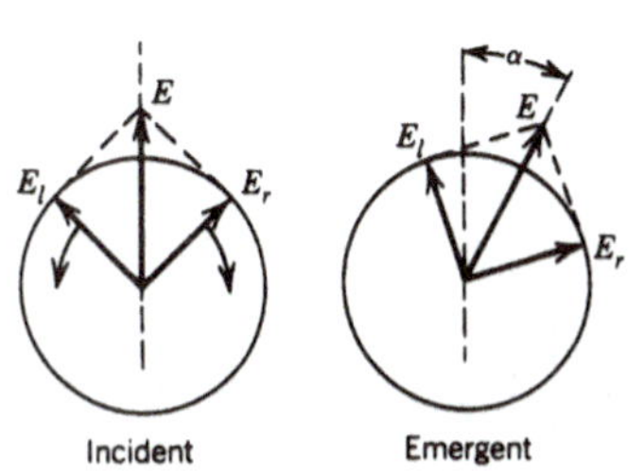


**Figure 5.49 Plane-polarized light before entering and after emerging from an optically active substance.** (From Douglas, B., McDaniel, D. H., and Alexander, J. J. *Concepts and Models of Inorganic Chemistry*, 3rd ed. Wiley, New York, 1994. (Reprinted by permission of John Wiley & Sons, Inc., New York.)

## The Polarimeter

The polarimeter measures the amount of rotation caused by an optically active compound (in solution) placed in the beam of the plane polarized light. The principal parts of the instrument are diagrammed in Figure 5.48. Two Nicol prisms are used in the instrument. The first prism, which polarizes the original light source, is called the polarizer. The second prism, called the analyzer, is used to examine the polarized light after it passes through the solution being analyzed.

When the axes of the analyzer and polarizer prisms are parallel (0°) and no optically active substance is present, the maximum amount of light is transmitted. If the axes of the analyzer and polarizer are at right angles to each other (90°), no transmission of light is observed. Placing an optically active solution into the path of the plane-polarized light causes one of the circularly polarized components to be slowed more than the other. The refractive indices are, therefore, different in the two circularly polarized beams. Figure 5.48 represents a case in which the left-hand component has been affected the most.

*NOTE. In the simplified drawing, Figure 5.48, the effect on only one of the circularly polarized waves is diagrammed. See Figure 5.49 for a more accurate description (view from behind the figure).*

This tilts the plane of polarization. The analyzer prism must be rotated to the left to maximize the transmission of light. If rotation is counterclockwise, the angle of rotation is defined as (−) and the enantiomer that caused the effect is called levorota-tory (*l*). Conversely, clockwise rotation is defined as (+), and the enantiomer is dextrorotatory (*d*). Tilting the plane of polarization is called *optical activity.* Note that if a solution of equal amounts of a *d* and an *l* enantiomeric pair is placed in the beam of the polarimeter, no rotation is observed. Such a solution is *racemic;* it is an equimolar mixture of enantiomers.

The magnitude of optical rotation depends on several factors: (1) the nature of the substance, (2) the path length through which the light passes, (3) the wavelength of light used as a source, (4) the temperature, (5) the concentration of the solution used to make the measurement of optical activity, and (6) the solvent used in making the measurement.

The results obtained from the measurement of the observed angle of rotation, $\alpha_{obs}$, are generally expressed in terms of *specific rotation* $[\alpha]$. The sign and magnitude of $[\alpha]$ are dependent on the specific molecule and are determined by complex features of molecular structure and conformation; they cannot be easily explained or predicted. The specific rotation is a physical constant characteristic of a substance. The relationship of $[\alpha]$ to $\alpha_{obs}$ is as follows:

$$[\alpha]_\lambda^T = \frac{\alpha_{obs}}{lc}$$

*where*

$T$ = temperature of the sample in degrees Celsius (°C),

$l$ = the length of the polarimeter cell in decimeters (1 dm = 0.1 m = 10 cm),

$c$ = concentration of the sample in grams per milliliter (g/mL),

$\lambda$ = the wavelength of light in nanometers (nm) used in the polarimeter.

These units are traditional, though most are esoteric by contemporary standards. The specific rotation for a given compound depends on both the concentration and the solvent, and thus both the solvent and concentration used must be specified. For example, $[\alpha]_D^{25}$ ($c$ = 0.4, $CHCl_3$) = 12.3° implies that the measurement was recorded in a $CHCl_3$ solution of 0.4 g/mL at 25 °C using the sodium D line (589 nm) as the light source.

For increased sensitivity, many simple polarimeters have an optical device that divides the viewed field into three adjacent parts (triple-shadow polarimeter; Fig. 5.50). A very slight rotation of the analyzer will cause one portion to become dimmer and the other lighter (Fig. 5.50*a* and 5.50*c*). The angle of rotation reading ($\alpha$) is recorded when the field sections all have the same intensity. An accuracy of ± 0.1° can be obtained.

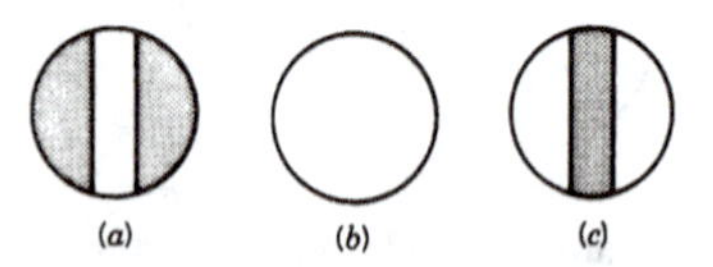


**Figure 5.50 View through the eyepiece of the polarimeter. The analyzer should be set so that the intensity in all parts of the field is the same (b). When the analyzer is displaced to one side or the other, the field will appear as in (a) or (c).**

***Inaccurate Measurements.*** Several conditions may lead to inaccurate measurements, including trapped air bubbles in the cell, and solid particles suspended in the solution. Filter the solution, if necessary.

***High-Performance Polarimeters and Optical Rotary Dispersion.*** For details of these two related topics refer to online discussion, Technique 8.

www

***Applications to Structure Determination in Natural Products.*** Natural products provide interesting opportunities for measuring optical activity. An excellent example is the lichen metabolite, usnic acid, which can be easily isolated from its native source,"Old Man's Beard" lichens, as golden crystals.

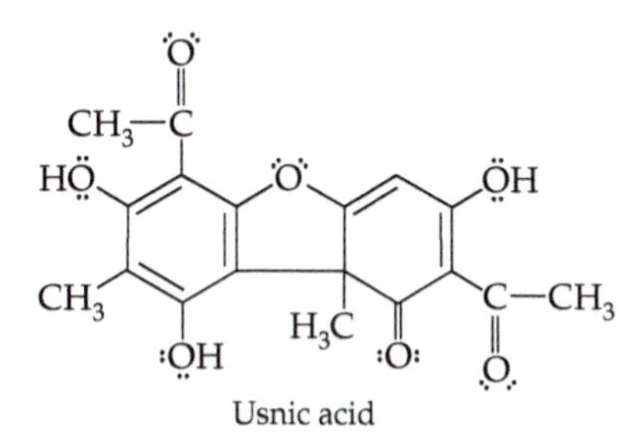


Usnic acid

Usnic acid contains a single stereocenter (stereogenic center, or chiral center) and, therefore, has the possibility of existing as an enantiomeric pair of stereoisomers. Generally, in a given lichen, only one of the stereoisomers (*R* or *S*) is present. Usnic acid has a very high specific rotation (~ ± 460°) which makes it an ideal candidate for optical rotation measurements at the microscale level.

## QUESTIONS

**5-42.** A solution of 300 mg of optically active 2-butanol in 10 mL of water shows an optical rotation of −0.54°. What is the specific rotation of this molecule?

**5-43.** Draw the structure of usnic acid and locate its stereocenter.

**5-44.** After drawing all stereoisomers of 3-amino-2-butanol, identify the enantiomeric pairs.

**5-45.** If a solution of an equimolar mixture of an enantiomeric pair is placed in the beam path of the polarimeter, what would you observe?

**5-46.** The specific rotation of (+)Q is + 12.80°. At identical concentration, solvent, pathlength, and light wavelength, the observed rotation of a solution containing both enantiomers of Q is 6.40°. What are the relative concentrations of each enantiomer in the solution?

*NOTE. The following experiment uses Technique 8: Experiment [11A]*

# Sublimation

Sublimation is especially suitable for purifying solids at the microscale level. It is useful when the impurities present in the sample are nonvolatile under the conditions used. Sublimation is a relatively straightforward method; the impure solid need only be heated.

Sublimation has additional advantages: (1) It can be the technique of choice for purifying heat-sensitive materials—under high vacuum it can be effective at low temperatures; (2) solvents are not involved and, indeed, final traces of solvents are effectively removed; (3) impurities most likely to be separated are those with lower vapor pressures than the desired substance and often, therefore, lower solubilities, exactly those materials very likely to be contaminants in a recrystallization; (4) solvated materials tend to desolvate during the process; and (5) in the specific case of water of solvation, it is very effective even with substances that are deliquescent. The main disadvantage of the technique is that it can be less selective than recrystallization when the vapor pressure of the desired material being sublimed is similar to that of an impurity.

Some materials sublime at atmospheric pressure ($CO_2$, or dry ice, is a well-known example), but most sublime when heated below their melting points under reduced pressure. The lower the pressure, the lower the sublimation temperature. Substances that do not have strong intermolecular attractive forces are excellent candidates for purification by sublimation. Napthalene, ferrocene, and *p*-dichlorobenzene are examples of compounds that are readily sublimed.

## Sublimation Theory

Sublimation and distillation are closely related. Crystals of solid substances that sublime, when placed in an evacuated container, will gradually generate molecules in the vapor phase by the process of evaporation (i.e., the solid has a vapor pressure). Occasionally, one of the vapor molecules will strike the crystal surface or the walls of the container and be held by attractive forces. This process, condensation, is the reverse of evaporation.

Sublimation is the complete process of evaporation from the solid phase to condensation from the gas phase to directly form crystals without passing through the liquid phase.

A typical single-component phase diagram is shown in Figure 5.51, which relates the solid, liquid, and vapor phases of a substance to temperature and pressure. Where two of the areas (solid, liquid, or vapor) touch, there is a line, and along each line the two phases exist in *equilibrium*. Line *BO* is the sublimation–vapor pressure curve of the substance in question; only along line *BO* can solid and vapor exist together in equilibrium. At temperatures and pressures along the *BO* curve, the liquid state is thermodynamically unstable. Where the three lines representing pairs of phases intersect, all three phases exist together in equilibrium. This point is called the *triple point*.

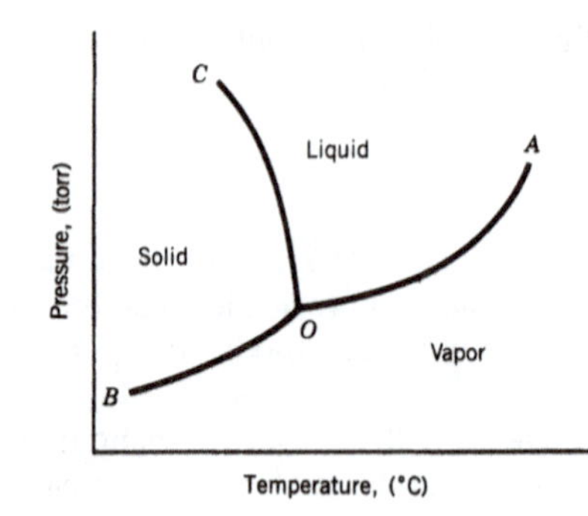


**Figure 5.51 Single-component phase diagram.**

Many solid substances have a sufficiently high vapor pressure near their melting point that allows them to be sublimed easily under reduced pressure in the laboratory. Sublimation occurs when the vapor pressure of the solid equals the pressure of the sample's environment.

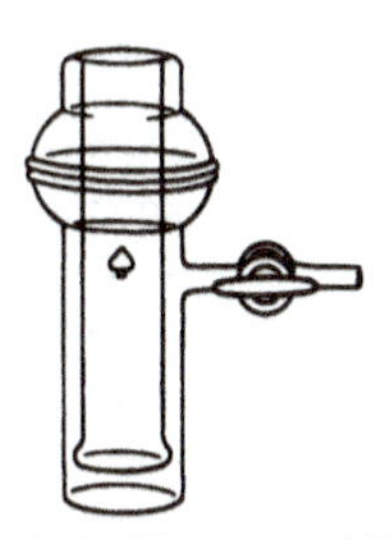

**Figure 5.52 Vacuum sublimator.** *(Courtesy of ACE Glass Inc., Vineland, NJ.)*

## Experimental Setup

Heating the sample with a microburner or a sand bath to just below the melting point of the solid causes sublimation to occur. Vapors condense on the cold-finger surface, whereas any less volatile residue will remain at the bottom of the flask. Apparatus for sublimation of small quantities are now commercially available (Fig. 5.52). Two examples of simple, inexpensive apparatus suitable for sublimation of small quantities of material in the microscale organic laboratory are shown in Figure 5.53.

An example of the purification of a natural product, where the sublimation technique at the microscale level is effective, is the case of the alkaloid caffeine. This substance can be isolated by extraction from tea.

## Precautions

Several precautions should be observed when performing a sublimation:

1. If you use the first setup in Figure 5.53, make sure you attach the hose connections to the cold finger in the proper manner. *The incoming cold water line is attached to the center tube.*
2. If you generate a vacuum using a water aspirator, make sure you place a water trap in the line. *Apply the vacuum to the system before you turn on the cooling water to the condenser.* This will keep moisture in the air in the flask from being condensed on the cold finger. Let the cold finger warm up before releasing the vacuum.
3. After the sublimation is complete, release the vacuum *slowly* so as not to disturb the sublimed material.
4. When using either of the arrangements in Figure 5.53, be careful to avoid loss of purified product as you remove the cold finger from the assembly.
5. The distance between the tip of the cold finger and the bottom of the sublimator should be less than 1 cm in most cases.

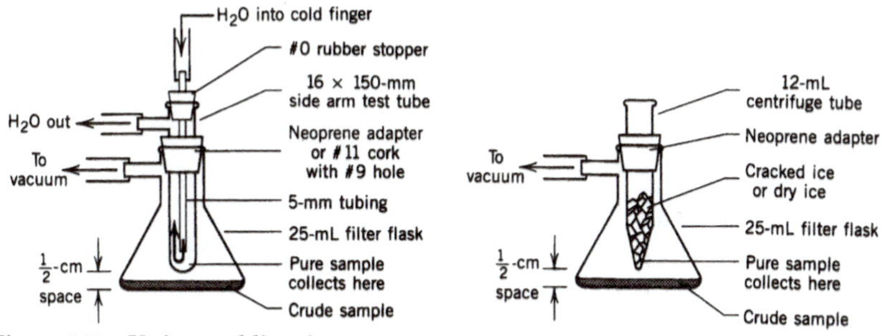

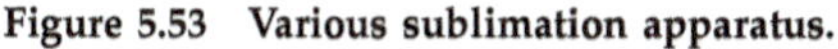

**Figure 5.53 Various sublimation apparatus.**

## QUESTIONS

**5-47.** List the advantages and disadvantages of sublimation as a purification technique.

**5-48.** For a solid compound to evaporate at atmospheric pressure it must have an unusually high vapor pressure. What molecular structural features contribute to this vapor pressure?

**5-49.** Why apply the vacuum to the sublimation system *before* you turn on the cooling water to the water condenser?

**5-50.** Why place a water trap in the vacuum line when using an aspirator to obtain the vacuum?

**5-51.** Why is sublimation particularly useful for purifying deliquescent compounds?

**5-52.** The 72% recovery after performing a sublimation translates to 32 mg of material. Calculate the amount of crude sample prior to performing the sublimation.

*NOTE. The following experiments use Technique 9: Experiments [11B], [25A], and [25B].*

EXPERIMENT 3105-1

# Distillation

In the following experiment, we will examine the applications of a variety of distillation techniques to the purification of liquid mixtures. In Experiment 3105-1 you will conduct a simple distillation and a fractional distillation. In particular, you will use the simple distillation apparatus pictured on page 62. Be sure to know what is meant by the "boiling point" of a pure compound, and what determines the boiling point of a mixture.

You must be aware of any safety considerations with regards to the chemicals and the techniques you are using. These must be recorded in your prelab write-up.

Methanol is flammable, with a low boiling point. The liquid and vapor cause blindness in small doses, and can be fatal at higher doses.

Ensure that you are using an alcohol thermometer (red or blue liquid) and not a mercury thermometer (silver liquid). Notify your TA immediately if you have a mercury thermometer.

## Simple and Fractional Distillation: Separation of Methanol and Water

***Purpose.*** To separate a mixture of methanol and water, two miscible liquids with somewhat different boiling points. To compare the difference between simple and fractional distillation methods.

*Prior Reading*

*Technique 2:* Distillation (pp. 61–64)

Distillation Theory (p. 61 and online)

Simple Distillation at the Semimicroscale Level (pp. 61–64)

*Technique 3:* Fractional Semimicroscale Distillation (pp. 64–67)

*Chapter 4:* Determination of Physical Properties

Ultramicro-Boiling Point (pp. 46–48)

Refractive Index (online)

## DISCUSSION

Methanol and water have boiling points approximately 35 °C apart. The liquid–vapor composition curve in Figure 5.6 on page 63 represents a similar system; it is apparent that a two-plate distillation should yield nearly pure components. The procedure to be outlined consists of two techniques. The first deals with a simple distillation. The second deals with a fractional distillation. Exercising careful technique during these distillations should provide a good comparison of the different distillation techniques.

## COMPONENTS

| $CH_3OH$ | $H_2O$ |
|---|---|
| Methanol | water |

## EXPERIMENTAL PROCEDURE

Estimated time for the experiment: 2.0 h.

| Physical Properties of Components | | | |
|---|---|---|---|
| Compound | MW | Amount | bp (°C) |
| Methanol ($CH_3OH$) | 32.05 | 20.0 mL | 65 |
| Water ($H_2O$) | 18.02 | 20.0 mL | 100 |

***Procedure.*** Work in pairs for this experiment (as is the norm for the rest of the semester) so you can make simultaneous measurements of volume and temperature. Your TA will assign your group the simple distillation procedure or the fractional distillation procedure. Each group will perform their respective distillation using their own methanol:water solution aliquot. You will share your groups' data with one other group performing the other distillation procedure. Remember to record your partner's name and the other group members' names.

***Simple Distillation Apparatus Set-Up.*** Clean your glassware if needed. Begin by setting up a simple distillation apparatus, as illustrated in Figure 5.4 on p. 62. Use a 100 mL round bottomed flask for your distilling flask, but use a 50 or 100 mL graduated cylinder for the receiving flask instead of the round-bottomed flask.

***Fractional Distillation Apparatus Set-Up.*** Groups assigned the fractional distillation will begin by setting up an apparatus similar to the simple distillation apparatus, as illustrated in Figure 5.4 on p 62, but also install a second condenser vertically between the distilling flask and three-way adapter as shown in Figure 5.9 on page 65. This condenser should have steel turnings in it (obtained from your TA). Do not connect this condenser to the water lines. The vapor will condense and re-boil on the turnings, producing the fractional distillation.

***Simple and Fractional Distillations.*** Clamp your apparatus securely (ensuring that the clamp grips the neck of the 100 mL round bottom flask), but be careful not to strain the glass, as it may break. Pay attention to details in the figures such as the position of the thermometer (the bulb should be BELOW the opening to the condenser) and where cooling water enters and leaves the condenser. You will heat your distillation flask directly with a heating mantle or a hot plate, not with a sand bath.

Add 40 mL of the distillation sample to the round bottom-distilling flask. This is a 50:50 mixture of methanol and water. Add a boiling chip and heat it so that it boils at a rate of two to three drops per second. Do not turn the hot plate on high! One person will watch the graduated cylinder and tell the other when 2 mL of distillate has been collected. The second person will record the temperature at that volume. Record the temperature at 2 mL intervals until the temperature evens off close to 100°C, then stop the distillation. Never boil a flask to dryness.

Graph the volume of distillate (mL) on the X-axis versus the temperature (°C) on the Y-axis. On the same graph, show the curves for both the simple and the fractional distillation. Was there any difference between the two curves? If so, what is the significance of the result you obtained?

## QUESTIONS

**6-1.** The boiling point of a liquid is affected by several factors. What effect does each of the following conditions have on the boiling point of a given liquid?
  **(a)** The pressure of the atmosphere
  **(b)** Use of an uncalibrated thermometer
  **(c)** Rate of heating of the liquid in a distillation flask

**6-2.** Explain why packed and spinning-band fractional distillation columns are more efficient at separating two liquids with close boiling points than are unpacked columns.

**6-3.** Explain what effect each of the following mistakes would have had on the simple distillation carried out in this experiment.
  **(a)** You did not add a boiling stone.
  **(b)** You heated the distillation flask at too rapid a rate.

## BIBLIOGRAPHY

**General references on distillation:**

Lodwig, S. N. *J. Chem. Educ.* **1989,** *66,* 77.

Stichlmair, J. G.; Fair, J. R. *Distillation: Principles and Practice;* Wiley: New York, 1998.

*Technique of Organic Chemistry,* Vol. IV, *Distillation,* 2nd ed., Perry, E. S.; Weissberger, A., Eds.; Interscience-Wiley: New York, 1967.

Vogel, A. I. *Vogel's Textbook of Practical Organic Chemistry,* 5th ed.; Furnis, B. S., et al. Eds.; Wiley: New York, 1989.

Zubrick, J. W. *The Organic Chem Lab Survival Manual,* 7th ed.; Wiley: New York, 2008.

# Solvent Extraction

EXPERIMENT
3105-2

## Separation of an Acid and a Hydrocarbon

***Purpose.*** To separate a mixture of a hydrocarbon and an acid, one from each pair shown in the "Components" section. This experiment utilizes the difference in water solubility between a neutral carboxylic acid and its anionic conjugate base, to separate it from a neutral hydrocarbon. Extraction is a useful technique in the organic laboratory. In the next experiment you will use melting points to identify the acid and hydrocarbon in your unknown mixture.

*Prior Reading*

*Chapter 3:* Experimental Apparatus
- Pasteur Filter Pipet (pp. 36–37)
- Automatic Pipet (pp. 37–38)
- Weighing of Solids in Milligram Quantities (p. 39)

*Technique 4:* Solvent Extraction
- Liquid–Liquid Extraction (p. 72)
- Separation of Acids and Bases (pp. 77–79)
- Salting Out (p. 79)
- Drying Agents (pp. 80–83)

## DISCUSSION

The solubility characteristics of organic acids in water can be shown to be highly dependent on the pH of the solution.

*NOTE. Refer to Technique 4, p. 78–79 for a flowchart outlining the procedure for an extraction similar to the one described in this experiment. Draw your own flowchart for the present experiment.*

The components of the mixture to be separated in this experiment are naphthalene, biphenyl, benzoic acid, and o-Chlorobenzoic Acid.

The reaction of interest is a classic Brønsted-Lowry acid-base reaction of the type you have studied previously, as shown in equation 1 for benzoic acid. Benzoic acid has many non-polar covalent bonds and just a few polar covalent ones, with no ionic bonds, so it is poorly soluble in water, but very soluble in non-polar solvents like diethyl ether (or ether). Non-polar hydrocarbons are also insoluble in water and soluble in ether. When benzoic acid is treated with a strong base like NaOH, the ionic salt of the conjugate base of the acid is formed. This ionic salt is much more soluble in water than in ether, so it will dissolve into the aqueous phase while the hydrocarbon stays in the ether phase, allowing for separation of the two components in the mixture.

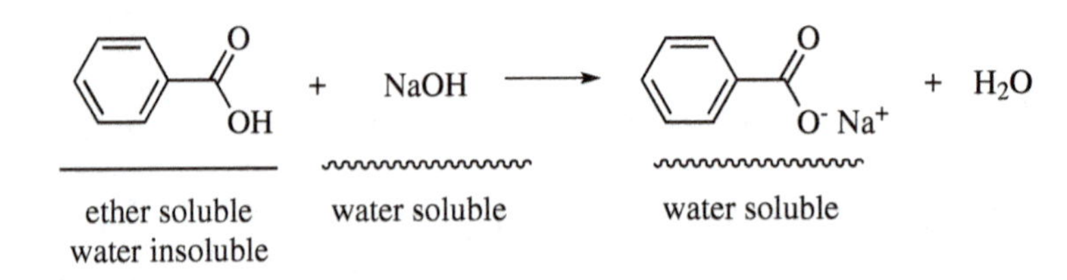


Equation 1

Treatment of the conjugate base with a strong acid, such as HCl, will regenerate the carboxylic acid as in equation 2. This will again be ether-soluble, but not water-soluble. Thus by adjusting the pH of the water, it is possible to shuttle the benzoic acid back and forth between the ether and water layers, while any non-polar hydrocarbons will stay put in the ether layer.

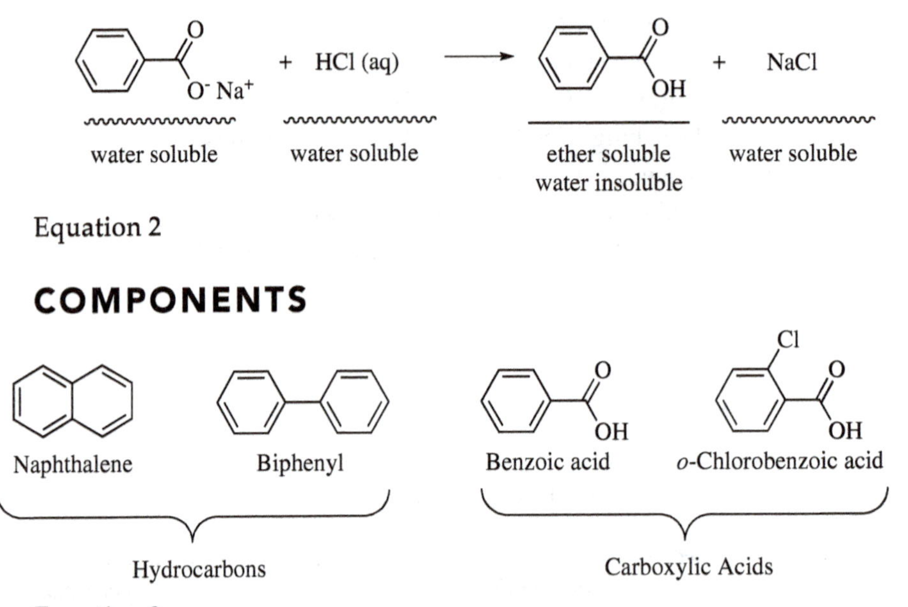


Equation 3

## EXPERIMENTAL PROCEDURE

Estimated time of experiment: 2.5 h.

**Physical Properties of Reactants**

| Compound | MW | Amount | mp (°C) | bp (°C) | *d* |
|---|---|---|---|---|---|
| Naphthalene | 128.17 | | 78-80 | | |
| Biphenyl | 154.21 | | 68-70 | | |
| Benzoic acid | 122.12 | | 120-122 | | |
| *o*-Chlorobenzoic Acid | 156.57 | | 140-142 | | |
| Diethyl ether | 74.12 | 40 mL | | 35 | 0.7184 |
| 2 M NaOH | | 40 mL | | | |
| 6 N HCl | | 8 mL | | | |
| sat'd NaCl | | 10 mL | | | |

*NOTE. In carrying out the separation, you should keep a record or flowchart of your procedure (as suggested in the prior reading assignment) in your laboratory notebook. You should also be particularly careful to label all flasks. You are heartily encouraged to ask questions and discuss the experiments with your colleagues in lab and your TA.*

***Reagents and Equipment.*** Clean your glassware if necessary. Put about 40 mL of a 2 M NaOH solution in an Erlenmeyer flask and cool it in an ice bath. Obtain a 0.50 g sample of unknown composition and record the sample number. Weigh your sample using the analytical balance. Save a few milligrams (just enough to see) for a melting point in the next experiment. Transfer the remainder to the separatory funnel (with the stopcock closed). Add about 30 mL of ether and swirl to dissolve the solid.

***Separation of the Acidic Component.*** Cautiously add the NaOH solution to the ether in 5-10 mL portions. (Acid-base reactions are exothermic, which means the reaction gets hot. Ether is a very low boiling solvent, so it may begin to boil.) Your TA will demonstrate the proper technique for shaking and venting a separatory funnel and separating the ether and water layers. Use pH paper or red litmus paper (which should turn blue) to check that the aqueous layer is basic. If not, add more NaOH, until the aqueous layer remains basic.

---

**WARNING:** Ether is volatile and flammable. Sodium hydroxide is a strong base and can quickly burn your skin and eyes. Wear your safety goggles. Be particularly careful not to spill any. If you do get sodium hydroxide on your skin, wash it immediately with cold running water and notify your TA.

---

**Hint:** It is easy to get confused in an extraction is by forgetting which layer is the organic one. Two immiscible liquids will separate by density, with the liquid of lowest density on top. Most organic solvents are less dense than water, and therefore will form the upper layer. A common exception is halogenated solvents, which are denser than water, and form the bottom layer (unless the aqueous layer has a sufficiently high salt concentration). If you're not sure, add a drop of one layer to a few drops of water in a test tube, and swirl. If it dissolves in the water, it's aqueous. If not, it's organic.

---

Remove the stopper and drain the aqueous layer into an Erlenmeyer flask labeled "basic extract". Wash the organic layer by adding about 10 mL of distilled water to the separatory funnel, shaking, venting, and draining the water into the basic extract.

**Hint:** Never discard any of your solutions until you have isolated the products, to make sure that you do not accidentally throw them away.

***Separation of the Hydrocarbon Component.*** Next, transfer the ether layer into another Erlenmeyer flask and label it as the ether wash. Some hydrocarbon may still be in the funnel or aqueous layer, so return the basic extract to the funnel and wash it by adding about 10 mL of fresh ether. Again, shake and vent the separating funnel, and drain the aqueous layer into the basic extract flask. You should still have ether in the separatory funnel. Add the contents of the ether wash flask to the ether in the separatory funnel. Finally, add about 10 mL of saturated aqueous NaCl to the ether wash. Shake and vent the combined ether layers in the separatory funnel. (The saturated aqueous NaCl removes most of the water that's still dissolved in the ether.) Drain the water layer into the basic extract flask, and drain the ether layer into the ether wash flask.

There will still be traces of water dissolved in the ether. To remove them, add a small amount of anhydrous magnesium sulfate (a drying agent) to the ether. If it all clumps up, it's forming a hydrate with the water, and you can add some more. If some crystals remain loose, then the ether is dry.

***Isolation of the Acidic Component.*** To recover the acid from the basic extract, you will use 6 N HCl. Add about 8 mL and check that the final pH is 1-2 (blue litmus paper turns pink). If not, add more HCl until the aqueous layer remains acidic. A voluminous precipitate should form. Collect this solid using vacuum filtration, using a Hirsch funnel. Wash the flask with 15 mL or less of cold distilled water to ensure that all the crystals get onto the funnel and that any water soluble material is washed off the filtrate.

Save this sample for next lab period. Allow it to dry on a piece of filter paper in your glassware drawer. The leftover aqueous phase can be washed down the sink. It only contains water, NaCl, and dilute HCl.

***Isolation of the Hydrocarbon Component.*** Pour the ether through a conical filter to separate it from the magnesium sulfate. Evaporate the ether by blowing a gentle stream of air over it, in the hood.

Store the remaining solid hydrocarbon in a covered shell vial until next lab period. The hydrocarbon will sublime away unless it is covered.

## QUESTIONS

**6-8.** Ether (MW = 46) evaporates much faster than water (MW = 18), yet ether's molecular weight is about 2.5 times than that of water. Explain.

**6-9.** Suggest a way to separate a mixture of these two compounds. **Hint:** Amines ($R-NH_2$) can act as bases. Answer can be written out or depicted in a flowchart.

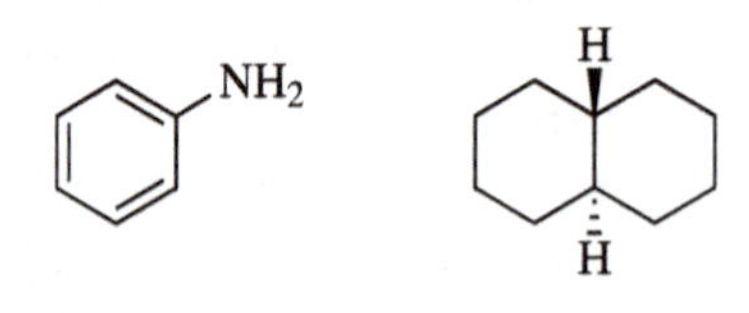


## BIBLIOGRAPHY


Lo, T. C.; Baird, M. H. I.; Hanson, C., Eds. *Handbook of Solvent Extraction;* Krieger: Melbourne, FL, 1991.

Rydberg, J.; Cox, M.; Musikas, C.; Choppin, G. R., Eds., *Solvent Extraction Principles and Practices,* 2nd ed.; C.H.I.P.S.: Weimar, Texas, 2004.

Shriner, R. L.; Hermann, C. K. F.; Morrill, T. C.; Curtin, D. Y.; Fuson, R. C. *The Systematic Identification of Organic Compounds,* 8th ed.; Wiley: New York, 2003.

Vogel, A. I. *Vogel's Textbook of Practical Organic Chemistry,* 5th ed.; Furnis, B. S., et al. Eds.; Wiley: New York, 1989.

Zubrick, J. W. *The Organic Chem Lab Survival Manual,* 7th ed.; Wiley: New York, 2008.

EXPERIMENT 3105-3

# Recrystallization

## Separation of an Acid and a Hydrocarbon

***Purpose.*** To identify the unknown compounds you isolated last lab period, using melting points of recrystallized materials.

*Prior Reading*

*Technique 5:* Crystallization
Introduction (pp. 85-87)
Use of the Hirsch Funnel (pp. 88-89)

*Chapter 4:* Solids (pp. 50-54)

## COMPONENTS

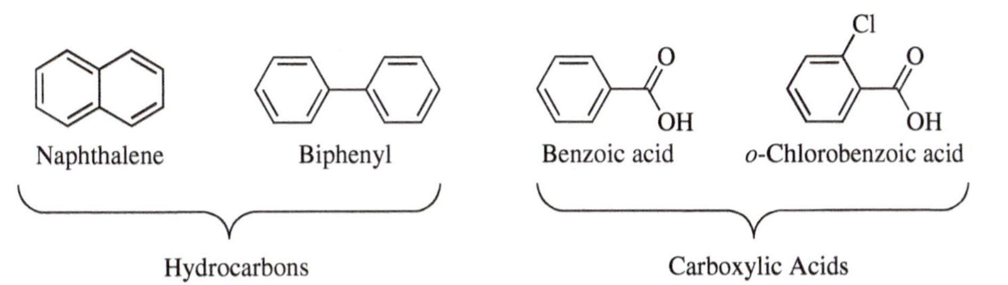


## EXPERIMENTAL PROCEDURE

**Physical Properties of Reactants**

| Compound | MW | Amount | mp (°C) | bp (°C) | *d* |
|---|---|---|---|---|---|
| Naphthalene | 128.17 | | 78-80 | | |
| Biphenyl | 154.21 | | 68-70 | | |
| Benzoic acid | 122.12 | | 120-122 | | |
| *o*-Chlorobenzoic Acid | 156.57 | | 140-142 | | |
| Ethanol | 46.07 | | | 78 | 0.789 |

Weigh your crude, dry product from last lab period. Then purify it by recrystallization. To recrystallize the hydrocarbon, weigh 50 mg of the solid into a shell vial. Add 1 mL of ethanol and gently heat to boiling. All the material should dissolve. To adjust the polarity of the solvent and the solubility of the hydrocarbon, add a few drops of water. With each drop, there will be a temporary cloudiness; stop adding water before you reach persistent cloudiness. Cool to room temperature or below using an ice bath. Collect the crystals using either vacuum or gravity filtration. Dry them, and take melting points.

To recrystallize the acid, heat 50 mg of the solid in 1 mL of water. Add just enough ethanol to dissolve it. Then cool and collect crystals again. Blot the crystals on filter paper to dry off most of the water.

Take the following melting point ranges:
1) Your unknown mixture.
2) Your crude hydrocarbon.
3) Your crude acid.
4) Your recrystallized hydrocarbon.
5) Your recrystallized acid.
6) A mixed melting point with an authentic sample of the hydrocarbon identified in 4.
7) A mixed melting point with an authentic sample of the acid identified in 5.

Note that steps 1, 2, and 3 can be done before you have done your recrystallization.

## QUESTIONS

**6-10.** What were the hydrocarbon and acid components (and sample number) of your unknown?

**6-11.** What two effects can impurities have on a melting point range? Did you see these consequences in your results?

# The Isolation of Natural Products

EXPERIMENT 3105-4

This experiment is designed to acquaint you with the procedures used to isolate naturally occurring and often biologically active organic compounds. These substances are known as *natural products* because they are produced by living systems. The particular natural product you are going to study comes from the plant kingdom. At the end of the nineteenth century more than 80% of all medicines in the Western world were natural substances found in roots, barks, and leaves. There was a widespread belief at that time that in plants there existed cures for all diseases. As Kipling wrote, "Anything green that grew out of the mold/ Was an excellent herb to our fathers of old." Even as the power of synthetic organic chemistry has grown during this century, natural materials still constitute a significant fraction of the drugs used in modern medicine. For example, in the mid-1960s when approximately 300 million new prescriptions were written each year, nearly half were for substances of natural origin. These materials have played a major role in successfully combating the worst of human illnesses, from malaria to high blood pressure; diseases that affect hundreds of millions of people.

Unfortunately, during the latter half of this century a number of very powerful natural products that subtly alter the chemistry of the brain have been used in vast quantities by our society. The ultimate impact on civilization is of grave concern. Evidence clearly demonstrates that these natural substances disrupt the exceedingly complex and delicate balance of biochemical reactions that lead to normal human consciousness. How well the brain is able to repair the damage from repetitive exposure is unknown. We are currently conducting experiments to answer that question.

The natural product that you will isolate in the following experiment is a white crystalline alkaloid that acts as a stimulant in humans.

## Isolation and Characterization of a Natural Product: Extraction of Caffeine from Tea

### *Product*

Common names: caffeine, 1,3,7-trimethyl-2,6-dioxopurine

CA number: [58-08-2]

CA name as indexed: 1*H*-purine-2,6-dione, 3,7-dihydro-1,3,7-trimethyl-

***Purpose.*** To extract the active principle, an alkaloid, caffeine, from a native source, tea leaves. Caffeine is a metabolite (a product of the living system's biochemistry) found in a variety of plants. We will use ordinary tea bags as our source of raw material. This experiment illustrates an extraction technique often used to isolate water-soluble, weakly basic natural products from their biological source. The isolation of caffeine will also give you the opportunity to use sublimation as a purification technique, since caffeine is a crystalline alkaloid that possesses sufficient vapor pressure to make it a good candidate for this procedure.

## ALKALOIDS

Caffeine belongs to a rather amorphous class of natural products called alkaloids. This collection of substances is unmatched in its variety of structures, biological response on nonhost organisms, and the biogenetic pathways to their formation.

The history of these fascinating organic substances begins at least 4000 years ago. They were incorporated into poultices, potions, poisons, and medicines, but no attempt was made to isolate and identify the substances responsible for the physiological response until the very early 1800s.

The first alkaloid to be obtained in the pure crystalline state was morphine. Friedrich Wilhelm Sertürner (1783–1841) isolated morphine in 1805. He recognized that the material possessed basic character and he, therefore, classified it as a vegetable alkali (that is, a base with its origin in the plant kingdom). Thus, compounds with similar properties ultimately became known as alkaloids. The term "alkaloid" was introduced for the first time by an apothecary, Meissner, in Halle in 1819.

Sertürner, also a pharmacist, lived in Hamelin, another city in Prussia. He isolated morphine from opium, the dried sap of the poppy. Since the analgesic and narcotic effects of the crude resin had been known for centuries, it is not surprising that, with the emerging understanding of chemistry, the interest of Sertürner became focused on this drug, which is still medicine's major therapy for intolerable pain. He published his studies in detail in 1816 and very quickly two French professors, Pierre Joseph Pelletier (1788–1842) and Joseph Caventou (1795–1877) at the Ecole de Pharmacie in Paris, recognized the enormous importance of Sertürner's work.

In the period from 1817 to 1820, these two men and their students isolated many of the alkaloids, which continue to be of major importance. Included in that avalanche of purified natural products was caffeine, which they obtained from the coffee bean. This substance is the target compound that you will be isolating directly from the raw plant in this experiment. A little more than 75 years later, caffeine was first synthesized by Fischer in 1895 from dimethylurea and malonic acid.

## THE CLASSIFICATION OF ALKALOIDS

These compounds are separated into three general classes of materials.

1. **True alkaloids:** These compounds contain nitrogen in a heterocyclic ring; are almost always basic (the lone pair of the nitrogen is responsible for this basic character); are derived from amino acids in the biogenesis of the alkaloid; invariably are toxic and possess a broad spectrum of pharmacological activity; are found in a rather limited number of plants (of the 10,000 known genera only 8.7% possess at least one alkaloid); and normally occur in a complex with an organic acid (this helps to make them rather soluble in aqueous media). As we will see, there are numerous exceptions to these rules. For example, there are several very well-known quaternary alkaloids. These are compounds in which the nitrogen has become tetravalent and positively charged (as in the ammonium ion). Thus, they are not actually basic.

2. **Protoalkaloids:** These compounds are simple amines, derived from amino acids, in which the basic nitrogen atom is not incorporated into a ring system; they are often referred to as *biological amines*. An example of a protoalkaloid is mescaline.
3. **Pseudoalkaloids:** These compounds contain nitrogen atoms usually *not* derived from amino acids. There are two main classes into which pseudoalkaloids are divided, the steroidal alkaloids and the *purines*. Caffeine has been assigned to this latter class of alkaloids.

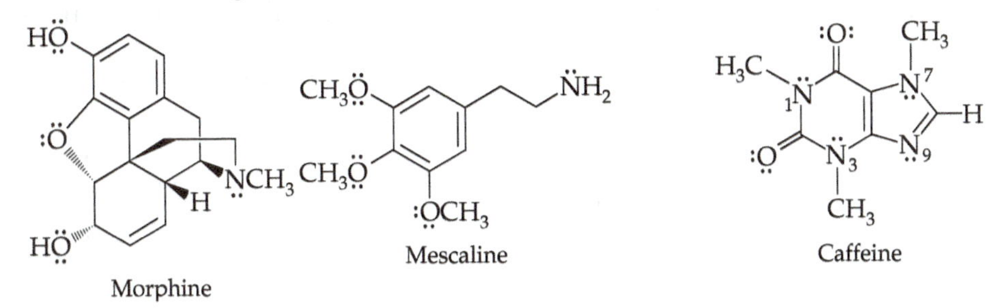

Morphine
Mescaline
Caffeine

*Prior Reading*

*Technique 4:* Solvent Extraction
Solid–Liquid Extraction (p. 79)
Liquid–Liquid Extraction (p. 72)
Microscale Extraction (pp. 73-75)

*Technique 9:* Sublimation
Sublimation Theory (pp. 112–113)

## DISCUSSION

Caffeine (1,3,7-trimethylxanthine) and its close relative theobromine (3,7-dimethyl-xanthine) both possess the oxidized purine skeleton (xanthine). These compounds are often classified as pseudoalkaloids, since only the nitrogen atom at the 7 position can be traced to an amino group originally derived from an amino acid (in this case glycine). This classification emphasizes the rather murky problem of deciding just what naturally occurring nitrogen bases are *true* alkaloids. We will simply treat caffeine as an alkaloid.

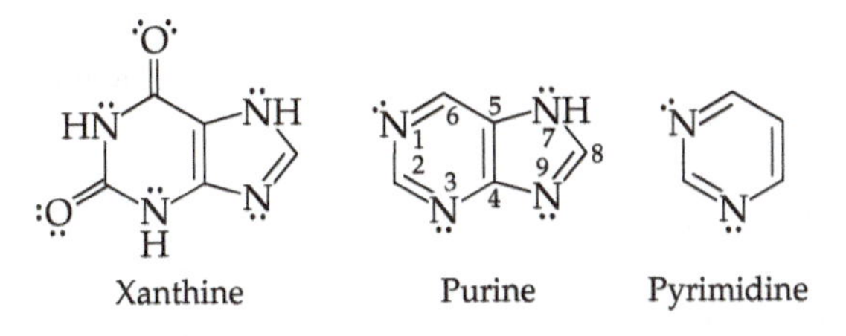

Xanthine
Purine
Pyrimidine

Although the pyrimidine ring (present in caffeine's purine system) is a significant building block of nucleic acids, it is rare elsewhere in nature.

These two methylated xanthines are found in quite a number of plants and have been extracted and widely used for centuries. Indeed, they very likely have been, and remain today, the predominant stimulant consumed by humans. Every time you make a cup of tea or coffee, you perform an aqueous extraction of plant material (tea leaves, *Camellia sinenis*, 1–4%, or coffee beans, *Coffea* spp., 1–2%) to obtain a dose of 25–100 mg of caffeine. Caffeine is also the active substance (~2%) in maté (used in Paraguay as a tea) made from the leaves of *Ilex paraguensis*. In coffee and tea, caffeine is the dominant member of the pair, whereas in *Theobroma cacao*, from which we obtain cocoa, theobromine (1–3%) is the primary source of the biological response. Caffeine acts to stimulate the central nervous system with its main impact on the cerebral cortex, and as it makes one more alert, it is no surprise that it is the chief constituent in No-Doz® pills.

Caffeine is readily soluble in hot water, because the alkaloid is often bound in thermally labile, partially ionic complexes with naturally occurring organic acids, such as with 3-caffeoylquinic acid in the coffee bean. For this reason it is relatively easy to separate caffeine from black tea leaves by aqueous extraction.

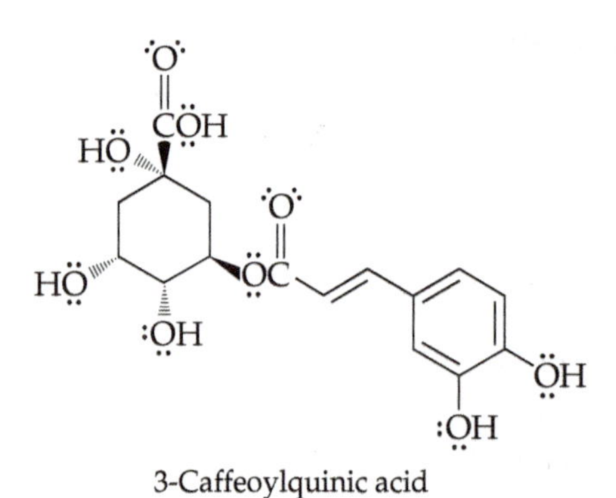

3-Caffeoylquinic acid

Other substances, mainly tannic acids, are also present in the tea leaves and are also water soluble. The addition of sodium carbonate, a base, during the aqueous extraction helps to increase the water solubility of these acidic substances by forming ionic sodium salts and liberating the free base.

Subsequent extraction of the aqueous phase with methylene chloride, in which free caffeine has a moderate solubility, allows the transfer of the caffeine from the aqueous extract to the organic phase. At the same time, methylene chloride extraction leaves the water-soluble sodium salts of the organic acids behind in the aqueous phase.

Extraction of the tea leaves directly with nonpolar solvents (methylene chloride) to remove the caffeine gives very poor results—since, as we have seen, the caffeine is bound in the plant in a partially ionic complex that will not be very soluble in nonpolar solvents. Thus, water is the superior extraction solvent for this alkaloid. The water also swells the tea leaves and allows for easier transport across the solid–liquid interface.

Following extraction and removal of the solvent, sublimation techniques are applied to the crude solid residues to purify the caffeine. This technique is especially suitable for the purification of solid substances at the microscale level, if they possess sufficient vapor pressure. Sublimation techniques are particularly advantageous when the impurities present in the sample are nonvolatile under the conditions used.

Sublimation occurs when a substance goes directly from the solid phase to the gas phase upon heating, bypassing the liquid phase. Sublimation is technically a straightforward method for purification in that the materials need only be heated and therefore, mechanical losses can be kept to a minimum (the target substance must, of course, be thermally stable at the required temperatures). Materials sublime only when heated *below* their melting points, and reduced pressure is usually required to achieve acceptable sublimation rates. Obviously, substances that lend themselves best to purification by sublimation are those that do not possess strong intermolecular attractive forces. Caffeine and ferrocene (used as a reactant in 3106-5) meet these criteria because they present large flat surfaces occupied predominantly with repulsive $\pi$ electrons. For other isolation methods, see the discussion of *solid-phase* extraction in Technique 4, where caffeine is extracted from coffee beans (pp. 83–84).

## EXPERIMENTAL PROCEDURE

Estimated time to complete the experiment: 2.5 h.

| Physical Properties of Constituents | | | | |
|---|---|---|---|---|
| Compound | MW | Amount | mmol | mp (°C) |
| Tea | | 1.0 g | | |
| Water | | 10 mL | | |
| Sodium carbonate | 105.99 | 1.1 g | 10 | 851 |

***Reagents and Equipment.*** Carefully open a commercial tea bag (2.0–2.5 g of tea leaves) and empty the contents. Weigh out 1.0 g of tea leaves and place them back in the empty tea bag. Close and secure the bag with staples.

Weigh, and add to a 50-mL Erlenmeyer flask, 1.1 g (0.01 mol) of anhydrous sodium carbonate followed by 10 mL of water. Heat the mixture with occasional swirling on a hot plate to dissolve the solid. Now add the 1.0 g of tea leaves (in the tea bag) to the solution. Place the bag in the flask so that it lies flat across the bottom.

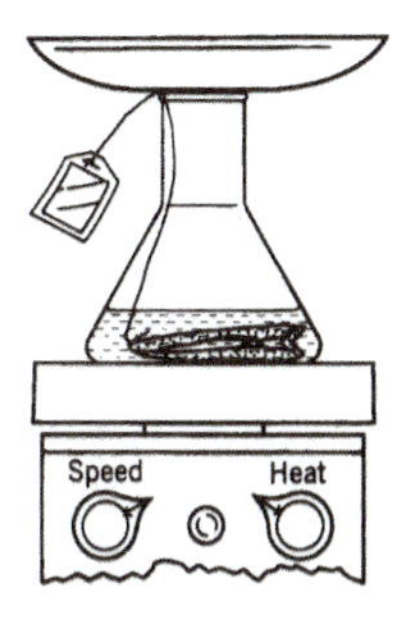

Anhydrous $Na_2CO_3$, 1.1 g + $H_2O$, 10 mL + tea bag with tea leaves, 1.0 g

***Reaction Conditions.*** Place a small watch glass over the mouth of the Erlenmeyer flask and then heat the aqueous suspension to **gentle** boiling for 30 min on the hot plate.

GENTLY

***Isolation of Product.*** Cool the flask and contents to room temperature. Transfer the aqueous extract from the Erlenmeyer flask to a 12- or 15-mL centrifuge tube using a Pasteur filter pipet. In addition, gently squeeze the tea bag by pressing it against the side of the Erlenmeyer flask to recover as much of the basic extract as possible. Set aside the tea bag and its contents.

Extract the aqueous solution with 2.0 mL of methylene chloride.

*NOTE. The tea solution contains some constituents that may cause an emulsion. If you obtain an emulsion during the mixing of the aqueous and organic solvent layers (by shaking or using a Vortex mixer), it can be broken readily by centrifugation. If you do this, after centrifugation, transfer the methylene chloride to a clean, dry test tube.*

---

CAUTION: Methylene chloride (dichloromethane) is a weak carcinogen. Avoid breathing its vapors; evaporate it from your sample in the hood.

---

Separate the lower (methylene chloride) layer (check to make sure that the lower layer is, indeed, the organic layer by testing the solubility of a few drops of it in a test tube with distilled water) using a Pasteur pipet. Add a small amount (about 0.1 to 0.2 g, or the end of a spatula) of magnesium sulfate to the test tube, swirl, and after a minute of drying, pipette the methylene chloride away from the drying agent and into a clean, dry shell vial or flask. (The organic phase will be saturated with water following the extraction, therefore it is referred to as "wet." It also may contain a few droplets of the aqueous phase, which become entrained during the phase separation; this can be particularly troublesome if an emulsion forms during the mixing.)

Extract the remaining aqueous phase with four additional 2.0-mL portions of methylene chloride (4 × 2 mL). Each extraction (an extraction is often referred to as a washing) is separated, dried as above, and transferred to the same shell vial or flask. Finally, rinse the magnesium sulfate with an additional 2.0 mL of methylene chloride and combine this wash with the earlier organic extracts.

Carefully evaporate the solvent in the hood. Do not walk away and leave your sample; evaporation should not take very long. If you heat it too long, or too hot, the caffeine will sublime away or turn to black gunk. The crude caffeine should be obtained as an off-white crystalline solid.

***Purification and Characterization.*** Purify the crude solid caffeine by sublimation.

Assemble a sublimation apparatus as shown in Figure 5.53, on page 113; either arrangement is satisfactory. Using an aspirator, apply a vacuum to the system through the filter flask (remember to install a water-trap bottle between the sidearm flask and the aspirator). After the system is evacuated, run cold water gently through the cold finger or add ice to the centrifuge tube. By cooling the surface of the cold finger *after* the system has been evacuated, you will minimize the condensation of moisture on the area where the sublimed sample will collect.

NOTE. *Less caffeine will be lost if the bottom of the cold finger is positioned less than 5 mm from the bottom of the filter flask.*

Once the apparatus is evacuated and cooled, begin the sublimation by **gently** heating the flask on a hot plate.

NOTE. *Be careful. Do not melt the caffeine. If the sample does begin to melt, remove from the heat source for a few seconds before heating is resumed. Overheating the crude sample will lead to decomposition and the deposition of impurities on the cold finger. High temperatures are not necessary since the sublimation temperature of caffeine (and of all solids that sublime) is below the melting point. It is generally worthwhile to carry out sublimations as slowly as possible, as the purity of the material collected will be enhanced.*

When no more caffeine will sublime onto the cold finger, remove the heat, shut off the aspirator and the cooling water to the cold finger, and allow the apparatus to cool to room temperature under reduced pressure. Once cooled, carefully vent the vacuum and return the system to atmospheric pressure. *Carefully* remove the cold finger from the apparatus.

*NOTE. If the removal of the cold finger is done carelessly, the sublimed crystals may be dislodged from the sides and bottom of the tube and drop back onto the residue left in the filter flask.*

Scrape the caffeine from the cold finger onto weighing paper using a microspatula and a sample brush. Weigh the purified caffeine and calculate its percent by weight in the original tea leaves.

Obtain an IR spectrum and compare it with that of an authentic sample.

## QUESTIONS

**6-4.** What was the % caffeine you isolated from the tea leaves? (NOTE: This is not a percent yield, just percent isolated. You do not have to convert to moles.)

**6-5.** Why was it helpful to use $Na_2CO_3$ in the extraction?

**6-6.** What would happen if you used HCl in the extraction?

**6-7.** Extra credit (1 point). There are four nitrogen atoms in caffeine. Only one acts as a base. Which one? (Hint: draw resonance structures.)

## BIBLIOGRAPHY

**Several extraction procedures for isolating caffeine from tea, some designed for the introductory organic laboratory, have been reported.**

Ault, A.; Kraig, R. *J. Chem. Educ.* **1969,** *46,* 767.
Mitchell, R. H.; Scott, W. A.; West, P. R. *J. Chem. Educ.* **1974,** *51,* 69.
Moyé, A. L. *J. Chem. Educ.* **1972,** *49,* 194.
Murray, S. D.; Hansen, P. J. *J. Chem. Educ.* **1995,** *72,* 851.
Onami, T.; Kanazawa, H. *J. Chem. Educ.* **1996,** *73,* 556.

**For additional information on the Fisher-Johns melting point apparatus see**

Morhrig, J. R.; Neckers, D. C. *Laboratory Experiments in Organic Chemistry,* 3rd ed.; Van Nostrand: New York, **1979,** p. 92.
Zubrick, J. W. *The Organic Chem Lab Survival Manual,* 7th ed.; Wiley: New York, **2008,** p. 94, Fig. 12.5.

# Thin-Layer Chromatography and Infrared Spectroscopy

EXPERIMENT 3105-5

***Purpose.*** To learn how to perform thin-layer chromatography (TLC), to apply it for the analysis of different organic compounds and to calculate the $R_f$ value of these compounds. You will use TLC to identify the composition of mixture. You will also learn the operation of the infrared spectrometer and apply infrared (IR) spectroscopy as a tool to identify different organic functional groups.

*Prior Reading*

*Technique 6A:* Thin-Layer Chromatography (pp. 92, 97–99)
*Chapter 8:* Infrared Spectroscopy
"Group Frequencies of Carbonyl groups: C=O" (pp. 545-547)
Table 8.7 (p. 546) for the frequencies of C=O functional group in various compounds
"Group Frequencies of the Heteroatom Functional Groups" (pp. 539-561)

## DISCUSSION

Estimated time to complete the experiment: two laboratory periods for 3105-5 and 3106-6.

In the reaction scheme shown below, benzyl alcohol can be oxidized to benzaldehyde. Further oxidation from benzaldehyde will result benzoic acid as the final product. TLC could be used to monitor the progress of such a reaction and IR spectroscopy could be used to characterize the product(s). In this laboratory, you will learn how to perform both of these analytical techniques (TLC and IR) and apply them to these three compounds.

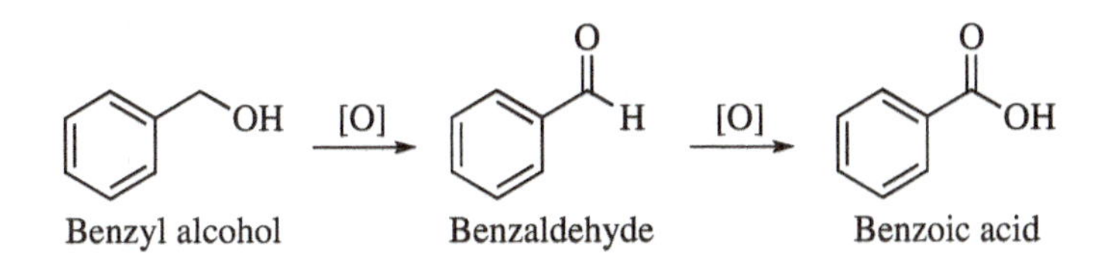


### Thin-Layer Chromatograpy (TLC) Procedure:

***Standards.*** Take a small amount of each chemical (benzyl alcohol, benzaldehyde and benzoic acid) separately into 3 different shell vials and dissolve them with a couple drops of dichloromethane.

(Hint: a solution with appropriate concentration is essential to achieve a good TLC result. For a liquid, just use 1 drop of the compound then a few drops of solvent; for a solid, just cover the tip of your spatula with the compound and dissolve it in the solvent. Too high concentration will result a "stripe" or a "smear" on the TLC plate and cause difficulty in $R_f$ value calculation).

Follow the procedure that is described on pages 97-99 for TLC development, observe the TLC result under UV light and perform the $R_f$ calculation. The developing liquid is a 4:1 mixture of hexanes and ethyl acetate that can be found in the labeled bottle.

(Hint: You can spot all three chemicals on one TLC plate and therefore run just one TLC!)

***Unknown Mixture.*** In the bottles labeled 1, 2, and 3 are mixtures of any two of these three compounds (benzyl alcohol, benzaldehyde and benzoic acid) dissolved in dichloromethane. Prepare and run a TLC with a sample from each bottle. Observe the results under UV light, and try to figure out what chemicals are present in each bottle. Report your results in the lab report. (Hint: Run the TLC and calculate the $R_f$ for each component, compare the $R_f$ value with the $R_f$ value of the pure compounds (standards) that you measured in the first step and make your judgment).

## INFRARED SPECTROSCOPY PROCEDURE

Based on your drawer location, run an IR spectrum for one of the 3 **pure** compounds (**not the mixture**) with your partner to learn the operation of the spectrometer. A simple operation menu can be found below. Obtain standard IR spectra of all three compounds from an online database such as the Spectral Database for Organic Compounds (SDBS, <http://riodb01.ibase.aist.go.jp/sdbs/cgi-bin/direct_frame_top.cgi>). Use these spectra and the spectra you obtained in lab to answer post-lab question 6-14. Turn in all 4 spectra (3 standards and your spectra) with your lab report.

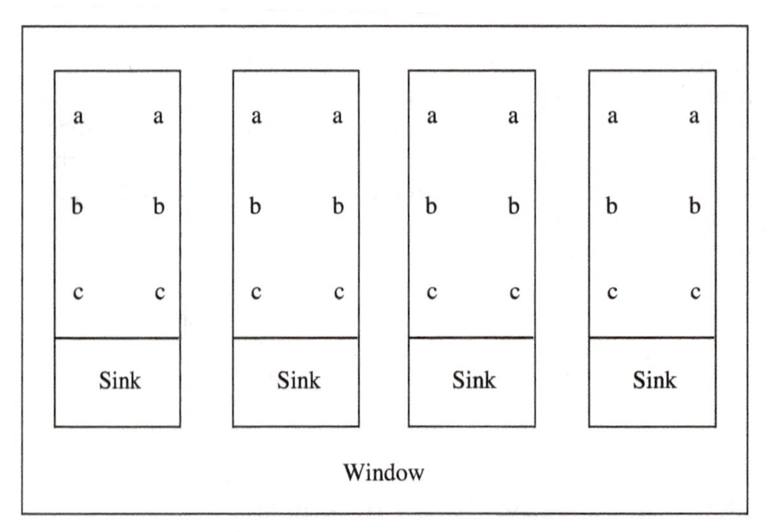


a: Benzyl alcohol b: Benzaldehyde c: Benzoic acid

### Simple Operation Menu for Infrared Spectroscopy

Infrared (IR) spectroscopy is a powerful tool for the identification of the functional groups commonly found in organic chemistry. A student should be able to run an IR spectrometer independently and interpret simple IR spectra after completing the organic chemistry laboratory and lecture courses.

We use a salt plate (NaCl) to hold the sample in the IR spectrometer because a NaCl crystal is transparent to IR radiation between 4000 and 500 $cm^{-1}$. Water, methanol, ethanol, and acetone will partially dissolve the salt and cause permanent damage to salt plate. These solvents should never come in contact with the salt plate. Fingerprints on the surface of salt plates should be avoided as well; when holding a salt plate, only hold it by the edge, never touch the center surface.

To obtain an IR spectrum:

1. Clean the salt plate by washing with dichloromethane and drying with a Kimwipe.
2 Load the empty salt plate into the sample well and click "collect" then "background" to obtain a background spectrum. This compensates for water and **carbon dioxide** in the air. Be careful not to breathe into the sample chamber.
3. Take the salt plate out of the spectrometer and add 1-2 drops of a liquid sample or 2-3 drops of a solution containing a solid sample (e.g. a solution for TLC) onto the center part of salt plate. For a solution, wait 1-2 mins to let the solvent evaporate.
4. If the sample is solid (such as benzoic acid), you are ready to run the IR after air dry; if the sample is liquid (such as benzyl alcohol and benzaldehyde), take another clean salt plate and sandwich your sample between the two salt plates.
5. Load the salt plate/plates into the well and click "collect" then "sample" to obtain the sample's spectrum.
6. You should be able to see the sample's spectrum upon the collection completion; if the spectrum did not show-up, click "view" then "full display"
7. Use the mouse and select the whole spectrum in a box; click in the box to enlarge the spectrum.
8. Click "analyze" then "find peaks" for annotation, adjust the bar to the appropriate height to select major peaks
9. Click "file" then "print" to get a print out of the spectrum. Make sure the page settings are "Landscape" and click "Ok."
10. Select the collected IR spectrum by clicking the spectrum, click "view" then "clear spectrum" to delete this spectrum; click "view" then "show report" then "clear report" to delete the report for the list of peaks. The spectrometer is ready for the next user.
11. Clean the salt plate(s) by washing with dichloromethane and you are done!

NOTE: Ensure that the TA has set the proper display options. Only one spectrum should be displayed on the screen at a time. Notify the TA if this is not the case.

## QUESTIONS

**6-12.** Rank the 3 chemicals (benzyl alcohol, benzaldehyde and benzoic acid) in order of increasing $R_f$ value.

**6-13.** List the **names** of the functional groups that are present in each of the 3 chemicals mentioned above.

**6-14.** Compare and contrast the OH peaks that are present in the benzyl alcohol and benzoic acid IR spectra; compare and contrast the C=O peak that are present in the benzaldehyde and benzoic acid IR spectra. How does the spectrum you obtained in lab compare to the standard spectra for your assigned compound?

# Oxidation of Alcohols; Trends in Infrared Spectroscopy

EXPERIMENT 3105-6

***Prior to lab:*** Prior to your laboratory period, you will be assigned an alcohol from the table at the end of this experiment. The alcohols shown on the table that already have the structure drawn and IR frequency listed will not be used as reagents in lab. You are to calculate the number of grams that will be required for you to run the reaction with 2.5 mmol of your alcohol. If your alcohol is a liquid, you should calculate the number of mL required. The molecular weights and densities can be found in various online sources. Predict the structure of the aldehyde or ketone that is the expected product from every reaction.

***Purpose:*** To prepare a library of aldehydes and ketones by oxidation of the corresponding 1° and 2° alcohols. To compare the infrared (IR) spectra obtained for all the compounds in the lab section and to explore trends in the carbonyl absorptions.

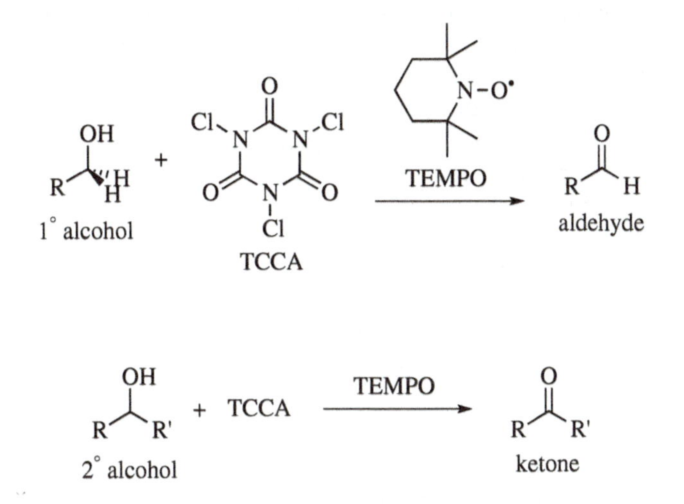


***Introduction:*** The oxidation of alcohols is an important reaction, both as a fundamental reaction of organic chemistry and as central process in biochemical processes. Many oxidation recipes call for the use of carcinogenic chromium (IV) salts. However, this procedure uses less toxic reagents that also are less persistent in the environment. TCCA is one of the chemicals used to "chlorinate" swimming pools.

The presence of a carbonyl functionality in a molecule is readily identified by a strong IR absorption in the region of 1800 – 1650 $cm^{-1}$. The specific absorption depends on the structure of the molecule. In this experiment you will have the opportunity to examine some of the factors that influence the carbonyl absorption frequency.

***Background.*** As you will recall from general chemistry, oxidation means the removal of electrons. Thus an electrochemical oxidation of an alcohol may be carried out; the mechanism is shown in Figure 1. The initial oxidation step (removal of an electron, $e^-$) followed by the loss of a proton ($H^+$) gives a radical. Note that the loss of a proton from the carbon rather than the oxygen is favored because the radical is resonance stabilized. The radical can easily lose one electron to be oxidized again; loss of another proton gives the corresponding carbonyl compound. Most oxidations in organic chemistry follow different mechanisms, but because the overall result is the same, the conversion of an alcohol to a carbonyl is always called an oxidation. (It is easy to recognize such an oxidation – there is one more carbon-oxygen bond in the product than in the reactant!)

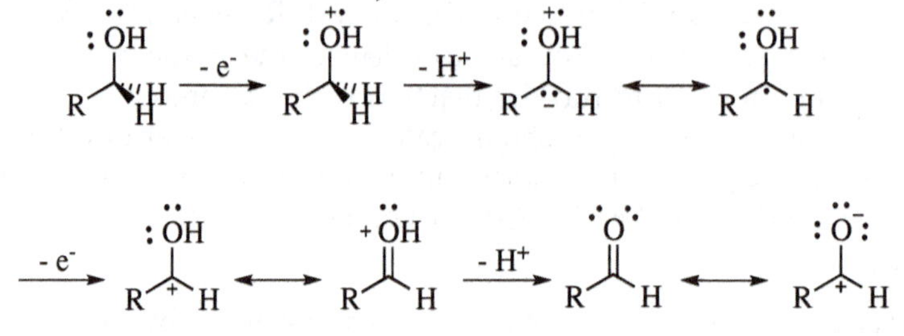


Figure 1. Mechanism for the electrochemical oxidation of a primary (1°) alcohol to an aldehyde.

In this experiment, TEMPO is used as a catalytic oxidant and TCCA is used as the stoichiometric (or terminal) oxidant. The oxidized form of TEMPO reacts selectively with 1° and 2° alcohols to produce the corresponding carbonyl compound (aldehyde or ketone, respectively). In the process, TEMPO is reduced to the hydroxylamine. TCCA does not react directly with alcohols, but will reoxidize the hydroxylamine to the TEMPO cation. The overall catalytic scheme is shown in Figure 2.

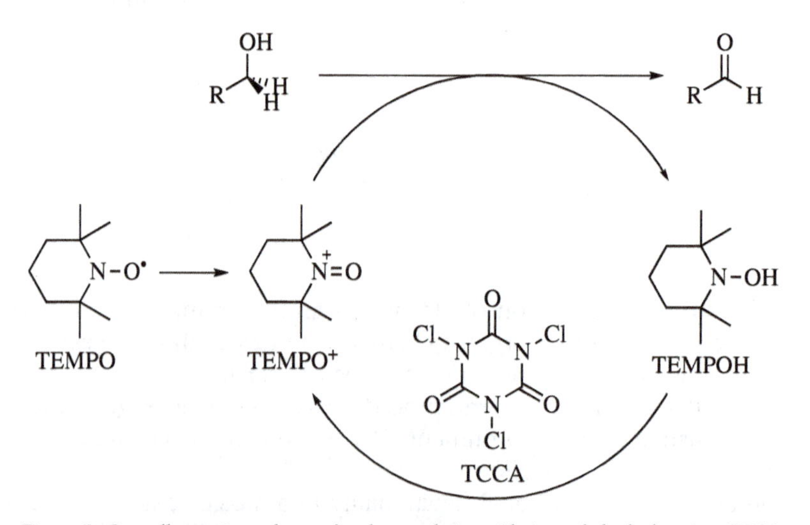


Figure 2. Overall reaction scheme for the catalytic oxidation of alcohols using TCCA.

Oxidation of a primary alcohol to an aldehyde is a particularly challenging reaction, because of the possibility of over oxidation to a carboxylic acid. Even if the oxidation initially forms an aldehyde in high yield there can be problems. In the presence of even a small amount of water, aldehydes exist in equilibrium with their hydrates. From the point of view of the oxidant, this is just another alcohol that can again be oxidized. This leads to the carboxylic acid, as shown in Figure 3.

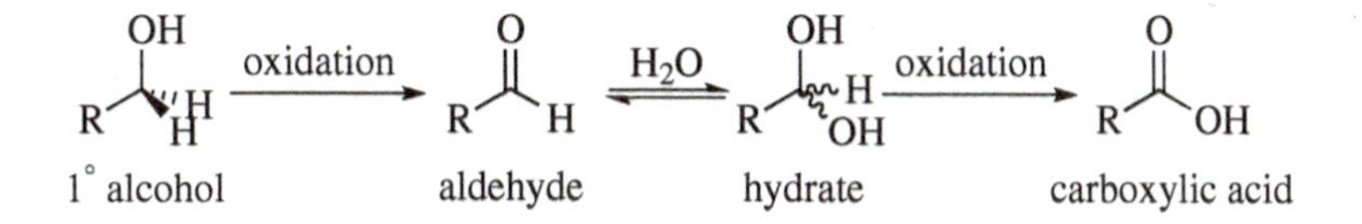


Figure 3. Oxidation of a primary alcohol, in the presence of water, via an aldehyde and hydrate, to a carboxylic acid.

***Background Reading:*** Oxidations are also discussed in your lecture textbook in various sections.

You should also review infrared (IR) spectroscopy, Ch 8: Introduction to Group Frequencies (pp. 290-304), as you will use this technique during this laboratory.

# EXPERIMENTAL PROCEDURE

## REACTION

1. Weigh (tare) an empty round-bottomed flask (50 mL), then add the appropriate mass or volume of your alcohol for 2.5 mmol. Add a magnetic stirrer and 10 mL of dichloromethane as solvent. Note: For a solid alcohol, weigh out the calculated mass. Mass of alcohol (g) = molecular weight of alcohol × 0.0025 mol For a liquid alcohol, measure the calculated volume. Volume of alcohol (mL) = mass of alcohol (g) / density (g/mL)

   154.25 · 0.0025 = 0.3856g ÷ .8894 = .4336mL

2. Add 0.60 g (2.6 mmol) of TCCA to the above solution. Cool the flask in an ice water bath in a beaker.
3. Weigh out 0.0040 g (0.0025 mmol) TEMPO in a small conical vial. Dissolve the TEMPO in about 1-2 mL of dichloromethane and add this to the reaction mixture with stirring.
4. Remove the reaction mixture from the ice bath and allow it to warm to room temperature with stirring.
5. Allow the reaction to stir at room temperature for 30 minutes. The reaction should be complete after this time, regardless of the alcohol you are using.

### Product isolation

→ centrifuge

6. After 30 minutes, filter the solution in the reaction flask through a small pad of Celite using a Buchner funnel and filtration flask connected to the water aspirator. Your TA can show you the proper filtration technique. Wash any remaining product off the Celite using 2 mL of dichloromethane.

   Note: First add a filter paper on the Buchner funnel. Wet the paper with dichloromethane. Then add Celite powder onto the paper with an amount enough to cover the paper. Then filter your solution. Celite is a filter aid which promotes a more efficient filtration by removing more suspended solid than filter paper alone.

7. Then wash the filtrate solution with 15 mL of 0.5 M $Na_2CO_3$ using a separatory funnel. Collect both the organic layer (on the bottom because the solvent $CH_2Cl_2$ is more dense than water) and aqueous layer (on the top) in two separate beakers.
8. Wash the organic layer solution from step 7 with 15 mL of dilute HCl using a separatory funnel. Collect both the organic layer (on the bottom) and aqueous layer (on the top) in two separate beakers.
9. Wash the organic layer solution from step 8 with 15 mL of brine (saturated NaCl) using the same separatory funnel. Collect both the organic layer (on the bottom) and aqueous layer (on the top) in two separate beakers.
10. Add $MgSO_4$ to the organic layer solution from step 9 to remove any dissolved water (drying it).
11. Filter the dried organic layer solution into a clean, dry beaker using a conical funnel.
12. Transfer some of the organic solution into a tared (weighed) shell vial.
13. Evaporate the dichloromethane solvent from the shell vial in the hood on a hot plate using a stream of air. Transfer more of the solution from 11 into the shell vial and evaporate; repeat until all the solvent is gone.
14. Weigh the beaker to obtain the final mass of the product.

### Characterization by IR spectroscopy

Ask your TA if you still have questions about how to use the IR spectrometer.

1. In the hood, clean the two salt plates with dichloromethane.
2. Add 2-3 drops of your product (if it is a solid, then dissolve it in 1-2 mL dichloromethane) onto the center of one salt plate.

   Note: The salt plates are made of NaCl, which is soluble in water. Do not touch the center of salt plate with your fingers. Always hold the edges of it. Do not wash it with water or acetone, only use dichloromethane.
3. Allow 1 or 2 min for the solvent to evaporate on the salt plate, if necessary. Sandwich your sample between the two salt plates.
4. Obtain the IR spectrum of your sample by following the procedure outlined in Experiment 3105-5.

   Depending on how quickly everyone completes their experiment, you may need to obtain an IR spectrum of your compound outside of your regular lab period. If so, clearly label your vial, put your name on it, and give it to your TA. Schedule a time with your TA to take the IR spectrum.

**Lab Report: To write your lab report, you will need to obtain the following information from every other student in your lab section:**

1) the structure of their aldehyde or ketone
2) the IR frequency of the carbonyl absorption

3) whether the IR shows evidence of an OH stretch.

Recall that the frequency of an IR stretch is related to the mass of the atoms and the force constant of the bond (i.e. the bond strength). Thus one expects that for a carbonyl stretch, where the mass of the oxygen atom is always the same, that **any variations in the frequency of the bond stretch would be due to differences in the bond strength.**

The equation below shows how vibrational frequency is related to the mass of the atoms in the compound. Think of each bond as a spring that is affected by the atoms on each end.

$$\bar{\nu} = \frac{1}{2\pi c}\left[\frac{f}{M_x \cdot M_y / (M_x + M_y)}\right]^{1/2}$$

| | |
|---|---|
| $\nu$ | vibrational frequency ($cm^{-1}$) (strange units, I know!) |
| c | speed of light |
| f | force constant (the strength of the bond) |
| $M_x$, $M_y$ | mass of the atoms on the end of the "spring." |

## QUESTIONS

**Remember you need the IR frequencies of the carbonyl absorptions from everyone else in your lab section.**

**6-15.** Bonus (1 point): What is the systematic name of the aldehyde or ketone you made?

**6-16.** Explain the role of each of the three aqueous washes in the workup procedure.

**6-17.** Look for trends in the carbonyl absorption frequencies of the aldehydes and ketones that your lab section prepared. Offer hypotheses for relationships between the structures and the absorptions.

Think about the differences in the structures of the products:

Aldehyde vs. ketone?

How close is the carbonyl group to a double bond?

If the product contains an aromatic ring: What substituents are attached to the ring? What position are they attached on the ring (*ortho, meta, para*)?

If you think some of the data do not fit the trends, explain your reasoning.

| Reactant Alcohol Structure | Product Carbonyl Structure | Observed IR Frequency ($cm^{-1}$) | Observed Residual –OH in IR? |
|---|---|---|---|
| OH / Cl<br>1-(2-chlorophenyl)-1-propanol | O / Cl | 1710 | n/a |
| OH / O<br>2-methoxy-alpha-methylbenzyl alcohol<br>1-(2-methoxyphenyl)ethanol | O / O | 1700 | n/a |
| OH / F<br>4-fluoro-alpha-methylbenzyl alcohol<br>1-(4-fluorophenyl)ethanol | O / F | 1685 | n/a |
| OH<br>1-phenyl-2-propanol | | | |
| OH<br>4-methylphenethyl alcohol<br>2-(4-methylphenyl)ethanol | O | 1720 | n/a |
| OH / O<br>4-methoxyphenethyl alcohol<br>2-(4-methoxyphenyl)ethanol | | | |
| OH / O<br>4-methoxybenzyl alcohol | | | |

| Reactant Alcohol Structure | Product Carbonyl Structure | Observed IR Frequency ($cm^{-1}$) | Observed Residual –OH in IR? |
|---|---|---|---|
| OH, F<br>2-fluorobenzyl alcohol | O, F | 1700 | n/a |
| OH<br>2-methylbenzyl alcohol | | | |
| OH<br>benzyl alcohol | | | |
| O, OH<br>3-methoxybenzyl alcohol | O, O | 1710 | n/a |
| OH<br>cinnamyl alcohol | | | |
| OH<br>3-cyclohexene-1-methanol | | | |
| OH<br>3-heptanol | | | |
| OH<br>1-hepten-3-ol | | | |
| * OH<br>geraniol | | | |

| Reactant Alcohol Structure | Product Carbonyl Structure | Observed IR Frequency ($cm^{-1}$) | Observed Residual –OH in IR? |
|---|---|---|---|
| OH<br>cyclohexanol | | | |
| OH<br>borneol | | | |

EXPERIMENT
3105-7

# The Separation of a 25-μL Mixture of Heptanal (bp 153 °C) and Cyclohexanol (bp 160 °C) by Gas Chromatography

***Purpose.*** This experiment illustrates the separation of a heptanal and cyclohexanol mixture, into the pure components. The materials boil within 7 °C of each other. This mixture would be difficult, if not impossible, to separate by the best distillation techniques available.

*Prior Reading*

*Technique 1:* Microscale Separation of Liquid Mixtures by Preparative Gas Chromatography (pp. 55–61)

## DISCUSSION

The efficacy of GC separations is highly dependent on the experimental conditions. For example, two sets of experimental data on the heptanal–cyclohexanol mixture are given below to demonstrate the effects of variations in oven temperature on retention times.

In Data Set A, the oven temperature was allowed to rise slowly from 160 to about 170 °C during a series of sample collections. The retention time of heptanal dropped from about 3 min to close to 2 min, whereas the retention time of cyclohexanol was reduced from about 5.5 min to nearly 4 min. The significant decrease in resolution over this series of collections is reflected in the number of theoretical plates calculated, which was over 300 for heptanal and about 500 for cyclohexanol in the first trial, but declined to below 200 for both compounds toward the last run (see Data Set A).

## COLLECTION YIELD

Cyclohexanol

Density of cyclohexanol = 0.963 mg/μL.

In 25 μL of 1:1 cyclohexanol–heptanal, we have 12.5 μL of cyclohexanol.

Therefore, 125 μL × 0.963 mg/μL = 12 mg of cyclohexanol injected.

Percent recovered = (8.3 mg/12.0 mg) × 100 = 69% cyclohexanol collected.

Heptanal

Density of heptanal = 0.850 mg/μL.
Therefore, 12.5 μL × 0.85 mg/μL = 10.6 mg of heptanal injected.
Percent recovered = (8.1 mg/10.6 mg) × 100 = 76% heptanal.

In the Data Set B collections, stable oven temperatures and flow rates were maintained, and the data exhibit excellent reproducibility. Oven temperature was held at 155 °C throughout the sampling process. The retention time of heptanal was observed to be slightly longer than 3 min, with a variance of 6 s, whereas the cyclohexanol retention time was found to be slightly longer than 6 min, with a variance of 12 s. The resolution remained essentially constant throughout the series, and the number of theoretical plates calculated was about 350 for heptanal and about 500 for cyclohexanol (see Data Set B).

## COLLECTION YIELD

Cyclohexanol

Density of cyclohexanol = 0.963 mg/μL.
In 25 μL of 1:1 cyclohexanol–heptanal, there are 12.5 μL of cyclohexanol.
Therefore, 125 μL × 0.963 mg/μL = 12 mg of cyclohexanol injected.
Percent recovered = (8.5 mg/12.0 mg) × 100 = 71% cyclohexanol collected.

Heptanal

Density of heptanal = 0.850 mg/μL.
Therefore, 12.5 μL × 0.85 mg/μL = 10.6 mg of heptanal injected.
Percent recovered = (8.3 mg/10.6 mg) × 100 = 78% heptanal.

**Data Set A**

| | Heptanal | | | | Cyclohexanol | | | |
|---|---|---|---|---|---|---|---|---|
| Trial No. | Retention Time (min) | Baseline Width (min) | Number of Theoretical Plates | Recovery (mg) | Retention Time (min) | Baseline Width (min) | Number of Theoretical Plates | Recovery (mg) |
| 1 | 3.1 | 0.7 | 314 | 8.0 | 5.6 | 1.0 | 502 | 8.0 |
| 2 | 2.9 | 0.7 | 275 | 8.0 | 5.3 | 1.0 | 449 | 8.0 |
| 3 | 3.0 | 0.7 | 294 | 7.0 | 5.7 | 1.0 | 520 | 8.0 |
| 4 | 2.8 | 0.7 | 256 | 8.0 | 5.1 | 1.1 | 344 | 8.0 |
| 5 | 2.5 | 0.6 | 278 | 8.0 | 4.3 | 1.1 | 244 | 9.0 |
| 6 | 2.7 | 0.5 | 467 | 7.0 | 4.6 | 1.0 | 339 | 10.0 |
| 7 | 2.5 | 0.6 | 278 | 10.0 | 4.2 | 1.0 | 282 | 8.0 |
| 8 | 2.2 | 0.5 | 310 | 9.0 | 3.5 | 1.0 | 196 | 8.0 |
| 9 | 1.8 | 0.5 | 207 | 8.0 | 3.0 | 1.0 | 144 | 8.0 |
| 10 | 2.3 | 0.7 | 173 | 8.0 | 3.9 | 1.0 | 243 | 8.0 |
| Av | 2.5 ± 0.4 | 0.6 ± 0.09 | 285 ± 78.1 ± 0.8 | | 4.5 ± 0.9 | 1.0 ± 0.05 | 326 ± 1298.3 ± 0.7 | |

**Data Set B**

| | Heptanal | | | | Cyclohexanol | | | |
|---|---|---|---|---|---|---|---|---|
| Trial No. | Retention Time (min) | Baseline Width (min) | Number of Theoretical Plates | Recovery (mg) | Retention Time (min) | Baseline Width (min) | Number of Theoretical Plates | Recovery (mg) |
| 1 | 3.5 | 0.7 | 400 | 8.0 | 6.6 | 1.1 | 576 | 8.0 |
| 2 | 3.2 | 0.7 | 334 | 9.0 | 6.0 | 1.1 | 476 | 7.0 |
| 3 | 3.5 | 0.7 | 400 | 7.0 | 6.6 | 1.2 | 484 | 10.0 |
| 4 | 3.2 | 0.7 | 334 | 9.0 | 6.1 | 1.0 | 595 | 9.0 |
| 5 | 3.1 | 0.6 | 427 | 8.0 | 6.0 | 1.1 | 476 | 8.0 |
| 6 | 3.2 | 0.7 | 334 | 9.0 | 6.0 | 1.1 | 476 | 9.0 |
| 7 | 3.3 | 0.8 | 272 | 9.0 | 6.1 | 1.1 | 492 | 8.0 |
| 8 | 3.1 | 0.7 | 313 | 8.0 | 6.0 | 1.1 | 476 | 10.0 |
| 9 | 3.2 | 0.7 | 334 | 8.0 | 6.1 | 1.1 | 492 | 8.0 |
| 10 | 3.2 | 0.7 | 334 | 8.0 | 6.2 | 1.1 | 508 | 8.0 |
| Av | 3.2 ± 0.1 | 0.7 ± 0.05 | 348 ± 47 | 8.3 ± 0.7 | 6.2 ± 0.2 | 1.1 ± 0.05 | 505 ± 44 | 8.5 ± 1.0 |

The results just described demonstrate that the resolution of GC peaks may be very sensitive to changes in retention time resulting from instability in oven temperatures. Since the number of theoretical plates is related to resolution values, significant degradation in column plate values can occur with variations in oven temperatures. When you compare the time and effort required to obtain a two-plate fractional distillation on a 2-mL mixture (see Experiment 3105-1 and Technique 2) with the speed and ease used to obtain a 500 plate separation on 12.5 μL of cyclohexanol in this experiment, it is hard not to be impressed with the enormous power of this technique.

## COMPONENTS

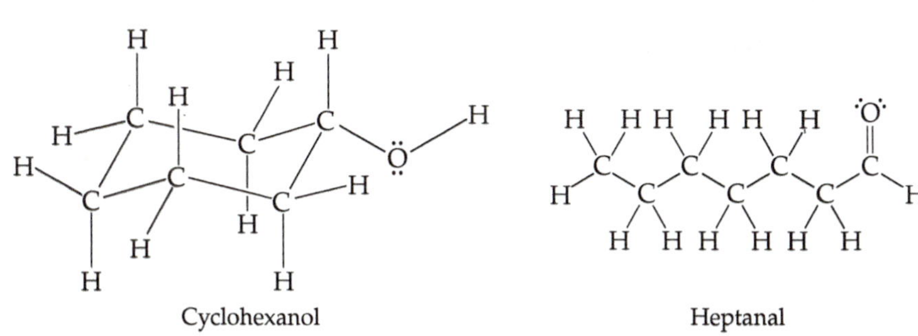

Cyclohexanol        Heptanal

## EXPERIMENTAL PROCEDURE

Estimated time for the experiment: 2.0 h.

***Sample Preparation.*** Your TA will divide the section into three groups. Each group will prepare one of the following samples to analyze by gas chromatography. The general method for preparing a sample for the Organic Chemistry Instructional Laboratory GC is to make a 1 % solution of the sample to be analyzed in some volatile solvent (e.g. – dichloromethane or pentane).

Group 1, Cyclohexanol Standard Sample

Prepare a 1 % solution of cyclohexanol in dichloromethane by adding 20 µL of cyclohexanol into a shell vial and then 2 mL of dichloromethane into the same shell vial (about ½ the shell vial). Mix the solution by drawing the solution into a disposable Pasteur pipet and then adding the solution back to the shell vial. Fill a GC vial about ¾ full with the standard solution. Cap the GC vial & label it with a permanent marker.

Group 2, Heptanal Standard Sample

Prepare a 1 % solution (of each component) of heptanal in dichloromethane by adding 20 µL of heptanal into a shell vial and then 2 mL of dichloromethane into the same shell vial (about ½ the shell vial). Mix the solution by drawing the solution into a disposable Pasteur pipet and then adding the solution back to the shell vial. Fill a GC vial about ¾ full with the standard solution. Cap the GC vial & label it with a permanent marker.

Group 3, Cyclohexanol/Heptanal Mixture Sample

Prepare a 1 % solution of cyclohexanol and heptanal in dichloromethane by adding 20 µL of cyclohexanol and 20 µL of heptanal into a shell vial. Then 2 mL of dichloromethane into the same shell vial (about ½ the shell vial). Mix the solution by drawing the solution into a disposable Pasteur pipet and then adding the solution back to the shell vial. Fill a GC vial about ¾ full with the standard solution. Cap the GC vial & label it with a permanent marker.

*Sample Analysis.* Your TA introduce you to GC analysis.

*Chromatogram Analysis.* Student should always label the important peaks on the chromatogram with the compound name. Turn chromatograms in with your lab report.

Chromatographic Parameters.

Column Temperature: 80 °C

Injector Temperature: 200 °C

Detector Temperature: 250 °C

Time: 5 minutes

## QUESTIONS

**6-18.** Based on the data presented in the Data Set A chromatographic separation, can you explain why there is such a steep decline in column efficiency with temperature change?

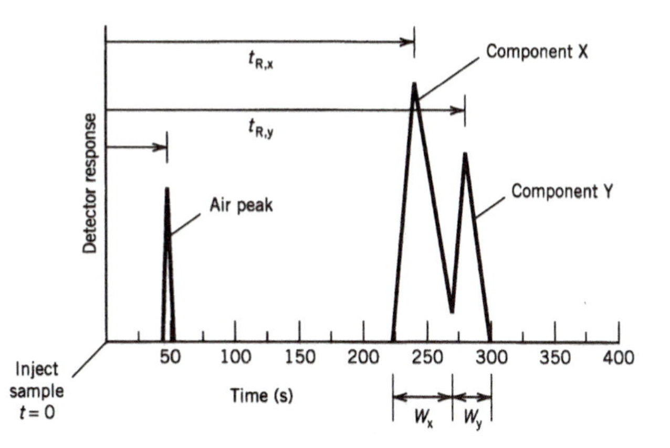


**6-19.** Consider the above gas chromatogram for a mixture of analytes X and Y:

**(a)** Calculate the number of theoretical plates for the column in reference to the peaks of each component (X and Y).

**(b)** If the column is 12 m long, calculate the height equivalent theoretical plate (HETP, in plates per cm) for this column.

**6-20.** Retention times for several organic compounds separated on a GC column are given below.

| Compound | $t_R$ (s) |
|---|---|
| Air | 75 |
| Pentane | 190 |
| Heptane | 350 |
| 2-Pentene | 275 |

**(a)** Calculate the relative retention of 2-pentene with respect to pentane.

**(b)** Calculate the relative retention of heptane with respect to pentane.

## BIBLIOGRAPHY

**Selected references on gas chromatography:**


Grob, R. L.; Barry, E. F., Eds.; *Modern Practices of Gas Chromatography;* Wiley: New York, 2004.

Jennings, W.; Mittiefehidt, E.; Stremple, P., Eds.; *Analytical Gas Chromatography;* 2nd ed., Academic Press: New York, 1997.

McNair, H. M.; Miller, J. M. *Basic Gas Chromatography;* Wiley: New York, 1997.

Sadek, P. C. *Illustrated Pocket Dictionary of Chromatography;* Wiley: *New York, 2004.*

# Williamson Synthesis of Ethers

EXPERIMENT 3105-8

Common names: propyl *p*-tolyl ether, 4-propoxytoluene
CA number: [5349-18-8]
CA name as indexed: benzene, 1-methyl-4-propoxy-

Common names: methyl *p*-ethylphenyl ether, *p*-ethylanisole
CA number: [1515-95-3]
CA name as indexed: benzene, 1-ethyl-4-methoxy-

Common names: butyl *p*-nitrophenyl ether
CA number: [7244-78-2]
CA name as indexed: benzene, 1-butoxy-4-nitro-

***Purpose.*** The conditions under which ethers are prepared are explored by the well-known Williamson ether synthesis. You will prepare alkyl aryl ethers by $S_N2$ reactions of alkyl halides with substituted phenoxide anions. The use of phase-transfer catalysis is demonstrated.

*Prior Reading*

*Technique 4:* Solvent Extraction
Liquid–Liquid Extraction (p. 72)
*Technique 6:* Chromatography
Column Chromatography (pp. 92–95)

## REACTION

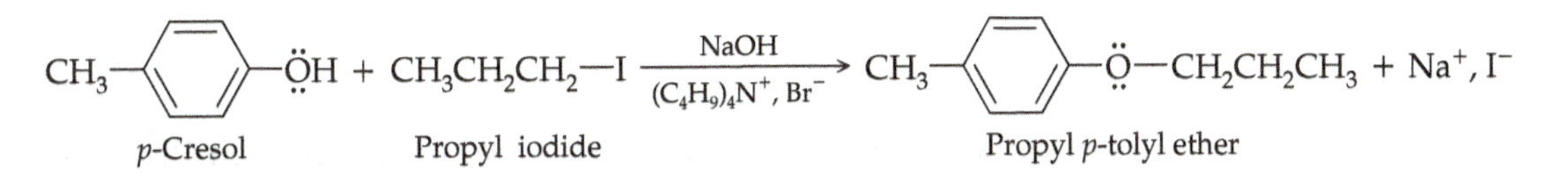

*p*-Cresol   Propyl iodide   Propyl *p*-tolyl ether

## DISCUSSION

A general method of preparing ethers is the Williamson synthesis, an $S_N2$ reaction specifically between a phenoxide ion ($ArO^-$) nucleophile and an alkyl halide. This reaction is often used for the synthesis of symmetrical and unsymmetrical ethers where at least one of the ether carbon atoms is primary or methyl, and thus amenable to an $S_N2$ reaction. Elimination (E2) is generally observed if secondary or tertiary halides are used, since phenoxide ions are also bases.

The conditions under which these reactions are conducted lend themselves to the use of phase-transfer catalysis. The reaction system involves two phases: the aqueous phase and the organic phase. In the present case, the alkyl halide reactant acts as the organic solvent, as does the product formed. The phase-transfer catalyst plays a very important role. In effect, it carries the phenoxide ion, as an ion-pair, from the aqueous phase, across the phase boundary into the organic phase, where the $S_N2$ reaction then occurs. The ether product and the corresponding halide salt of the catalyst are produced in this reaction. The halide salt then migrates back into the aqueous phase, where the halide ion is exchanged for another phenoxide ion, and the process repeats itself. The catalyst can play this role, since the large organic groups (the four butyl groups) allow the solubility of the ion-pair in the organic phase, while the charged ionic center of the salt renders it soluble in the aqueous phase.

In the reactions described below, the mechanism is a classic $S_N2$ process, and involves a backside nucleophilic attack of the phenoxide anion on the alkyl halide.

$$CH_3-C_6H_4-\ddot{O}H + NaOH \rightleftharpoons CH_3-C_6H_4-\ddot{O}:^-, Na^+ + H_2O$$

$$CH_3-C_6H_4-\ddot{O}:^-, Na^+ + CH_3CH_2-CH_2-I \longrightarrow CH_3-C_6H_4-\ddot{O}-CH_2CH_2CH_3 + Na^+, I^-$$

It is of interest to contrast the acidity of phenols with that of simple alcohols. A phenol is more acidic than an alcohol. In a typical aliphatic alcohol (e.g., ethanol) loss of the proton forms a strong anionic base, alkoxide ion (ethoxide ion).

$$R-CH_2-OH \rightleftharpoons H^+ + R-CH_2-O^-$$

Alkoxide ion

The strongly basic characteristics of the alkoxide species are due to the fact that the negative charge is localized on the oxygen atom. Ethanol has a $pK_a = 16$. In contrast, the conjugate base of a phenol can delocalize its negative charge.

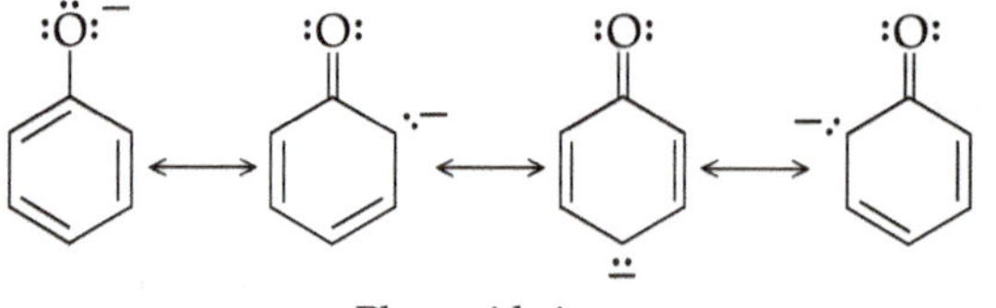

Phenoxide ion

The phenoxide ion is stabilized by this resonance delocalization; therefore, it is a weaker base than the alkoxide ion. Conversely, the phenol is a stronger acid than a typical aliphatic alcohol. Phenol has a $pK_a = 10$ and is thus 1 million times more acidic than ethanol.

## EXPERIMENTAL PROCEDURE

Estimated time of the experiment: 2.5 h.

**Physical Properties of Reactants**

| Compound | MW | Amount | mmol | mp (°C) | bp (°C) | $d$ | $n_D$ |
|---|---|---|---|---|---|---|---|
| *p*-Cresol | 108.15 | 160 μL | 1.56 | 32–34 | 202 | 1.02 | 1.5312 |
| 25% NaOH solution | | 260 μL | | | | | |
| Tetrabutylammonium bromide | 322.38 | 18 mg | 0.056 | 103–104 | | | |
| Propyl iodide | 169.99 | 150 μL | 1.54 | | 102 | 1.75 | 1.5058 |

***Reagents and Equipment.*** Weigh and place 160 μL (168 mg, 1.56 mmol) of *p*-cresol in a 5.0-mL conical vial containing a magnetic spin vane. Now add 260 μL of 25% aqueous sodium hydroxide (TA will dispense) and thoroughly mix the resulting solution. To this solution weigh and add the tetrabutylammonium bromide ($Bu_4{}^+Br^-$) catalyst (18 mg), followed by 150 μL (262 mg, 1.54 mmol) of propyl iodide. Immediately attach the vial to a reflux condenser.

*NOTE. Warm the cresol in a hot water bath to melt it. Dispense this reagent and the propyl iodide in the* **hood** *using an automatic delivery pipet. Never lay automatic delivery pipets flat; return to holder/stand after use.*

HOOD

CAUTION: Propyl iodide is a cancer suspect agent. Sodium hydroxide is a strong base and can quickly burn your skin and eyes. Wear your safety goggles and gloves. Be particularly careful not to spill any. If you do get sodium hydroxide on your skin, wash it immediately with cold running water and notify your TA.

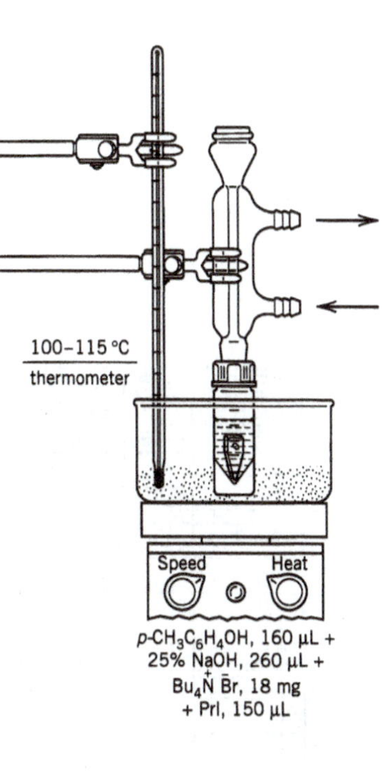


***Reaction Conditions.*** Place the reaction vessel on a hot plate and stir vigorously at 110–115 °C for 45–60 min.

***Isolation of Product.*** Cool the resulting two-phase mixture to room temperature, and remove the spin vane with forceps. Rinse the spin vane with 1.0 mL of diethyl ether, adding the rinse to the two-phase mixture. Cap the vial, agitate, vent, and transfer the bottom aqueous layer, using a Pasteur filter pipet, to a 3.0-mL conical vial. A Vortex mixer, if available, can be used to good advantage in this extraction step. Wash this aqueous fraction with 1.0 mL of diethyl ether. Save this and all subsequent aqueous fractions together in a small Erlenmeyer flask until your final product has been isolated and characterized. Now transfer this diethyl ether wash to the 5-mL conical vial containing the ether solution of the product. Extract the resulting ether solution with a 400-μL portion of 5% aqueous sodium hydroxide solution. Cap the vial, agitate, vent, and remove and save the bottom aqueous layer, using a Pasteur filter pipet. Wash the product–ether solution with 200 μL of water. Remove, and save, the aqueous phase to obtain the crude, wet ether solution of the product. Add a boiling stone to the vial and concentrate the solution on a hot plate under a gentle stream of air to isolate the crude product.

***Purification and Characterization.*** The crude product is purified by chromatography on silica gel. Prepare a microchromatographic column by placing a small amount of sand (approximately 1 cm height in the pipet), in a Pasteur filter pipet (see p 36-37), followed by 500 mg of activated silica gel, and then 50 mg of anhydrous magnesium sulfate (⬅). Dissolve the crude product in 250 μL of methylene chloride and transfer the resulting solution to the dry column by use of a Pasteur pipet. Elute the material with 2.0 mL of methylene chloride and collect the eluate in a tared 3.0-mL conical vial or shell vial containing a boiling stone. Evaporate the solvent by placing the vial in an aluminum block on a hot plate in the hood maintained at a temperature of 60–65 °C.

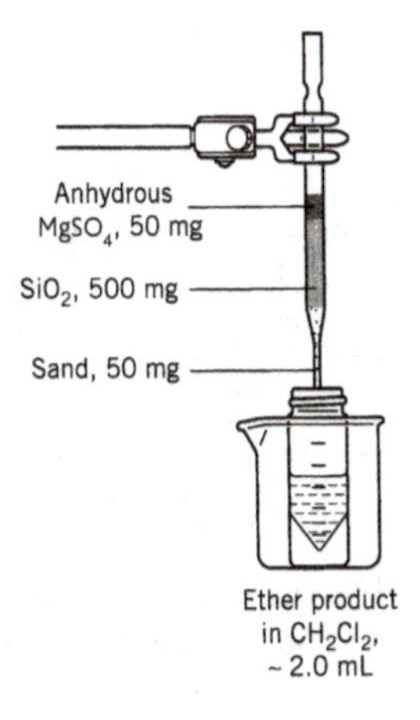


Weigh the pure propyl *p*-tolyl ether and calculate the percent yield.

Obtain an IR spectrum of the compound and compare it to that shown in Figure 6.36 for 4-propoxytoluene (propyl *p*-tolyl ether).

***Nuclear Magnetic Resonance Analysis.*** Look at the $^1$H and $^{13}$C NMR spectra of propyl *p*-tolyl ether in $CDCl_3$ in Figures 6.37 and 6.38. There are two extraneous peaks in the $^1$H spectrum: the small singlet at 7.24 ppm is due to residual $CHCl_3$ in the $CDCl_3$, and the small singlet at 1.55 ppm is probably due to a trace amount of $H_2O$ in either the sample or the NMR solvent. The 1:1:1 triplet at 77 ppm in the $^{13}$C spectrum is from the $CDCl_3$ solvent.

Since the $^1$H spectrum is entirely first order, it can be readily interpreted. You should be able to use the splitting patterns to assign peaks to each of the different groups of protons in the molecule. The integration can assist you. Include a copy of the $^1$H spectrum and the peak assignments in your lab report.

***Chemical Tests.*** Qualitative chemical tests can also be used to assist in characterizing this compound as an ether.

A key factor to investigate is the solubility characteristics of this material (see Experiments 3106-7 - 3106-9). Determine its solubility in water, 5% sodium hydroxide, 5% hydrochloric acid, concentrated sulfuric acid, and 85% phosphoric acid (your TA will dispense the concentrated acids). Do the results place this compound in the solubility class of an ether containing less than 8, or more than 8, carbon atoms?

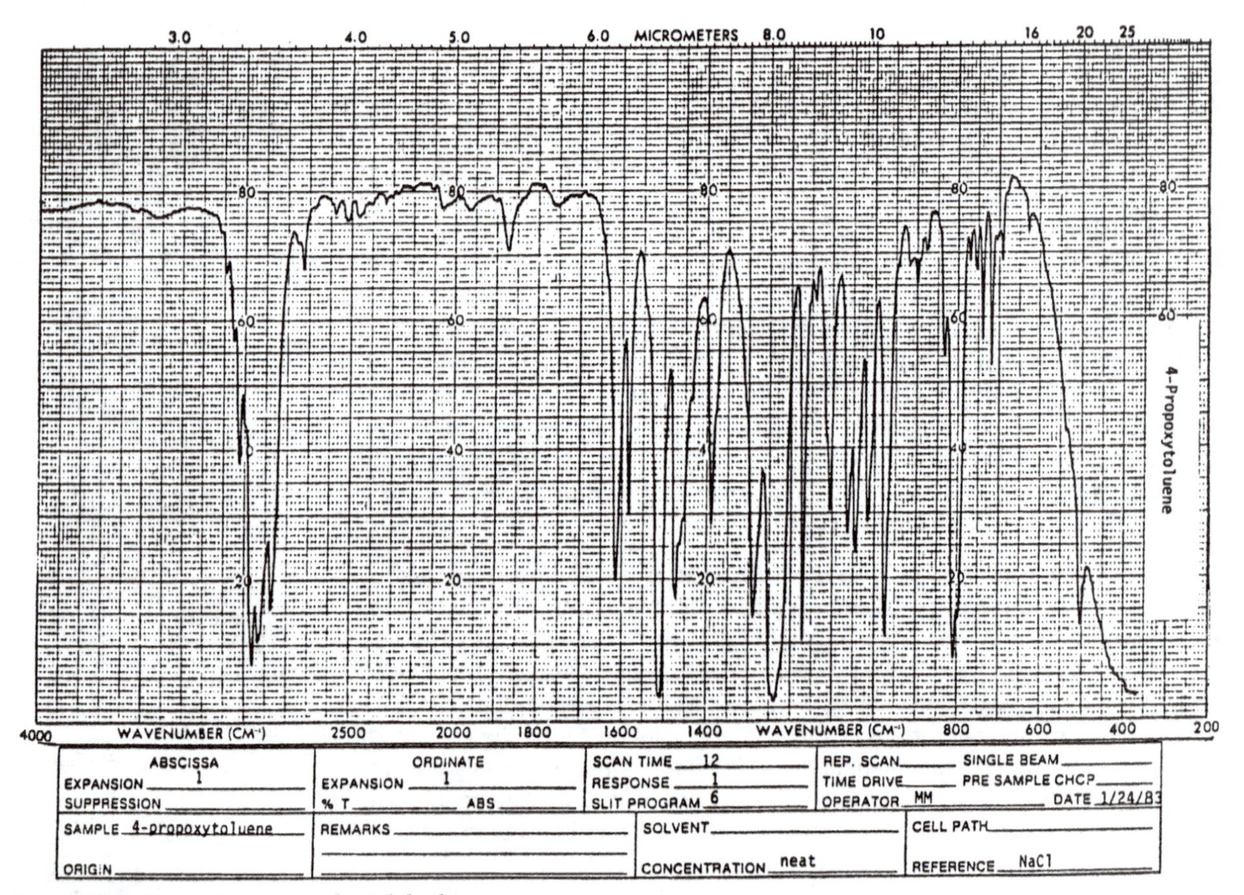


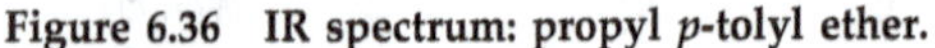

**Figure 6.36 IR spectrum: propyl *p*-tolyl ether.**

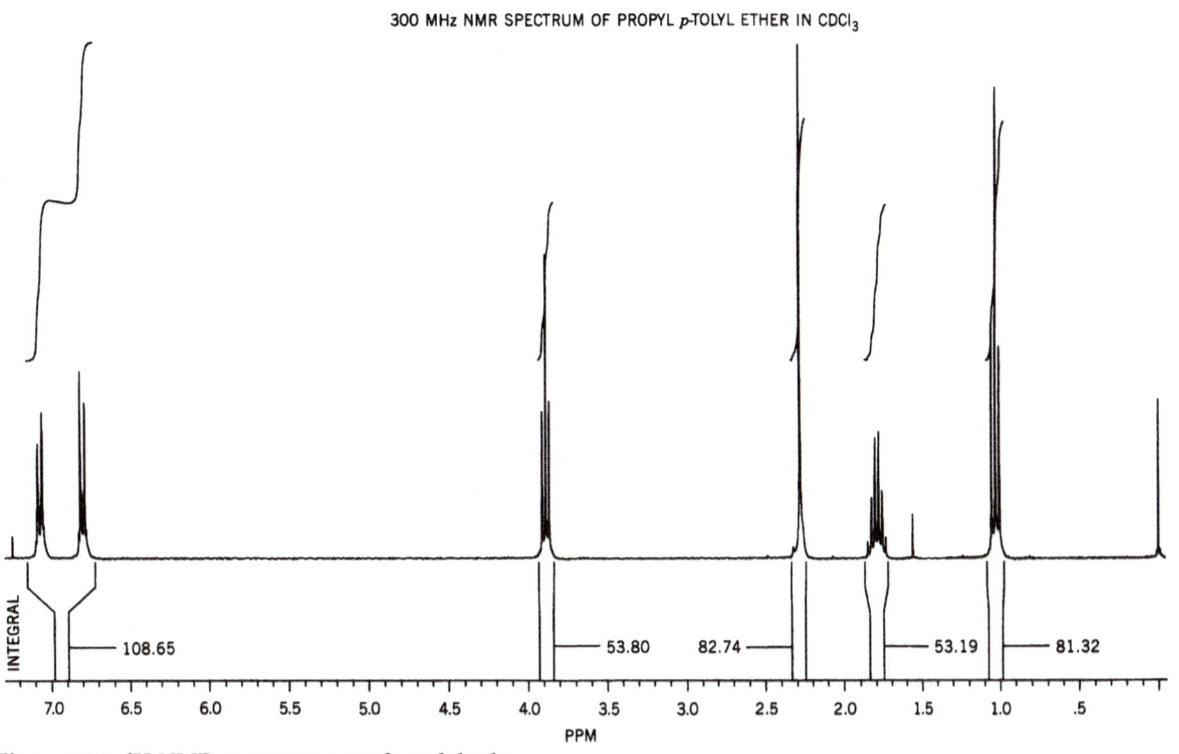


**Figure 6.37** $^1$H-NMR spectrum: propyl-*p*-tolyl ether.

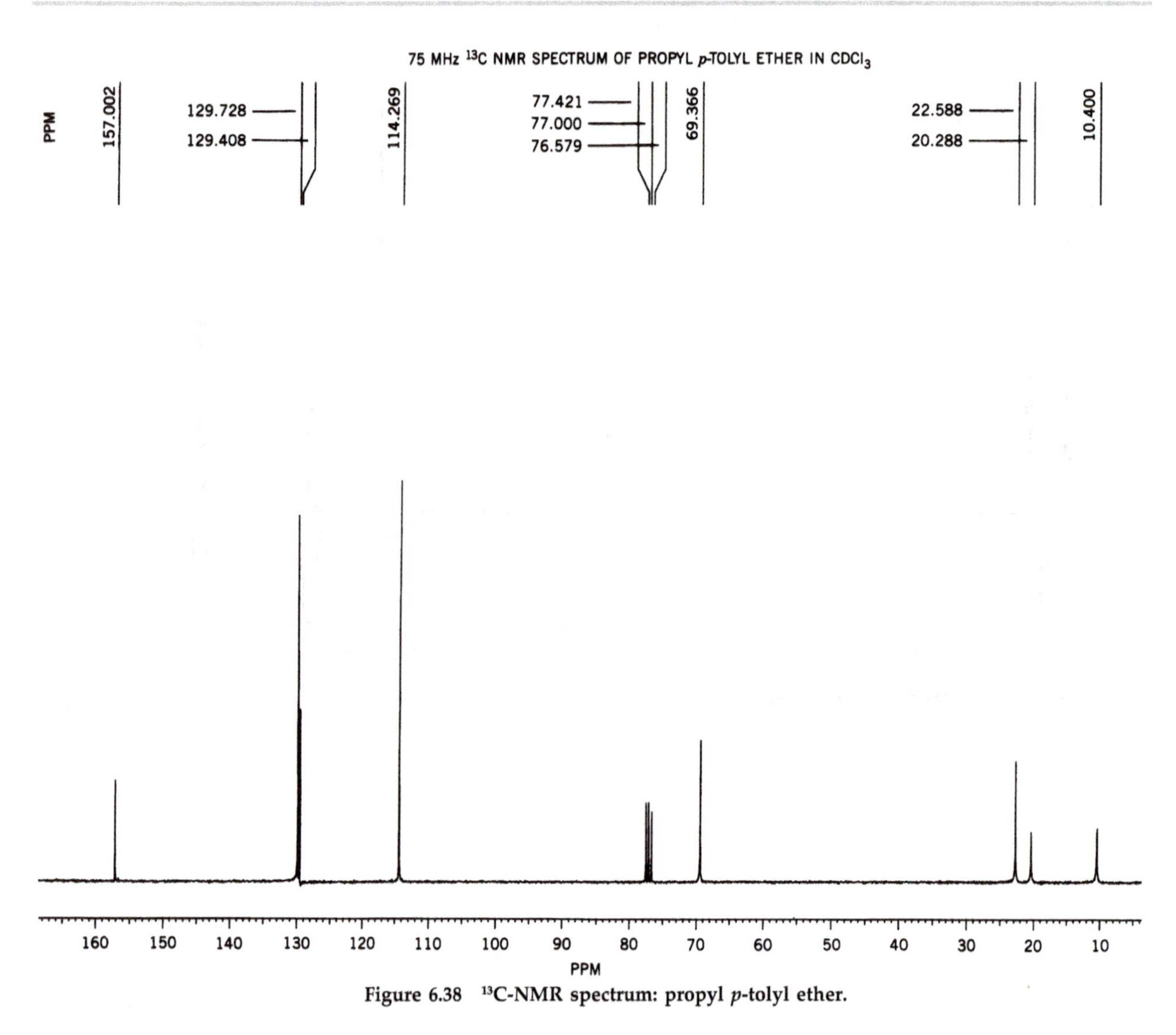


**Figure 6.38** $^{13}$**C-NMR spectrum: propyl** ***p*****-tolyl ether.**

# QUESTIONS

**6-21.** There are at least four potential nucleophiles in the reaction mixture. What are they? Why does the reaction work to give primarily one isolated product?

**6-22.** Based on your IR spectrum, is there any evidence for unreacted starting material?

**6-23.** What is the purpose of washing the product with 5% NaOH (aq)?

**6-24.** Could you make the same product using *p*-iodotoluene and propanol as shown below? Explain.

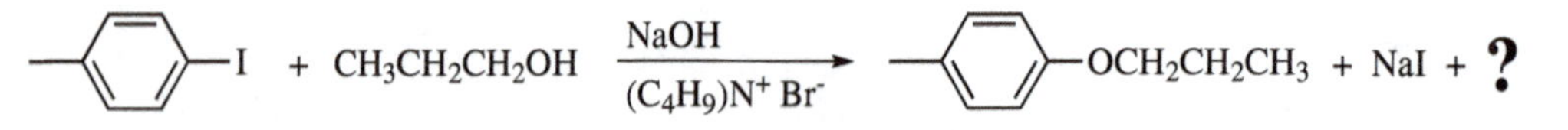

**6-25.** *trans*-2-Chlorocyclohexanol reacts readily with NaOH to form cyclohexene oxide, but the cis isomer will not undergo this reaction. Explain.

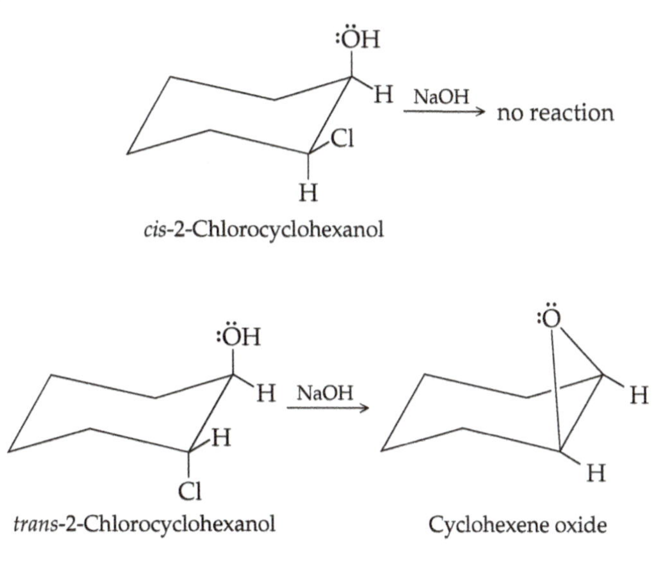


*cis*-2-Chlorocyclohexanol

*trans*-2-Chlorocyclohexanol

Cyclohexene oxide

## BIBLIOGRAPHY

**Selected review articles on the Williamson synthesis.**

Dermer, O. C. *Chem. Rev.* **1934,** *14,* 409.

Feuer, H.; Hooz, J. in *The Chemistry of the Ether Linkage;* Patai, S., Ed.; Wiley: NewYork, 1967, p. 446.

Smith, M. B.; March J. *Advanced Organic Chemistry,* 6th ed.; Wiley-Interscience: New York, 2007, Chap. 10, p. 529.

**Examples of the Williamson reaction in *Organic Syntheses* include**

Allen, C. F. H.; Gates, J. W., Jr. *Organic Syntheses;* Wiley: New York, 1955; Collect. Vol. III, p. 140. *ibid.,* p. 418.

Boehme, W. R. *Organic Syntheses;* Wiley: New York, 1963; Collect. Vol. IV, p. 590.

Fuson, R. C.; Wojcik, B. H. *Organic Syntheses;* Wiley: New York, 1943; Collect. Vol. II, p. 260.

Gassman, P. G.; Marshall, J. L. *Organic Syntheses;* Wiley New York, 1973; Collect. Vol. V, p. 424.

Gokel, G. W.; Cram, D. J.; Liotta, C. L.; Harris, H. P.; Cook, F. L. *Organic Syntheses;* Wiley: New York, 1988; Collect. Vol. VI, p. 301.

Kuryla, W. C.; Hyve, J. E. *Organic Syntheses;* Wiley: New York, 1973; Collect, Vol. V, p. 684.

Mirrington, R. N.; Feutrill, G. I. *Organic Syntheses;* Willey: New York, 1988; Collect. Vol. VI, p. 859.

Pedersen, C. J. *Organic Syntheses;* Wiley: New York, 1988; Collect. Vol. VI, p. 395.

Vyas, G. N.; Shah, N. M. *Organic Syntheses;* Wiley: New York, 1963; Collect. Vol. IV, p. 836.

**Review articles on phase transfer catalysis:**

Dehmlow, E. V.; Dehmlow, S. S. *Phase Transfer Catalysis,* 3rd ed.; VCH: New York, 1993.

Gokel, G. W.; Weber, W. P. *J. Chem. Educ.* **1978,** *55,* 350; *ibid.,* 429.

Smith, M. B.; March. J. *Advanced Organic Chemistry,* 6th ed.; Wiley-Interscience: New York, 2007, Chap. 10, p. 508.

**The procedures used in these experiments for the preparation of the ethers were adapted from the work of**

McKillop, A.; Fiaud, J. C.; Hug, R. P. *Tetrahedron* **1974,** *30,* 1379.

Rowe, J. E. *J. Chem. Educ.* **1980,** *57,* 162.

# Kinetics of an $S_N1$ Reaction: Solvolysis of t-Butyl Chloride

EXPERIMENT 3105-9

***Purpose.*** To measure the rate of the $S_N1$ reaction of *t*-butyl chloride (2-chloro-2-methylpropane) with water as the nucleophile and solvent.

*Prior Reading*

*Read the $S_N1$ section in your Organic Chemistry lecture textbook.*

## REACTION

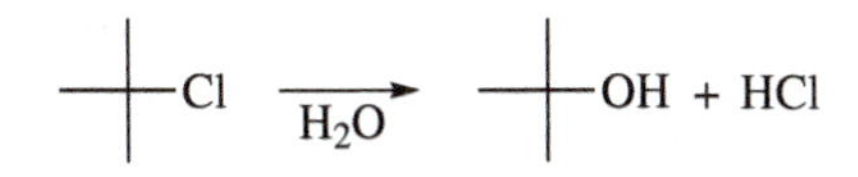


## DISCUSSION

The determination of reaction rates is important in industrial chemistry, because the ability to predict how long a reaction will take is important for optimizing yields. On a multi-ton scale, small improvements in yield can translate into significantly higher profits. Rates also provide clues to understanding the mechanism of a reaction.

The "rate of a reaction" means the rate at which a reactant is consumed (or product is formed) as a function of time. Thus:

$$\text{rate} = \Delta c/\Delta t = (c-c_0)/(t-t_0)$$

where c = the concentration of a reactant (or product) at time t
$c_0$ = the initial concentration
$t_0$ = the initial time

A first-order reaction has the simplest reaction kinetics, where the rate is directly proportional to the concentration of the reactant:

$$\text{rate (first order)} = -kc$$

The half-life of a reactant (the time necessary for half of the reactant to disappear) in a first order reaction is constant. After one half-life, half of the reactant will remain, and after two half-lives one fourth of the reactant will remain. Plots of concentration and ln(concentration) of reactant versus time (in half-lives) are shown below.

**An arbitrary example of a first order decay, plotting concentration vs. half-lives:**

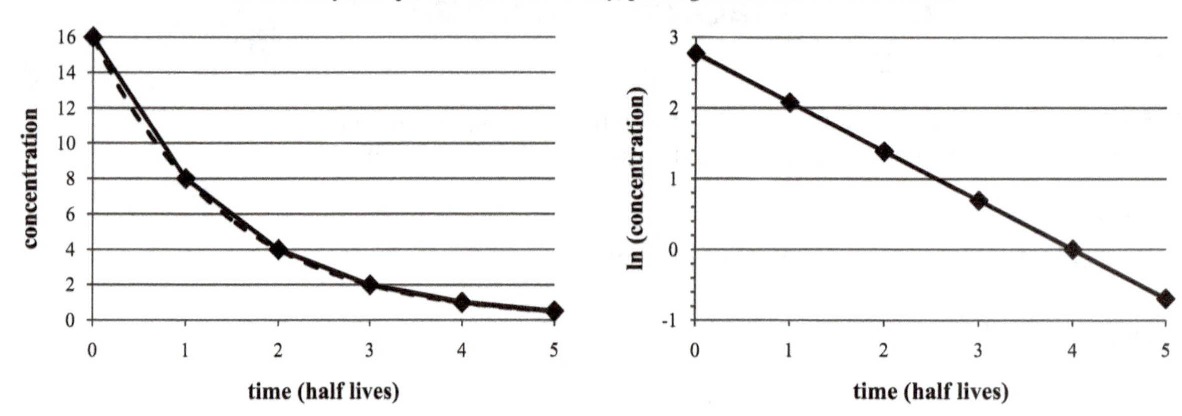


The symbols (◆) represent each half-life; they are connected by solid straight lines. The dashed line is an exponential fit to the data.

The same data, plotted as ln(concentration) *vs.* half-lives. Note that the ln plot gives a straight line. The slope of this line is -*k*, the rate constant.

***$S_N$1 reaction of t-butyl chloride.*** When *t*-butyl chloride (2-chloro-2-methylpropane) partially dissolves in water (a polar solvent), a small fraction of the molecules will ionize, because the 3° cation is stabilized by hyperconjugation, and the solvation energy of the ions in water partially compensates for the energy needed to break the C-Cl bond. The resulting carbocation reacts rapidly with water to give *t*-butyl alcohol, pulling the equilibrium for the first step to the right. The byproduct is HCl. Since the solvent is one of the reactants, this is called a **solvolysis** reaction.

$$(CH_3)_3C{-}Cl \xrightarrow[\text{(solvent)}]{H_2O} \left[(CH_3)_3C^{\oplus}\ Cl^-\right] \xrightarrow[\text{(reactant)}]{H_2O} \left[(CH_3)_3C{-}O^+H_2 + Cl^-\right] \longrightarrow (CH_3)_3C{-}OH + HCl$$

3° alkyl halide → 3° cation → 3° alcohol + HCl (used for analysis)

***Experimental Design.*** The goal of the experiment is to follow the disappearance of starting material (*t*-butyl chloride) over time. The quantitative analysis of the concentration of an alkyl halide is difficult, but one mole of HCl is produced for every mole of *t*-butyl chloride that reacts. Therefore acid/base chemistry can be used to follow the reaction.

A known amount of NaOH is added to the reaction mixture, along with an indicator, bromphenol blue. The NaOH added to the reaction is not involved in the $S_N$1 reaction. The $S_N$1 reaction will produce HCl and then neutralize the known amount of NaOH. Bromphenol blue changes from blue to yellow when the HCl that is produced has neutralized the NaOH. One mole of HCl will neutralize one mole of NaOH. As stated above, one mole of HCl is produced for every one mole of *t*-butyl chloride that reacts. The molar concentrations for the NaOH solution and the *t*-butyl chloride solution are equal to each other (0.1 M each). All three of these are directly comparable. Therefore, if one mole of NaOH is neutralized, then one mole of *t*-butyl chloride must have reacted to produce one mole of HCl.

Example:

If 0.60 mL of 0.10 M NaOH is neutralized by the HCl produced by the $S_N1$ reaction, then 0.60 mL of 0.10 M *t*-butyl chloride reacted. If you started with 3.00 mL of *t*-butyl chloride, then only 20% of the *t*-butyl chloride reacted (0.60 mL / 3.00 mL x 100% = 20%). At this time in the reaction, 80% of the initial amount of *t*-butyl chloride remains unreacted.

By measuring the time between adding the *t*-butyl chloride to water, and the color change, you are measuring the time for the number of moles of t-butyl chloride equal to the number of moles of NaOH that you added to react.

You will perform several experiments, starting with the same concentration of *t*-butyl chloride each time, and measuring the completion times with varying quantities of NaOH. This will give you the multiple data points you need to calculate the rate constant.

Reaction rates are often strongly dependent on temperature. In this lab, the temperature is kept constant by using a water bath.

**Bring to lab a stopwatch or a watch with a second hand.**

## EXPERIMENTAL PROCEDURE

| Physical Properties Table | | | |
|---|---|---|---|
| Compound | MW | Amount | mmol |
| 0.10 M *t*-butyl chloride in acetone | 92.7 | 1.50-3.00 mL | 0.15-0.30 |
| 0.10 M NaOH (aq) | 322.38 | 0.15-0.60 mL | 0.02-0.06 |
| bromphenol blue solution | | | |

1) Fill an 800 mL beaker with lukewarm water and put a thermometer in it. This water bath should sit on the bench during the prelab lecture to equilibrate to room temperature.

2) Equipment and chemicals:

- two 10.00 mL graduated pipettes
- two 25 mL Erlenmeyer flasks
- 0.10 M *t*-butyl chloride in acetone
- distilled water
- one digital pipette (150 μL)
- watch
- 0.1 M NaOH (aqueous)
- bromphenol blue solution

---

CAUTION: Sodium hydroxide is a strong base and can quickly burn your skin and eyes. Wear your safety goggles. Be particularly careful not to spill any. If you do get sodium hydroxide on your skin, wash it immediately with cold running water and notify your TA.

---

3) **Measure the volumes carefully.** The accuracy of your results depends on how accurately you measure the reagents. Use separate pipettes for each reagent. In one 25 mL flask, measure 3.0 mL of the *t*-butyl chloride solution. In the second, measure 0.60 mL of the NaOH solution and 6.40 mL of distilled water, and add 2 drops of the bromphenol blue solution.

NOTE: Use the digital pipette for measuring the NaOH solution (150 μL = 0.150 mL). Do not reset the delivery quantity of the digital pipettes.
To avoid contact between the NaOH and the mechanism, NEVER lay the digital pipettes down. ALWAYS return them to their holders.

Clamp both flasks in the water bath and allow them to equilibrate for 3 minutes. (Swirl occasionally.) Have one person mix the contents of the two flasks while the other starts the timing. Measure the time to the disappearance of the blue color. Recreate Table 3105-9 in your lab notebook and record all data in the table.

4) Clean the flasks, rinse with acetone and allow to dry. Then repeat the experiment, varying the amount of NaOH solution as below. Record your data in this table as well.

Table 3105-9: Time Trials for $S_N1$ Reactions

| Exp. | *t*-BuCl solution (mL) | NaOH solution (mL) | Water (mL) | % completion | % *t*-BuCl remaining | Time Trial 1 (sec) | Temp Trial 1 (° C) | Time Trial 2 (sec) | Temp Trial 2 (° C) |
|---|---|---|---|---|---|---|---|---|---|
| a | 3.00 | 0.60 | 6.40 | 20 | 80 | | | | |
| b | 3.00 | 0.45 | 6.55 | 15 | | | | | |
| c | 3.00 | 0.30 | 6.70 | 1 0 | | | | | |
| d | 3.00 | 0.15 | 6.85 | 5 | | | | | |

***Analysis.*** Plot your data for Experiments a-d as ln(% t-butyl chloride remaining) vs. time. A straight line indicates that our proposed equation for a first order reaction is correct. The slope of this line is -k, where k is the rate constant. Determine k, remembering that slope = $(y_2 - y_1) / (x_2 - x_1)$.

## QUESTIONS

**6-26.** What effect on the rate would you expect if you raised the temperature? What if you doubled the concentration of *t*-butyl chloride?

**6-27.** From what you know from lecture, another organic product is almost certainly formed in addition to t-butyl alcohol. What is this product? In the equation we used for the reaction we did not take this into account. Should we have done so?

**6-28.** a. The *t*-BuCl solution is in acetone. How would the rate of product formation change if you used more acetone and less water? (Hint: consider the nature of the solvents)

b. How would the rate of product formation change if you used only acetone and no water? (Hint: there are two steps in the overall reaction)

**6-29.** How would the rate change with 2-chloropropane instead of *t*-butyl chloride?

# The E2 Elimination Reaction: Dehydrohalogenation of 2-Bromoheptane to Yield 1-Heptene, trans-2-Heptene, and cis-2-Heptene

EXPERIMENT 3105-10

Common name: 1-heptene
CA number: [592-76-7]
CA name as indexed: 1-heptene

Common name: trans-2-heptene
CA number: [14686-13-6]
CA name as indexed: 2-heptene, (E)-

Common name: cis-2-heptene
CA number: [229-242-7]
CA name as indexed: 2-heptene, (Z)-

***Purpose.*** This reaction illustrates the base-induced dehydrohalogenation of alkyl halides with strong base and is used extensively for the preparation of alkenes. The stereo- and regiochemical effects of the size of the base is investigated, and the product mixture can be analyzed by the use of gas chromatography.

*Prior Reading*

*Technique 1:* Gas Chromatography (pp. 55–61)

## REACTION

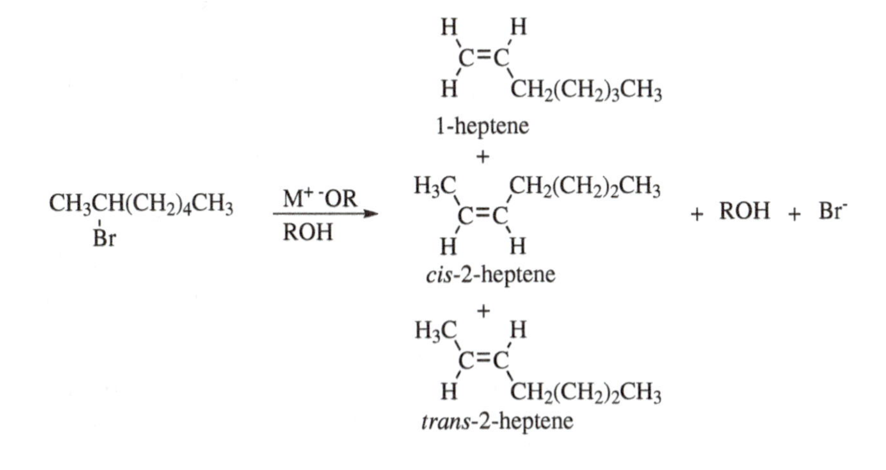


## DISCUSSION

Base-induced elimination (dehydrohalogenation) of alkyl halides is a general reaction and is an excellent method for preparing alkenes. This process is often referred to as β *elimination,* since a hydrogen atom is always removed β to the halide (leaving group):

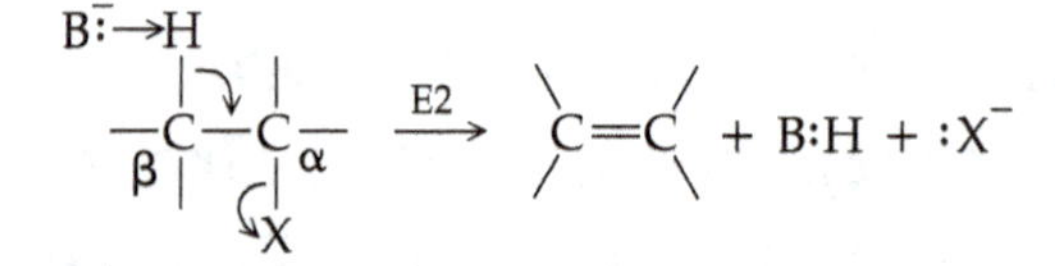

A high concentration of a strong base in a relatively nonpolar solvent is used to carry out the dehydrohalogenation reaction. Such combinations as sodium methoxide in methanol, sodium ethoxide in ethanol, potassium isopropoxide in isopropanol, and potassium *tert*-butoxide in *tert*-butanol or dimethyl sulfoxide (DMSO) are often used.

Elimination reactions almost always yield an isomeric mixture of alkenes, where this is possible. Under the reaction conditions, the elimination is *regioselective* and follows the Zaitsev rule when more than one route is available for the elimination of HX from an unsymmetrical alkyl halide. That is, the reaction proceeds in the direction that yields the most highly substituted alkene. For example,

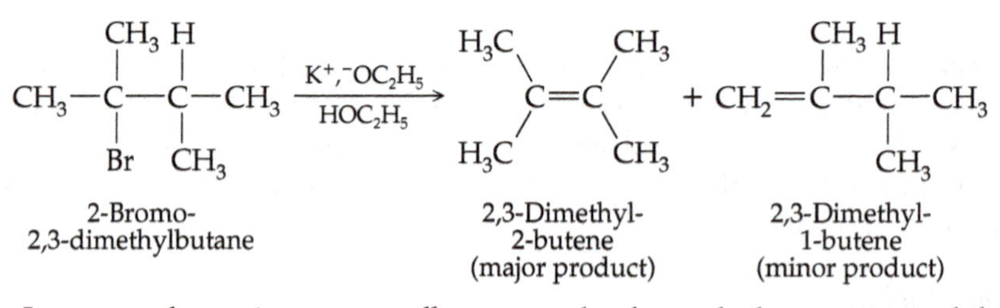

In cases where cis or trans alkenes can be formed, the reaction exhibits stereo-selectivity, and the more stable trans isomer is the major product.

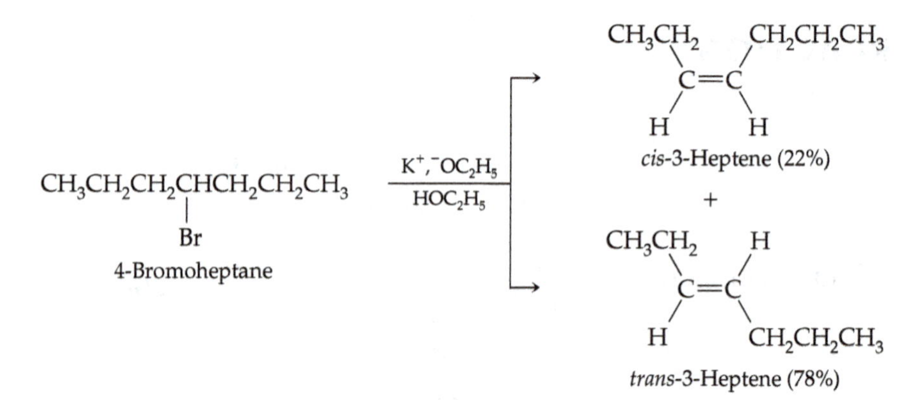

Experimental evidence indicates that the five atoms involved in the E2 elimination reaction must lie in the same plane; the anti-periplanar conformation is preferred. This conformation is necessary for the orbital overlap that must occur for the $\pi$ bond to be generated in the alkene. The $sp^3$-hybridized atomic orbitals on carbon that comprise the C—H and C—X $\sigma$ bonds broken in the reaction develop into the p orbitals comprising the $\pi$ bond of the alkene formed:

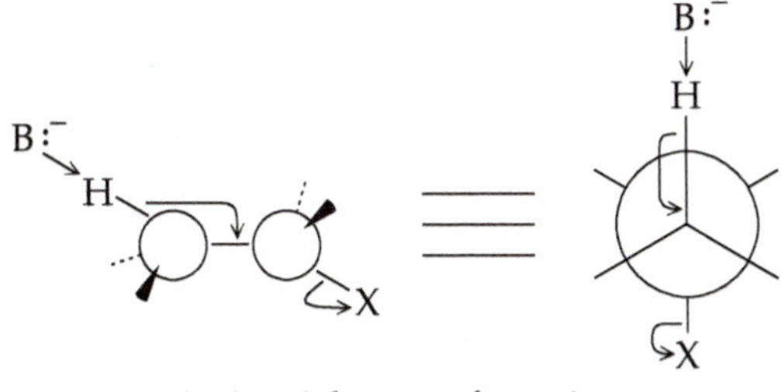

Anti-periplanar conformation

There is a smooth transition between reactant and product. Analogous to the $S_N2$ reaction, no intermediate has been isolated or detected. Furthermore, no rearrangements occur under E2 conditions. This situation is in marked contrast to E1 elimination reactions, where carbocation intermediates are generated and rearrangements are frequently observed (see Experiment 3105-11).

The alkyl halide adopts the anti-periplanar conformation in the transition state, and experimental evidence demonstrates that if the size of the base is increased, then it must be difficult for the large base to abstract an internal β-hydrogen atom. In such cases, the base removes a less hindered β-hydrogen atom, leading to a predominance of the thermodynamically **less stable (terminal) alkene** in the product mixture. This type of result is often referred to as **anti-Zaitsev** or **Hofmann elimination.** Thus, in the reaction of 2-bromo-2,3-dimethylbutane given above, the 2,3-dimethyl-1-butene would be the major product (anti-Zaitsev) if the conditions used a bulkier base. The anti-periplanar arrangements are illustrated in the Newman projections below.

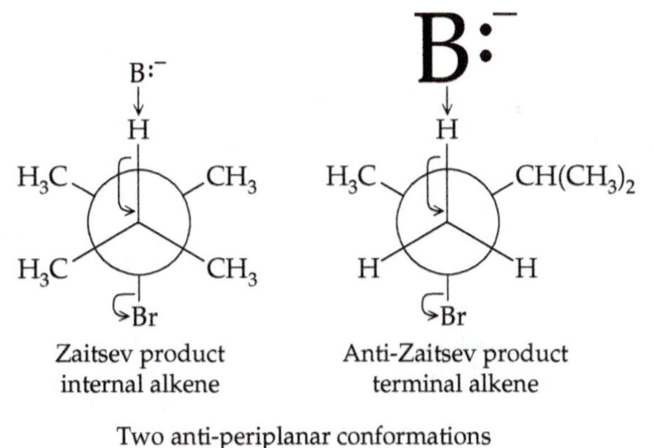

Two anti-periplanar conformations of 2-bromo-2,3-dimethylbutane

Dehydrohalogenation of alkyl halides in the presence of strong base (E2) is often accompanied by the formation of substitution ($S_N2$) products. The extent of the competitive substitution reaction depends on the structure of the alkyl halide. Primary alkyl halides give predominantly substitution products (the corresponding ether), secondary alkyl halides give predominantly elimination products, and tertiary alkyl halides give exclusively elimination products. For example, the reaction of 2-bromopropane with sodium ethoxide proceeds as follows:

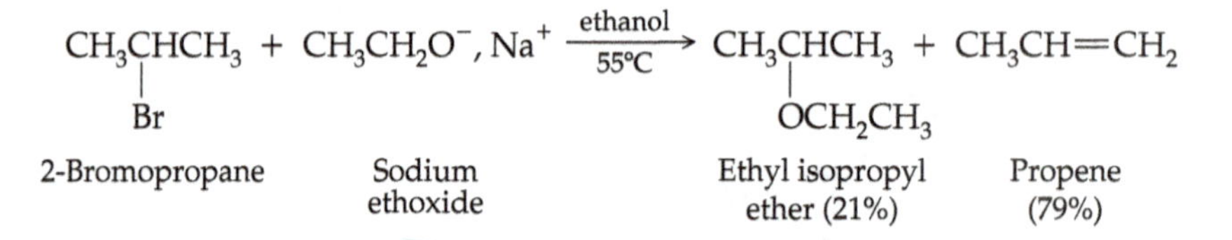


In general, for the reaction of alkyl halides with strong base,

1° 2° 3°

ease of $S_N2$ reaction ←

→ ease of E2 reaction

## EXPERIMENTAL PROCEDURE

Estimated time of experiment: 2.5 h.

**Physical Properties of Reactants**

| Compound | MW | Amount | mmol | bp (°C) | $d$ | $n_D$ |
|---|---|---|---|---|---|---|
| 2-Bromoheptane | 179.10 | 0.15 mL | 0.92 | 165-167 | 1.128 | |
| 25% (w) soln sodium methoxide | 53.98 | 1.10 mL | 5.09 | | | |
| Methanol | 32.04 | 1.90 mL | | 64.9 | 0.791 | 1.3288 |
| 2-Methyl-2-propanol (*tert*-butanol) | 74.12 | 3.0 mL | | 82–830 | .786 | 1.3838 |
| Potassium tert-butoxide | 112.21 | 0.56 g | 5.00 | 256-258 | | |

***Table 6.5*** **Reagent Combinations**

| Alcohol Solvent | Alkoxide Base Produced |
|---|---|
| Methanol | Sodium methoxide |
| 2-Methyl-2-propanol (*tert*-butanol) | Potassium 2-methyl-2-propoxide (potassium *tert*-butoxide) |

***Reagents and Equipment.*** Each group will be assigned one of the solvent/base combinations in Table 6.5. Students will compare results to observe a total picture of the effect. Sodium methoxide is a small base and potassium tert-butoxide is a bulky base.

***Reaction Conditions—Sodium Methoxide Reaction.*** Begin heating a hot plate to 100-110 °C. Measure and add to a 5.0-mL conical vial containing a magnetic spin vane 1.10 mL of a 25% (weight) solution of sodium methoxide in methanol. Add 1.90 mL of dry methanol and 0.15 mL of 2-bromoheptane and immediately attach the vial to a reflux condenser protected by a calcium chloride drying tube. Begin stirring and heat the reaction for 30 min.

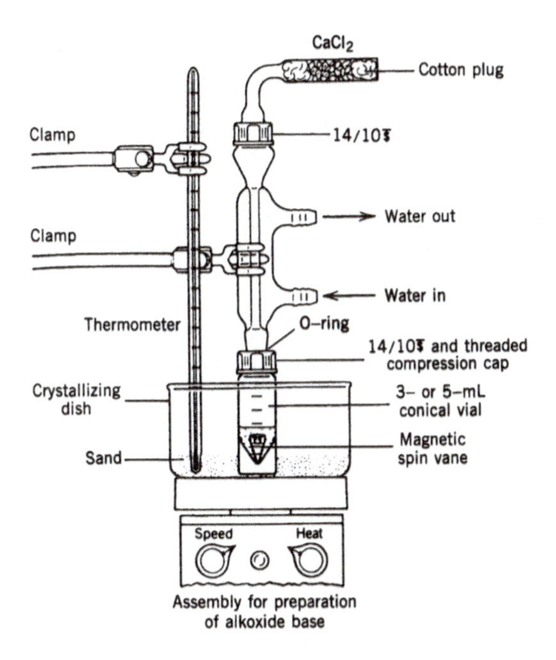

Assembly for preparation of alkoxide base

***Reaction Conditions - Potassium tert-Butoxide reaction*** Begin heating a hot plate to 140-150 °C. Measure and add to a 5.0-mL conical vial containing a magnetic spin vane 3.00 mL of dry 2-methyl-2-propanol (tert-butanol). Add 0.56 g of potassium tert-butoxide and 0.15 mL of 2-bromoheptane and immediately attach the vial to a reflux condenser protected by a calcium chloride drying tube. Begin stirring and heat the reaction for 30 min.

***Isolation of Product Mixture for Both Reactions.*** Cool the vial and contents to room temperature. Remove the spin vane using forceps and add 1 mL of pentane and 0.5 mL of water to the reaction vial. Transfer the mixture to a 12- or 15-mL centrifuge tube using a Pasteur pipet. Rinse the conical vial with a small portion of pentane and then water; transfer each rinse to the centrifuge tube. Cap the tube tightly and shake the mixture until the white solid dissolves. If needed, add additional drops of water to completely dissolve the solid. Separate the lower aqueous layer using a Pasteur pipet. For potassium tert-butoxide reactions, extract the remaining organic layer with 2 x 1.0 mL KOH and then two additional 1.0-mL portions of water (2 x 1 mL). For sodium methoxide reactions, extract the remaining organic layer with two additional 1.0-mL portions of water (2 x 1 mL). Each extraction is separated and transferred to the same flask. Finally, dry the organic layer in the centrifuge tube with magnesium sulfate. Transfer the dried product solution to a shell vial using a Pasteur filter pipet.

***Purification and Characterization.*** The collected product is analyzed by gas chromatography using a nonpolar column. (see Experiment 3105-10 Handout)

Assuming that the amount of each substance in the product mixture is proportional to the areas of its corresponding peak, determine the ratio of the three components in the sample.

The order of elution of the heptenes is 1-heptene, *trans*-2-heptene, and *cis*-2-heptene. Record the literature values of their physical properties. Compare your results with another group that used a different base/solvent combination. How does the nature of the different bases affect this E2 reaction? Answer this question in the conclusions section of your lab report.

## QUESTIONS

**6-30.** Outline a complete mechanistic sequence for the reaction of 2-bromoheptane with potassium 2-methyl-2-propoxide in 2-methyl-2-propanol solvent to form the three alkenes generated in the reaction (1-heptene, *trans*-2-heptene, and *cis*-2-heptene). Include a clear drawing of the anti-periplanar transition state for the formation of each alkene.

**6-31.** Predict the predominant alkene product that would form when 2-bromo-2-methylpentane is treated with sodium methoxide in methanol. If the base were changed to $KOC(CH_2CH_3)_3$ would the same alkene predominate? If not, why? What would be the structure of this alternate product, if it formed?

**6-32.** Predict the more stable alkene of each of the following pairs:

**(a)** 1-Hexene or *trans*-3-hexene

**(b)** *trans*-3-Hexene or *cis*-3-hexene

**(c)** 2-Methyl-2-hexene or 2,3-dimethyl-2-pentene

## BIBLIOGRAPHY

**Several dehydrohalogenation reactions of alkyl halides using alkoxide bases are given in *Organic Syntheses:***

Allen, C. F.; Kalm, M. J. *Organic Syntheses;* Wiley: New York, 1963; Collect. Vol. IV, p. 398.

McElvain, S. M.; Kundiger, D. *Organic Syntheses;* Wiley: New York, 1955; Collect. Vol. III, p. 506.

Paquette, L. A.; Barrett, J. H. *Organic Syntheses;* Wiley: New York, 1973; Collect. Vol. V, p. 467.

Schaefer, J. P.; Endres, L. *Organic Syntheses;* Wiley: New York, 1973; Collect. Vol. V, p. 285.

**For an overview of elimination reactions:**

March, J. *Advanced Organic Chemistry,* 4th ed.; Wiley: New York, 1992, p. 982.

Smith M. B.; March, J. *Advanced Organic Chemistry*, 6th ed.; Wiley: New York, 2007, Chap. 17.

**Experiment 3105-10 is adapted from the method given by:**

Leone, S. A.; Davis, J. D. *J. Chem. Educ.* **1992,** 69, A175.

# *The E1 Elimination Reaction: Dehydration of Cyclohexanol to Yield Cyclohexene*

EXPERIMENT 3105-11

Common name: cyclohexene
CA number: [110-83-8]
CA name as indexed: cyclohexene

***Purpose.*** To carry out an E1 elimination reaction, the dehydration of cyclohexanol, and to isolate and characterize the product of this reaction, cyclohexene.

## THE DEVELOPMENT OF CARBOCATION THEORY

The dehydration reaction that you are about to study is representative of the large collection of reactions that are classified as E1 elimination reactions. These reactions all form an intermediate in which one of the carbon atoms bears, if not a *full* positive charge, at least a significant fractional positive charge. It is this fleeting, high-energy intermediate, an aliphatic carbocation, that makes these reactions so interesting.

The development of bonding theory in organic chemistry during the late nineteenth and early twentieth centuries did not accept the existence of carbocations, except in a few esoteric instances. This position was reasonable because aliphatic compounds show little ionic character. The first proposal that these nonpolar substances might actually form cations came from Julius Stieglitz (1867–1937), of the University of Chicago, in a paper published in 1899. Eight years later, James F. Norris (1871–1940) at Massachusetts Institute of Technology produced compelling evidence that these substances might be intermediates in reactions of certain *tert*-butyl halides.

These two papers were the origin of the concept that organic carbon cations, which in those days became known as *carbonium ions*, were far more widespread in organic reactions than previously anticipated. Indeed, these early investigations caused more controversy and more experimental work to unambiguously prove the existence of what are now called *carbocations* than any other single problem in American chemistry.

It was, however, the English chemists Arthur Lapworth, Sir Robert Robinson, C. K. Ingold (University of London), and E. D. Hughes who, from 1920 to 1940, undertook a massive effort to develop the experimental and theoretical data to place these early postulates on a solid scientific foundation. Between 1920 and 1922, Hans Meerwein in Bonn, Germany, demonstrated that carbon rearrangements in the camphene series could be best explained by postulating the presence of carbocation intermediates. Perhaps the most important contribution to the entire subject was published in 1932 by Frank C. Whitmore of Pennsylvania State College, where he was Dean of the School of Chemistry and Physics. His paper, "The Common Basis of Intramolecular Rearrangements," brought together a vast array of data in a beautifully consistent interpretation that essentially cemented the carbocation into contemporary organic chemical theory.

A reaction not too distant from the one that is to be carried out below, but one that also included a rearrangement along with the dehydration, was studied by Dorothy Bateman and C. S. "Speed" Marvel at the University of Illinois as early as 1927.

Within 2 years of the publication of the Whitmore paper, Robinson proposed the formation of the steroids (including cholesterol) from squalene (a $C_{30}$ polyunsaturated polyisoprene molecule) via an incredible series of intermediates and rearrangements. Later, following the elucidation of the structure of lanosterol, R. B. Woodward and K. Bloch made a brilliant proposal that at once rationalized the biosynthetic origin of both lanosterol and cholesterol and implicated lanosterol as an intermediate in cholesterol biosynthesis. Their mechanism involved the concerted (bonds made and broken simultaneously) cyclization of four rings, as well as four rearrangements following the generation of the initial carbocation intermediate, to ultimately yield lanosterol.[9]

The elucidation of these biochemical pathways is further evidence of the major impact that our understanding of carbocation chemistry has had on related fields, such as biochemistry.

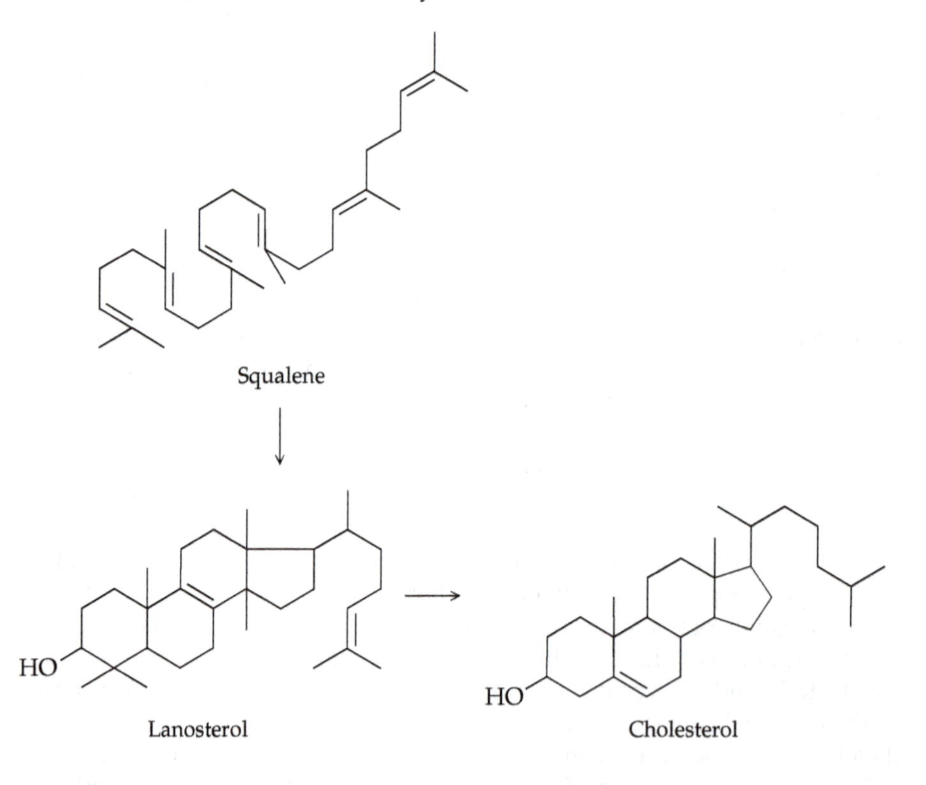


[9]Stieglitz, J. *Am. Chem. J.* **1899,** *21,* 101. Norris, J. F. *Am. Chem. J.* **1907,** *38,* 627. Meerwein, H.; van Emster, K. *Berichte* **1922,** *55,* 2500. Whitmore, F. C. *J. Am. Chem. Soc.* **1932,** *54,* 3274. Bateman, D. E.; Marvel, C. S. *J. Am. Chem. Soc.* **1927,** *49,* 2914. Robinson, R. *Chem. Ind.* **1934,** *53,* 1062. Woodward, R. B.; Bloch, K. *J. Am. Chem. Soc.* **1953,** *75,* 2023. See also: Tarbell, D. S.; Tarbell, T. *The History of Organic Chemistry in the United States, 1875–1955;* Folio: Nashville, TN, 1986.

# REACTION

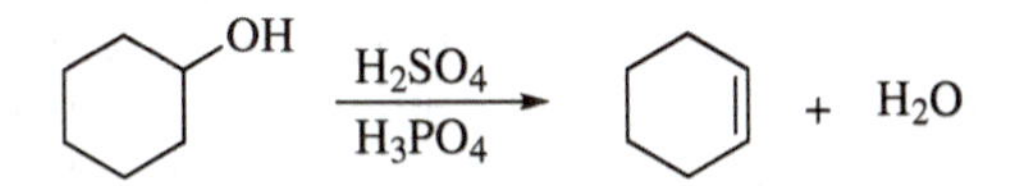


# DISCUSSION

The formation of an alkene (or alkyne) frequently involves loss of a proton and a leaving group from adjacent carbon atoms. The generalized reaction scheme is shown below.

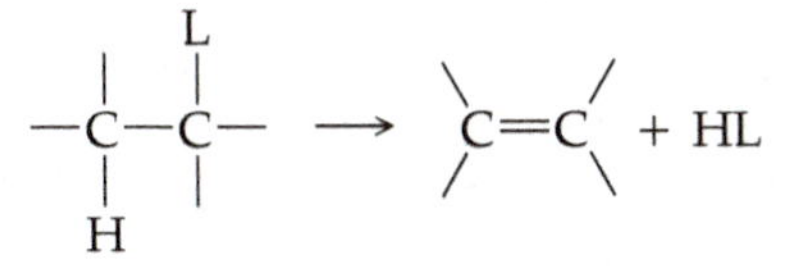


Such a reaction is called an *elimination reaction,* because a small molecule (HL) is *eliminated* from the organic molecule. If the molecule eliminated is water, the reaction may also be referred to as a *dehydration*. One of the common synthetic routes to alkenes is the dehydration of an alcohol.

Dehydration of alcohols is an acid-catalyzed elimination reaction. Experimental evidence shows that alcohols react in the order tertiary (3°) > secondary (2°) > primary (1°); this reactivity relates directly to the stability of the carbocation intermediate formed in the reaction. Generally, sulfuric or phosphoric acid is used as the catalyst in the research laboratory. A Lewis acid, such as aluminum oxide or silica gel, is usually the catalyst of choice at the fairly high temperatures used in industrial scale reactions.

The mechanism for this reaction is classified as E1 (elimination, unimolecular). The elimination, or dehydration reaction, proceeds in several steps:

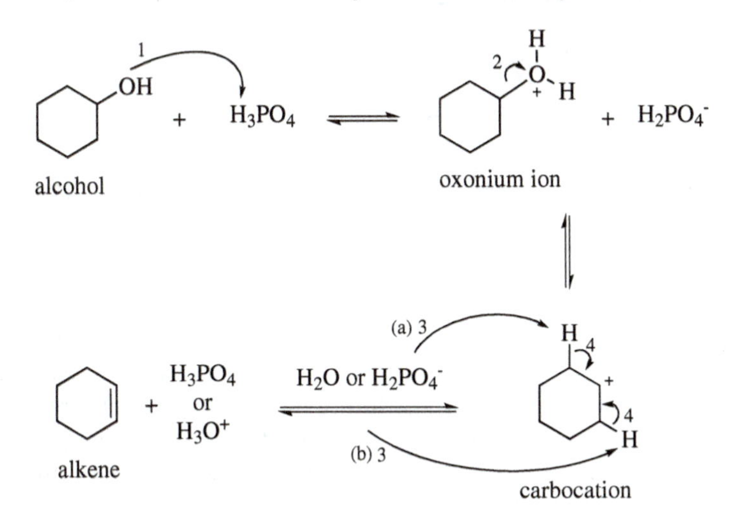


notice the carbocation is symmetric,
either route (a or b) will lead to the same product

The first step (1) involves the very rapid, though reversible, protonation of the oxygen atom of the alcohol to form an *oxonium ion* (an oxygen cation with a full octet of electrons). This protonation step is important because it produces a good leaving group, water. Without acid, the only available leaving group is hydroxide ion ($HO^-$) which, as a strong base, is a poor leaving group. This finding is the reason why acid plays such an important role in the mechanism of this dehydration. The second step (2) of the reaction is the dissociation of the oxonium ion to form an intermediate *carbocation* and water. This step is the rate-determining (and therefore the slowest) step of the reaction. In the third step (3), the carbocation is deprotonated by a ubiquitous water molecule (or other base, such as biphosphate ion, present in the system) in another rapid equilibration. The carbocation gains stability (lower energy) by releasing a proton ($H^+$) from a carbon atom adjacent ($\alpha$) to the carbocation (route (a) or (b) shown) to the attacking base. Thus, in step 4 the catalyst is regenerated as a protonated molecule of water ($H_3O^+$), and the electron pair, previously comprising the C—H bond adjacent to the carbocation, flows toward the positive charge, generating a stable and neutral alkene. For asymmetric reactants, a variety of isomeric products may be formed, since different protons adjacent to the carbocation may be removed, and different conformations of the intermediate carbocation are also possible.

E1 elimination reactions, as we have just seen, involve equilibrium conditions and, thus, to maximize the yield (drive the reaction to completion), the alkene is usually removed from the reaction while it is in progress. A convenient technique for accomplishing this task is distillation, which is often used because alkenes *always* have a lower boiling point than the corresponding alcohol.

Many 1° (primary) alcohols also undergo dehydration, but usually by a different route, the E2 (elimination, bimolecular) mechanism. This step is governed mainly by the fact that the 1° carbocation that would be required in an E1 process is a relatively unstable (very high energy) intermediate. In this case, attack by the base occurs directly on the oxonium ion:

$$CH_3CH_2CH_2\ddot{O}H \xrightleftharpoons{H_2SO_4} CH_3CH_2CH_2\overset{+}{O}H_2 + HSO_4^-$$

Oxonium ion

$$HSO_4^- + CH_3-\underset{H}{\overset{H}{C}}-CH_2-\overset{+}{\ddot{O}}H_2 \rightleftharpoons CH_3CH{=}CH_2 + H_2SO_4 + H_2O$$

Oxonium ion

E1 elimination is usually accompanied by a competing $S_N1$ substitution reaction that involves the *same* carbocation intermediate. Since both reaction mechanisms are reversible in this case, and since the alkene products are easily removed from the reaction, the competing substitution reaction (which predominantly regenerates the starting alcohol by attack of water, as a nucleophile, at the carbocation) is not troublesome.

Many alcohols can dehydrate to yield more than one isomeric alkene. In the reaction below involving the dehydration of 2-butanol, at least three alkenes are usually formed—1-butene, *trans*-2-butene, and *cis*-2-butene:

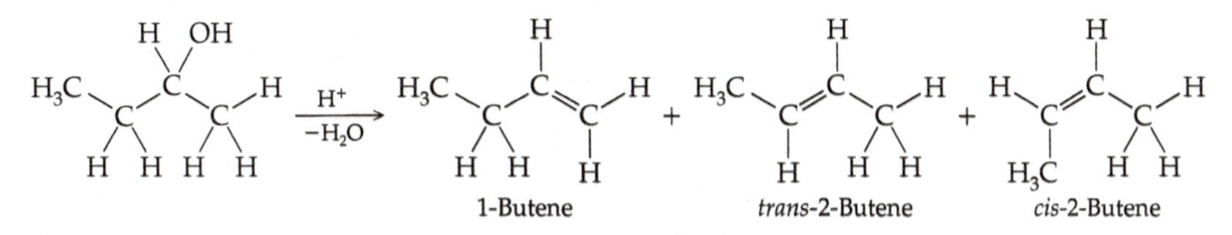


We would expect the alkene generated in the largest amount to be the one possessing the highest degree of substitution, since it is the most stable product (lowest energy). This is exactly what is observed: more than 90% of the products are isomers of 2-butene. Because trans alkenes are thermodynamically more stable than their cis counterparts, and since the reaction is reversible, one might expect the dominant isomer in the 2-butene mixture to be trans. Indeed, nearly twice as much trans as cis isomer is formed. An empirical rule, originally formulated by the Russian chemist Alexander Zaitsev (or Saytzeff), for base-catalyzed E2 eliminations states that the alkene with the largest number of alkyl substituents on the double bond will be the major product. This rule can also correctly predict the relative ratio of substituted alkenes to be expected from a given E1 elimination reaction, and it obviously applies in the case of 2-butanol. The present experiment uses cyclohexanol, which forms a symmetrical intermediate carbocation, and therefore undergoes no rearrangement and can generally form only one C=C (in other words, we don't worry about Zaitsev's rule here).

Rearrangements of alkyl groups (such as methide, $^{-}:CH_3$ and ethide, $^{-}:CH_2CH_3$) and hydrogen (as hydride, $:H^{-}$) are often observed during the dehydration of alcohols, especially in the presence of very strong acid where carbocations can exist for longer periods of time. For example, when 3,3-dimethyl-2-butanol is treated with sulfuric acid, the elimination reaction yields the mixture of alkenes shown below. Can you predict which alkene is the principal product?

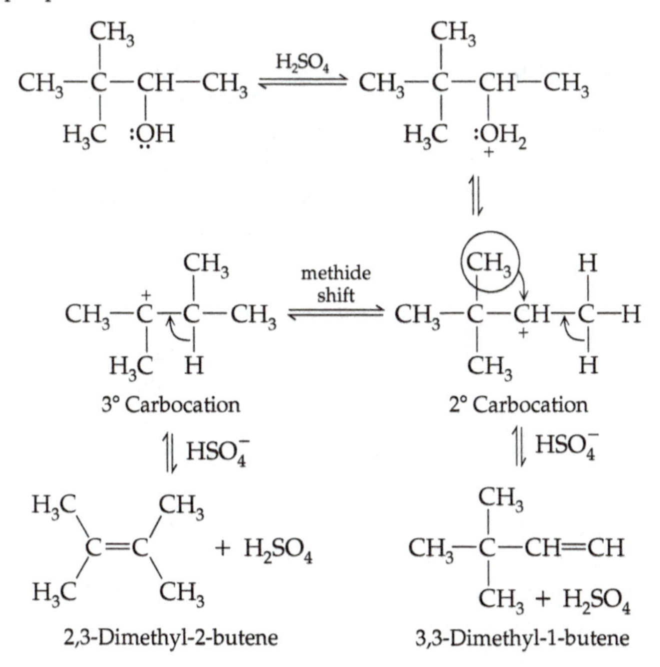

Reactions carried out under these conditions are very susceptible to alkide or hydride shifts if a more stable carbocation intermediate can be formed. In the example above, a 2° carbocation rearranges into the more stable 3° carbocation. This intramolecular rearrangement involves the transfer of an entire alkyl group (in this case a methyl substituent), together with its bonding pair of electrons, to an adjacent carbon atom. This migration, or shift, of the methyl group to an adjacent carbocation is called a *1,2-methide shift*. It commonly occurs in aliphatic systems involving carbocation intermediates that have alkyl substituents adjacent to the cation. Hydride shifts are also frequently observed and actually appear in a wider range of molecules.

Phosphoric acid is often used as the protonating acid even though it is a weak acid. Sulfuric acid used alone can lead to excess polymerization, tar formation, and evolution of sulfur dioxide when heated too strongly. Concentrated acid is used to minimize the presence of water since this is a reversible reaction. The product will also be removed by distillation as it is formed to drive the reaction toward completion by Le Chatlier's principle.

You will also use a qualitative test for the presence of unsaturation to confirm the presence of the alkene in the product. The test uses molecular bromine ($Br_2$) in $CCl_4$ or $CH_2Cl_2$ as a test for unsaturation since bromine will add across the double bond and its characteristic deep orange color will disappear.

## EXPERIMENTAL PROCEDURE

**Physical Properties of Reactants**

| Compound | MW | Amount | mmol | bp (°C) | *d* |
|---|---|---|---|---|---|
| Cyclochexanol | 100.6 | 8.0 mL | 7.7 | 160.8 | 0.96 |
| Concd sulfuric acid | 98.08 | 5-8 drops | | | |
| Concd phosphoric acid | 98.00 | 2.5 mL | | | |

***Reagents and Equipment.*** Place 8.0 mL of cyclohexanol (d = 0.96 g/mL) in a 50 mL round bottom flask. Your TA will dispense 2.5 mL of concentrated phosphoric acid and 5-8 drops of concentrated sulfuric acid (CAUTION) into your reaction flask. Mix the solution well, add boiling chips, and set up a simple distillation apparatus with the 50 mL flask as the reaction flask and a 25 mL round bottom flask as the receiving flask immersed in an ice bath (use the diagram on page 62 as a guide).

CAUTION: Bromine is toxic. Concentrated phosphoric acid and sulfuric acid are strong, corrosive materials. They can quickly burn your skin and eyes. Wear your safety goggles. Be particularly careful not to spill any. If you do get any of these chemicals on your skin, wash it immediately with cold running water and notify your TA.

***Reaction Conditions.*** Heat the reaction mixture carefully (you are allowing the alcohol to react and distilling the product away at the same time). Keep the temperature of the distilling vapor below 102°C. The distillate will be a mixture of cyclohexene and water. You must be patient in allowing the reaction/distillation to proceed. Stop the distillation when most of the mixture has been distilled (at the proper temperature). Never heat a distillation flask to dryness.

***Isolation of Product.*** Pour the collected distillate into a small separatory funnel. Add 2 mL of saturated sodium chloride solution to the funnel and extract. Drain off the aqueous (lower) layer. Then wash the remaining organic layer with 3 mL of 0.5 M sodium carbonate solution (swirl carefully before capping as this could generate pressure due to carbon dioxide evolution). Drain off the lower aqueous layer and test with litmus paper to make sure it is not acidic. If it is acidic, repeat the sodium carbonate wash of the organic layer. If it is not acidic, pour the organic layer into a beaker, and add magnesium sulfate drying agent.

---

CAUTION: If water is drawn back into hot concentrated sulfuric acid, a very dangerous situation can occur, particularly if larger quantities of the reagents than recommended above are used in the experiment.

---

***Purification and Characterization.*** While your crude product is drying, rinse the distillation apparatus with water, then acetone, and dry it out. Decant the crude product away from the drying agent into a clean 25 mL round bottom flask, add boiling chips, and set the distillation apparatus back up as before, again with the (preweighed) receiving flask in an ice bath. Distill the product, collecting distillate that comes over between 78-84°C. Determine the mass of your product and calculate percent yield (density of cyclohexene: 0.81 g/mL). Obtain an IR of your product.

Place 4 drops of the starting material (cyclohexanol) in a shell vial and 4 drops of your cyclohexene product in another shell vial. To each vial, add bromine (in either $CCl_4$ or $CH_2Cl_2$) dropwise until the original color persists (that is, the color remains rather than disappearing). Count how many drops this takes for both starting material and product.

***Waste Cleanup.*** To the original reaction flask, add 15 mL of (tap) water, then add solid sodium carbonate a little at a time to neutralize the remaining acid (it will foam). You may then wash it down the sink (preferably in the hood) with water. The distillation flask from the final distillation should be rinsed with acetone in the hood, with all waste going into the waste jar. The residue from the bromine tests for unsaturation should also be placed in the halogenated waste jar in the hood. After you have weighed and tested your final product, you may pour it into the waste jar also, again rinsing with acetone in the hood.

## QUESTIONS

**6-33.** What major IR absorption would disappear in going from reactant to product in this reaction?

**6-34.** What new IR absorptions would you see?

**6-35.** When *tert*-pentyl bromide is treated with 80% ethanol, the following amounts of alkene products are detected on analysis:

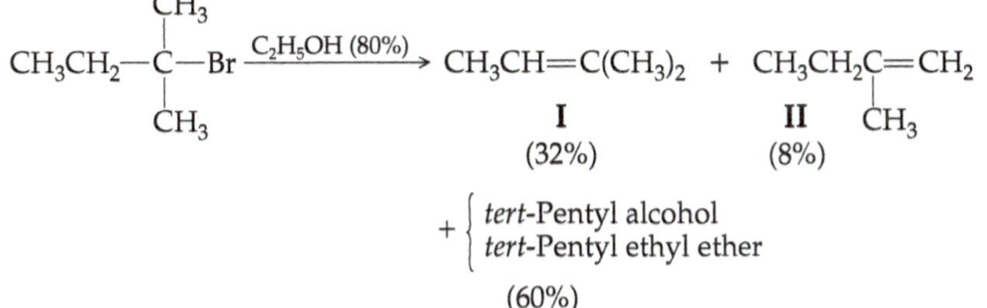


Offer an explanation of why compound **I** is formed in far greater amount than the terminal alkene.

**6-36.** The —$^{+}SR_2$ group is easily removed in elimination reactions, but the —SR group is not. Explain.

**6-37.** Why are sulfuric or phosphoric acids, rather than hydrochloric acid, used to catalyze the dehydration of alcohols?

## BIBLIOGRAPHY

**Several dehydration reactions of secondary alcohols using sulfuric acid as the catalyst are given in *Organic Syntheses:***

Adkins, H.; Zartman, W. *Organic Syntheses;* Wiley: New York, 1943; Collect. Vol. II, p. 606.

Bruce, W. F. *Organic Syntheses;* Wiley: New York, 1943; Collect. Vol. II, p. 12.

Coleman, G. H.; Johnstone, H. F. *Organic Syntheses;* Wiley: New York, 1941; Collect.Vol. I, p. 183.

Grummitt, O.; Becker, E. I. *Organic Syntheses;* Wiley: New York, 1963; Collect. Vol. IV, p. 771.

Norris J. F. *Organic Syntheses;* Wiley: New York, 1941; Collect. Vol. I, p. 430.

Wiley, R. H.; Waddey, W. E. *Organic Syntheses;* Wiley: New York, 1955; Collect. Vol. III, p. 560.

**For an overview of elimination reactions:**

March, J. *Advanced Organic Chemistry,* 4th ed.; Wiley: New York, 1992, p. 982.

Smith, M. B.; March, J. *Advanced Organic Chemistry,* 6th ed.; Wiley: New York, 2007, Chap 17.

**For details on carbonium ion formation:**

Olah, G. A.; Schleyer, P. von R., Eds. *Carbonium Ions;* Wiley: New York, Vol. 1, 1968; Vol. II, 1970; Vol. III, 1972.

Olah, G. A.; Prakash, G. K. S. *Carbocation Chemistry;* Wiley: New York, 2004.

# Photochemical Isomerization of an Alkene: cis-1,2-Dibenzoylethylene

EXPERIMENT 3105-12

Common names: *cis*-1,2-dibenzoylethylene
*cis*-1,4-diphenyl-2-butene-1,4-dione
CA number: [959-27-3]
CA name as indexed: 2-butene-1,4-dione, 1,4-diphenyl-, (Z)-

***Purpose.*** This exercise illustrates the ease of cis–trans isomerization in organic molecules and, specifically in this case, demonstrates the isomerization of a trans alkene to the corresponding cis isomer via photochemical excitation.

## BIOLOGICALLY IMPORTANT PHOTOCHEMICAL REACTIONS

A number of important biochemical reactions are promoted by the adsorption of UV-vis radiation.Vitamin $D_3$, which regulates calcium deposition in bones, is biosynthesized in just such a photochemical reaction. This vitamin is formed when the provitamin, 7-dehydrocholesterol, is carried through fine blood capillaries just beneath the surface of the skin and exposed to sunlight. The amount of radiation exposure, which is critical for the regulation of the concentration of this vitamin in the blood stream, is controlled by skin pigmentation and geographic latitude. Thus, the color of human skin is an evolutionary response to control the formation of vitamin $D_3$ via a photochemical reaction.

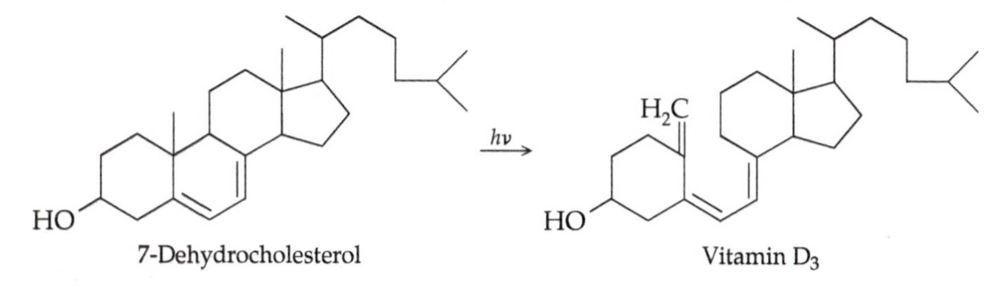


Another set of significant photochemical reactions in human biochemistry is contained in the chemistry of vision. These reactions involve vitamin $A_1$ (retinol), which is a $C_{20}$ compound belonging to a class of compounds known as diterpenes. These compounds are molecules formally constructed by the biopolymerization of four isoprene, $CH_2{=}C(CH_3){-}CH{=}CH_2$, molecules. Retinol (an all-trans pentaene) is first oxidized via liver enzymes (biological catalysts) to vitamin A aldehyde (*trans*-retinal). The *trans*-retinal, which is present in the light-sensitive cells (the retina) of the eye, undergoes further enzymatic transformation (retinal isomerase) to give *cis*-retinal (a second form of vitamin A aldehyde) in which one of the double bonds of the all-trans compound is isomerized.

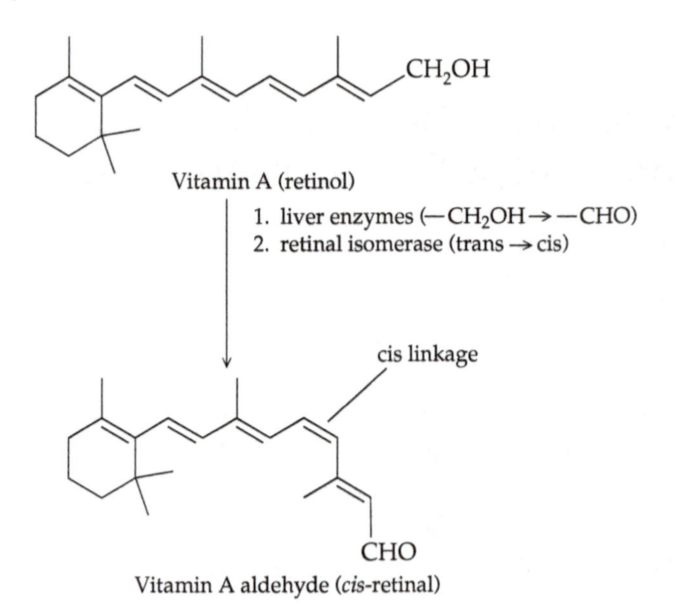


The cis isomer of vitamin A aldehyde (retinal) possesses exactly the correct dimensions to become coupled to opsin, a large protein molecule (MW ~38,000) (coupling involves a reaction of the retinal aldehyde group, —C(H)=O, with an amine group [$-NH_2$] of the protein to form an imine linkage [RCH=NR]), to generate a light-sensitive substance, rhodopsin. This material is located in the rodlike structures of the retina. When the protonated form of rhodopsin ($-CH=N^{+}HR$) which absorbs in the blue-green region of the visible spectrum near 500 nm, is exposed to radiation of this wavelength, isomerization of the lone cis double bond of the diterpene group occurs and *trans*-rhodopsin is formed:

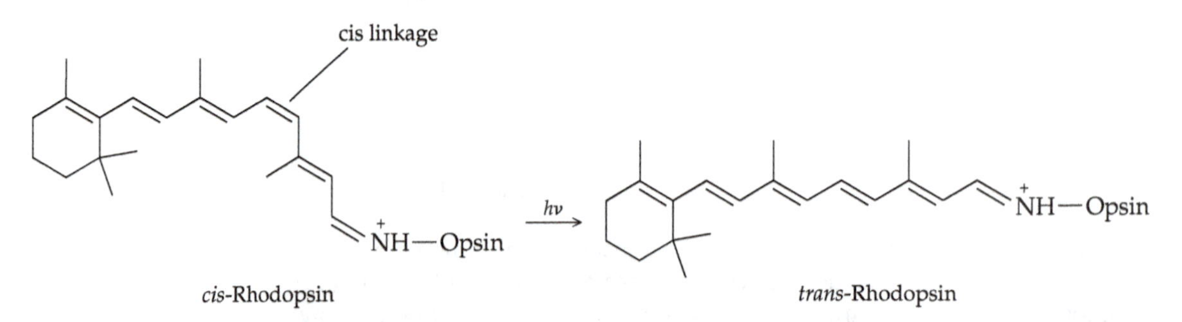


This photoreaction (a fast reaction, $10^{-12}$ s) involves a significant change in the geometry of the diterpene group, which eventually ($10^{-9}$ s) results in both a nerve impulse and the separation of *trans*-retinal from the opsin. The trans isomer is then enzymatically reisomerized back to the cis compound, which then starts the initial step of the visual cycle over again.

There are two interesting facts about this reaction: (1) This reaction is incredibly sensitive. A single photon will cause the visual nerve to fire. (2) All known visual systems use *cis*-retinal, regardless of their evolutionary trail.

The photoreaction that we study next is very similar to the cis–trans double-bond isomerism found in the vitamin A visual pigments. The only difference is that in our case we will be photochemically converting a trans double-bond isomer to the cis isomer.

*Prior Reading*

*Technique 5:* Crystallization
Introduction (pp. 85–87)
Use of the Hirsch Funnel (pp. 88–89)

*Technique 6A:* Thin-Layer Chromatography (pp. 97–99)

## REACTION

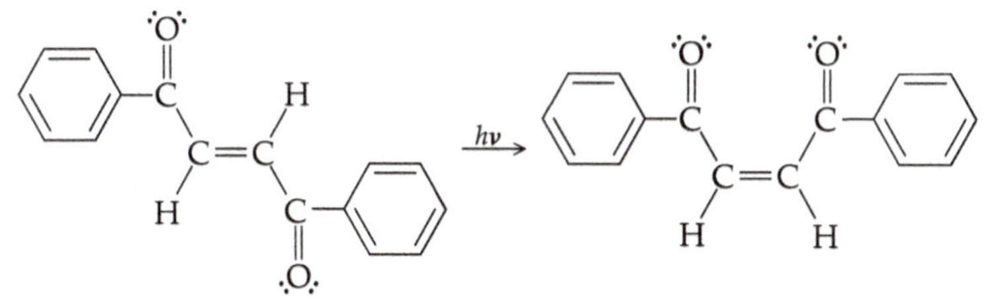


*trans*-1,2-Dibenzoylethylene    *cis*-1,2-Dibenzoylethylene

## DISCUSSION

***Cis-Trans Isomerizations.*** Experiment 3105-12 studies the photochromic properties of *trans*-dibenzoylethylene. In this case, the highly conjugated bright-yellow trans diastereomer is rapidly isomerized under intense sunlamp visible radiation, via excited electronic states, to the colorless cis alkene. A $\pi$ electron is promoted to an anti-bonding $\pi^*$ molecular orbital, which destroys the $\pi$ bond, and thus permits facile rotation about the remaining $\sigma$ bond and formation of the cis alkene:

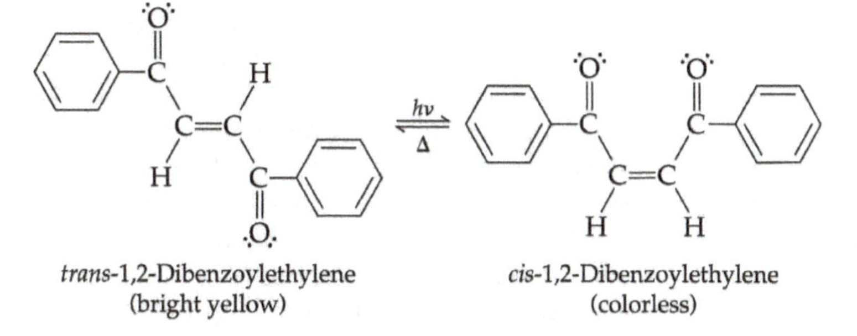


*trans*-1,2-Dibenzoylethylene (bright yellow)    *cis*-1,2-Dibenzoylethylene (colorless)

This is an endothermic reaction that does not reach equilibrium, but goes to completion because the cis isomer's electronic transitions are shifted to the ultraviolet and do not absorb visible radiation. The cis alkene is a structurally shorter chromophore than its trans isomer, and it is also likely to experience steric crowding with resultant distortion of the $\pi$ system. The cis isomer, therefore, absorbs at shorter wavelengths (higher energies) and has a lower molecular extinction coefficient (weaker) than the trans isomer. Thus, once formed, the cis isomer is trapped. Upon heating, however, the cis isomer undergoes exothermic isomerization back to the original, more stable, trans alkene under equilibrium conditions.

A good example of photochromic behavior is the highly colored *cis-* and *trans*-azobenzene. In this case the $\pi \rightarrow \pi^*$ transition is promoted by ultraviolet light so that nonequilibrium isomerization to the cis isomer requires UV irradiation. The cis isomer is considerably less stable, however, and it undergoes relatively easy reversion back to the trans isomer by other mechanisms (the thermal conversion of visible radiation absorbed by the colored cis compound is an additional isomerization route apparently open to the cis isomer):

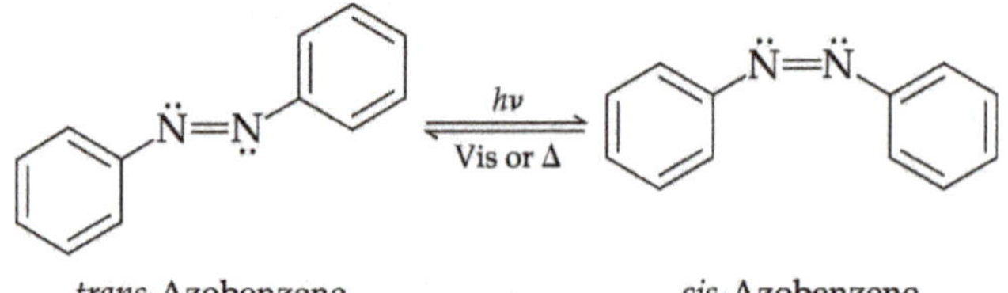


*trans*-Azobenzene *cis*-Azobenzene

The π bond of an alkene (C=C) is created by overlap of the $sp^2$ hybridized carbons' p orbitals. Rotation about the axis of the C=C requires a good deal of energy because it destroys the p orbital overlap, and thus the π bond. Unless the material is irradiated with light of the appropriate wavelength, absorption of the radiation does not take place, and the isomers do not interconvert unless sufficient thermal energy (60–65 kcal/mol, typically >200 °C) is supplied to break the π bond. If one of the electrons in the π bond (π molecular orbital) is photochemically excited into an antibonding, π*, molecular orbital, as occurs in this experiment, the π bond is weakened significantly. Rotation about this bond can then occur rapidly at room temperature.

It is the high-energy barrier to rotation about the C=C that gives rise to the possibility of alkene stereoisomers. The cis and trans isomers of an alkene system are called *diastereoisomers*, or *diastereomers*. Like all stereoisomers, these isomers differ only in the arrangement of the atoms in space. These isomers have all the same atoms bonded to each other. Diastereomers are not mirror images of each other and, of course, are not superimposable (identical). This particular type of diastereomer often was referred to in the older literature as a "geometric isomer." Diastereomers, therefore, would be expected to possess different physical properties, such as melting points, boiling points, dipole moments, densities, and solubilities, as well as different spectroscopic properties. Because of these differences in physical properties, diastereomers are amenable to separation by chromatography, distillation, crystallization, and other separation techniques. We use both chromatographic and crystallization methods in the present experiment.

The course of the isomerization will be followed using *thin-layer chromatography*

## Isomerization of an Alkene: Thin-Layer Chromatographic Analysis

## EXPERIMENTAL PROCEDURE

Estimated time to complete the experiment: 2.5 h of laboratory time. The reaction requires approximately 1 h of irradiation; the actual time to completion is quite sensitive to both radiation flux and temperature. These factors are largely determined by the distance the reaction vessel is positioned from the source of radiation.

You will monitor the progress of your reaction about every half-hour. If much of the ethanol solvent has evaporated, add more to bring it back to the original volume.

**Physical Properties of Reactants**

| Compound | MW | Amount | mmol | mp (°C) | bp (°C) |
|---|---|---|---|---|---|
| *trans*-1,2-Dibenzoylethylene | 236.27 | 25 mg | 0.11 | 111 | |
| Ethanol (95%) | | 3.5 mL | | | 78.5 |

***Reagents and Equipment.*** To a 50 mL beaker weigh and add 25 mg (0.11 mmol) *trans*-1,2-dibenzoylethylene and 3.0 mL of 95% ethanol. Be sure to **label** your beaker.

GENTLY

***Reaction Conditions.*** Warm the mixture ***gently*** until a homogeneous solution is obtained. Place the beaker directly under the sunlamp (this step will be assisted by your TA). Irradiate the solution for approximately 1 h.

*NOTE. If a lower wattage lamp is used, longer irradiation times or shorter distances will be necessary. The photolysis will probably take about 1 to 1.5 hours. However, you will monitor the progress of your reaction about every half-hour. If much of the ethanol solvent has evaporated, add more to bring it back to the original volume.*

---

Reminder: There are periods of waiting in this lab. But do not take off your goggles in the laboratory.

---

The progress of the isomerization will be followed by thin-layer chromatography (TLC) analysis every 30 minutes.

*INFORMATION. The TLC analysis is carried out using Eastman Kodak silica gel–polyethylene terephthalate plates with a fluorescent indicator.* ***Activate the plates at an oven temperature of 100 °C for 30 min*** *and then place them in a desiccator to cool until needed. After spotting, elute the plates using methylene chloride as the solvent. Visualize the spots with a UV lamp. The course of the reaction is followed by removing small samples (2–3 drops) of solution from the* ***hot*** *test tube or beaker at set time intervals with a Pasteur pipet and placing them in separate shell vials. See Technique 6A for the method of TLC analysis and the determination of $R_f$ values. Approximate $R_f$ values: trans = 0.72; cis = 0.64.*

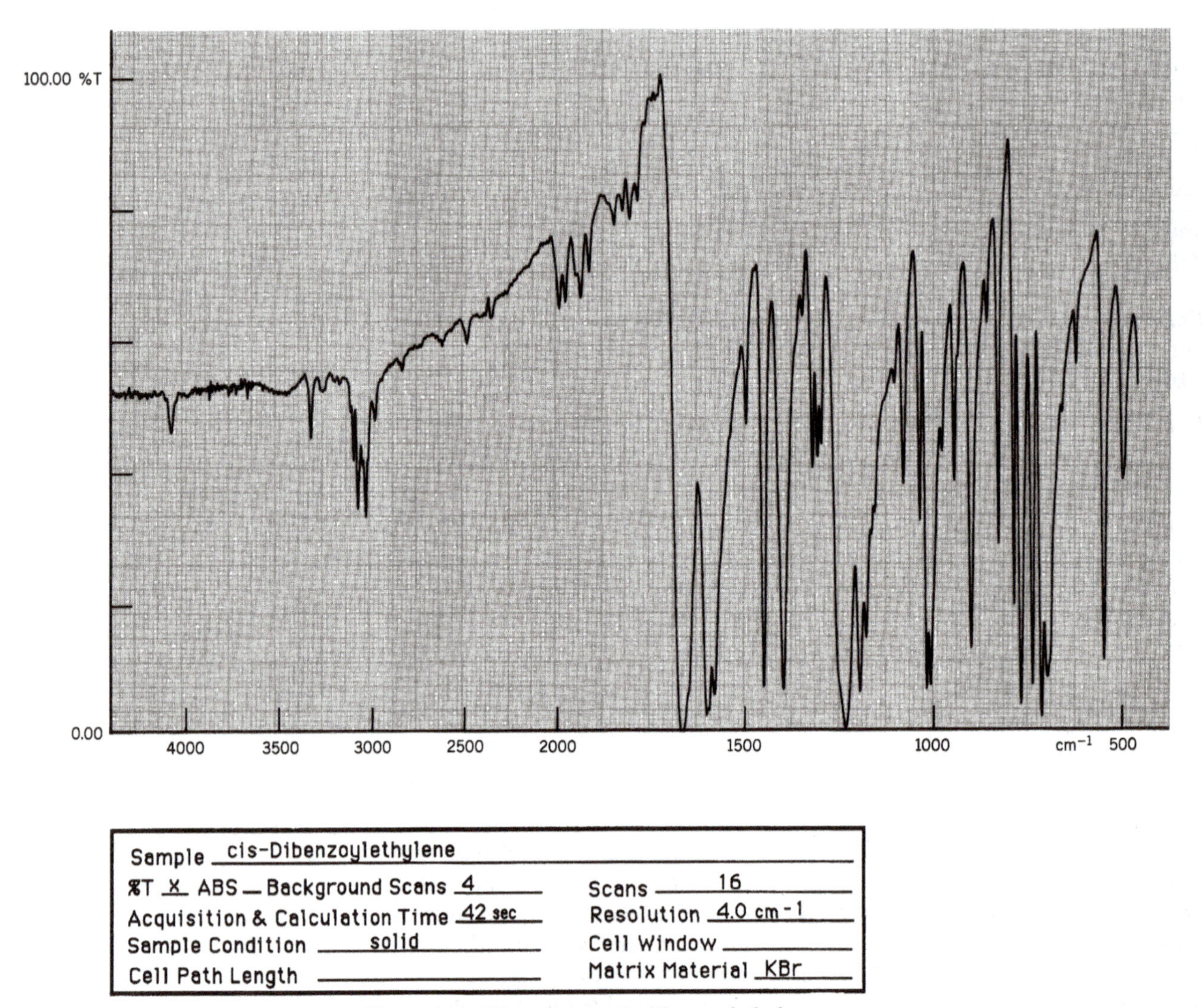


**Figure 6.11** **IR spectrum: *cis*-dibenzoylethylene.**

***Isolation of Product.*** Remove the **hot** beaker from the light source and allow the solution to cool to room temperature.

Place the resulting mixture in an ice bath to complete crystallization of the *colorless cis*-1,2-dibenzoylethylene product. Collect the crystals by vacuum filtration using a Hirsch funnel, wash them with 0.5 mL of cold 95% ethanol, and then air-dry them on a porous clay plate or on filter paper.

***Purification and Characterization.*** Weigh the dried product and calculate the percent yield. Determine the melting point and compare your result with the literature value. The purity of the crude isolated product may be further determined by TLC analysis (if not used above).

Obtain IR spectra of the cis and trans isomers and compare them to Figures 6.11 and 6.12. The UV-vis spectra in methanol solution are shown in Figures 6.13 and 6.14. These spectra will not be obtained in the lab.

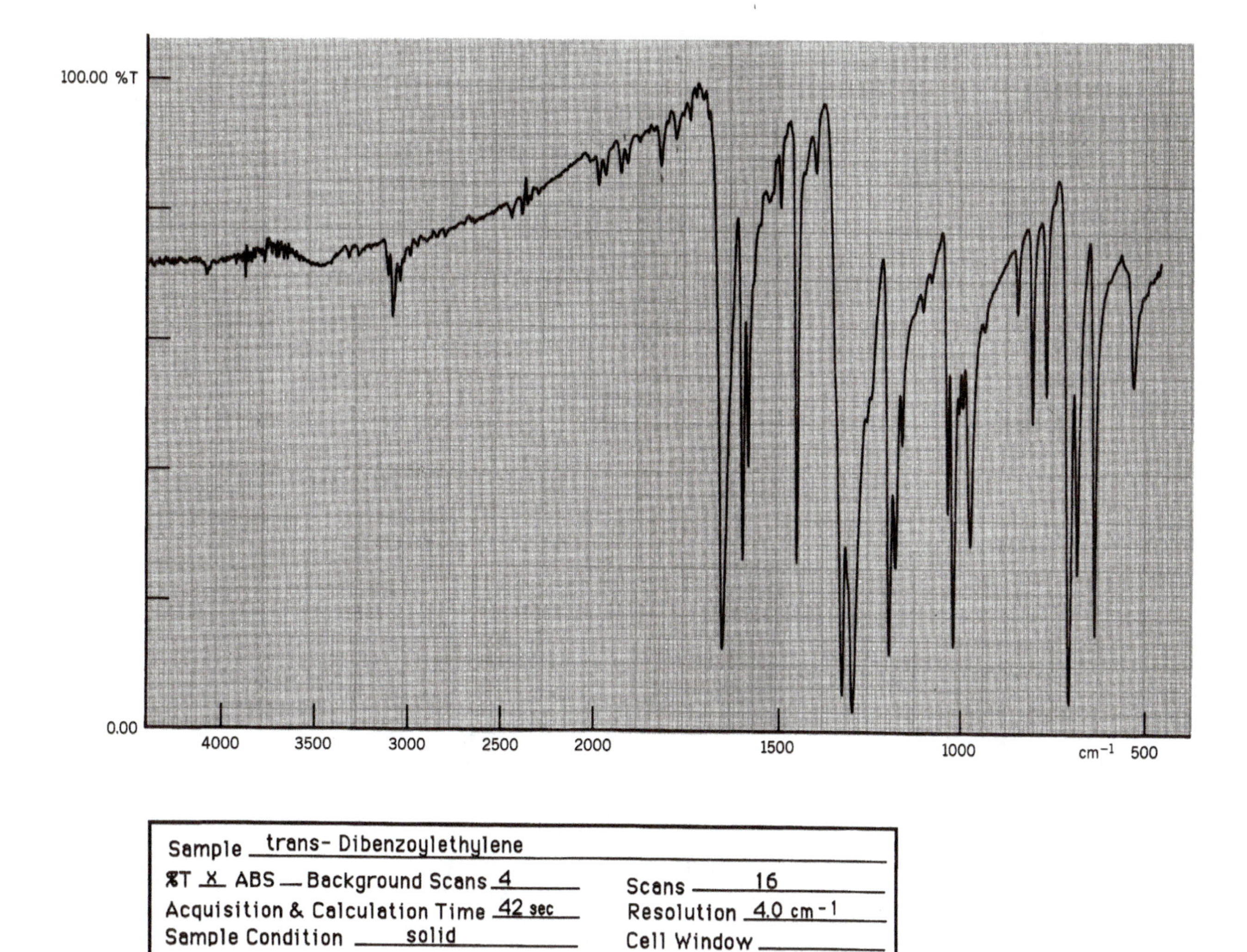


**Figure 6.12 IR spectrum: *trans*-dibenzoylethylene.**

***Ultraviolet-Visible Analysis.*** The bright yellow color of the *trans*-dibenzoylethylene rapidly fades as the conversion to the colorless cis compound progresses under irradiation. This visual observation may be supported by an examination of the absorption spectra of the isomers in a methanol solution from 225 to 400 nm (see Figs. 6.13 and 6.14). The $\lambda_{max}$ of the trans isomer drops from 268 to 259 nm in the cis compound. This shift to shorter wavelengths is just enough to move the long-wavelength end of the absorption band in the trans isomer out of the visible region and into the near-ultraviolet (thus, the cis compound does not absorb light to which the eye is sensitive and the compound appears colorless). This observation is consistent with the theory that the spatial contraction of extended $\pi$ systems moves the associated electronic transitions to higher energy gaps (higher frequencies or shorter wavelengths), which is exactly what takes place during the trans–cis isomerization.

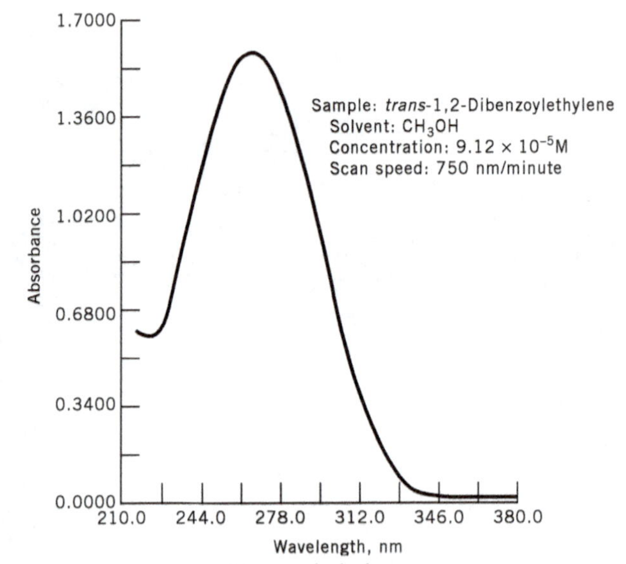


**Figure 6.13 UV spectrum: *trans*-dibenzoylethylene.**

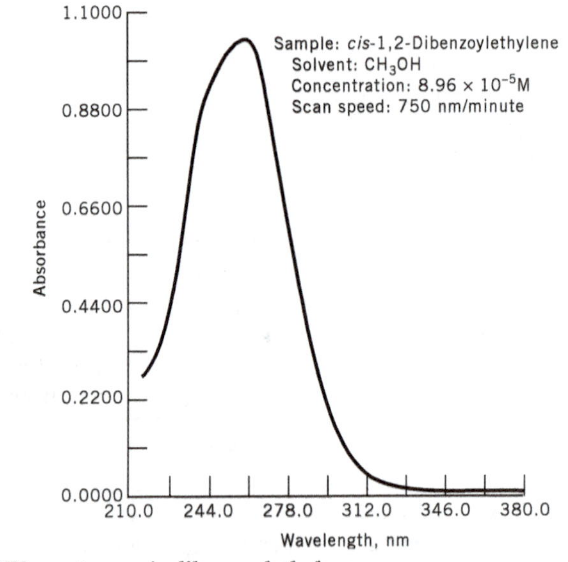


**Figure 6.14 UV spectrum: *cis*-dibenzoylethylene.**

***Infrared Analysis.*** The infrared spectral changes are consistent with the proposed reaction product. Consider the spectrum of the trans starting material (Fig. 6.12). It contains the following:

**1.** The macro group frequency train (see Strategies for Interpreting Infrared Spectra, p. 542, for conjugated aromatic ketones is 3080 and 3030 (C—H, aromatic), 1652 (doubly conjugated carbonyl), 1599, and 1581 ($\nu_{8a}$, $\nu_{8b}$ degenerate ring stretch, strong intensity of 1581-wavenumber peak confirms ring conjugation), 1495 and 1450 ($\nu_{19a}$, $\nu_{19b}$ degenerate ring stretch, $\nu_{19a}$ weak) $cm^{-1}$.

**2.** The monosubstituted phenyl ring macro group frequency train is 1980(d), 1920(d),[4] 1820, 1780 (mono combination-band pattern), 708 (C—H out-of-plane bend, C═O conjugated), 686 (ring puckering) $cm^{-1}$.

**3.** The presence of the trans double bond is indicated by the single medium intensity 970 $cm^{-1}$ band, as the C—H stretching region is overlapped by the aromatic ring C—H stretches.

The spectrum of the cis photoproduct (Fig. 6.11) possesses the same macro group frequencies as the starting material:

**1.** The conjugated aromatic ketone frequency train is 3335 (overtone of C═O stretch), 3080 and 3040, 1667, 1601, 1581, 1498 (weak), and 1450 $cm^{-1}$.

**2.** The monosubstituted phenyl ring macro frequency train is assigned peaks of 1990, 1920, 1830, 1795, 710, and 695 $cm^{-1}$.

**3.** The cis double bond is clearly present and utilizes a macro frequency train of 3030 (overlapped by aromatic C—H stretch), 1650 (C═O stretch, shoulder on the low wavenumber side of conjugated carbonyl; resolved in spectra run on thin samples), 1403 (═C—H out-of-phase, in-plane bending mode, strong band not present in trans compound), 970 (trans, in-phase, out-of-plane bend missing), 820 (cis, in-phase out-of-plane bend; band not found in spectrum of trans isomer) $cm^{-1}$.

Discuss the similarities and differences of the experimentally derived spectral data to the reference spectra (Figs. 6.11 and 6.12).

## QUESTIONS

**6-38.** What are some clues that tell you that you have made a different isomer product in this reaction?

**6-39.** Suggest a reason why the starting material (*trans*-1,4-diphenyl-2-butene-1,4-dione) should be recrystallized (or pure) prior to use (NOTE: you did not perform this recrystallization, but were provided with the recrystallized material).

**6-40.** *Cis*- and *trans*-1,4-diphenyl-2-butene-1,4-dione are stereoisomers. What is the specific relationship between the two of them?

**6-41.** Explain how this photochemical isomerization allows the production of the thermodynamically less stable cis isomer. In other words, why is the trans isomer exclusively converted to the cis isomer during short reaction periods and not vice versa? Is it possible, under these conditions, that the trans and cis isomers are in equilibrium with one another?

## BIBLIOGRAPHY

**The photochemical isomerization was adapted from the following references:**

Pasto,D. J.; Ducan, J. A.; Silversmith, E. F. *J. Chem. Educ.* 1974, 51, 277.

Silversmith, E. F.; Dunsun, F. C. *J. Chem. Educ.* 1973, 50, 568.

**Reviews on photochemical isomerization reactions may be found in the following references.**

Arai,T.; Tokumaru, K. *Chem. Rev.* 1993, 93, 23.

Coxton, J. M.; Halton, B. *Organic Photochemistry*, 2nd ed.; Cambridge University Press: New York, 1987.

Coyle, J. D. *Introduction to Organic Photochemistry;* Wiley: New York, 1991.

Kagan, J. *Organic Photochemistry: Principles and Applications;* Academic Press: Orlando, FL, 1993.

Kopecky, J. *Organic Photochemistry: A Visual Approach;* Wiley/VCH, New York, 1992.

Prasad, P. N. *Introduction to Biophotonics;* Wiley: New York, 2003.

# Reductive Catalytic Hydrogenation of an Alkene: Octane

EXPERIMENT 3106-1

Common name: octane

CA number: [1111-65-9]

CA name as indexed: octane

***Purpose.*** This experiment shows you how to reduce the carbon–carbon double bond of an alkene by addition of molecular hydrogen ($H_2$). You will gain an understanding of the important role that metal catalysts play in the stereospecific reductions of alkenes (and alkynes), to form the corresponding alkanes, by the activation of molecular hydrogen. You can observe the powerful influence on column chromatography of heavy metal ions, such as silver ($Ag^+$), which lead to effective separation of mixtures of alkenes from alkanes. Finally, you will appreciate the enormous importance and breadth of application of these reduction reactions in both industrial synthesis and basic biochemistry (for important examples see Experiment [3106-12])

## REACTION

$$\underset{\text{1-Octene}}{CH_3{-}(CH_2)_5{-}CH{=}CH_2} \xrightarrow[\substack{NaBH_4 \\ C_2H_5OH \\ HCl(6\ M)}]{H_2PtCl_2} \underset{\text{Octane}}{CH_3{-}(CH_2)_5{-}CH_2{-}CH_3}$$

*Prior Reading*

*Technique 4:* Solvent Extraction
Liquid–Liquid Extraction (p. 72)

*Technique 6:* Chromatography
Column Chromatography (pp. 92–95)
Concentration of Solutions (pp. 101–104)

## DISCUSSION

The addition of hydrogen to an alkene (or to put it another way, the saturation of the double bond of an alkene with hydrogen) to produce an alkane is an important reaction in organic chemistry. Alkanes are also called saturated hydrocarbons, because the carbon skeletons of alkanes contain the greatest possible number of hydrogen atoms permitted by tetravalent carbon atoms; alkenes are thus unsaturated hydrocarbons. Hydrogenation reactions have widespread use in industry. For example, we all consume vegetable fats hardened by partial hydrogenation (see Experiment [3106-12]) of the polyunsaturated oils that contained several carbon–carbon double bonds per molecule as isolated from their original plant sources. Partially hydrogenated fats represented a consumer market of over 2.9 billion pounds in 1992. Since that time major restaurants, fast food chains, and food producers have greatly lowered the amount of trans fats in their produts. Furthermore, batch processes involving hydrogenation reactions are nondiscriminating when considering the geometric isomers cis and trans.

The hydrogenation reaction is exothermic; the energy released is approximately 125 kJ/mol for most alkenes, but on the kinetic side of the ledger, this reductive pathway requires a significant activation energy to reach the transition state.Thus, alkenes can be heated in the presence of hydrogen gas at high temperatures for long periods without any measurable evidence of alkane formation. However, when the reducing reagent ($H_2$) and the substrate (the alkene) are in intimate contact with each other in the presence of a finely divided metal catalyst, rapid reaction does occur at room temperature. Under these conditions, successful reduction is generally observed at pressures of 1–4 atm. For this reason, this reaction is often referred to as low-pressure catalytic hydrogenation. These reactions are called heterogeneous reactions, since they occur at the boundary between two phases—in this case a solid and a liquid.

The main barrier to the forward progress of the reaction is the very strong H—H bond that must be broken. Molecular hydrogen, however, is adsorbed by a number of metals in substantial quantities; indeed, in some instances the amount of hydrogen contained in the metal lattice can be greater than that in an equivalent volume of pure liquid hydrogen! In this adsorption process the H—H bond is broken or severely weakened. (This adsorption process, which necessarily involves a large exchange of chemical energy between the metal lattice and the adsorbed hydrogen, may in some, as yet unexplained way, be related to the "cold fusion" problem in which palladium saturated with a heavy isotope of hydrogen [deuterium] allegedly exhibits apparent excess thermal energies on electrolysis.)

The $\pi$ system of the alkene is also susceptible to adsorption onto the metal surface and when this occurs the barrier to reaction between the alkene and the activated hydrogen drops dramatically.

Catalytic hydrogenations are also a representative example of a class of organic reactions known as *addition reactions,* which are reactions in which two new substituents are *added* to a molecule (the alkene substrate in this case) across a $\pi$ system. Usually addition is 1,2, but in extended $\pi$ systems such as 1,3 dienes, the addition may occur 1,4. In catalytic hydrogenations, formally, one hydrogen atom of a hydrogen molecule adds to each carbon of the alkene linkage, C═C. It is not at all clear that both hydrogen atoms must come from the same original hydrogen molecule even though they are added stereospecifically in syn (cis) fashion while both systems are coordinated with the metal surface. A representation of the stereochemistry of this addition is given below.

The metals most often used as catalysts in low-pressure (1–4 atm) hydrogenations in the laboratory are nickel, platinum, rhodium, ruthenium, and palladium. In industry, high-pressure, large-scale processes are more likely to be found. For example, Germany had little or no access to naturally formed petroleum deposits during World War II, but did possess large coal mines. The Germans mixed powdered coal with heavy tar (from previous production runs) and 5% iron oxide and heated this in the presence of $H_2$ at a pressure of 3000 lb/in$^2$, at about 500 °C for 2 h to yield synthetic crude oil. Thirteen German plants operating in 1940 produced 24 million barrels that year, with an average of 1.5–2 tons of coal ultimately converted to about 1 ton of gasoline.

In the present experiment the metal catalyst, platinum, is generated in situ by the reaction of chloroplatinic acid with sodium borohydride. The reduced platinum metal is formed in a colloidal suspension, which provides an enormous surface area, and therefore excellent conditions, for heterogeneous catalysis. The molecular hydrogen necessary for the reduction can also be conveniently generated in situ by the reaction of sodium borohydride with hydrochloric acid:

$$4\ NaBH_4 + 2\ HCl + 7\ H_2O \longrightarrow Na_2B_4O_7 + 2\ NaCl + 16\ H_2$$

This reduction technique does not require equipment capable of safely withstanding high pressures. The use of chloroplatinic acid, therefore, is particularly attractive for saturating easily reducible groups, such as unhindered alkenes or alkynes, in the laboratory. The potential limitation to the use of this reagent is that other reducible functional groups, such as aldehydes and ketones, normally inert to catalytic hydrogenations of alkenes and alkynes, may be reduced by the sodium borohydride. Thus, with chloroplatinic acid and sodium borohydride, we accept, as a compromise, a more limited set of potential reactants (substrates) for the convenience inherent in the reagent.

The platinum catalyst generated in the reaction medium adsorbs both the internally generated molecular hydrogen and the target alkene on its surface. The addition of the hydrogen molecule ($H_2$) (evidence strongly suggests that it is actually atomic hydrogen that attacks the alkene $\pi$ system) to the alkene system while they are both adsorbed on the metal surface results in the reduction of the substrate and the formation of an alkane. The addition is, as mentioned earlier, syn (or cis), since both hydrogen atoms add to the same face of the alkene plane. The mechanistic sequence is outlined below:

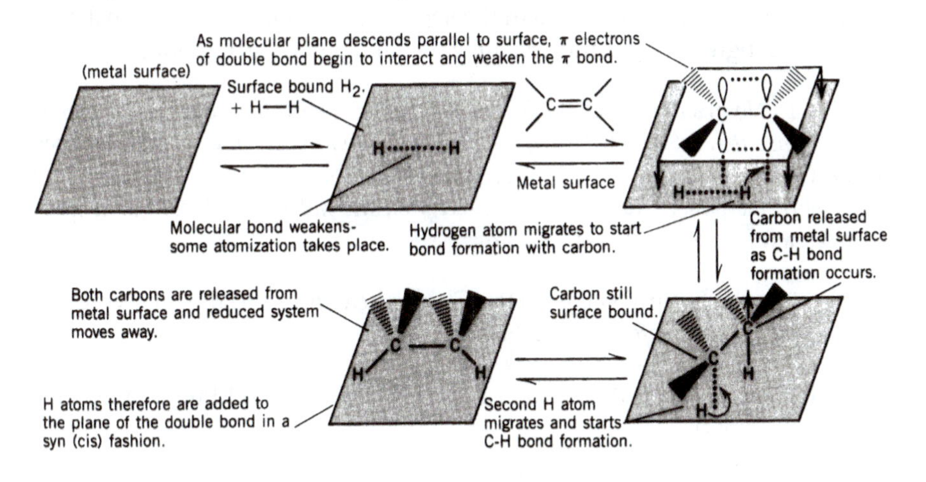


This experiment also provides an opportunity for you to study a powerful aspect of column chromatography in which heavy metal ions have a particularly important role in the purification of the product. Unreacted alkene has the potential to be a problem contaminant during the isolation and purification of the relatively low-boiling saturated reaction product.The successful removal of the remaining 1-octene from the desired *n*-octane in the product mixture is achieved by using column chromatography with silver nitrate/silica gel as the stationary phase. Complex formation between the silver ion (on the silica gel surface) and the $\pi$ system of the unreacted alkene acts to retard the rate of elution of the alkene relative to that of the alkane.

The ability of alkenes to form coordination complexes with certain metal ions having nearly filled d orbitals was established some time ago. In the case of the silver ion complex with alkenes, the orbital nature of the bonding is believed to involve a $\sigma$ bond formed by overlap of the filled $\pi$ orbital of the alkene with the free s orbital of the silver ion plus a $\pi$ bond formed by overlap of the vacant antibonding $\pi^*$ orbitals of the alkene together with the filled d orbitals of the metal ion.

## EXPERIMENTAL PROCEDURE

Estimated time to complete the experiment: 2.5 h.

**Physical Properties of Reactants**

| Compound | MW | Amount | mmol | bp (°C) | $d$ | $n_D$ |
|---|---|---|---|---|---|---|
| 1-Octene | 112.22 | 120 μL | 0.76 | 121.3 | 0.72 | 1.4087 |
| Ethanol (absolute) | 46.07 | 1.0 mL | | 78.5 | | |
| Chloroplatinic acid (0.2 M) | 517.92 | 50 μL | | | | |
| Sodium borohydride (1 M) | 37.83 | 125 μL | | | | |
| Dilute HCl (6 M) | | 100 μL | | | | |

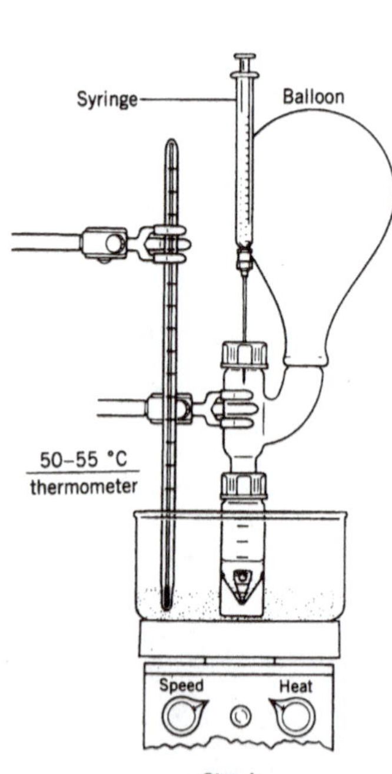


**Step I:**
0.2 *M* $H_2PtCl_6$ solution, 50 μL +
$CH_3CH_2OH$, 1.0 mL
+ $NaBH_4$ solution, 125 μL
**Step II:**
6 *M* HCl, 100 μL
+ 1-octene, 120 μL
+ $NaBH_4$ solution, 1.0 mL

***Reagents and Equipment.*** Equip a 5.0-mL conical vial containing a magnetic spin vane with a Claisen head fitted with a rubber balloon and Teflon-lined rubber septumcap (good GC *septa* work best—this is an important point, because several injections through the septum are required and the seal must remain gas-tight). Mount the assembly on a magnetic stirring hot plate.

*NOTE. 1. No residual acetone (perhaps from cleaning the equipment) can be present since it reacts with the $NaBH_4$. 2. The balloon must make a gas-tight seal to the Claisen head, so be sure to secure it with copper wire or a rubber band. Make sure the balloon does not rest directly on the hot plate. It will be deflated at first then inflate as the reaction progresses*

Remove the 5.0-mL vial from the Claisen head and add the following reagents. *Recap the vial after each addition.*

**a.** Add 50 μL of a 0.2 M solution of chloroplatinic acid ($H_2PtCl_6$) (automatic delivery pipet).

**b.** Add 1.0 mL of absolute ethanol.

**c.** Add 125 μL of the sodium borohydride reagent (automatic delivery pipet).

*NOTE. Reattach the vial* ***immediately*** *to the Claisen head after the $NaBH_4$ solution is added.*

Stir the mixture vigorously. The solution should turn black immediately as the finely divided platinum catalyst is formed.

*INSTRUCTOR PREPARATIONS. 1. The 0.2 M $H_2PtCl_6$ solution is prepared by adding 41 mg (0.1 mmol) of the acid to 0.5 mL of deionized water. 2. The sodium borohydride reagent is prepared by adding 0.38 g (0.01 mol) of $NaBH_4$ to a solution of 0.5 mL of 2.0 M aqueous NaOH in 9.5 mL of absolute ethanol.*

After 1 min, use a syringe to add 100 μL of 6 M HCl solution through the septum cap. In a like manner using a fresh syringe, add *immediately* to the acid solution, a solution of 120 μL (86 mg, 0.76 mmol) of 1-octene dissolved in 250 μL of absolute ethanol. (This solution is conveniently prepared in a shell vial; the reagents are best dispensed using automatic delivery pipets.)

Now add dropwise (clean syringe) 1.0 mL of the $NaBH_4$ reagent solution over a 2-min interval.

*NOTE. At this point the balloon should inflate and remain inflated for at least 30 min. If it does not, the procedure must be repeated.*

***Reaction Conditions.*** Stir the reaction mixture vigorously at a temperature of 50 °C for 45 min.

***Isolation of Product.*** Cool the reaction to ambient temperature and dropwise add 1 mL of water. Extract the resulting mixture in the reaction vial with three 1.0-mL portions of pentane. Transfer each pentane extract to a stoppered 25-mL Erlenmeyer flask containing 0.5 g of anhydrous magnesium sulfate.

The reaction mixture is extracted as follows. Upon addition of each portion of pentane, cap the vial, shake, vent carefully, and then allow the layers to separate. A Vortex mixer may be used if available. The transfers must be made using a Pasteur filter pipet because the pentane solvent is particularly volatile.

Using a Pasteur filter pipet, transfer the dried solution to a second 25-mL Erlenmeyer flask. Rinse the drying agent with an additional 1.0 mL of pentane and add the rinse to this second flask. Add a boiling stone and concentrate the solution to a volume of about 1.0–1.5 mL by warming it gently on a hot plate in the **hood.**

HOOD

***Purification and Characterization.*** The saturated product, *n*-octane, is purified by column chromatography. In a Pasteur filter pipet place about 50 mg of sand, 500 mg of 10% silver nitrate on activated silica gel (200 mesh), and then 50 mg of anhydrous magnesium sul fate (➡).

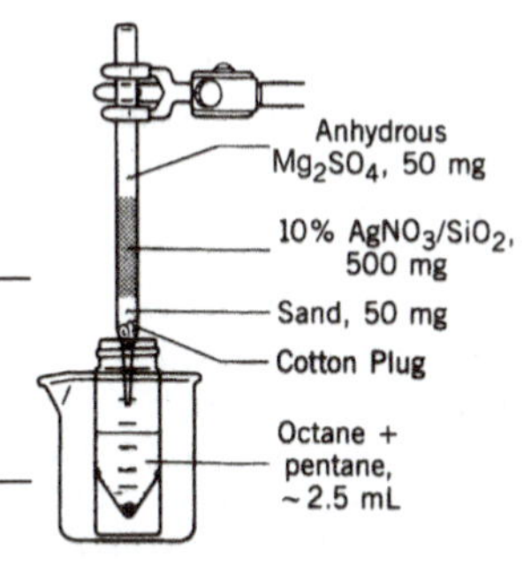


CAUTION: Silver nitrate stains the skin. Protective gloves should be worn during this operation. The silver nitrate/silica gel used in this separation is commercially available.

Wet the column with 0.5 mL of pentane (calibrated Pasteur pipet) and then transfer the concentrated crude product, as obtained above, to the column by Pasteur filter pipet. Elute the octane from the column using 1.5 mL of pentane and collect the eluate in a tared 5.0-mL conical vial containing a boiling stone.

Fit the vial with an air condenser and then place the assembly in the **hood** on a hot plate maintained at a temperature of 90–100 °C to evaporate the pentane solvent.

HOOD

*OPTIONAL. The evaporation is continued until a constant weight of product is obtained. This procedure is the best approach, but has to be done very carefully or a considerable amount of product can be lost.*

Record the weight of product and calculate the percent yield. Obtain an IR spectrum of your product. Compare your results with those reported in the literature (*Aldrich Library of IR Spectra* and/or SciFinder Scholar). Also, compare your IR spectrum to that of the 1-octene starting material. Can you establish from the above data if your sample is contaminated by traces of the pentane extraction solvent? If not, how would you go about determining the presence of this potential impurity?

## QUESTIONS

**7-1.** Squalene, first isolated from shark oil and a biological precursor of cholesterol, is a long-chain aliphatic alkene ($C_{30}H_{50}$).The compound undergoes catalytic hydrogenation to yield an alkane of molecular formula $C_{30}H_{62}$. How many double bonds does a molecule of squalene have?

**7-2.** A chiral carboxylic acid A ($C_5H_6O_2$) reacts with 1 mol of hydrogen gas on catalytic hydrogenation. The product is an achiral carboxylic acid B ($C_5H_8O_2$). What are the structures of compounds A and B?

**7-3.** Two hydrocarbons, A and B, each contain six carbon atoms and one C=C. Compound A can exist as both *E* and *Z* isomers but compound B cannot. However, both A and B on catalytic hydrogenation give only 3-methylpentane. Draw the structures and give a suitable name for compounds A and B.

**7-4.** What chemical test would you use to distinguish between the 1-octene starting material and the octane product? (Hint: look in "The Classification Tests; Alkenes & Alkynes" section of Experiment 3106-8, p. 240)

**7-5.** Give the structure and names of five alkenes having the molecular formula $C_6H_{12}$ that produce hexane on catalytic hydrogenation.

## BIBLIOGRAPHY

**Selected references on catalytic hydrogenation:**

Birch, A. J.; Williamson, D. H. *Org. React.* **1976,** *24,* 1.

Carruthers, W. *Some Modern Methods of Organic Synthesis,* 3rd ed.; Cambridge Univ. Press: New York, 1986; p. 411.

De, S; Gambhir, G.; Krishnamurty, H. G. *J. Chem. Educ.* **1994,** *71,* 922.

Kabalka, G. W.; Wadgaonkar, P. P.; Chatla, N. *J. Chem. Educ.* **1990,** *67,* 975.

Parker, D. in *The Chemistry of the Metal-Carbon Bond,* Hartley, F. R., Ed.; Wiley: New York, 1987, Vol 4, Chapter 11, p. 979.

Pelter, A.; Smith, K.; Brown, H. C. *Borane Reagents (Best Synthetic Methods);* Academic Press: Orlando, FL, 1988.

Smith, M. B.; March, J. *Advanced Organic Chemistry,* 6th ed.; Wiley: New York, 2007, Chap. 15, p. 1065.

**Selected examples of catalytic hydrogenation of alkenes in *Organic Syntheses:***

Adams, R.; Kern, J. W.; Shriner, R. L. *Organic Syntheses;* Wiley: New York, 1941; Collect. Vol. I, p. 101.

Bruce, W. F.; Ralls, J. O. *Organic Syntheses;* Wiley: New York, 1943; Collect. Vol. II, p. 191.

Cope, A. C.; Herrick, E. C. *Organic Syntheses;* Wiley: New York, 1963; Collect. Vol. IV, p. 304.

Herbst, R. M.; Shemin, D. *Organic Syntheses;* Wiley: New York, 1943; Collect. Vol. II, p. 491.

Ireland, R. E.; Bey, P. *Organic Syntheses;* Wiley: New York, 1988; Collect. Vol. VI, p. 459.

McMurry, E. *Organic Syntheses;* Wiley: New York, 1988; Collect. Vol. VI, p. 781.

Meyers, A. I.; Beverung, W. N.; Gault, R. *Organic Syntheses;* Wiley: New York, 1988; Collect. Vol. VI, p. 371.

# Polymerization

EXPERIMENT 3106-2

## REACTION

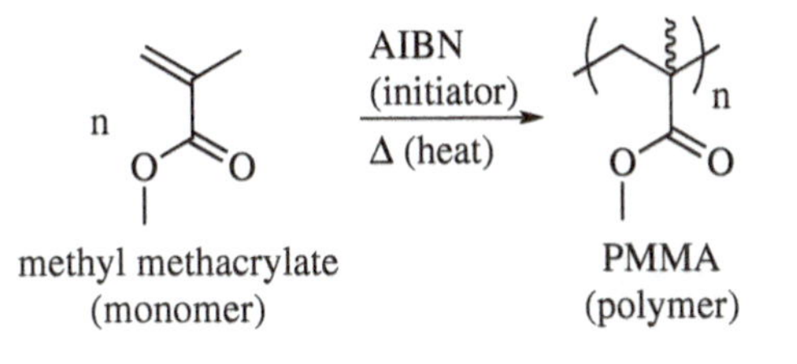


## DISCUSSION

Organic glass (PMMA) is a transparent thermoplastic. This material was developed in the late 19th century to the early 20th century in various research labs and commercialized by Rohm and Hass Company in 1933 and sold under the trade name of "Plexiglas". It is widely used as a glass substitute because of its transparency, strength, and lack of brittleness as compared to glass, as well as its ease of handling and processing and relatively low cost.

Step 1: Radical initiation: Heating AIBN leads to the breaking of the weak C-N single bonds and the simultaneous formation of nitrogen ($N_2$) along with two radicals, as shown below. A single headed (fishhook) arrow is used to show the movement of a single electron.

Reaction 1: CN–N=N–CN —Δ (heat)→ CN• N≡N •CN

Mechanism - Step 2: Propagation: The radical from AIBN adds to the π-bond of methyl methacrylate, forming a new σ-bond and another radical. This newly generated radical can add to another molecule of methyl methacrylate. The same process can happen numerous times to lead to polymeric radicals. The number of repeating units can as high as $10^5$.

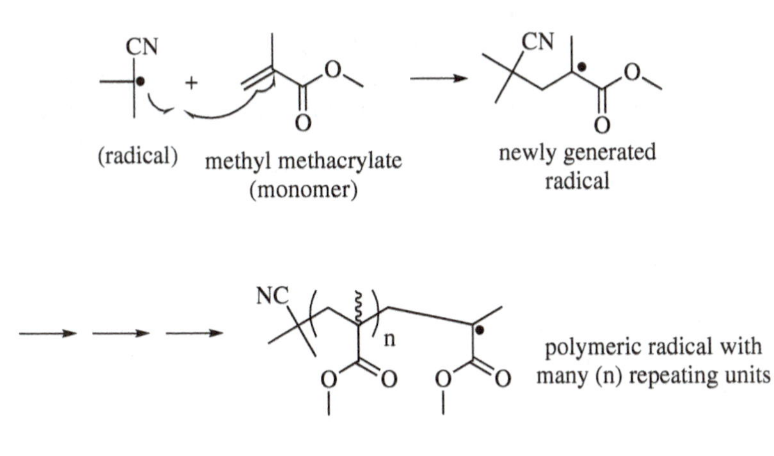

Mechanism - Step 3: Termination: Radicals are very reactive and therefore short lived. If two collide, they will recombine to form a single bond. One way this can happen is shown below. Notice that while there are CN groups at one or both ends of the polymer, it is written as just the repeating unit, since there are so many more of them than the CN groups.

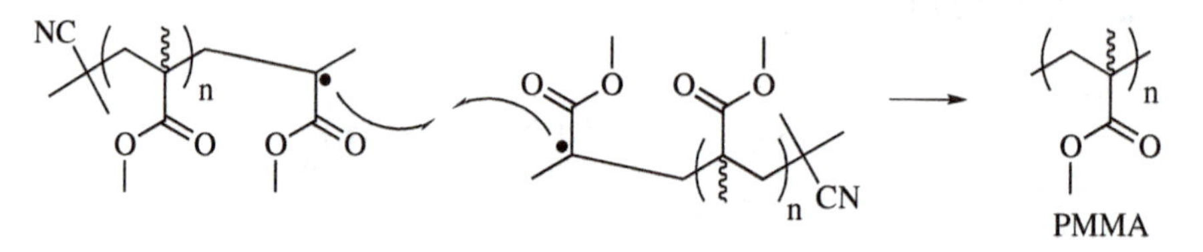


Combination of two polymeric radicals results in one PMMA molecule.

## EXPERIMENTAL PROCEDURE

CAUTION: Methyl methacrylate is a flammable/toxic liquid. It will cause irritation to skin and eyes. The reaction must be performed in the hood.

**Physical Properties of Reactants**

| Compound | MW | Amount | mmol | bp (°C) | *d* |
|---|---|---|---|---|---|
| Methyl methacrylate | 100.2 | 5.0 mL | 47 | 101 | .94 |
| AIBN | 164.21 | 0.05 g | 0.3 | | |

### *Reagents and Equipment*

To a medium sized screw-capped vial provided in the hood (do not use a conical vial), add 5 mL (47 mmol) of methyl methacrylate monomer and 0.05g (0.3 mmol) of the AIBN initiator. Mix well in the vial. Loosely cap the top of the vial. The reaction needs to be carried out at 75-80° C; use a water-bath for heating. Secure the vial with a clamp and heat it by water bath for approximately 40 min. (Caution: radical reactions are generally sensitive to water and other impurities, so the reactant in the vial should avoid contact with water in the water bath.) Check the status of your vial every 10 minutes, noting in particular any increase in the viscosity. The solution in the vial should go from a water-like liquid to a viscous liquid and finally end up as a solid. After formation of the solid in the vial, take the vial out of the water bath and leave it under the hood to cool down to room temperature. Thoroughly wash the product in the vial with water and take the product home. You have made your own polymer!

**TA's demonstration experiment:** In a separate vial with 5 mL of methyl methacrylate monomer without addition of AIBN, what will happen after heating the monomer under the same temperature for same amount of time? This shows how important the role of the small amount of AIBN is in this reaction.

## QUESTIONS

**7-6.** What is a radical?

**7-7.** In the PMMA product that is produced in this experiment, there will be bubbles trapped in the solid. Where do the bubbles come from? (Hint, see Reaction 1)

**7-8.** The density of methyl methacrylate monomer is 0.94 g/mL, the density for poly(methyl methacrylate) is 1.19 g/mL. In an ideal case (do not count the volume of the bubbles), 5 mL of methyl methacrylate monomer will give how many mL of poly(methyl methacrylate)?

# Hydroboration–Oxidation of an Alkene: Octanol

EXPERIMENT 3106-3

Common name: octanol
CA number: [111-87-5]
CA name as indexed: 1-octanol

***Purpose.*** The oxidation of an alkene to an alcohol is investigated via the in situ formation of the corresponding trialkylborane, followed by the oxidation of the carbon–boron bond with hydrogen peroxide. The conditions required for hydroboration (a reduction) of unsaturated hydrocarbons are explored. Alkylboranes are particularly useful synthetic intermediates for the preparation of alcohols. The example used in this experiment is the conversion of 1-octene to 1-octanol in which an anti-Markovnikov addition to the double bond is required to yield the intermediate, trioctylborane. Since it is this alkyl borane that subsequently undergoes oxidation to the alcohol, hydroboration offers a synthetic pathway for introducing substituents at centers of unsaturation that are not normally available to the anti-Markovnikov addition reactions that are based on radical intermediates.

*Prior Reading*

*Technique 4:* Solvent Extraction
Liquid–Liquid Extraction (p. 72)
*Technique 1:* Gas Chromatography (pp. 55-61)

## REACTION

$$3\,CH_3{-}(CH_2)_5{-}CH{=}CH_2 \xrightarrow{THF\cdot BH_3} [CH_3(CH_2)_7]_3B \xrightarrow[OH^-]{H_2O_2} 3\,CH_3{-}(CH_2)_7{-}OH$$

1-Octene — Trioctylborane — 1-Octanol

## DISCUSSION

The course of this reaction depends (1) on the *stereospecific* reductive addition of diborane ($B_2H_6$, introduced as the borane • tetrahydrofuran complex ($BH_3 \cdot THF$)) to an alkene to form an intermediate trialkylborane and (2) on oxidation of the borane with alkaline hydrogen peroxide to yield the corresponding alcohol.

The first step in the reaction sequence is generally called a *hydroboration.* The addition of diborane is a rapid, quantitative, and general reaction for all alkenes (as well as alkynes) when carried out in a solvent that can act as a Lewis base. The ether solvation of the diborane, for example, is the key to the success of this reaction. In the absence of a Lewis base, borane ($BH_3$) exists as a dimer ($B_2H_6$), which is much less reactive than the monomer ($BH_3$). Borane, however, does exist in coordination with ether type solvents. It is the monomer ($BH_3$) that functions as the active reagent in the reductive addition.

As depicted in the following mechanism, the boron hydride rapidly adds successively to three molecules of the alkene to form a trialkylborane.

$$CH_3-(CH_2)_5-HC{=}CH_2 \longrightarrow CH_3-(CH_2)_5-CH_2-CH_2-BH_2$$

$$\overset{\delta-}{H}-\overset{\delta+}{B}H_2$$

$$\xrightarrow[\text{repeat}]{2\,CH_3-(CH_2)_5-CH{=}CH_2} [CH_3(CH_2)_5-CH_2-CH_2]_3B$$

Note that the transition state of this addition reaction is generally considered to be a *four-center* one, and that the 1-octene substrate is oriented such that the boron becomes bonded to the least-substituted carbon atom of the double bond. Thus, the reaction can be classified as *regioselective,* and it will be sensitive to substitution on the carbon–carbon double bond.

In the developing transition state, the alkene $\pi$ electrons (the least tightly held, and most nucleophilic) flow to the electron-deficient boron atom (the vacant p orbital is the electrophile). The formation of the *transition state* is controlled in large part by the polarization of the alkene $\pi$ system during the early stages of formation of the transition state. At this point a partial positive charge begins to form on the more highly substituted carbon (the more stable carbocation), and a partial negative charge on the least substituted carbon. The orientation of the polarization, therefore, is to a large extent controlled by the electron-releasing effects of the alkyl substituents on the alkene, which enables the more highly substituted of the $sp^2$ carbon atoms to better accommodate the positive charge. As the reaction proceeds, the boron acquires a partial negative charge in response to the incoming electron density. The ease of hydride ($:H^-$) transfer from the boron to the more highly substituted carbon atom of the alkene, therefore, increases. Thus, hydroboration involves simultaneous hydride release and boron–carbon bond formation, and is a concerted reaction. The reaction can be conveniently considered as passing through a four-centered transition state, wherein the atoms involved undergo simultaneous changes in bonding (i.e., electron redistribution [see below]).

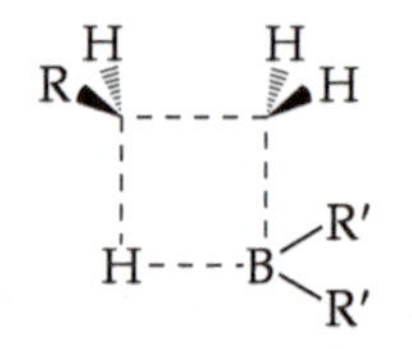

Hydroboration, as we have seen, can be classified as a **concerted addition reaction** in which no intermediate is formed. The mechanism is characteristic of a group of reactions called **pericyclic** (from the Greek, meaning *around the circle*) **reactions,** which involve a cyclic shift of electrons in and around the *transition state.* The mechanism proposed is further supported by the fact that rearrangements are not normally observed in hydroboration reactions, which implies that there are no carbocationic intermediates.

When alkenes with varying degrees of substitution undergo hydroboration, the boron ends up on the least substituted $sp^2$ carbon atom. While it might appear from the products that the regioselectivity is controlled by steric factors, this assumption is probably too simplistic. Steric and electronic factors both favor, and are both likely responsible for, the observed regioselectivity in hydroboration reactions.

Accumulated evidence demonstrates that the reaction occurs by **syn** addition, which is a consequence of the four-centered transition state. Therefore, the new C—B and C—H bonds are necessarily formed on the same face of the C═C bond, as shown in the following example:

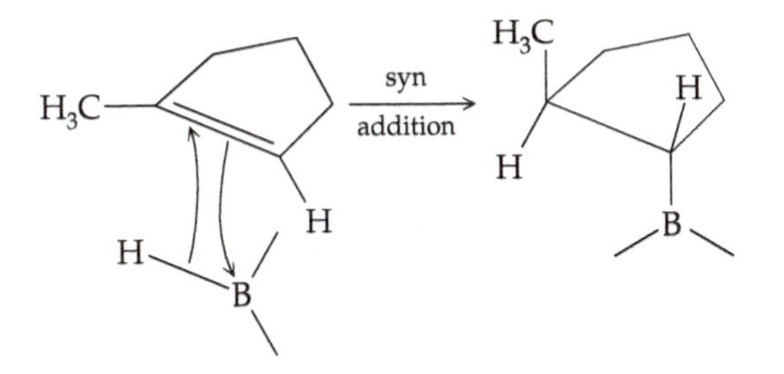


Organoboranes are important in organic synthesis as reactive intermediates. Reactions have been developed by which the boron atom may be replaced by a wide variety of functional groups, such as —H, —OH, —$NH_2$, —Br, —I, and —COOH. The present experiment demonstrates the conversion of an organoborane to an alcohol by oxidation with alkaline hydrogen peroxide. It is not necessary to isolate the organoborane prior to its oxidation. This simplification is particularly fortuitous in this case, since most alkylboranes, when not in solution, are pyrophoric (spontaneously flammable in air).

With regard to the second stage of the hydroxylation process, there is now conclusive evidence that oxidation of the C—B bond proceeds with retention of configuration at the carbon atom bearing the boron. That is, the hydroxyl group that replaces the boron atom has the identical orientation in the molecule as the boron:

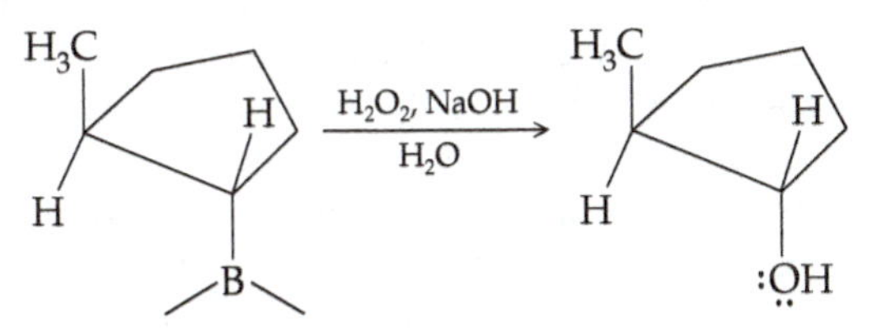


Thus, in unsymmetrical alkenes the hydroboration–oxidation sequence of reactions leads to the addition of the elements of H—OH to the original C═C in an anti-Markovnikov manner.

In the oxidation step a hydroperoxide anion ($HOO^-$) is generated in the alkaline medium. This species makes a nucleophilic attack on the boron atom to form a boron hydroperoxide. A 1,2 migration of an alkyl group from boron to oxygen occurs to yield a boron monoester (a borate). Hydrolysis of the boron triester, generated by successive rearrangement of all three alkyl groups, produces the desired alcohol. The mechanism of the oxidation sequence is given below.

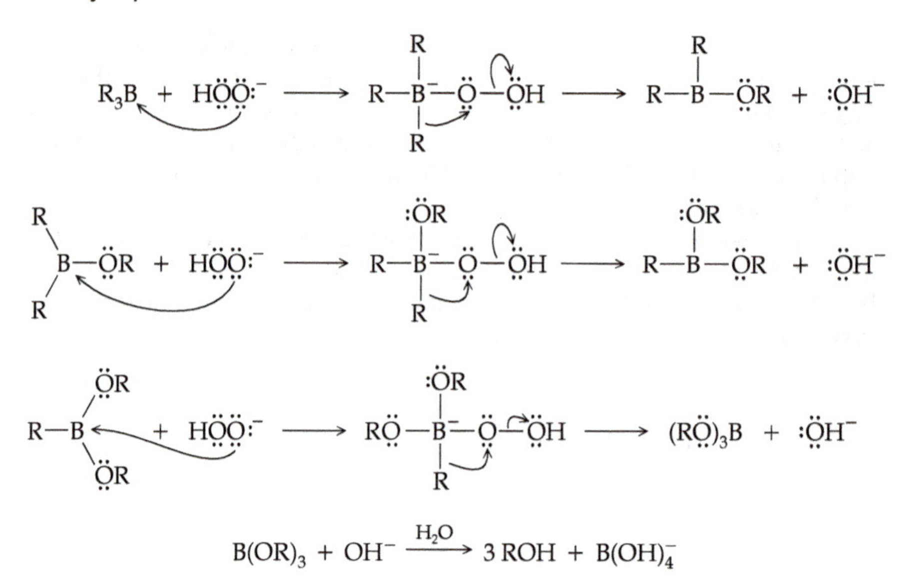

In the final step, alkaline hydrolysis of the trialkyl borate ester yields 3 moles of the alcohol.

The effective use of $B_2H_6$ in the hydroboration reaction was discovered in 1955 by H. C. Brown, and is just one of the many important hydride reagents developed by Professor Brown and his coworkers at Purdue University.

**Herbert Charles Brown (1912)** Brown obtained his B.S. in chemistry from the University of Chicago (1936) and his Ph.D. from the same institution in 1938. He later became a Professor of Chemistry at Wayne State and Purdue Universities.

Working with H. I. Schlesinger at the University of Chicago, Brown developed practical routes for the synthesis of diborane ($B_2H_6$). He discovered that diborane reacted rapidly with LiH to produce lithium borohydride ($LiBH_4$), discovering and opening a synthetic route to the metal borohydrides. These compounds proved to be powerful reducing agents. Later, he developed an effective route to $NaBH_4$, which led to the commercial production of this material. Metal borohydrides, particularly $LiAlH_4$ (developed by Schlesinger and Albert Finholt), have revolutionized how organic functional groups are reduced in both the research laboratory and the industrial plant.

In 1955, Brown discovered that alkenes can be converted to organoboranes by reaction with diborane (actually the monomer in ether solution) and with organoboranes containing a B—H bond (the hydroboration reaction). The organoboranes are valuable intermediates in organic synthesis because the boron substituent can be quickly and quantitatively replaced by groups such as —OH, —H, $—NH_2$, or —X (halogen). Thus, organoboranes have become an attractive pathway for the preparation of alcohols, alkanes, amines, and organohalides.

Brown's investigation of the addition compounds of trimethyl borane, diborane, and boron trifluoride with amines has provided a quantitative estimation for steric strain effects in chemical reactions. He also investigated the role of steric effects in solvolytic, displacement, and in elimination reactions. His results demonstrate that steric effects can assist, as well as hinder, the rate of a chemical reaction.

Brown has published over 700 scientific papers and is the author of several texts. For his extensive work on organoboranes, Brown (with G. Wittig [organophosphorus compounds]) received the Nobel Prize in Chemistry in 1979.[13]

## EXPERIMENTAL PROCEDURE

Estimated time to complete the experiment: 4.0 h, performed over the course of two lab periods.

**Physical Properties of Reactants**

| Compound | MW | Amount | mmol | bp (°C) | *d* | $n_D$ |
|---|---|---|---|---|---|---|
| 1-Octene | 112.22 | 420 μL | 2.68 | 121 | 0.72 | 1.4087 |
| Borane • THF (1M) | | 2.0 mL | 2.00 | | | |
| Sodium hydroxide (3 M) | 40.00 | 600 μL | | | | |
| Hydrogen peroxide (30%) | 34.01 | 600 μL | | | | |

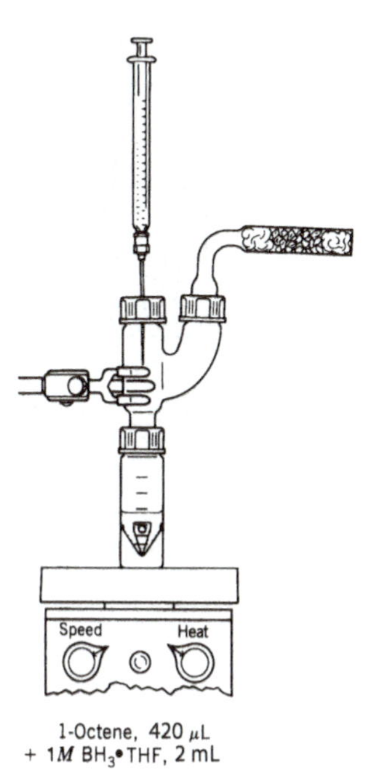

1-Octene, 420 μL + 1*M* $BH_3$•THF, 2 mL

***Lab Period 1: Reagents and Equipment.*** Equip a 5.0-mL conical vial, containing a spin vane, with a Claisen head fitted with a rubber septum and calcium chloride drying tube. Through the rubber septum add 420 μL (300 mg, 2.68 mmol) of 1-octene (in one portion) with a 1.0-$cm^3$ syringe.

*NOTE. This reaction is moisture sensitive. Instead of drying the glassware, two equivalents of borane will be used.*

Cool the reaction vessel in an ice bath and, using a 5 mL syringe, add 2.00 mL (2.0 mmol) of the 1 M borane · THF solution through the septum over a 5-min period.

CAUTION: The $BH_3$ • THF reagent reacts *violently* with water. Never recap a used needle, you may accidentally stick yourself if you do this.

***Reaction Conditions.*** Allow the reactants to warm to room temperature and then stir for 45 min. Using a Pasteur pipet, carefully and slowly add eight drops of water to hydrolyze any unreacted borane complex.

*NOTE. At this stage of the procedure the vial may be removed from the Claisen head, capped, and allowed to stand until the next laboratory period.*

***Lab Period 2:*** Complete the Experiment 3106-3 by adding 600 μL of 3 M NaOH solution to the reaction vial, followed by the dropwise addition of 600 μL of 30.0% hydrogen peroxide solution over a 10-min period. Stir the reaction vial gently after each addition.

CAUTION: 30% hydrogen peroxide is caustic and blisters the skin. It is ten times stronger than the 3% hydrogen peroxide often used as a household antiseptic, and it will rapidly burn your skin. If you spill some on your skin, immediately wash it off with cold water and notify your TA. Concentrated solutions of hydrogen peroxide can explode!

[13]See *McGraw-Hill Modern Scientists and Engineers;* S. P. Parker, Ed.; McGraw-Hill: New York, 1980, Vol. 1, p. 150.

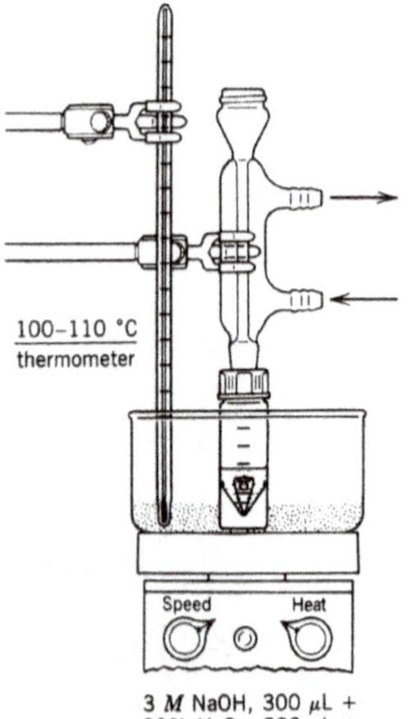


Attach the vial to a reflux condenser and warm the reaction mixture with stirring, for 1 h on a hot plate at 40–50 °C (➡). Cool the resulting two-phase mixture to room temperature and use forceps to remove the spin vane. Add 1.0 mL of diethyl ether to establish a reasonable volume for extraction of the organic phase.

***Isolation of Product.*** Using a Pasteur filter pipet, separate the bottom aqueous layer and transfer it to a 3.0-mL reaction vial. *Save* the organic phase in a 10 mL Erlenmeyer flask.

Extract the aqueous phase placed in the 3.0-mL conical vial with two 2.0-mL portions of diethyl ether. Upon the addition of each portion of ether, cap, shake, and carefully vent the vial and allow the layers to separate. The top ether layer is then separated using a Pasteur filter pipet and the ether extracts are combined with the previously saved organic phase in a 10 mL Erlenmeyer flask. Rinse the conical vial with 1.0 mL of ether and combine the rinse with the ether solution in the Erlenmeyer flask. *If a solid forms during the extraction, add a few drops of 0.1 M HCl.*

Extract the combined organic phases with 1.5 mL of 0.1 M HCl solution, followed by extraction with several 1.0-mL portions of distilled water or until the aqueous extract is neutral to pH paper. Add granular anhydrous magnesium sulfate ($MgSO_4$) (~200 mg or the tip of a spatula) to the combined organic phases and let it stand with occasional swirling for 20 min. If large clumps of drying agent form, add an additional ~100 mg of $MgSO_4$. The solution should be clear at the end of the drying period. Then transfer the solution to a tared 3.0-mL conical vial in 1-mL aliquots, add a preweighed boiling stone to the vial, and concentrate by carefully evaporating on a hot plate (60–65 °C) in the

HOOD

**hood** to yield the crude product residue.

Weigh the vial and calculate the crude yield. Obtain an IR spectrum of the crude alcohol and compare your result to that recorded in the literature (Aldrich Library of IR Spectra and/or SciFinder Scholar).

***Gas Chromatographic Analysis.*** This crude product is easily analyzed by gas chromatography. Prepare a sample for GC analysis as described in your Experiment 3106-3 Handout.

The liquid components elute in the order: unreacted 1-octene, a small amount of 2-octanol, and the major product, 1-octanol.

***Chemical Tests.*** A positive ceric nitrate test (Experiment 3106-8, pgs. 231-232 should confirm the presence of the alcohol grouping.

It might also be of interest to determine the solubility characteristics of this $C_8$ alcohol in water, ether, concentrated sulfuric acid, and 85% phosphoric acid (Experiment 3106-7, pgs. 223-225, concentrated acids will be dispensed by your TA). Do your results agree with what you would predict for this alcohol?

What other chemical tests could you perform to determine the difference between the starting alkene and the alcohol product?

## QUESTIONS

**7-9.** Using the hydroboration-oxidation reaction, outline a reaction sequence for each of the following conversions:

**(a)** 1-Butene to 1-butanol

**(b)** 1-Methylcyclohexene to *trans*-2-methylcyclohexanol

**(c)** 2-methylpropene to 2-methyl-1-propanol

**7-10.** When diborane ($B_2H_6$) dissociates in ether solvents, such as tetrahydrofuran (THF), a complex between borane ($BH_3$) and the ether is formed. For example,

$$B_2H_6 + 2\ \text{THF} \longrightarrow 2\ \text{THF}{:}\overset{+}{O}{-}\overset{-}{B}H_3$$

**(a)** In the Lewis sense, what is the function of $BH_3$ (acid or base) as it forms the complex? Explain.

**(b)** Write the Lewis structure for $BH_3$. Diagram its expected structure indicating the bond angles in the molecule.

**7-11.** Hydroboration/oxidation usually gives the anti-Markovnikov product. What would be the product of Markovnikov addition in the case of this experiment?

**7-12.** An advantage of the hydroboration reaction is that rearrangement of the carbon skeleton does not occur. This lack of migration contrasts with results obtained upon the addition of hydrogen chloride to the double bond. For example,

$$CH_3{-}CH(CH_3){-}CH{=}CH_2 \xrightarrow{HCl} CH_3{-}CH(CH_3){-}CH(Cl){-}CH_3 + CH_3{-}C(CH_3)(Cl){-}CH_2CH_3$$

$$CH_3{-}CH(CH_3){-}CH{=}CH_2 \xrightarrow{BH_3} CH_3{-}CH(CH_3){-}CH_2CH_2BH_2$$

Offer an explanation for the difference in these results.

## BIBLIOGRAPHY


Brown, H. C.; Subba, B. C. *J. Am. Chem. Soc.* **1956,** *78,* 5694.

Matteson, D. S. in *The Chemistry of the Metal-Carbon Bond,* Hartley, F. R., Ed.; Wiley: New York, 1987, Vol: 4, Chapter 3, p. 307.

Pelter, A.; Smith, K.; Brown, H. C. *Boron Reagents: Best Synthetic Methods;* Academic Press: New York, 1988.

Smith, K. *Organoboron Chemistry in Organometallics in Synthesis —A Manual;* Schlosser, M., Ed.; Wiley: New York, 1994.

**Also see**

Kabalka, G. W.; Maddox, J. T.; Shoup, T.; Bowers, K. R. *Organic Syntheses* **1996,** *73,* 116.

Experiment 3106-3 was based on the work reported by

Kabalka, G. W. J. *Chem. Educ.* **1975,** *52,* 745; also see Kabalka, G. W; Wadgaonkar, P. P.; Chatla, N. *J. Chem. Educ.* **1990,** *67,* 975.

Smith, M. B.; March, J. *Advanced Organic Chemistry,* 6th ed., Wiley-Interscience: New York, 2007, Chap. 15, p. 1075.

# Diels–Alder Reaction: 9,10-Dihydroanthracene-9,10-α,β-succinic Acid Anhydride

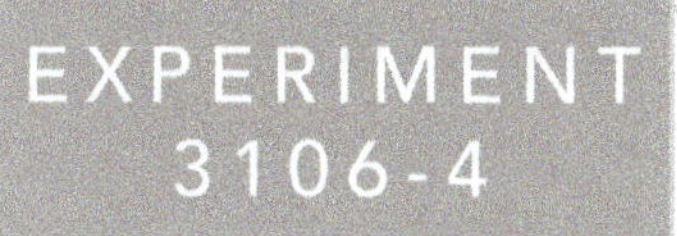


Common name: 9,10-dihydroanthracene-9,10-α,β-succinic acid anhydride
CA number: [85-43-8]
CA name as indexed: 1,3-isobenzofurandione, 3a,4,7,7a-tetrahydro-

***Purpose.*** The Diels–Alder reaction is investigated. You will explore the role of an aromatic ring system as the diene substrate in this addition reaction. The reaction studied in this experiment is an example of a 1,4 addition by an activated alkene dienophile across the 9,10 positions of anthracene. These addition products are often called adducts.

*Prior Reading*

*Technique 5:* Crystallization
Use of the Hirsch funnel (pp. 88–89)

## REACTION

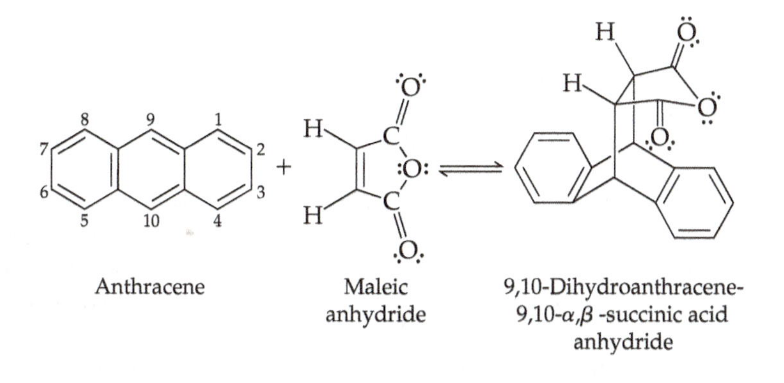


This famous class of reactions are named for **Otto Paul Hermann Diels** and **Kurt Alder,** who were primarily responsible for its development. Diels and Alder received the Nobel Prize in 1950 for this work.

**Otto Paul Hermann Diels (1876–1954)** Diels obtained his Ph.D. in 1899 while studying with Emil Fischer at the University of Berlin. He later became Associate Professor of Chemistry at the University of Berlin, and in 1916 he moved to the University of Kiel. In 1906 Diels discovered carbon suboxide gas ($C_3O_2$), obtained from the dehydration of malonic acid. He did extensive studies on saturated fats and fatty acids. Diels also developed the use of selenium as a mild dehydrogenation agent. This latter work led to the commercial production of polyunsaturated oils.

In the same year that he identified carbon suboxide, Diels began to investigate cholesterol with E. Abderhalden. The structure of this lipid had not yet been determined. He was the first to study the products of the selenium dehydrogenation of cholesterol and isolated a hydrocarbon ($C_{18}H_{16}$), which became known as "Diels' hydrocarbon." This substance proved to possess the basic steroidal ring structure; its subsequent synthesis by Diels in 1935 (almost 30 years after he started this work) led to the rapid elucidation of the structures of a vast array of steroidal sex hormones, saponins, cardiac glycosides, bile pigments, and adrenal cortical hormones, such as cortisone.

Diels, however, is best known for his discovery (with his student Kurt Alder) of the reaction that now bears his name (it is now known as the *Diels–Alder cycloaddition reaction*), which was first published in 1928 (see Diels, O.; Alder, K. *Liebigs Ann. Chem.* **1928,** *460,* 98). This reaction involves the 1,4 addition of dienophile reagents to diene substrates to produce six-membered cycloalkenes. The reaction has found extensive application in the synthesis of terpenes and other natural products, since six-membered rings abound in the metabolites of living systems, and because for some time it was one of the few methods available for the synthesis of these cyclic structures. Because of the impact of their work in the field of organic synthesis, Diels shared the 1950 Nobel Prize (in chemistry) with Alder. He also was the author of a popular textbook (*Einfuhrung in die organische Chemie*), first published in 1907, which went through 19 editions by 1962.[14]

**Kurt Alder (1902–1958)** Alder obtained his Ph.D. in 1926 while studying with Otto Diels at the University of Kiel. His dissertation was titled *Causes of the Azoester Reaction*. He became Professor of Chemistry at the University of Kiel in 1934 and later became head of the Chemical Institute at the University of Cologne (1940). For a few years (1936–1940) he was Research Director of the Baeyer dye works.

Together with Diels, Alder was responsible for the development of what came to be known as the Diels–Alder reaction. This reaction typically involves the reaction of a 1,3-conjugated diene with an activated alkene (dienophile) to form a six-membered cycloalkene. While reactions of this type had been reported as early as 1893, Diels and Alder were the first to recognize their great versatility. Alder continued to focus his academic research in this area following his graduate work with Diels. Over a number of years, Alder carried out a systematic study of the reactivity of a large number of dienes and dienophiles and established the structure and stereochemistry of many new adducts. He also expanded his doctoral research, studying the condensation of azoesters with dienes to yield the corresponding heterocyclic adducts.

Alder demonstrated that successful addition required that the diene double bonds possess an *s*-cis conformation (*s* refers to the single bond connecting the two double bonds). Furthermore, he realized that the bridged ring adducts formed by using cyclic dienes were closely related to natural products, such as camphor, and that this reaction offered a powerful route for the synthesis of a wide variety of naturally occurring compounds, particularly the terpenes. The Diels–Alder reaction also has been invaluable in the industrial synthesis of thousands of new organic materials from insecticides and dyes to lubricating oils and pharmaceuticals.

---

[14]See Newett, L. C. *J. Chem. Educ.* **1931,** *8,* 1493; *Dictionary of Scientific Biography*, C. C. Gillespie, Ed.; Scribner's: New York, 1971, Vol. IV, p. 90; *McGraw-Hill Modern Scientists and Engineers*, S. P. Parker, Ed.; McGraw-Hill: New York, 1980, Vol.1, p. 289.

Alder investigated autooxidation and polymerization processes particularly during his industrial years. For example, he was involved in an extensive study of polymerizations related to the formation of Buna-type synthetic rubbers.[16]

## DISCUSSION

The Diels–Alder reaction is one of the most useful synthetic reactions in organic chemistry because, in a *single* step, it produces two new carbon–carbon bonds and up to four stereocenters. It is an example of a [4 + 2] cycloaddition reaction (4 π electrons + 2 π electrons) between a *conjugated* 1,3-diene and an alkene (dienophile; to have an affinity for dienes, from the Greek *philos,* meaning loving), which leads to the formation of cyclohexenes. Alkynes may also be used as dienophiles, in which case the reaction produces 1,4-cyclohexadienes. The reaction proceeds faster if the dienophile bears electron-withdrawing groups and if the diene bears electron-donating groups. Thus, α,β-unsaturated esters, ketones, nitriles, and so on, make excellent dienophiles, which are often used in the Diels–Alder reaction. By varying the nature of the diene and dienophile, a very large number of compounds can be prepared. Unsubstituted alkenes, such as ethylene, are poor dienophiles and react with 1,3-butadiene only at elevated temperatures and pressures. These high activation energies (slow reactions) pose a particular problem for the Diels–Alder reaction. The Diels–Alder reaction is a reversible, equilibrium reaction that is not very exothermic. Since the equilibrium constant ($K_{eq}$) is temperature dependent [$K_{eq} = e^{-\Delta G/RT}$], $K_{eq}$ decreases with increasing temperature and eventually can become quite small at the high temperature needed for the reaction of an unactivated dienophile to proceed at a reasonable rate. Elevating the temperature will increase the rate of the reaction, but this will also reduce the amount of product formed, and therefore the lowest possible temperature must often be used.

The reaction is a thermal *cycloaddition* (a ring is formed), which occurs in one step and is thus a *concerted* reaction. Both new C—C single bonds and the new C=C π bond are formed simultaneously, as the three π bonds in the reactants break. The electron flow for the reaction is shown below. The reaction is thus classified as a *pericyclic* reaction (from the Greek meaning "around the circle").

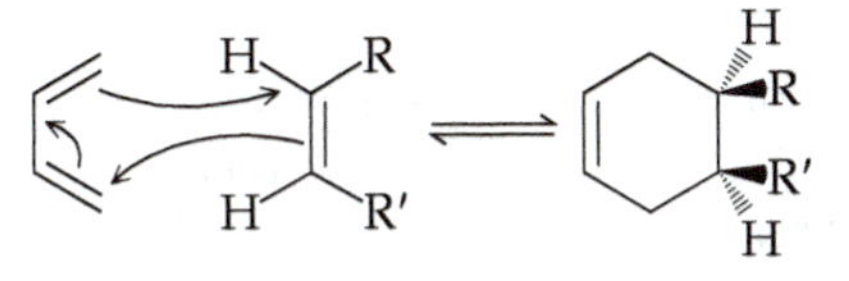


[16]See Allen, C. F. H. *J. Chem. Educ.* **1933,** *10,* 494; *Dictionary of Scientific Biography,* C. C. Gillespie, Ed.; Scribner's: New York, 1970,Vol. I, p. 105; *McGraw-Hill Modern Scientists and Engineers,* S. P. Parker, Ed., McGraw-Hill: New York, 1980, Vol. 1, p. 8.

The diene component must be in the *s-cis* (the "*s*" refers to the conformation about a single bond) conformation to yield the cyclic product with the cis C═C required by the six-membered ring. For this reason, cyclic dienes usually react more readily than acyclic species. For example, 1,3-cyclopentadiene, which is locked in the *s-cis* configuration, reacts with maleic anhydride about 1000 times faster than 1,3-butadiene, which prefers an *s-trans* conformation.

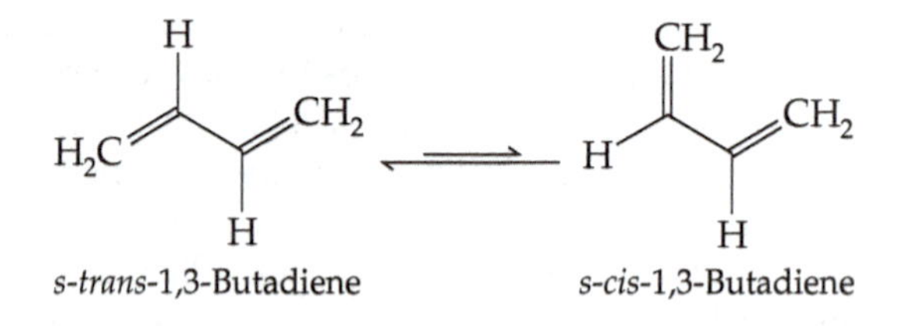


The reaction is highly stereospecific and the orientation of the groups on the dienophile are *retained* in the product; thus, the addition must be suprafacial–suprafacial. That is, by having the stereochemical information preserved, both new bonds are formed on the same face of the diene and on the same face of the dienophile. Thus, two groups that are cis on the dienophile will be cis in the product (and trans will give a trans product).

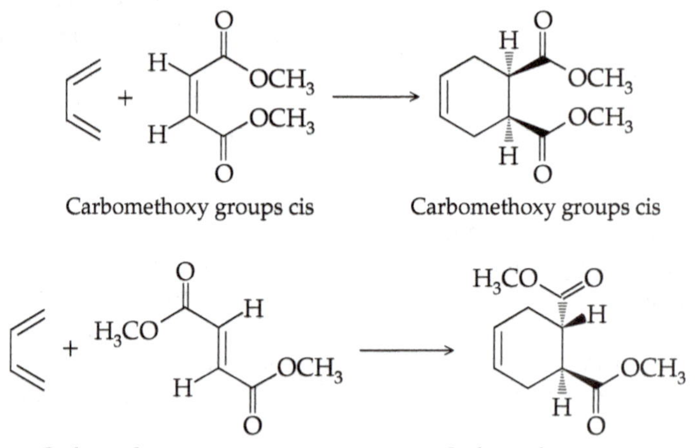


The reaction of cyclopentadiene with maleic anhydride demonstrates the further stereochemical consequence of the relative orientation of the reactants in Diels–Alder reactions. In this situation, there are two possible ways in which the reactants may bond. This reaction leads to the formation of two products: the endo and exo stereoisomers:

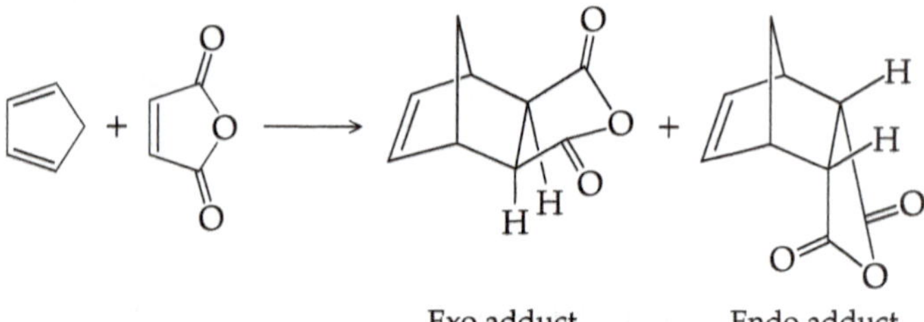

Generally, the endo form of the product predominates (endo-rule which is a result of secondary orbital overlap between the diene and dienophile), but endo/exo ratios may vary, depending on several steric and electronic factors and with reaction conditions.

In the present case, the central ring of anthracene is shown to possess the characteristic properties of a diene system. Thus, this aromatic compound reacts to form stable Diels–Alder adducts with many dienophiles at the 9 and 10 positions (the two positions on the central ring where new bonds can be made without destroying the aromaticity of the other two rings). Maleic anhydride, a very reactive dienophile, is used here in the reaction with anthracene. Note, that as this reaction is reversible, it is usually best carried out at the lowest possible temperatures consistent with an acceptable reaction rate.

Higher molecular weight polynuclear aromatic hydrocarbons (PAHs) containing the anthracene nucleus have also been found to react with maleic anhydride. These ring systems, however, can differ widely in reaction rates. Typical examples of those systems that undergo the Diels–Alder reaction are 1,2,5,6-dibenzanthracene (a), 2,3,6,7-dibenzanthracene (pentacene) (b), and 9,10-diphenylanthracene (c).

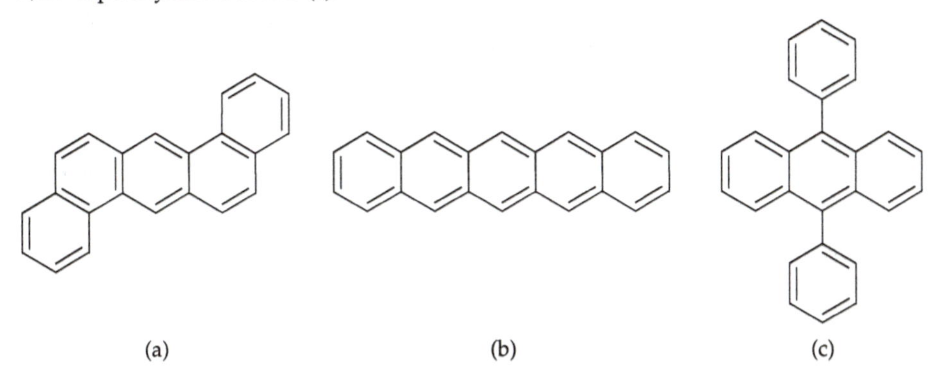

(a) (b) (c)

## EXPERIMENTAL PROCEDURE

Estimated time to complete the experiment: 1.5 h.

| Physical Properties of Reactants and Product | | | | | |
|---|---|---|---|---|---|
| **Compound** | **MW** | **Amount** | **mmol** | **mp (°C)** | **bp (°C)** |
| Anthracene | 178.24 | 80 mg | 0.44 | 216 | |
| Maleic anhydride | 98.06 | 40 mg | 0.40 | 60 | |
| Xylenes | | 1.0 mL | | | 137–140 |
| 9,10-Dihydroanthracene-9,10-α,β-succinic acid anhydride | 276 | | | 261–262 | |

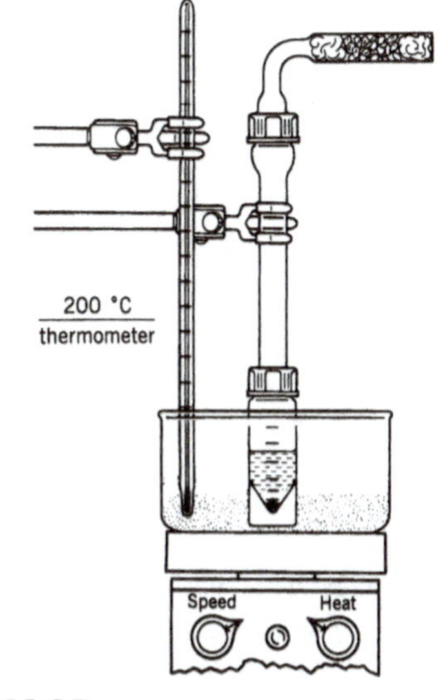


Anthracene, 80 mg + maleic anhydride, 40 mg + xylene, 1.0 mL

***Reagents and Equipment.*** Weigh and place 80 mg (0.44 mmol) of anthracene and 40 mg (0.40 mmol) of maleic anhydride in a 3.0-mL conical vial containing a boiling stone and equipped with an air condenser protected by a calcium chloride drying tube. Now, while in the **hood,** add 1.0 mL of xylene to the solid mixture using an automatic delivery pipet (➡).

*NOTE. High-purity grades of anthracene and maleic anhydride are strongly recommended. Anthracene may be recrystallized from 95% ethanol. A mixture of xy-*

*lenes with a boiling-point range of 137–140 °C is sufficient, but the solvent (xylenes) should be dried over molecular sieves before use.*

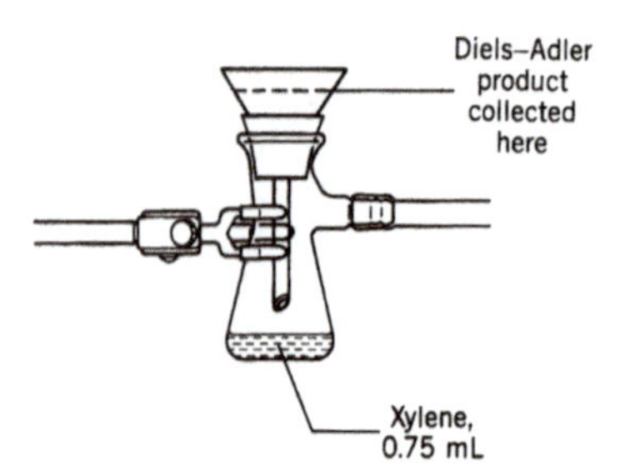


***Reaction Conditions.*** Heat the reaction mixture at reflux for 30 min at about 200 °C. During this time the initial yellow color of the reaction mixture gradually disappears. (Why?) Allow the resulting bleached solution to cool to room temperature and then place it in an ice bath for 10 min to complete the crystallization of the product.

***Isolation of Product.*** Collect the crystals by vacuum filtration using a Hirsch funnel and wash the filter cake with two 300-μL portions of cold ethyl acetate (➡). Partially dry the filter cake under suction using plastic food wrap (see Prior Reading). Transfer the filtered, washed, and partially dried product to a porous clay plate or filter paper to complete the drying.

***Purification and Characterization.*** The product is often of sufficient purity for direct characterization at this point. Weigh the dried Diels–Alder product and calculate the percent yield. Determine the melting point and compare your result with the value listed above. Obtain an IR spectrum of the product.

*NOTE. It is suggested that a hot-stage melting point apparatus, such as the Fisher–Johns, be used due to the high melting point of the product.*

## QUESTIONS

**7-13.** There are four reasonable resonance structures for anthracene. Draw them.

**7-14.** Offer an explanation of why anthracene preferentially forms a Diels–Alder adduct at the 9,10 positions.

**7-15.** Predict the structure of the product formed in the following reactions.

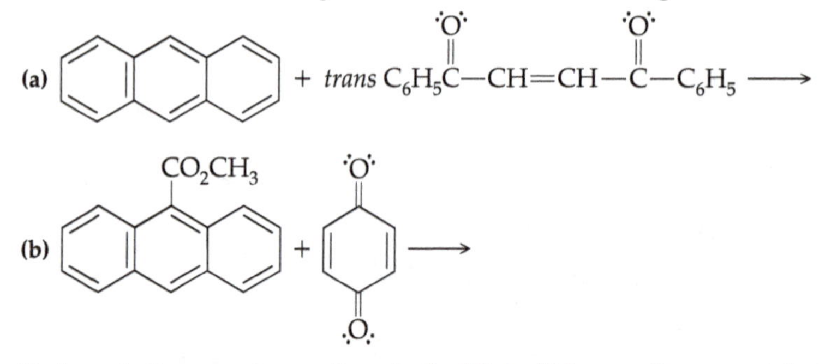


**7-16.** Cyclopentadiene reacts as a diene in the Diels–Alder reaction a great deal faster than does 1,3-butadiene. Explain.

## BIBLIOGRAPHY

**For a review of diastereoselectivity in the reaction see**

Coxon, J. M. et al., *Diastereofacial Selectivity in the Diels–Alder Reaction* in *Advances in Detailed Reaction Mechanisms* **1994,** 3, 131.

Waldmann, H. *Synthesis* **1994,** 535.

# Friedel–Crafts Acylation: Acetylferrocene and Diacetylferrocene

EXPERIMENTS 3106-5 & 6

Common name: acetylferrocene
CA number: [1271-55-2]
CA name as indexed: ferrocene, acetyl-

Common names: diacetylferrocene, 1,1′-diacetylferrocene
CA number: [1273-94-5]
CA name as indexed: ferrocene, 1,1′-diacetyl-

***Purpose.*** To investigate the conditions under which the synthetically important Friedel–Crafts acylation (alkanoylation) reaction is carried out. The reaction described here illustrates electrophilic aromatic substitution on an aromatic ring contained in an organometallic compound. The highly colored products are easily separated by both thin-layer and dry-column chromatography.

*Prior Reading*

*Technique 4:* Solvent Extraction
Liquid–Liquid Extraction (p. 72)
Drying of the Wet Organic Layer (pp. 80–83)

*Technique 6:* Chromatography
Column Chromatography (pp. 92–95)
Thin-Layer Chromatography (pp. 97–99)
Concentration of Solutions (pp. 101–104)

## REACTION

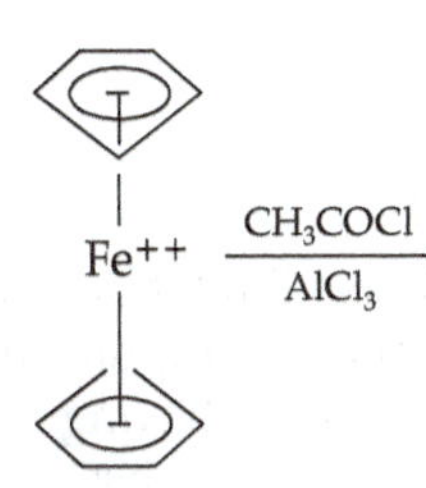


Ferrocene

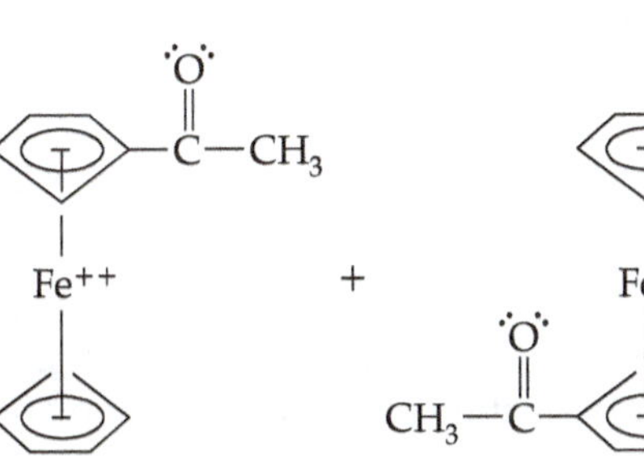


1-Acetylferrocene

+

O
C—CH3
Fe++
O
CH3—C

1,1′-Diacetylferrocene

## DISCUSSION

Electrophilic aromatic substitution is an important reaction and is typically applied to benzene and benzene derivatives. Ferrocene is a stable bright orange organometallic aromatic compound, discovered serendipitously in 1954. Sometimes called a "sandwich" compound, or a "metallocene," it consists of an iron (II) cation surrounded by two cyclopentadienyl anions, like a piece of meat between two slices of bread. The cyclopentadienyl anion is aromatic (with 6 $\pi$ electrons), just like benzene, so it is perhaps not surprising that ferrocene undergoes many of the same reactions as benzene.

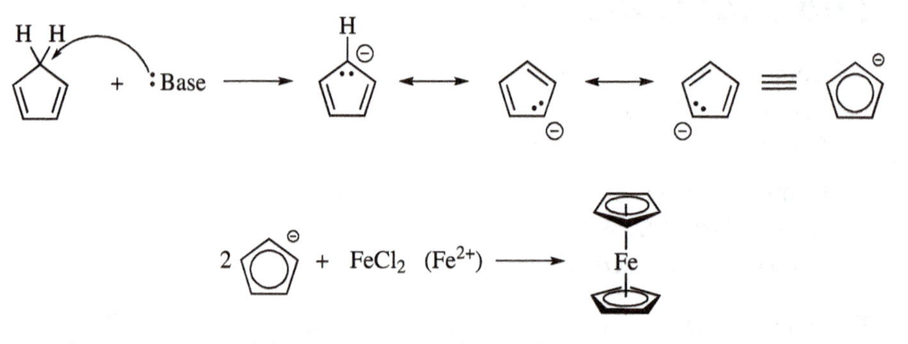


The generation of the appropriate electrophile (carbocation, carbocation complex, or acylium ion) in the presence of an aromatic ring system (nucleophile) can lead to alkylation or acylation of the aromatic ring. This set of reactions, discovered by Charles Friedel and James Crafts in 1877, originally used aluminum chloride as the catalyst. The reaction is now known to be catalyzed by a wide range of Lewis acids, including ferric chloride, zinc chloride, boron trifluoride, and strong acids, such as sulfuric, phosphoric, and hydrofluoric acids.

Alkylation is accomplished by use of haloalkanes, alcohols, or alkenes; any species that can function as a carbocation precursor. The alkylation reaction is accompanied by two significant and limiting side reactions: polyalkylation, due to ring activation by the added alkyl groups, and rearrangement of the intermediate carbocation. These lead to diminished yields, and mixtures of products that can be difficult to separate as shown here:

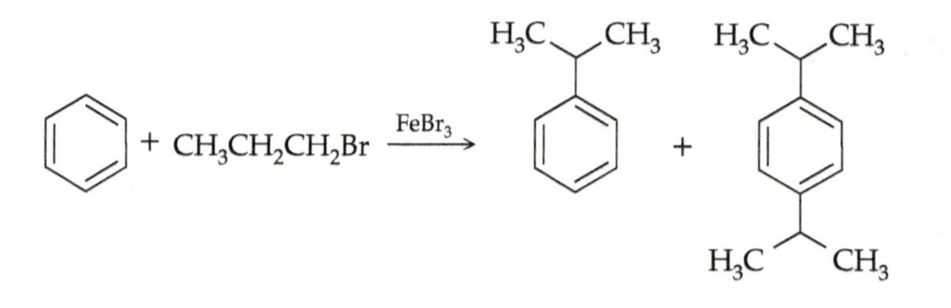


**Acylation** reactions generally do not suffer from these limitations, and can be conducted using acid chlorides or anhydrides as the electrophilic reagents. Since the introduction of a carbonyl group onto the aromatic ring in an acylation reaction deactivates the ring, the problem of multiple substitution is avoided. The acylium cation, since it is resonance stabilized, is unlikely to rearrange.

The mechanism involves three steps: (1) formation of a cationic electrophile, (2) nucleophilic attack on this electrophile by an aromatic ring, and (3) loss of a proton from the resulting cation to regenerate the aromatic ring system. The mechanism shown here represents the $AlCl_3$ catalyzed generation of the acylium ion electrophile from acetyl chloride (ethanoyl chloride), followed by subsequent nucleophilic attack by the ferrocene ring system:

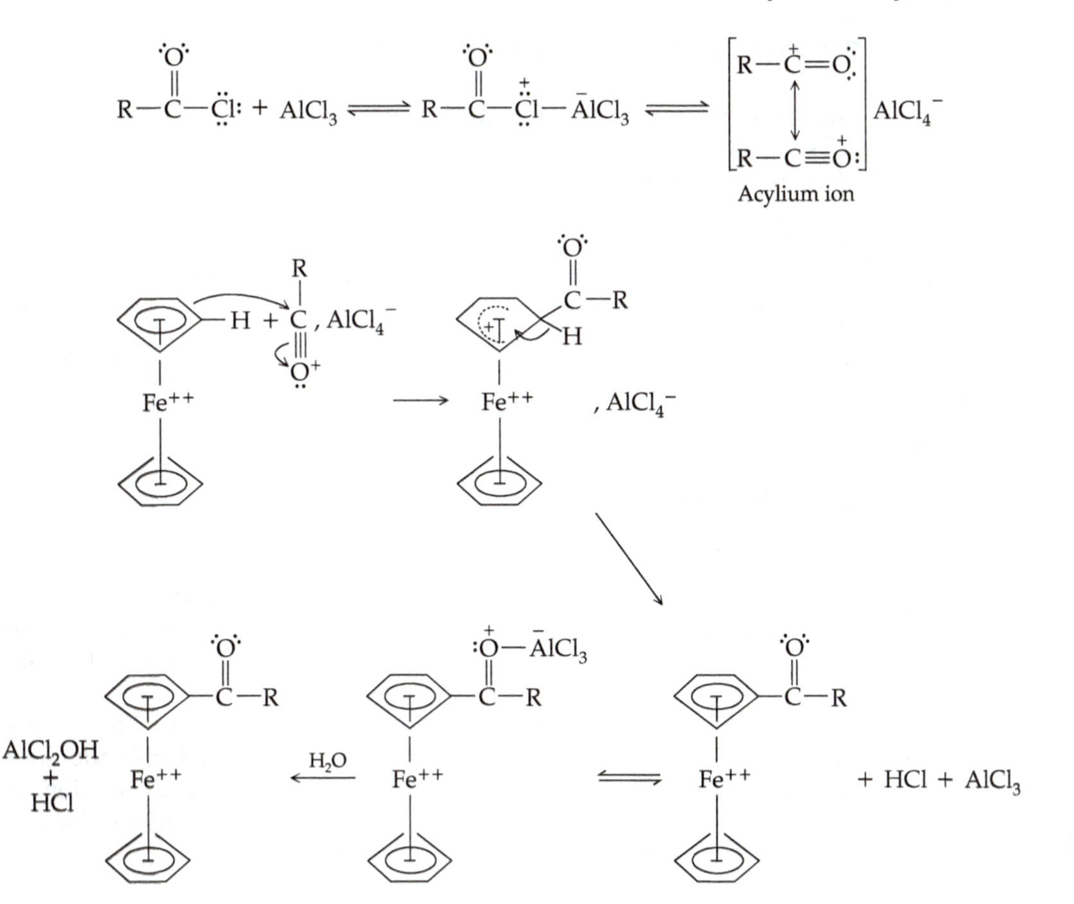


The present experiment also demonstrates the practical value of monitoring reaction progress by TLC analysis.

**Charles Friedel (1832–1899)** Friedel was Professor of Chemistry at the Sorbonne. He did extensive work on ketones, lactic acid, and glycerol and he discovered isopropyl alcohol. He is best known for his studies of the use of aluminum chloride in the synthesis of aromatic products (**Friedel–Crafts reaction,** 1877). Friedel prepared a series of esters of silicic acid and demonstrated the analogy between the compounds of carbon and silicon, meanwhile confirming the atomic weight of silicon. He determined the vapor densities and molecular weights of the chlorides of aluminum, iron, and gallium.[23]

**James Mason Crafts (1839–1917)** Crafts was Professor of Chemistry at Cornell University and later at Massachusetts Institute of Technology, where he eventually became President. Crafts studied with Bunsen (Germany) and Wurtz (France) and also worked on the organic compounds of silicon. Crafts was, of course, the codiscoverer of the **Friedel–Crafts reaction.** He also carried out investigations in the area of thermochemistry, catalytic effects in concentrated solutions, and determination of the densities of the halogens at high temperatures.[24]

## EXPERIMENTAL PROCEDURE

Estimated time to complete the experiment: two 3.0-h laboratory periods. You will submit two lab reports: one for each lab period. You will include the post-lab questions in the Experiment 6 lab report.

| Physical Properties of Reactants | | | | | | | |
|---|---|---|---|---|---|---|---|
| Compound | MW | Amount | mmol | mp (°C) | bp (°C) | $d$ | $n_D$ |
| Aluminum chloride | 133.34 | 150 mg | 1.12 | 190 | | | |
| Acetyl chloride | 78.50 | 80 μL | 1.12 | | 51 | 1.11 | 1.3898 |
| Ferrocene | 186.04 | 100 mg | 0.54 | 173 | | | |
| Methylene chloride | | 4.0 mL | | | 40 | | |

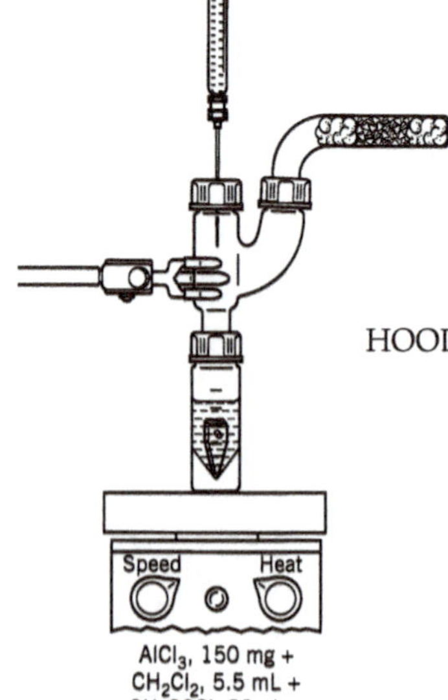


### *Reagents and Equipment*

#### *LAB PERIOD 1 (Experiment 5):*

Equip a tared 5.0-mL conical vial containing a magnetic spin vane with a Claisen head protected by a calcium chloride drying tube and a septum cap. Weigh and add 150 mg (1.12 mmol) of fresh, anhydrous aluminum chloride ( ⬅).

Add 2.5 mL of methylene chloride to the reaction vial. In the hood, with swirling, add 80 μL (1.12 mmol) of acetyl chloride from an automatic delivery pipet. Prepare in a shell vial a solution of 100 mg (0.54 mmol) of ferrocene dissolved in 1.5 mL of methylene chloride. Use a syringe to add this solution to the resulting mixture.

HOOD

*NOTE. Use a capped conical vial and recap it between the addition of each reagent. After addition of the acetyl chloride, attach the vial to the Claisen head and add the ferrocene solution through the septum as shown in the figure. Do this in one or two portions, depending on whether a 1- or 2-mL syringe is used.*

At this stage, the reaction mixture turns a deep-violet color.

---

CAUTION: It is important to minimize the exposure to moist air during these transfers. Both the aluminum chloride and the acetyl chloride are highly moisture sensitive so that rapid, yet accurate, manipulations are necessary to minimize deactivation of these reagents, which leads to poor results. In addition, both chemicals are irritants. Avoid breathing the vapors or allowing the reagents to come in contact with skin. These reagents must be dispensed in the **hood.**

HOOD

---

***TLC Sample Instructions.*** Obtain an aliquot for TLC analysis by removing a small amount of the reaction mixture by touching the open end of a Pasteur pipet to the surface of the solution. First, remove the cap from the straight neck of the Claisen head, and then insert the pipet down the neck so as to touch the surface

[23] See *Berichte* **1899,** *32,* 372; Crafts, J. M. *J. Chem. Soc.* **1900,** *77,* 993; *Bull. Soc. Chim. Fr.* **1900,** *23,* 1; Béhal, A. *ibid.,* **1932,** *51,* 1423; Willemart, A. *J. Chem. Educ.* **1949,** *26,* 3.

[24] Ashdown, A. A. *J. Chem. Educ.* **1928,** *5,* 911; Talbot, H. P. *J. Am. Chem. Soc.* **1917,** *39,* 171; Richards, T. W. *Proc. Am. Acad.* **1917–1919,** 53, 801; Cross, C. R. *J. Natl. Acad. Sci.* **1914,** *9,* 159.

of the solution. Dissolve this aliquot in about 10 drops of cold methylene chloride in a small capped vial. Mark the vial and *save* it for TLC analysis.

***Reaction Conditions.*** Following the addition of the ferrocene solution, note the time, and begin stirring. Allow the reaction to proceed at room temperature for 15 min.

***Isolation of Product.*** Quench the reaction by transferring the mixture by Pasteur pipet to a 15-mL capped centrifuge tube containing 5.0 mL of ice water. Cool the tube in an ice bath and neutralize the resulting solution by dropwise addition (calibrated Pasteur pipet) of about 0.25 mL of 25% aqueous sodium hydroxide.

*NOTE. Avoid an excess of base. Use litmus or pH paper to confirm the neutralization.*

Now extract the mixture with three 3-mL portions of methylene chloride. Cap the tube, shake, vent, and allow the layers to separate (a Vortex mixer may be used in this step). Remove the lower (methylene chloride) layer using a Pasteur filter pipet. Combine the methylene chloride extracts in a 25-mL Erlenmeyer flask, and dry the wet solution over about 200 mg of granular anhydrous magnesium sulfate for 20 min. Transfer the dried solution to a tared 10-mL Erlenmeyer flask, using a Pasteur filter pipet, in aliquots of 4 mL each. After each transfer, concentrate the solution, in the **hood,** under a stream of air on a warm hot plate to a volume of about 0.5 mL. Rinse the drying agent with an additional 2.0 mL of methylene chloride and combine this rinse with the concentrate. Remove several drops of this solution by Pasteur pipet and place them in a capped vial containing 10 drops of *cold* methylene chloride. Mark the vial and *save* it for TLC analysis. Remove the remaining solvent by warming on a hot plate in the **hood** to yield the crude, solid product (~130 mg). Weigh the residue.

HOOD

HOOD

*If the reaction is performed over 2 lab periods, this is a convenient point at which to stop. However, if time permits, perform the TLC analysis now.*

***Thin-Layer Chromatographic Analysis.*** Use TLC to analyze the two samples saved above. Also analyze a standard of the starting ferrocene at the same time. Use the developed TLC plates as a guide to determine the product mixture obtained in the reaction and as an aid in determining the appropriate elution solvent required for separation of the mixture by dry-column chromatography.

*INFORMATION. Good results have been achieved by conducting the TLC analysis with Eastman Kodak silica gel–polyethylene terephthalate plates (#13179). Activate the plates at an oven temperature of 100 °C for 30 min. Place them in a desiccator for cooling and storing until used. Elute the plates using pure methylene chloride as the elution solvent. Visualization of unreacted ferrocene can be enhanced with iodine vapor. See Prior Reading for methods of TLC analysis and determination of* $R_f$ *values.*

***LAB PERIOD 2 (Experiment 6):***

***Purification and Characterization.*** Now purify the reaction products formed in the reaction by dry-column chromatography. The term *dry-column chromatography* refers to the fact that the column is packed with dry alumina, rather than with a slurry (see Prior Reading). Dissolve the solid product residue isolated above in 0.5 mL (calibrated Pasteur pipet) of methylene chloride in a small vial. Mix this solution with 300 mg of alumina (activity III, see Glossary) in a tared vial, and evaporate the solvent under a stream of air in the **hood** to give a product–alumina mixture. Assemble a chromatographic buret column (provided in the hood) in the following order (bottom to top): prewashed cotton plug, 5 mm of sand, 60–80 mm of alumina (~5.0 g, activity III), *one-half* of the product–alumina mixture, and 10 mm of alumina (⬅).

HOOD

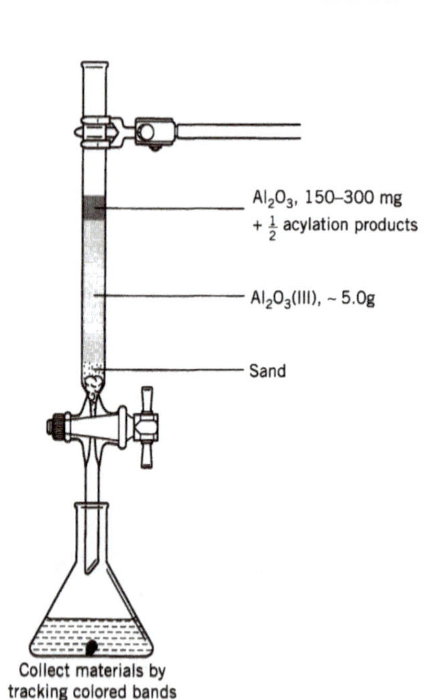


*NOTE. This procedure prevents overloading of the chromatographic column during the separation of the reaction products. If the yield of crude reaction products exceeds 75 mg (the usual case), introduce one-half of the alumina–product mixture to the column. If the crude products, however, are obtained in quantities of less than 75 mg, add the entire alumina–product mixture to the top of the column. If only one-half of the alumina–crude ferrocene acylation product mixture is placed on the column, it is important to reweigh the tared vial to establish a reasonably accurate estimate of the overall yields obtained in the reaction.*

Given the polar nature of the alumina, the products will elute in order of increasing polarity: ferrocene followed by acetylferrocene, followed by diacetylferrocene. Begin elution of the column with pure hexane if TLC analysis indicates that unreacted ferrocene is present in the product mixture. Be sure to add the initial solvent down the side of the column so as not to disturb the alumina bed. During elution, the ferrocenes will separate into two or three bands of different colors on the column. The volume of each eluted fraction should be in the range of 2–5 mL if the band is carefully tracked down the column. Once elution has begun, do not let the column go dry. The elution solvent level should remain about 20 mm above the alumina. Once the ferrocene band has been collected, continue the elution with a 1:1 mixture of $CH_2Cl_2$/hexane to obtain the monosubstituted product. The polarity of this elution solvent may need to be increased to collect the monosubstituted product. If a 1:1 mixture of $CH_2Cl_2$:hexane does not move the monosubstituted product band, try pure $CH_2Cl_2$. Further elution with a 9:1 mixture of $CH_2Cl_2$/$CH_3OH$ will elute the disubstituted material. The column can then be stripped by eluting with pure $CH_3OH$. Collect and save each chromatographic band separately. Store the solvent that elutes without color in an Erlenmeyer flask or beaker until you have isolated all your product. In the **hood,** remove the solvent under a stream of air, using a warm hot plate. During concentration of the solvent, spot each fraction on a TLC plate to verify the separation and purity of its contents.

HOOD

Determine the melting point of each of the isolated products and compare your results to those reported in the literature.

Obtain an IR spectrum of each material and compare the results with an authentic sample or spectra found in the literature (*The Aldrich Library of IR Spectra* and/or SciFinder Scholar). Interpretation of the spectra allows an unambiguous determination of substitution based on the presence or absence of absorption in the 1100- to 900-wavenumber region of the spectrum.

***Characterization of the Fractions (Total Sample) Isolated From the Reaction Workup***

Acetylferrocene: mp ________ °C; ________ mg, ________ mmol

Diacetylferrocene: mp ________ °C; ________ mg, ________ mmol

**Total:** ________ mmol, ________ % yield

## QUESTIONS

**7-17.** In the formation of diacetylferrocene, the product is always the one in which each ring is monoacetylated. Why is no diacetylferrocene produced in which both acetyl groups are on the same aromatic ring?

**7-18.** Predict the major product(s) in each of the following Friedel–Crafts reactions. If the reaction does not occur, offer a reasonable explanation for that fact.

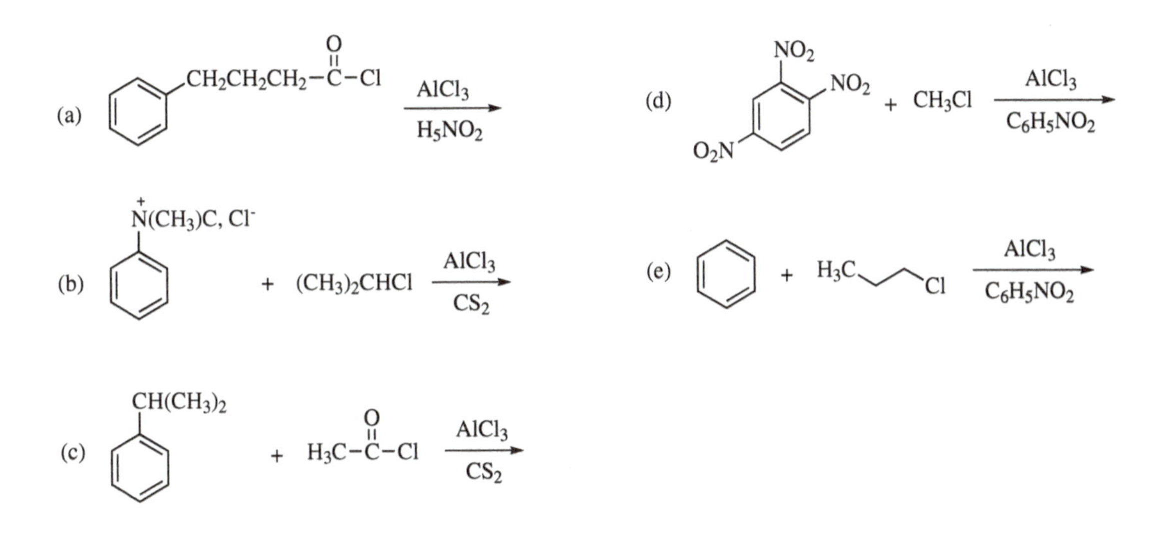


## BIBLIOGRAPHY

**The acetylation of ferrocene has been monitored using chromatographic techniques. Several references are listed:**

Amenta, D. S.; DeVore, T. C.; Gallaher, T. N.; Zook, C. M. *J. Chem. Educ.* **1996,** *73,* 575.

Bohen, J. M.; Joullié, M. M.; Kaplan, F. A. *J. Chem. Educ.* **1973,** *50,* 367.

Nerwith, T. L.; Srouji, N. *J. Chem. Educ.* **1995,** *72;* 455.

**Several reviews on the Friedel–Crafts reaction, selected from many cited in the literature:**

Gore, P. H. *Chem. Rev.* **1955,** *55,* 229.

Harada, T; Ohno, T.; Kobayashi, S.; Mukaiyama, T. *Synthesis* **1991,** 1216.

Heaney, H. *Comp. Org. Syn.* **1991,** *2,* 733.

Olah, G. A., Ed., *Friedel–Crafts and Related Reactions;* Interscience: New York, 1963–1965; Vols. I–IV.

Smith, M. B.; March. J. *Advanced Organic Chemistry,* 6th ed.; Wiley-Interscience: New York, 2007, Chap. 11, p. 719.

**Several of the large number of Friedel–Crafts acetylation reactions given in *Organic Syntheses:***

Adams, R.; Noller, C. R. *Organic Syntheses;* Wiley: New York, 1941; Collect. Vol. I, p. 109.

Arsenijivic, L; Arsenijivic, V.; Horeau, A.; Jacques, J. *Organic Syntheses;* Wiley: New York, 1988; Collect. Vol. VI, p. 34.

Fieser, L. F. *Organic Syntheses;* Wiley: New York, 1941; Collect. Vol. I, p. 517.

Fieser, L. F. *Organic Syntheses;* Wiley: New York, 1955; Collect. Vol. III, p. 6.

Marvel, C. S.; Sperry, W. M. *Organic Syntheses;* Wiley: New York, 1941; Collect. Vol. I, p. 95.

Olson, C. E.; Bader, A. F. *Organic Syntheses;* Wiley: New York, 1963; Collect. Vol. IV, p. 898.

Sims, J. J.; Selman, L. H.; Cadogan, M. *Organic Syntheses;* Wiley: New York, 1988; Collect. Vol.VI, p. 744.

# *Qualitative Identification of Organic Compounds*

EXPERIMENTS 3106-7–9

***Purpose.***

1) To determine the identity of two unknown compounds (a solid and a liquid) using the classical chemical methods of functional group identification and the preparation of derivatives.
2) To confirm those identifications by spectroscopic means.

***Historical Background.*** The chemical tests you will perform were once the primary method of determining the identity of an unknown compound. In the older literature, (prior to the 1960's) when a new compound was reported for the first time (whether is was isolated from natural sources or synthesized in lab), it was of critical importance to report not only the melting or boiling point, but also those of derivatives. This was the only way that someone else could confirm that they subsequently made the same compound. With the advent of spectroscopic methods (IR and NMR) these data are now reported in lieu of derivatives, but the purpose remains the same, to enable someone else to confirm that they have made or isolated the same compound. Nevertheless, the chemical tests serve a useful pedagogical purpose, as they illustrate characteristic reactivities of many functional groups. This can also be one of the most fun and memorable experiments in organic lab. Enjoy it!

## DISCUSSION

One of the exciting challenges that a chemist faces on a regular basis is identifying organic compounds. This challenge is an excellent way for a student to be initiated into the arena of chemical research. Millions of organic compounds are recorded in the chemical literature. At first glance it may seem a bewildering task to attempt to identify one certain compound from this vast array, but most of these substances can be grouped, generally by functional groups, into a comparatively small number of classes. In addition, chemists have an enormous database of chemical and spectroscopic information, which has been correlated and organized over the years, at their disposal. Determination of the physical properties of a molecule, the functional groups present, and the reactions the molecule undergoes has allowed the chemist to establish a systematic, logical identification scheme.

Forensic chemistry, the detection of species causing environmental pollution, the development of new pharmaceuticals, progress in industrial research, and development of polymers all depend to a large extent on the ability of the chemist to isolate, purify, and identify specific chemicals. The objective of organic qualitative analysis is to place a given compound, through screening tests, into one of a number of specific classes, which in turn greatly simplifies the identification of the compound.This screening is usually done by using a series of preliminary observations and chemical tests, in conjunction with the instrumental data that developments in spectroscopy have made available to the analyst. The advent of infrared (IR) and nuclear magnetic resonance

(NMR) spectroscopy and mass spectrometry (MS) have had a profound effect on the approach taken to identify a specific organic compound. Ultraviolet (UV) spectra may also be utilized to advantage with certain classes of materials.

The identification of your unknowns will be completed over the course of three experiments. The systematic approach taken in these experiments for the identification of an unknown organic compound is as follows:

**Experiment 3106-7:** Preliminary tests are performed to determine the physical nature of the compound.

The solubility characteristics of the unknown species are determined. This identification can often lead to valuable information related to the structural composition of an unknown organic compound.

**Experiment 3106-8:** The spectroscopic method of analysis is utilized. As your knowledge of chemistry develops, you will appreciate more and more the revolution that has taken place in chemical analysis over the past 25 -30 years and the powerful tools now at your disposal for the identification of organic compounds. After obtaining IR spectra of your solid and liquid unknowns, you will confirm your functional group by performing the classification tests.

Classification tests are carried out to detect *common functional groups* present in the molecule. Most of these tests may be done using a few drops of a liquid or a few milligrams of a solid. An added benefit to the student, especially in relation to the chemical detection of functional groups, is that a vast amount of chemistry can be *observed* and *learned* in performing these tests. The successful application of these tests requires that you develop the ability to think in a logical manner and to interpret the significance of each result based on your observation.

**Experiment 3106-9:** Preparation of derivatives will be carried out to identify your unknown once the functional group has been successfully identified. You will prepare at least one and up to three derivatives of your unknown.

It is important to realize that *negative* findings are often as important as positive results in identifying a given compound. Cultivate the habit of following a *systematic pathway or sequence* so that no clue or bit of information is lost or overlooked along the way. It is important also to develop the *attitude* and *habit* of planning ahead. Outline a logical plan of attack, depending on the nature of the unknown, and follow it. As you gain more experience in this type of investigative endeavor, the planning stage will become easier.

At the *initial* phase of your training, the unknowns to be identified will be pure materials and will all be known compounds. The properties of these materials are recorded in the literature, and in the "blue books" located in your laboratory.

*Record all observations and results of the tests in your laboratory notebook.* Review these data as you execute the sequential phases of your plan. This method serves to keep you on the path to success.

A large number of texts have been published on organic qualitative analysis. Several references are cited at the end of these experiments.

## DISCUSSION

Because everyone has a different unknown, with different functional groups, everyone will do a different series of tests and get different answers. Therefore, it is important to realize that this is not a cookbook laboratory; you will need to read ahead of lab and to think carefully about what reactions/tests you should perform during the lab period.

The description is fairly comprehensive, allowing for the identification of a wide range of functional groups. You will have one of these functional groups in your unknown:

carboxylic acid
ketone
phenol
aldehyde
alcohol
amine

In addition, you may have one of these functional groups in your molecule:

aromatic ring
ether

You will not have acid chlorides, halides, nitro groups, anhydrides, esters or amides as functional groups.

All of the unknowns are taken from those listed in supplemental tables for the functional groups listed above. These tables are available in the "blue book" found in your laboratory. The melting points or boiling points of the compounds are listed, along with derivative information.

***Timing.*** It is anticipated that in one lab period you will be able to carry out one experiment for both unknowns. You should use your time wisely.

You are encouraged to discuss your results with your TA over the course of these experiments. You will turn in to your TA duplicate copies of your results at the end of each lab period, detailing your conclusions. If you have made an error in an identification, your TA can help you correct that error before you waste too much time pursuing it. (For example, if you thought you had an alcohol, but you did not, you could waste a lot of time trying unsuccessfully to make derivatives.)

---

CAUTION: Some of the unknown compounds are toxic and/or carcinogenic. Take care when identifying their odor, wafting the scent toward your nose. Do not expose yourself to them unnecessarily.

---

## EXPERIMENT 3106-7: PRELIMINARY TESTS

First, put a small sample of each of your two unknowns (one liquid and one solid) in a shell vial. Write down their ID numbers, label the vials, and keep the vials in a small beaker when working on them. This will prevent them from spilling. You should only need ~1/2 a shell vial full of sample over the three lab periods.

Preliminary tests help you select a route to follow to ultimately identify the unknown material at hand. These tests frequently consume material, so, given the amounts of material generally available at the micro- or semimicroscale level, judicious selection of the tests to perform must be made (in some tests, the material analyzed may be recovered). Each preliminary test that can be conducted with *little expenditure of time and material* can offer valuable clues as to which class a given compound belongs.

### Nonchemical Tests

***Physical State.*** Determine the melting point, using a small amount of the solid material. A narrow melting point range (1–2° C is a good indication that the material is quite pure.

If the material is a *liquid,* the boiling point is determined by using a clean Hickman stillhead (Figure 5.5, p. 63). The condensed liquid can be recovered and reused in the subsequent tests.

***Color.*** Since the majority of organic compounds are colorless, examination of the color can occasionally provide a clue as to the nature of the sample. Use caution, however, since tiny amounts of some impurities can color a substance. Aniline is a classic example. When freshly distilled it is colorless, but on standing a small fraction oxidizes and turns the entire sample a reddish-brown color.

Colored organic compounds contain a *chromophore,* usually extended conjugation in the molecule. For example, *trans*-1,2-dibenzoylethylene (Experiment 3105-12) is yellow; 5-nitrosalicylic acid is light yellow; tetraphenylcyclopentadienone is purple.

Can you identify the chromophores that cause these compounds to be colored? Note that a colorless liquid or white solid would not contain these units. Thus, compounds containing these groupings would be excluded from consideration as possible candidates in identification of a given substance.

***Odor.*** Detection of a compound's odor can occasionally be of assistance, since the vast majority of organic compounds have no definitive odor. You should become familiar with the odors of the common compounds or classes. For example, aliphatic amines have a fishy smell; benzaldehyde (like nitrobenzene and benzonitrile) has an almond odor; esters have fruity odors (Experiments 3106-12). Common solvents, such as acetone, diethyl ether, and toluene, all have distinctive odors. Butyric and caproic acids have rancid odors. Low molecular weight mercaptans (—SH) have an intense smell of rotten eggs. In many cases, extremely small quantities of certain relatively high molecular weight compounds can be detected by their odor. For example, a $C_{16}$ unsaturated alcohol released by the female silk worm moth elicits a response from male moths of the same species at concentrations of 100 molecules/$cm^3$. Odors are an important facet of chemical communication between plants and animals and often result in a spectacular behavioral response (see also Experiment 3106-12).

Odor detection in humans involves your olfactory capabilities and thus can be a helpful lead, *but very rarely can this property be used to strictly classify or identify a substance.* As mentioned above, contamination by a small amount of an odorous substance is always a possibility.

---

CAUTION: You should be very cautious when detecting odors. Any odor of significance can be detected several inches from the nose. Do not place the container closer than this to your eyes, nose, or mouth. Open the container of the sample and gently waft the vapors toward you.

---

## EXPERIMENT 3106-7 (CONT'D): SOLUBILITY CHARACTERISTICS

Determination of the solubility characteristics of an organic compound can often give valuable information as to its structural composition. It is especially useful when correlated with spectral analysis. Several schemes have been proposed that place a substance in a definite group according to its solubility in various solvents. The scheme presented below is similar to that outlined in-Shriner et al.[8]

There is no sharp dividing line between soluble and insoluble, and an arbitrary ratio of solute to solvent must be selected. We suggest that a compound be classified as soluble if its solubility is greater than 15 mg/500 μL of solvent.

Carry out the solubility determinations, at ambient temperature, in 10 × 75-mm test tubes or shell vials. For the solid samples, place the sample (~15 mg) in the test tube and add a total of 0.5 mL of solvent in three portions from a graduated or calibrated Pasteur pipet. Between addition of each portion, stir the sample vigorously with a glass stirring rod for 1.5–2 min. If the sample is water soluble, test the solution with litmus paper to assist in classification according to the solubility scheme that follows. For liquid samples, determine the solubility by adding a few drops of the liquid unknown to 0.5 mL solvent in a shell vial or test tube. Observe whether the layers are miscible or immiscible. For example, a few drops of ether in water would not be miscible, therefore ether is not soluble in water.

*NOTE. To test with litmus paper, dip the end of a small glass rod into the solution and then gently touch the litmus paper with the rod.* ***Do not dip the litmus paper into the test solution.*** *Amines are the only unknowns that are basic. Carboxylic acids and phenols are acidic. Some aldehydes and ketones may also test acidic for various reasons.*

The solubility tests can be influenced by the size of the unknown molecule. If the molecule contains more than 5 or 6 carbon atoms, it may not be soluble in aqueous solutions.

In doing the solubility tests follow the scheme in the order given. *Keep a record of your observations.*

**Step I** Test for water solubility. If soluble, test the aqueous solution with litmus paper.
**Step II** If water soluble, determine the solubility in diethyl ether. This test further classifies water-soluble materials.
**Step III** Water-insoluble compounds are tested for solubility in a 5% aqueous NaOH solution. If soluble, determine the solubility in 5% aqueous $NaHCO_3$.The use of the $NaHCO_3$ solution aids in distinguishing between strong (soluble) and weak (insoluble) acids. This test may seem ambiguous. Some acids and phenols will dissolve in $NaHCO_3$, some will not. Don't worry about this; there will be enough evidence from other experiments to identify your unknown.
**Step IV** Compounds insoluble in 5% aqueous NaOH are tested for solubility in a 5% HCl solution.

*NOTE: If you get to this point on the flow chart, assume your unknown is soluble in both concentrated sulfuric acid and 85% phosphoric acid. If you think about the classes of unknowns discussed above, it should be clear that every unknown will have an oxygen, a nitrogen, or at least a carbon-carbon double bond, and thus all should be soluble in concentrated sulfuric acid; therefore you can skip this step and the $H_3PO_4$ test.*

To classify a given compound, it may not be necessary to test its solubility in every solvent. Do only those tests that are required to place the compound in one of the solubility groups. Make your observations with care, and proceed in a logical sequence as you make the tests.

If time permits, you will obtain an IR spectrum for each of your unknowns at the end of this experiment's lab period. You can obtain the spectra at the beginning of the next lab period, if necessary. You will turn these spectra in with your 3106-8 lab report.

[8] Shriner, R. L.; Fuson, R. C.; Morrill, T. C. *The Systematic Identification of Organic Compounds*, 6th ed.; Wiley: New York, 1980.

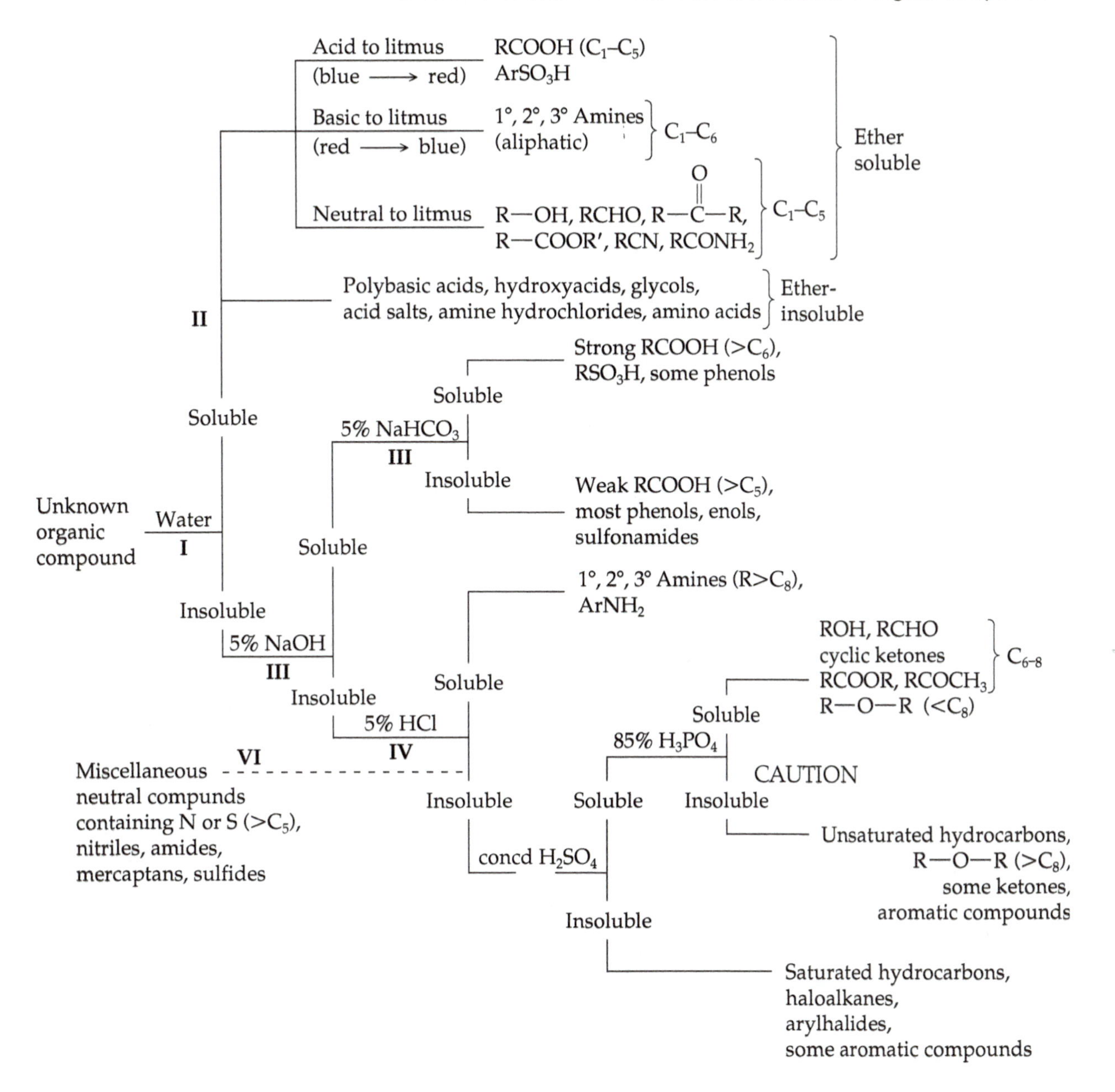

Acid to litmus (blue ⟶ red)
RCOOH ($C_1$–$C_5$)
$ArSO_3H$
Basic to litmus (red ⟶ blue)
1°, 2°, 3° Amines (aliphatic)
$C_1$–$C_6$
Neutral to litmus
R—OH, RCHO, R—C(=O)—R, R—COOR′, RCN, $RCONH_2$
$C_1$–$C_5$
Ether soluble
Polybasic acids, hydroxyacids, glycols, acid salts, amine hydrochlorides, amino acids
Ether-insoluble
II
Soluble
Unknown organic compound
Water
I
Insoluble
5% NaOH
III
Soluble
5% $NaHCO_3$
III
Soluble
Strong RCOOH (>$C_6$), $RSO_3H$, some phenols
Insoluble
Weak RCOOH (>$C_5$), most phenols, enols, sulfonamides
Insoluble
5% HCl
IV
Soluble
1°, 2°, 3° Amines (R>$C_8$), $ArNH_2$
VI
Miscellaneous neutral compunds containing N or S (>$C_5$), nitriles, amides, mercaptans, sulfides
Insoluble
concd $H_2SO_4$
Soluble
85% $H_3PO_4$
Soluble
ROH, RCHO
cyclic ketones
RCOOR, $RCOCH_3$
R—O—R (<$C_8$)
$C_{6-8}$
CAUTION
Insoluble
Unsaturated hydrocarbons, R—O—R (>$C_8$), some ketones, aromatic compounds
Insoluble
Saturated hydrocarbons, haloalkanes, arylhalides, some aromatic compounds

| | |
|---|---|
| Organic Chemistry Laboratory II | Name ____________________ |
| CHEM 3106 | TA ____________________ |
| Solid Unknown | Date ____________________ |
| Experiment 3106-7, Preliminary Classification | Unknown ID ____________________ |

Melting point range ____________________

Physical description of your unknown (crystal form, color, other observations):

Solubility (miscible, slightly soluble or insoluble):

| | |
|---|---|
| water | 5% NaOH |
| 5% $NaHCO_3$ | 5% HCl |
| Ether | |

pH or Litmus test: Is the compound acidic, neutral or basic?
*NOTE: blue litmus paper → red for acids; red litmus paper → blue for bases*

***Tentative conclusions***. What elements might be present in your unknown? What functional groups could be present? What compounds, found in the blue book, are possibilities for your unknown based on this information? Briefly summarize your reasoning

Organic Chemistry Laboratory II — Name ______________________________

CHEM 3106 — TA ______________________________

Liquid Unknown — Date ______________________________

Experiment 3106-7, Preliminary Classification — Unknown ID ________________________

Boiling point range ____________________

Physical description of your unknown (color, other observations):

Solubility (miscible, slightly soluble or insoluble):

water — 5% NaOH

5% $NaHCO_3$ — 5% HCl

Ether

pH or Litmus test: Is the compound acidic, neutral or basic?
*NOTE: blue litmus paper → red for acids; red litmus paper → blue for bases*

***Tentative conclusions***. What elements might be present in your unknown? What functional groups could be present? What compounds, found in the blue book, are possibilities for your unknown based on this information? Briefly summarize your reasoning.

# EXPERIMENT 3106-8: THE CLASSIFICATION TESTS[9]

A series of functional group classification tests are described. Depending on the results of your preliminary tests and solubility characteristics, you may make use of the tests for the following functional groups:

Alcohols (p ?)
Aldehydes and Ketones (p ?)
Amines (p ?)
Carboxylic Acids (p ?)
Methyl Ketones and Methyl Carbinols (p ?)
Phenols and Enols (p ?)

Use your list of possible compounds and the IR spectra to determine which functional group tests to run. For example, if the liquid sample is basic, start with the test for amines. Amines are the only unknowns in this experiment that are basic. You do not need to run every functional group test. If you do not think your unknown is an amine, run the 2,4-DNP test for Aldehydes and Ketones. This is a very fast test and doing it early can save you time. If you get a positive result, do not check for alcohols. If you get a negative result, your unknown is likely an alcohol. Run the Purpald (make sure to shake the vial to see the purple color) and/or Tollen's test only on unknowns that give a positive 2,4-DNP result. If the 2,4-DNP test is positive, keep the solid precipitate for next lab period's derivative tests. If the unknown is not basic, and the 2,4-DNP test is negative, run alcohol and/or phenol tests. The next option after alcohols is carboxylic acids or ethers.

*NOTE. For all tests given in this section, drops of reagents are measured out using Pasteur pipets.*

*NOTE. All of the unknowns in this experiment are soluble in ethanol, you may have to wait a few minutes and/or shake it for it to dissolve.*

## Alcohols

### *Ceric Nitrate Test*

*INSTRUCTOR PREPARATION. The reagent is prepared by dissolving 4.0 g of ceric ammonium nitrate [$(NH_4)_2Ce(NO_3)_6$] in 10 mL of 2 M $HNO_3$. Warming may be necessary.*

Primary, secondary, and tertiary alcohols with fewer than 10 carbon atoms give a positive test as indicated by a change in color from *yellow* to *red:*

$$\underset{\text{Yellow}}{(NH_4)_2Ce(NO_3)_6} + RCH_2OH \rightarrow \underset{\text{(Red complex)}}{[\text{alcohol + reagent}]}$$

[9]For a detailed discussion of classification tests see (a) Shriner, R. L.; Fuson, R. C.; Morrill, T. C. *The Systematic Identification of Organic Compounds,* 6th ed.; Wiley: New York, 1980, p. 138; (b) Pasto, D. J.; Johnson, C. R.; Miller, M. J. *Experiments and Techniques in Organic Chemistry;* Prentice Hall: Englewood Cliffs, NJ, 1992.

Place 5 drops of test reagent on a white spot plate. Add 1–2 drops of the unknown sample (5 mg if a solid). Stir with a thin glass rod to mix the components and observe any color change.

**1.** If the alcohol is water insoluble, 3–5 drops of dioxane may be added, but run a blank to make sure the dioxane is pure. Efficient stirring gives positive results with most alcohols.

**2.** Phenols, if present, give a brown color or precipitate.

### *Chromic Anhydride Test: The Jones Oxidation*

*INSTRUCTOR PREPARATION. The reagent is prepared by slowly adding a suspension of 1.0 g of $CrO_3$ in 1.0 mL of concentrated $H_2SO_4$ to 3 mL of water. Allow the solution to cool to room temperature before using.*

The Jones oxidation test is a rapid method to distinguish primary and secondary alcohols from tertiary alcohols. A positive test is indicated by a color change from *orange* (the oxidizing agent, $Cr^{6+}$) while the oxidizing agent is itself reduced to the *blue green* ($Cr^{3+}$):

$$\left.\begin{array}{c} RCH_2OH \\ \text{or} \\ R_2CHOH \end{array}\right\} + \underset{\text{Orange}}{H_2Cr_2O_7} \xrightarrow{H_2SO_4} \underset{\text{Green}}{Cr_2(SO_4)_3} + \begin{array}{c} RCO_2H \\ \text{or} \\ R_2C{=}O \end{array}$$

The test is based on oxidation of a primary alcohol to an aldehyde or acid, and of a secondary alcohol to a ketone.

On a white spot plate, place 1 drop of the liquid unknown (10 mg if a solid). Add 10 drops of acetone and stir the mixture with a thin glass rod. Add 1 drop of the test reagent to the resulting solution. Stir and observe any color change within a 2-second time period.

**1.** Run a blank to make sure the acetone is pure.

**2.** Tertiary alcohols, unsaturated hydrocarbons, amines, ethers, and ketones give a negative test within the 2-s time frame for observing the color change. Aldehydes, however, give a positive test, since they are oxidized to the corresponding carboxylic acids.

### *The HCl/ZnCl$_2$ Test: The Lucas Test*

*INSTRUCTOR PREPARATION. The Lucas reagent is prepared by dissolving 16 g of anhydrous $ZnCl_2$ in 10 mL of concd HCl while it is cooling in an ice bath.*

The Lucas test is used to distinguish between primary, secondary, and tertiary monofunctional alcohols having fewer than six carbon atoms:

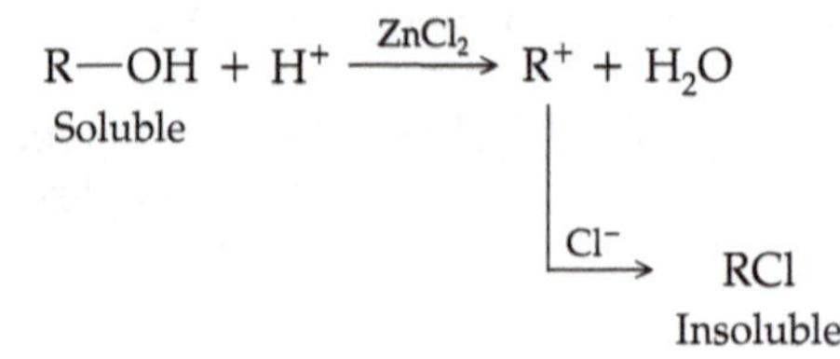


The test requires that the alcohol initially be soluble in the Lucas test reagent solution. As the reaction proceeds, the corresponding alkyl chloride is formed, which is insoluble in the reaction mixture. As a result, the solution becomes cloudy. In some cases a separate layer may be observed.

**1.** Tertiary, allyl, and benzyl alcohols react to give an immediate cloudiness to the solution. You may be able to see a separate layer of the alkyl chloride after a short time.

**2.** Secondary alcohols generally produce a cloudiness within 3–10 min. The solution may have to be heated to obtain a positive test.

**3.** Primary alcohols having less than six carbon atoms dissolve in the reagent but react very, very slowly. Those having more than six carbon atoms do not dissolve to any significant extent, no reaction occurs, and the aqueous phase remains clear.

**4.** A further test to aid in distinguishing between tertiary and secondary alcohols is to run the test using concentrated hydrochloric acid. Tertiary alcohols react immediately to give the corresponding alkyl halide, whereas secondary alcohols do not react under these conditions.

In a small test tube, place 2 drops of the unknown (10 mg if a solid) followed by 10 drops of the Lucas reagent.

Shake or stir the mixture with a thin glass rod and allow the solution to stand. Observe the results. Based on the times given above, classify the alcohol.

***Additional points to consider:***

**1.** Certain polyfunctional alcohols also give a positive test.

**2.** If an alcohol having three or fewer carbons is expected, a 1-mL conical vial equipped with an air condenser should be used to prevent low molecular weight alkyl chlorides (volatile) from escaping and thus remaining undetected.

***The Iodoform Test.*** This test is positive for compounds that on oxidation generate methyl ketones (or acetaldehyde) under the reaction conditions. For example, methyl carbinols (secondary alcohols having at least one methyl group attached to the carbon atom to which the —OH is attached), acetaldehyde, and ethanol give positive results.

For the test see Methyl Ketones and Methyl Carbinols **(p. 651).**

### Aldehydes and Ketones

#### *The 2,4-Dinitrophenylhydrazine Test*

*INSTRUCTOR PREPARATION. The reagent solution is prepared by dissolving 1.0 g of 2,4-dinitrophenylhydrazine in 5.0 mL of concentrated sulfuric acid. This solution is slowly added, with stirring, to a mixture of 10 mL of water and 35 mL of 95% ethanol. After mixing, filter the solution.*

Aldehydes and ketones react rapidly with 2,4-dinitrophenylhydrazine to form 2,4-dinitrophenylhydrazones. These derivatives range in color from *yellow* to *red*, depending on the degree of conjugation in the carbonyl compound:

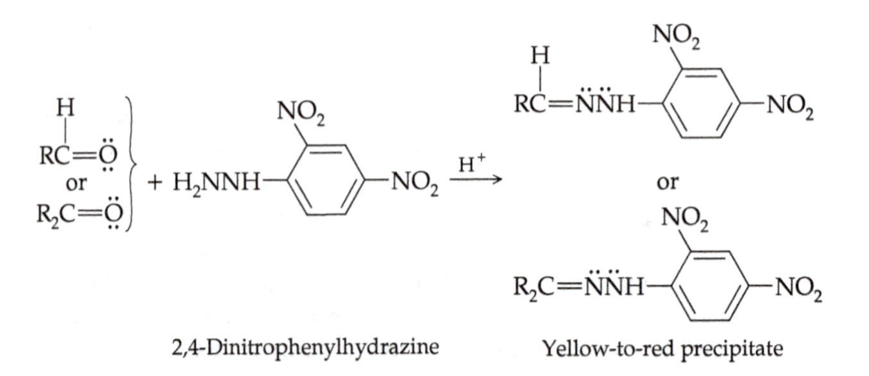


On a white spot plate place 7–8 drops of 2,4-dinitrophenylhydrazine reagent solution.

Then add 1 drop of a liquid unknown. If the unknown is a solid, add 1 drop of a solution prepared by dissolving 10 mg of the material in 10 drops of ethanol. The mixture is stirred with a thin glass rod. The formation of a red-to-yellow precipitate is a positive test.

*NOTE. The reagent, 2,4-dinitrophenylhydrazine, is orange-red and melts at 198 °C (dec). Do not mistake it for a derivative!*

Reactive esters or anhydrides react with the reagent to give a positive test. Allylic or benzylic alcohols may be oxidized to aldehydes or ketones, which in turn give a positive result. Amides do not interfere with the test. Be sure that your unknown is pure and does not contain aldehyde or ketone impurities.

Phenylhydrazine and *p*-nitrophenylhydrazine are often used to prepare the corresponding hydrazones. These reagents also yield solid derivatives of aldehydes and ketones.

***Silver Mirror Test for Aldehydes: Tollens Reagent.*** The Tollen's test can be a beautiful test, as it will make a silver mirror under favorable conditions. However, it often gives less diagnostic results, such as a black precipitate. Feel free to try it and enjoy the results if you get a nice mirror. However, if you have a carbonyl compound, you should also try the Purpald test.

This reaction involves the oxidation of aldehydes to the corresponding carboxylic acid, using an alcoholic solution of silver ammonium hydroxide. A positive test is the formation of a *silver* mirror, or a black precipitate of finely divided silver:

$$\mathrm{RC(H){=}O + 2\,Ag(NH_3)_2OH \longrightarrow 2\,Ag\downarrow + R{-}C({=}O){-}O^-,\,NH_4^+ + H_2O + 3\,NH_3}$$

The test should be run only after the presence of an aldehyde or ketone has been established.

In a small test tube place 1.0 mL of a 5% aqueous solution of $AgNO_3$, followed by 1 drop of aqueous 10% NaOH solution. Now add concentrated aqueous ammonia (also called ammonium hydroxide), drop by drop (2–4 drops) with shaking, until the precipitate of silver oxide just dissolves. Add 1 drop of the unknown (10 mg if a solid), with shaking, and allow the reaction mixture to stand for 10 min at room temperature. If no reaction has occurred, place the test tube on a hot plate at 40 °C for 5 min. Observe the result.

***Additional points to consider:***

1. Avoid a large excess of ammonia.

2. Reagents must be well mixed. Stirring with a thin glass rod is recommended.

3. *This reagent is freshly prepared for each test. It should not be stored since decomposition occurs with the formation of $AgN_3$, which is explosive.*

4. This oxidizing agent is very mild and thus alcohols are not oxidized under these conditions. Ketones do not react. Some sugars, acyloins, hydroxylamines, and substituted phenols do give a positive test.

### *Purpald Test for Aldehydes*

*INSTRUCTOR PREPARATION. The reagent is prepared by dissolving 5-10 mg Purpald 1 mL of 1N NaOH.*

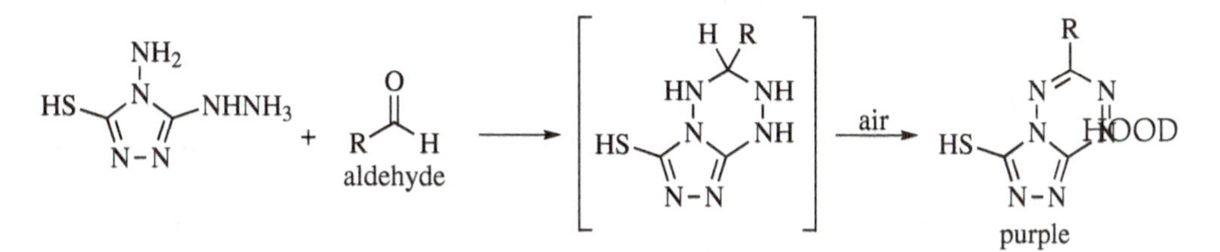


In a small test tube, add one drop of the unknown to 2 mL of the Purpald reagent. If the unknown is a solid, dissolve it first in a small amount of ethanol. If your unknown is an aldehyde, a reaction will take place, but no change should be evident. Shaking the solution will aerate it and air oxidation will then give an intense purple color, indicating an aldehyde.

### *Chromic Acid Test*

*INSTRUCTOR PREPARATION. The reagent is prepared by dissolving 1 g of chromium trioxide in 1 mL of concd $H_2SO_4$, followed by 3 mL of $H_2O$.*

Chromic acid in acetone rapidly oxidizes aldehydes to carboxylic acids. Ketones react very slowly, or not at all.

In a 3-mL vial or small test tube, place 2 drops of a liquid unknown (~10 mg if a solid) and 1 mL of spectral-grade acetone. Now add several drops of the chromic acid reagent (same as the Jones Reagent).

A green precipitate of chromous salts is a positive test. Aliphatic aldehydes give a precipitate within 30 s; aromatic aldehydes take 30–90 s.

The reagent also reacts with primary and secondary alcohols (see Chromic Anhydride Test: Jones Oxidation, **p. 640**).

### *Bisulfite Addition Complexes*

*INSTRUCTOR PREPARATION. The reagent is prepared by mixing 1.5 mL of ethanol and 6 mL of a 40% aqueous solution of sodium bisulfite. Filter the reagent before use, if a small amount of the salt does not dissolve.*

Most aldehydes react with a saturated sodium bisulfite solution to yield a crystalline bisulfite addition complex:

$$\text{>C=O} \underset{\text{H}^+ \text{ or HO}^-}{\overset{\text{NaHSO}_3}{\rightleftharpoons}} \text{>C(OH)SO}_3^-\text{, Na}^+$$

The reaction is reversible and thus the carbonyl compound can be recovered by treatment of the complex with aqueous 10% $NaHCO_3$ or dilute HCl solution.

Place 50–75 μL of the liquid unknown in a small test tube and add 150 μL of the sulfite reagent and mix thoroughly.

A crystalline precipitate is a positive test.

Alkyl methyl ketones and unhindered cyclic ketones also give a positive test.

### Amines

***Copper Ion Test.*** Amines will give a blue-green coloration or precipitate when added to a copper sulfate solution. In a small test tube, place 0.5 mL of a 10% copper sulfate solution. Now add 1 drop of an unknown (~10 mg if a solid). The blue-green coloration or precipitate is a positive test. Ammonia will also give a positive test.

***Hinsberg Test.*** The Hinsberg test is useful for distinguishing between primary, secondary, and tertiary amines. The reagent used is *p*-toluenesulfonyl chloride in alkaline solution.

*Primary* amines with fewer than seven carbon atoms form a sulfonamide that is soluble in the alkaline solution. Acidification of the solution results in the precipitation of the insoluble sulfonamide:

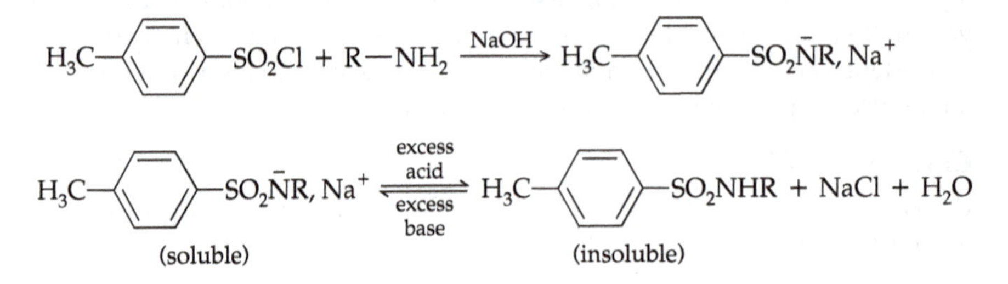


*Secondary* amines form an insoluble sulfonamide in the alkaline solution:

$$\text{H}_3\text{C–C}_6\text{H}_4\text{–SO}_2\text{Cl} + \text{R}_2\text{NH} \xrightarrow{\text{NaOH}} \underset{\text{(insoluble)}}{\text{H}_3\text{C–C}_6\text{H}_4\text{–SO}_2\text{NR}_2} + \text{NaCl} + \text{H}_2\text{O} \xrightarrow{\text{excess base}} \text{no change}$$

*Tertiary* amines normally give no reaction under these conditions:

$$\text{H}_3\text{C–C}_6\text{H}_4\text{–SO}_2\text{Cl} + \text{R}_3\text{N} \xrightarrow{\text{NaOH}} \underset{\text{(soluble)}}{\text{H}_3\text{C–C}_6\text{H}_4\text{–SO}_3^-} + \underset{\text{(oil)}}{\text{NR}_3} + 2\,\text{Na}^+ + \text{Cl}^- + \text{H}_2\text{O}$$

In a 1.0-mL conical vial containing a boiling stone, and equipped with an air condenser, place 0.5 mL of 10% aqueous sodium hydroxide solution, 1 drop of the sample unknown (~10 mg if a solid), followed by 30 mg of *p*-toluenesulfonyl chloride (in the **hood**). Heat the mixture to reflux for 2–3 min on a hot plate, and then cool it in an ice bath. Test the alkalinity of the solution using litmus paper. If it is not alkaline, add additional 10% aqueous sodium hydroxide dropwise.

HOOD

Using a Pasteur filter pipet, separate the solution from any solid that may be present. Transfer the solution to a clean 1.0-mL conical vial and **save.**

SAVE

*NOTE. If an oily upper layer is obtained at this stage, remove the lower alkaline phase using a Pasteur filter pipe and* ***save.*** *To the remaining oil add 0.5 mL of cold water and stir vigorously to obtain a solid material.*

SAVE

If a solid is obtained, it may be (1) the sulfonamide of a secondary amine; (2) recovered tertiary amine, if the original amine was a solid; or (3) the insoluble salt of a primary sulfonamide derivative, if the original amine had more than six carbon atoms.

***Additional points to consider:***

1. If the solid is a tertiary amine, it is soluble in aqueous 10% HCl.
2. If the solid is a secondary sulfonamide, it is insoluble in aqueous 10% NaOH.
3. If no solid is present, acidify the alkaline solution by addition of 10% aqueous HCl. If the unknown amine is primary, the sulfonamide will precipitate.

***Bromine Water.*** Aromatic amines, since they possess an electron-rich aromatic ring, can undergo electrophilic aromatic substitution with bromine, to yield the corresponding arylamino halide(s). Therefore, if elemental tests indicate that an aromatic group is present in an amine, treatment with the bromine water reagent may indicate that the amine is attached to an aromatic ring.

For the test, see Phenols and Enols **(p. 653).**

## Carboxylic Acids

The presence of a carboxylic acid is detected by its solubility behavior. An aqueous solution of the acid will be acidic to litmus paper (or pH paper may be used). A water-soluble phenol is acidic toward litmus paper but also would give a positive ferric chloride test.

Carboxylic acids also react with a 5% solution of sodium bicarbonate.

Place 1–2 mL of the bicarbonate solution on a watch glass and add 1–2 drops of the acid (~10 mg if a solid). Gas bubbles of $CO_2$ constitute a positive test.

## Ethers

CAUTION: Upon standing, ethers may form peroxides. Peroxides are very explosive. To test for the presence of these substances, use starch–iodide paper that has been moistened with 6 M HCl. Peroxides cause the paper to turn blue. To remove peroxides from ethers, pass the material through a short column of highly activated alumina (Woelm basic alumina, activity grade 1).[10] Always retest for peroxides before using the ether.

***Ferrox Test.*** The ferrox test is a color test sensitive to oxygen, which may be used to distinguish ethers from hydrocarbons that, like most ethers, are soluble in sulfuric acid.

In a dry 10 × 75-mm test tube using a glass stirring rod, grind a crystal of ferric ammonium sulfate and a crystal of potassium thiocyanate. The ferric hexathiocyanatoferrate that is formed adheres to the rod.

In a second clean 10 × 75-mm test tube, place 2–3 drops of a liquid unknown. If dealing with a solid, use about 10 mg and add toluene until a saturated solution is obtained. Now, using the rod with the ferric hexathiocyanatoferrate attached, stir the unknown. *If the unknown contains oxygen, the ferrate compound dissolves and a reddish-purple color is observed.*

Some high-molecular-weight ethers do not give a positive test.

***Bromine Water.*** Since the aromatic ring is electron rich, aromatic ethers can undergo electrophilic aromatic substitution with bromine to yield the corresponding aryl ether–halide(s). Therefore, if elemental tests indicate that an aromatic group is present in an ether, treatment with the bromine water reagent may substantiate the presence of an aryl ether.

For the test see Phenols and Enols **(p. 653)**.

## Methyl Ketones and Methyl Carbinols

### *Iodoform Test*

*INSTRUCTOR PREPARATION. Dissolve 3 g of KI and 1 g $I_2$ in 20 mL of water.*

The iodoform test involves hydrolysis and cleavage of methyl ketones to form a yellow precipitate of iodoform ($CHI_3$):

$$R{-}\overset{\overset{\displaystyle ..O..}{\|}}{C}{-}CH_3 + 3\,I_2 + 3\,KOH \longrightarrow R{-}\overset{\overset{\displaystyle ..O..}{\|}}{C}{-}CI_3 + 3\,KI + 3\,H_2O$$

$$\downarrow \text{KOH}$$

$$R{-}\overset{\overset{\displaystyle ..O..}{\|}}{C}{-}O^-, K^+ + \underset{\text{Yellow}}{CHI_3} \downarrow$$

It is also a positive test for compounds that, upon oxidation, generate methyl ketones (or acetaldehyde) under these reaction conditions. For example, methyl carbinols (secondary alcohols having at least one methyl group attached to the carbon atom to which the —OH unit is linked), acetaldehyde, and ethanol give positive results.

In a 3.0-mL conical vial equipped with an air condenser, place 2 drops of the unknown liquid (10 mg if a solid), followed by 5 drops of 10% aqueous KOH solution.

[11]Fuson, R. C.; Bull, B. A. *Chem. Rev.* **1934,** *15,* 275.

*NOTE. If the sample is insoluble in the aqueous phase, either mix vigorously or add dioxane (in the* ***hood****) or bis(2-methoxyethyl) ether to obtain a homogeneous solution.*

HOOD

Warm the mixture on a hot plate to 50–60 °C and add the $KI-I_2$ reagent dropwise until the solution becomes dark brown in color (~1.0 mL). Additional 10% aqueous KOH is now added (dropwise) until the solution is again colorless.

---

CAUTION: Iodine is highly toxic and can cause burns.

---

CAUTION

After warming for 2 min, cool the solution and determine whether a yellow precipitate ($CHI_3$, iodoform) has formed. If a precipitate is not observed, reheat as before for another 2 min. Cool and check again for the appearance of iodoform.

***Additional points to consider:***

1. The iodoform test is reviewed elsewhere.[11]

## Phenols and Enols

***Ferric Ion Test.*** Most phenols and enols form colored complexes in the presence of ferric ion, $Fe^{3+}$:

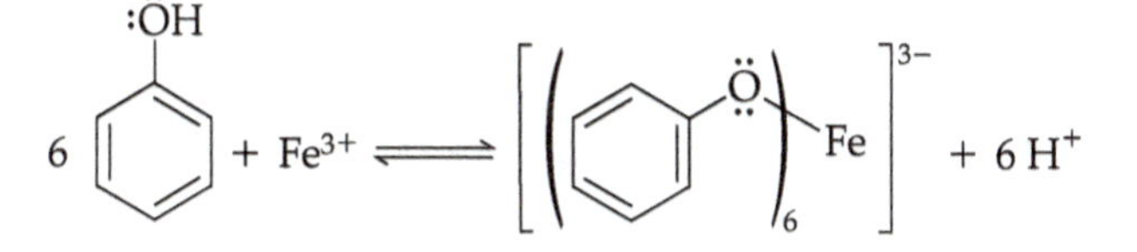


Phenols give red, blue, purple, or green colors. Sterically hindered phenols may give a negative test. Enols generally give a tan, red, or red-violet color.

On a white spot plate place 2 drops of water, or 1 drop of water plus 1 drop of ethanol, or 2 drops of ethanol, depending on the solubility characteristics of the unknown. To this solvent system add 1 drop (10 mg if a solid) of the substance to be tested. Stir the mixture with a thin glass rod to complete dissolution. Add 1 drop of 2.5% aqueous ferric chloride ($FeCl_3$) solution (light yellow in color). Stir and observe any color formation. If necessary, a second drop of the $FeCl_3$ solution may be added.

***Additional points to consider:***

1. The color developed may be fleeting or it may last for many hours. A slight excess of the ferric chloride solution may or may not destroy the color.
2. An alternative procedure using $FeCl_3–CCl_4$ solution in the presence of pyridine is available.[13]

***Bromine Water.*** Phenols, substituted phenols, aromatic ethers, and aromatic amines, since the aromatic rings are electron rich, undergo aromatic electrophilic substitution with bromine to yield substituted aryl halides. For example,

---

[12]Pasto, D. J.; Johnson, C. R.; Miller, M. J. *Experiments and Techniques in Organic Chemistry;* Prentice Hall: Englewood Cliffs, NJ, 1992, p. 321.

[13]Soloway, S.; Wilen, S. H. *Anal. Chem.* **1952,** 4, 979.

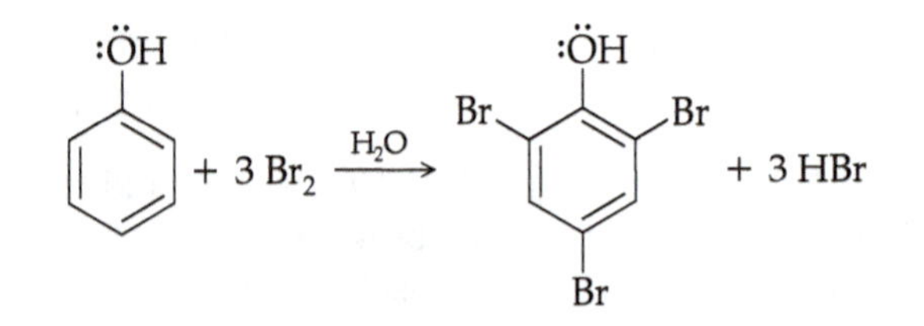


HOOD

---

CAUTION: The test should be run in the **hood.**

---

In a small test tube, place 1–2 drops of the unknown (~20 mg if a solid) and add 1–2 mL of water. Check the pH of the solution with pH paper. In the **hood,** add saturated bromine water dropwise until the bromine color persists. A precipitate generally forms.

A positive test is the decolorization of the bromine solution, and often the formation of an off-white precipitate. If the unknown is a phenol, this should cause the pH of the original solution to be less than 7.

### Alkenes and Alkynes: Unsaturated Hydrocarbons

***Bromine in Methylene Chloride.*** Unsaturated hydrocarbons readily add bromine ($Br_2$).

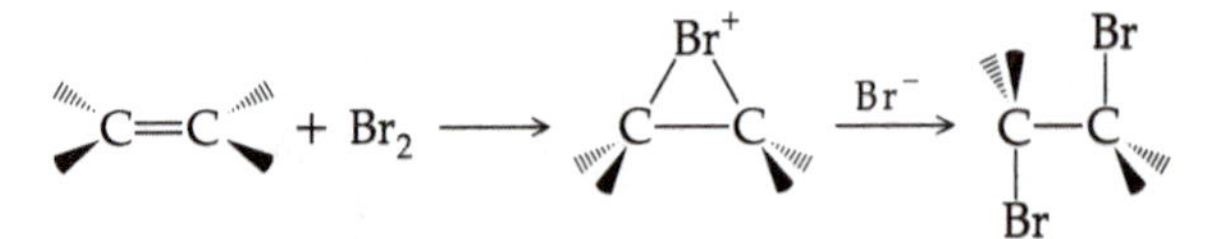


The test is based on the decolorization of a red-brown bromine–methylene chloride solution.

---

CAUTION: Bromine is highly toxic and can cause burns.

---

HOOD

In a 10 × 75-mm test tube, place 2 drops of a liquid unknown (~15 mg if a solid) followed by 0.5 mL of methylene chloride. Add dropwise, in the **hood** with shaking, a 2% solution of bromine in methylene chloride solvent. The presence of an unsaturated hydrocarbon will require 2–3 drops of the reagent before the reddish-brown color of bromine persists in the solution.

***Additional points to consider:***

**1.** Methylene chloride is used in place of the usual carbon tetrachloride ($CCl_4$) because it is less toxic.

**2.** Phenols, enols, amines, aldehydes, and ketones interfere with this test.

***Permanganate Test: Baeyer Test for Unsaturation.*** Unsaturation in an organic compound can be detected by the decolorization of permanganate solution. The reaction involves the cis hydroxylation of the alkene to give a 1,2 diol (glycol):

$$\mathrm{C{=}C} + 2\,\mathrm{MnO_4^-} + 4\,\mathrm{H_2O} \longrightarrow \underset{:\ddot{O}H\ :\ddot{O}H}{\mathrm{C{-}C}} + 2\,\mathrm{MnO_2} + 2\,\mathrm{OH^-}$$

On a white spot plate, place 0.5 mL of *alcohol-free* acetone, followed by 2 drops of the unknown compound (~15 mg if a solid). Now add dropwise (2–3 drops), with stirring, a 1% aqueous solution of potassium permanganate ($KMnO_4$). A positive test for unsaturation is the discharge of purple permanganate color from the reagent and the precipitation of brown manganese oxides.

Any functional group that undergoes oxidation with permanganate interferes with the test (phenols, aryl amines, most aldehydes, primary and secondary alcohols, etc.).

| | |
|---|---|
| Organic Chemistry Laboratory II | Name ______________________ |
| CHEM 3106 | TA ______________________ |
| Solid Unknown | Date ______________________ |
| Experiment 3106-8, Classification Tests | Unknown ID ______________________ |

You will identify the functional group(s) present in your unknown by performing various Classification Tests.

List (1) the tests you performed (2) the results you obtained and (3) the conclusions you made regarding the presence or absence of this functional group.

***Conclusion***. What functional group(s) is (are) present? What compounds, found in the blue book, are possibilities for your unknown based on this information? Briefly summarize your reasoning.

Organic Chemistry Laboratory II — Name ______________________________

CHEM 3106 — TA ______________________________

Liquid Unknown — Date ______________________________

Experiment 3106-8, Classification Tests — Unknown ID ________________________

You will identify the functional group(s) present in your unknown by performing various Classification Tests.

List (1) the tests you performed (2) the results you obtained and (3) the conclusions you made regarding the presence or absence of this functional group.

***Conclusion.*** What functional group(s) is (are) present? What compounds, found in the blue book, are possibilities for your unknown based on this information? Briefly summarize your reasoning.

# EXPERIMENT 3106-9: PREPARATION OF DERIVATIVES

Based on the preliminary and classification tests carried out to this point, you should have established the type of functional group (or groups) present (or lack of one) in the unknown organic sample. The next step in qualitative organic analysis is to consult a set of tables containing a listing of known organic compounds sorted by functional group and/or by physical properties or by both. Using the physical properties data for your compound, you can select a few possible candidates that appear to "fit" the data you have collected. On a chemical basis, the final step in the qualitative identification sequence is to prepare one or two *crystalline derivatives* of your compound. Selection of the specific compound, and thus final confirmation of its identity, can then be made from the extensive derivative tables that have been accumulated. With the advent of spectral analysis, the preparation of derivatives is often not necessary, but the wealth of chemistry that can be learned by the beginning student in carrying out these procedures is extensive and important. The preparation of selected derivatives for the most common functional groups are given below. Condensed tables of compounds and their derivatives are found in the blue books in your laboratory. For extensive tables and alternative derivatives that can be utilized, see the Bibliography section of this experiment.

Identify the functional group present in your unknowns by using the IR spectra and Classification Tests (attach the spectra to the Experiment 3106-8 lab report). You must perform at least one derivative for each unknown sample. Look at the blue notebook of possible unknowns and choose the derivative that gives the largest difference in melting points between your list of possibilities; make that derivative. Review all your data. This should allow you to accurately determine which unknown you were given.

Prepare appropriate derivatives and take melting points of them. Using this information and the tables in the blue book, you will make a final identification of your unknowns.

Carboxylic Acids (p. ?)
Alcohols (p. ?)
Aldehydes and Ketones (p. ?)
Amines (p. ?)
Phenols and Enols (p. ?)

*NOTE. In each of the procedures outlined below, drops of reagents are measured using Pasteur pipets.*

## CARBOXYLIC ACIDS[14]

### Preparation of Acid Chlorides

$$\mathrm{RC(=O)OH} + \mathrm{ClC(=O)C(=O)Cl} \xrightarrow{\text{DMF}} \mathrm{RC(=O)Cl} + CO_2\uparrow + CO\uparrow + HCl\uparrow$$

HOOD

Weigh and place 20 mg of the unknown acid in a dry 3.0-mL conical vial containing a boiling stone and fitted with a cap. Now, in the **hood,** add 4 drops of oxalyl chloride and 1 drop of *N,N*-dimethylformamide (DMF). Immediately attach the vial to a reflux condenser that is protected by a calcium chloride drying tube.

---

HOOD

CAUTION: This reaction is run in the **hood** since hydrogen chloride, carbon dioxide, and carbon monoxide are evolved. Oxalyl chloride is an irritant and is harmful to breathe. Immediately recap the vial after each addition until the vial is attached to the reflux condenser.

---

Allow the mixture to stand at room temperature for 10 min, heat it at gentle reflux on a hot plate for 15 min, and then cool it to room temperature. Dilute the reaction mixture with 5 drops of methylene chloride solvent.

The acid chloride is not isolated but is used directly in the preparations that follow.

### Amides

$$\mathrm{R{-}C(=\ddot{O})}{-}Cl + 2\,\ddot{N}H_3 \longrightarrow \mathrm{R{-}C(=\ddot{O})}{-}\ddot{N}H_2 + NH_4Cl$$

HOOD

Cool the vial in an ice bath and add 10 drops of concentrated aqueous ammonia (also called ammonium hydroxide), in the **hood** via Pasteur pipet, *dropwise,* with stirring. *It is convenient to make this addition down the neck of the air condenser.* The amide may precipitate during this operation. After the addition is complete, remove the ice bath and stir the mixture for an additional 5 min. Now add methylene chloride (10 drops) and stir the resulting mixture to dissolve any precipitate. Separate the methylene chloride layer from the aqueous layer using a Pasteur filter pipet and transfer it to another Pasteur filter pipet containing 200 mg of anhydrous magnesium sulfate. Collect the eluate in a flask containing a boiling stone. Extract the aqueous phase with an additional 0.5 mL of methylene chloride. Separate the methylene chloride layer as before and transfer it to the same column. Collect this eluate in the same flask.

HOOD

Evaporate the methylene chloride solution using a warm hot plate in the **hood** under a gentle stream of air. Recrystallize the solid amide product. Dissolve the material in about 0.5 mL of ethanol, add water (dropwise) until the solution becomes cloudy, cool in an ice bath, and collect the crystals in the usual manner. Dry the crystalline amide on a porous clay plate and determine the melting point.

---

WWW → [14]See Tables 9W.1 and 9W.2.

### Anilides

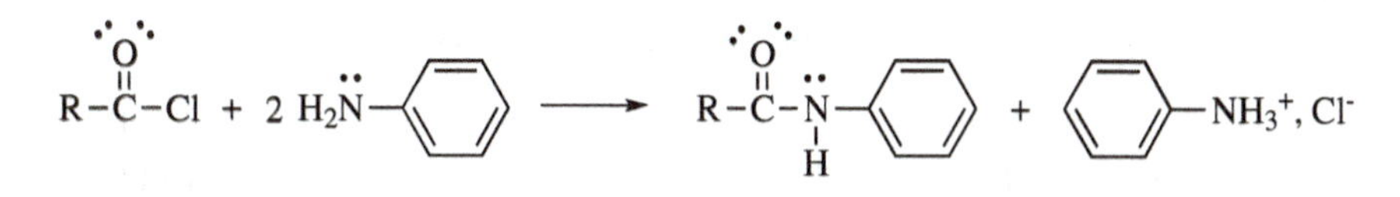


In a 3.0-mL conical vial containing a magnetic spin vane, and equipped with an air condenser, place 5 drops of aniline and 10 drops of methylene chloride. Cool the solution in an ice bath and transfer the acid chloride solution (prepared above) via Pasteur pipet, *dropwise,* with stirring, to the aniline solution in the **hood.** *It is convenient to make this addition down the neck of the condenser.* After the addition is complete, remove the ice bath and stir the mixture for an additional 10 min.

HOOD

Transfer the methylene chloride layer to a 10 × 75-mm test tube, and wash it with 0.5 mL of $H_2O$, 0.5 mL of 5% aqueous HCl, 0.5 mL of 5% aqueous NaOH, and, finally, 0.5 mL of $H_2O$. For each washing, shake the test tube and remove the top aqueous layer by Pasteur filter pipet. Transfer the resulting wet methylene chloride layer to a Pasteur filter pipet containing 200 mg of anhydrous magnesium sulfate. Collect the eluate in a flask containing a boiling stone. Rinse the original test tube with an additional 10 drops of methylene chloride. Collect this rinse and pass it through the same column. Both eluates are combined.

Evaporate the methylene chloride solvent on a warm hot plate under a gentle stream of air in the **hood.** Recrystallize the crude anilide from an ethanol–water mixture. Dissolve the material in about 0.5 mL of ethanol, add water (dropwise) to the cloud point, cool in an ice bath, and collect the crystals in the usual manner. Dry the purified derivative product on a porous clay plate, and determine its melting point.

HOOD

### Toluidides

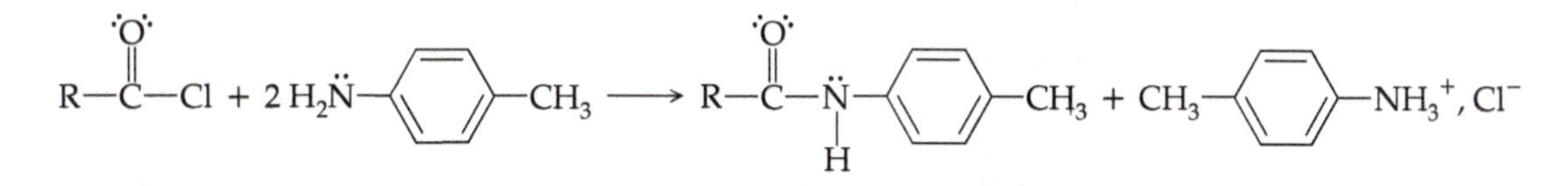


The same procedure described for the preparation of anilides is used, except that ~ 0.25 g *p*-toluidine replaces the aniline.

## ALCOHOLS[15]

### Phenyl- and α-Naphthylurethanes (Phenyl- and α-Naphthylcarbamates)

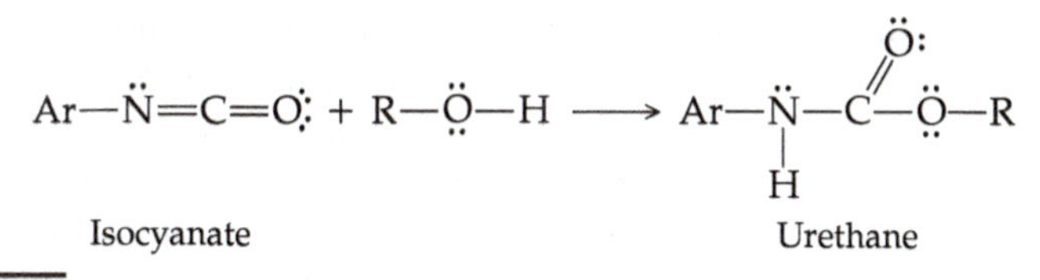


WW [15]See Table 9W.3.

*NOTE. For the preparation of these derivatives, the alcohols must be anhydrous. Water hydrolyzes the isocyanates to produce arylamines that react with the isocyanate reagent to produce high-melting, disubstituted ureas.*

In a 3.0-mL conical vial containing a boiling stone and equipped with an air condenser protected by a calcium chloride drying tube place 15 mg of an anhydrous alcohol or phenol. Remove the air condenser from the vial and add 2 drops of phenyl isocyanate or α-naphthyl isocyanate. Replace the air condenser immediately. If the unknown is a phenol, add 1 drop of pyridine in a similar manner.

HOOD

CAUTION: This addition must be done in the **hood.** The isocyanates are lachrymators! Pyridine has the characteristic strong odor of an amine.

If a spontaneous reaction does not take place, heat the vial at about 80–90 °C using a hot plate, for a period of 5 min. Then cool the reaction mixture in an ice bath. It may be necessary to scratch the sides of the vial to induce crystallization. Collect the solid product by vacuum filtration, using a Hirsch funnel, and purify it by recrystallization from ligroin. For this procedure, place the solid in a 10 × 75-mm test tube and dissolve it in 1.0 mL of warm (60–80 °C) ligroin. If diphenyl (or dinaphthyl) urea is present (formed by reaction of the isocyanate with water), it is insoluble in this solvent. Transfer the warm ligroin solution to a flask using a Pasteur filter pipet. Cool the solution in an ice bath and collect the resulting crystals in the usual manner. After drying the product on a porous clay plate, determine the melting point.

### 3,5-Dinitrobenzoates

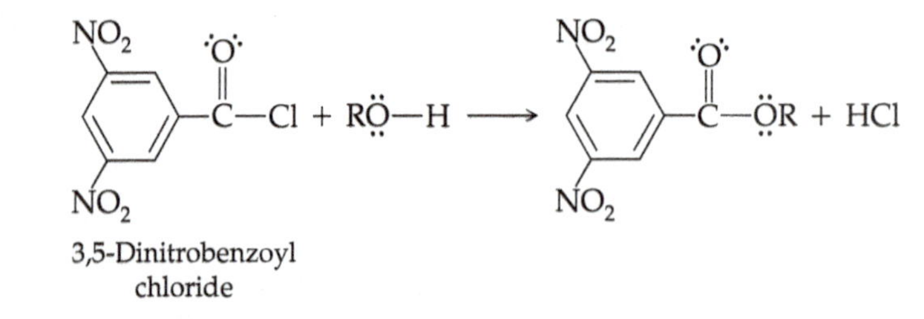


3,5-Dinitrobenzoyl chloride

*NOTE. The dinitrobenzoyl chloride reagent tends to hydrolyze on storage to form the corresponding carboxylic acid. Check its melting point before use (3,5-dinitrobenzoyl chloride, mp = 74 °C; 3,5-dinitrobenzoic acid, mp = 202 °C)*

In a 3.0-mL conical vial containing a boiling stone, and equipped with an air condenser protected by a calcium chloride drying tube, place 25 mg of pure 3,5-dinitrobenzoyl chloride and two drops of the unknown alcohol. Heat the mixture to about 10 °C below the boiling point of the alcohol (but not over 100 °C) on a hot plate for a period of 5 min. Cool the reaction mixture, add 0.3 mL of water, and then place the vial in an ice bath to cool. Collect the solid ester by vacuum filtration, using a Hirsch funnel, and wash the filter cake with three 0.5-mL portions of 2% aqueous sodium carbonate ($Na_2CO_3$) solution, followed by 0.5 mL of water. Recrystallize the solid product from an ethanol–water mixture. Dissolve the material in about 0.5 mL of ethanol, add water (dropwise) until the solution is just cloudy, cool in an ice bath, and collect the crystals in the usual manner. After drying the product on a porous clay plate, determine the melting point.

## ALDEHYDES AND KETONES[16]

### 2,4-Dinitrophenylhydrazones

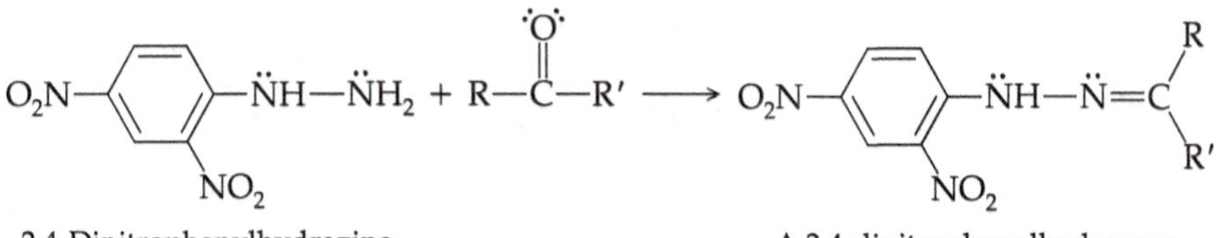


2,4-Dinitrophenylhydrazine A 2,4-dinitrophenylhydrazone

The procedure outlined in the Classification Test Section for aldehydes and ketones **(p. 642)** is used. Since the derivative to be isolated is a solid, it may be convenient to run the reaction in a 3-mL vial or in a small test tube. Double the amount of the reagents used. If necessary, the derivative can be recrystallized from 95% ethanol.

The procedure is generally suitable for the preparation of phenylhydrazone and *p*-nitrophenylhydrazone derivatives of aldehydes and ketones.

### Semicarbazones

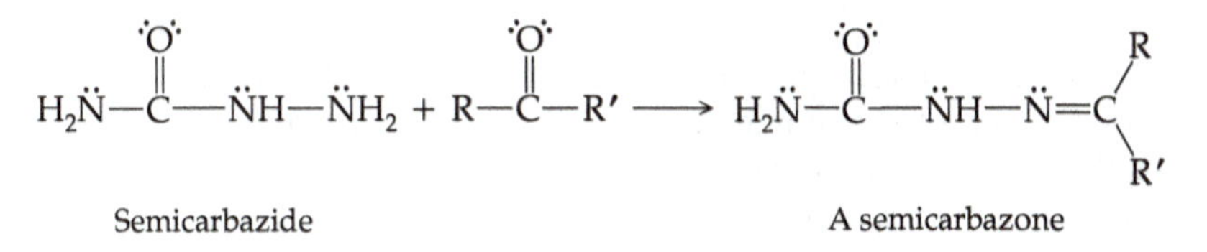


Semicarbazide A semicarbazone

In a 3.0-mL conical vial place 12 mg of semicarbazide hydrochloride, 20 mg of sodium acetate, 10 drops of water, and 12 mg of the unknown carbonyl compound. Cap the vial, shake vigorously, vent, and allow the vial to stand at room temperature until crystallization is complete (varies from a few minutes to several hours). Cool the vial in an ice bath if necessary. Collect the crystals by vacuum filtration, using a Hirsch funnel, and wash the filter cake with 0.2 mL of cold water. Dry the crystals on a porous clay plate. Determine the melting point.

## AMINES[17]

### Primary and Secondary Amines: Acetamides

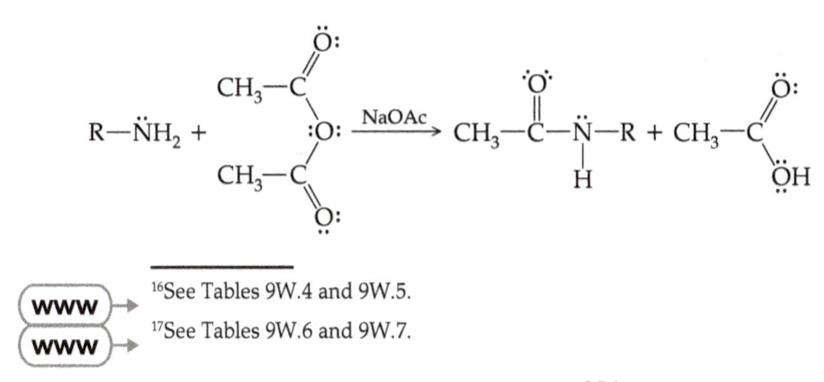


[16]See Tables 9W.4 and 9W.5.

[17]See Tables 9W.6 and 9W.7.

In a 3.0-mL conical vial equipped with an air condenser, place 20 mg of the unknown amine, 5 drops of water, and 1 drop of concentrated hydrochloric acid.

In a small test tube, prepare a solution of 40 mg of sodium acetate trihydrate dissolved in 5 drops of water. Stopper the solution and set it aside for use in the next step.

Warm the solution of amine hydrochloride to about 50 °C on a hot plate.

HOOD

Then cool it, and add 1 drop of acetic anhydride in one portion (in the **hood**) through the condenser by aid of a Pasteur pipet. In like manner, *immediately* add the sodium acetate solution (prepared previously). Swirl the contents of the vial to ensure complete mixing.

Allow the reaction mixture to stand at room temperature for about 5 min, and then place it in an ice bath for an additional 5–10 min. Collect the white crystals by vacuum filtration, using a Hirsch funnel, and wash the filter cake with two 0.1-mL portions of water. The product may be recrystallized from ethanol–water, if desired. Dry the crystals on a porous clay plate and determine the melting point.

### Primary and Secondary Amines: Benzamides

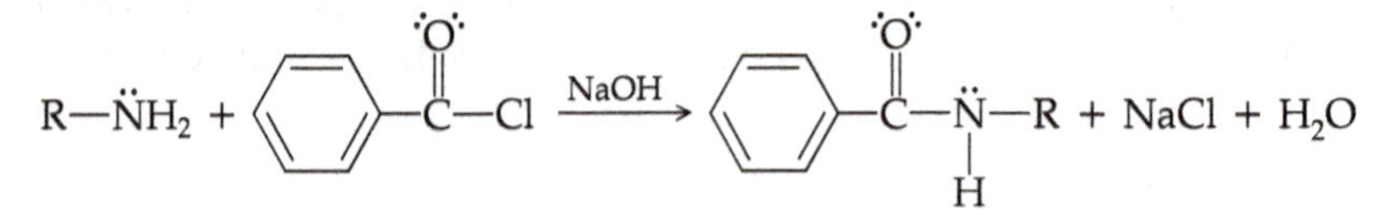


HOOD

In a 3.0-mL conical vial in the **hood** place 0.4 mL of 10% aqueous NaOH solution, 25 mg of the amine, and 2–3 drops of benzoyl chloride. Cap and shake the vial over a period of about 10 min. Vent the vial periodically to release any pressure buildup.

Collect the crystalline precipitate by vacuum filtration, using a Hirsch funnel, and wash the filter cake with 0.1 mL of dilute HCl followed by 0.1 mL of water. It is generally necessary to recrystallize the material from methanol or aqueous ethanol. Dry the product on a porous clay plate and determine the melting point.

## PHENOLS[21]

### α-Naphthylurethanes (α-Naphthylcarbamates)

The procedure outlined under Alcohols: Phenyl-, and α-Naphthylurethanes is used to prepare these derivatives (p. 656).

### Bromo Derivatives

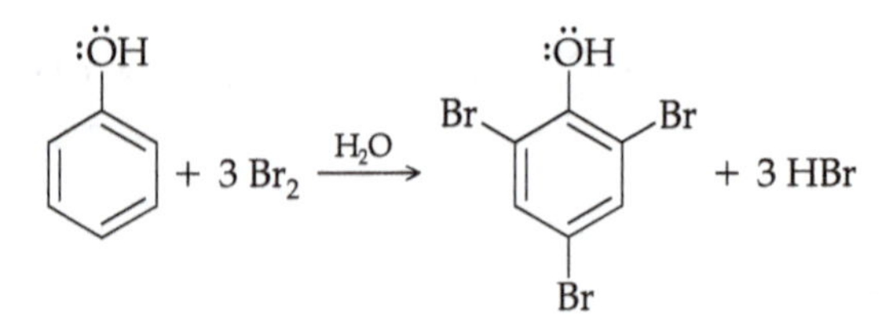

*INSTRUCTOR PREPARATION. The brominating reagent is prepared by adding 1.0 mL (3 g) of bromine in the **hood** to a solution of 4.0 g of KBr in 25 mL of water.*

In a 1.0-mL conical vial, place 10 mg of the unknown phenol followed by 2 drops of methanol and 2 drops of water. To this solution, in the **hood,** add 3 drops of brominating agent from a Pasteur pipet.

Continue the addition (dropwise) until the reddish-brown color of bromine persists. Now add water (4 drops), cap the vial, shake, vent, and then allow it to stand at room temperature for 10 min. Collect the crystalline precipitate by vacuum filtration using a Hirsch funnel and wash the filter cake with 0.5 mL of 5% aqueous sodium bisulfite solution. Recrystallize the solid derivative from ethanol, or from an ethanol–water mixture. Dissolve the material in about 0.5 mL of ethanol, add water until it becomes cloudy, cool in an ice bath, and collect the crystals in the usual manner. Dry the purified product on a porous clay plate and determine the melting point.

| | |
|---|---|
| Organic Chemistry Laboratory II | Name ____________________ |
| CHEM 3106 | TA ____________________ |
| Solid Unknown | Date ____________________ |
| Experiment 3106-9, Preparation of Derivatives | Unknown ID ____________________ |

Based on your functional group determinations and your IR spectrum, what functional group did you use for the formation of derivatives?

What derivative(s) did you make?

Briefly describe the procedure you followed and any problems you encountered.

| | observed | literature (from blue book) |
|---|---|---|
| Melting point of derivative 1: | | |
| Melting point of derivative 2: | | |
| Melting point of derivative 3: | | |

I conclude my unknown is:

| | |
|---|---|
| Organic Chemistry Laboratory II | Name ____________________ |
| CHEM 3106 | TA ____________________ |
| Liquid Unknown | Date ____________________ |
| Experiment 3106-9, Preparation of Derivatives | Unknown ID ________________ |

Based on your functional group determinations and your IR spectrum, what functional group did you use for the formation of derivatives?

What derivative(s) did you make?

Briefly describe the procedure you followed and any problems you encountered.

| | observed | literature (from blue book) |
|---|---|---|
| Melting point of derivative 1: | | |
| Melting point of derivative 2: | | |
| Melting point of derivative 3: | | |

I conclude my unknown is:

## BIBLIOGRAPHY

Cheronis, N. D.; Entriken, J. B. *Identification of Organic Compounds: A Student's Text Using Semimicro Techniques*; NewYork: Interscience, 1963.

Cheronis, N. D.; Ma, T. S. *Organic Functional Group Analysis by Micro and Semimicro Methods*; Interscience: NewYork, 1964.

Cheronis, N. D.; Entrikin, J. B.; Hodnett, E. M. *Semimicro Qualitative Organic Analysis*, 3rd ed.; Interscience: NewYork, 1965.

Feigl, F.; Anger,V. *Spot Tests in Organic Analysis*, 7th ed.; Elsevier: NewYork, 1966.

Pasto, D. J.; Johnson, C. R.; Miller, M. J. *Experiments and Techniques in Organic Chemistry*; Prentice Hall: Englewood Cliffs, NJ, 1992.

Schneider, F. L. In *Qualitative Organic Microanalysis,Vol. II of Monographien aus dem Gebiete der qualitativen Mikroanalyse*; Benedetti-Pichler, A. A., Ed.; Springer-Verlag:Vienna, 1964.

Shriner, R. L.; Hermann, C.K.F.; Morrill, T.C.; Curtin, D.Y.; Fuson, R.C. *The Systematic Identification of Organic Compounds*,

8th ed.; Wiley: NewYork, 2003.

Vogel, A. I. *Qualitative Organic Analysis, Part 2 of Elementary Practical Organic Analysis*; Wiley: NewYork, 1966.

______*Vogel's Textbook of Practical Organic Chemistry, Including Qualitative Organic Analysis*, 5th ed.; London: Longman Group, 1989.

Pasto, D. J.; Johnson, C. R.; Miller, M. J. *Experiments and Techniques in Organic Chemistry*; Prentice Hall: Englewood Cliffs, NJ, 1992.

Rappoport, Z. *Handbook of Tables for Organic Compound Identification*, 3rd ed.; CRC Press: Boca Raton, FL, 1967.

Shriner, R. L.; Hermann, C. K. F.; Morrill, T. C.; Fuson, R. C. *The Systematic Identification of Organic Compounds*, 8th ed.; Wiley: New York, 2003.

# Grignard Reaction with a Ketone: Triphenylmethanol

EXPERIMENT 3106-10

Common name: triphenylmethanol
CA number: [76-84-6]
CA name as indexed: benzenemethanol, α,α-diphenyl-

***Purpose.*** The reaction of Grignard reagents with ketones to form *tertiary* alcohols is investigated.

This illustrates the Grignard reaction, which is a versatile and useful reaction for the formation of carbon-carbon bonds.

*Prior Reading*

*Technique 4:* Solvent Extraction
Liquid–Liquid Extraction (p. 72)
Drying of the Wet Organic Layer (pp. 80–83)
*Technique 5:* Crystallization
General Crystallization Procedure (pp. 85–87)
Use of the Hirsch funnel (pp. 88–89)
*Technique 6A:* Chromatography
Thin-Layer Chromatography (pp. 97–99)

## REACTION

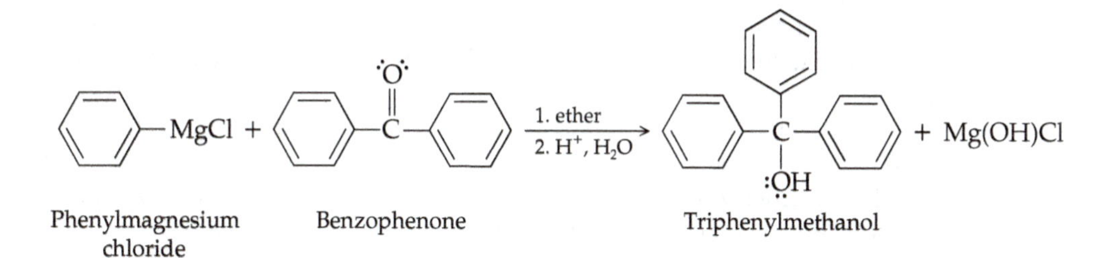


**François Auguste Victor Grignard (1871–1935)** Born in Cherbourg, Grignard was professor of Chemistry at the Universities of Lyons and Nancy. After studying for one year with Bouveault, Grignard became a graduate student of Phillippe Antoine Barbier, a professor at the University of Lyons. Barbier, who was working in the area of terpene chemistry, had found that magnesium could be used in place of zinc in the reaction of methyl iodide with an unsaturated ketone (methylheptenone) to yield the corresponding tertiary alcohol. This route was much preferred since the zinc reagents were difficult to work with because they were pyrophoric (spontaneously flammable in air). The use of magnesium in the formation of tertiary alcohols was reported in 1899. Barbier suggested to Grignard that it might be interesting to further investigate the reaction of magnesium with alkyl halides. This study was to form the basis of Grignard's doctoral dissertation. Grignard discovered that treatment of alkyl iodides with magnesium in *diethyl ether* produced an alkylmagnesium iodide by a spontaneous reaction at ambient temperatures. His initial results, reported in 1900, were followed by seven papers the following year. His doctoral thesis on organomagnesium compounds and their application to synthetic organic chemistry was presented in 1901, when Grignard was 30 years old.

Grignard continued this work, having recognized the enormous potential of the alkyl magnesium halides in organic synthesis. These species are now known as *Grignard reagents* and when the reagent is used in synthesis, the reaction is called a *Grignard reaction*. These reagents have found great utility in the preparation of many kinds of organic compounds, including alcohols, ketones, esters, and carboxylic acids. As mentioned above, these reagents contain a carbon–metal bond, and therefore they are classed in a large group of substances called *organometallic* compounds.

For this work, Grignard received the Nobel Prize in 1912. In 1919 Grignard returned to Lyons where he succeeded Barbier as chairman of the Department. By the end of his life, the scientific literature contained over 6000 papers dealing with Grignard reagents and their application.

Grignard also did extensive work in the areas of the terpenes, quantitative ozonolysis of alkenes, aldol reactions, catalytic hydrogenation, and dehydrogenation and cracking of hydrocarbons.[17]

## DISCUSSION

Grignard reagents possess significant nucleophilic character because of the highly polarized carbon–metal bond that results in considerable carbanionic character at carbon. Grignard discovered that these reactive materials readily attack the electrophilic carbon of a carbonyl group. It is this direct attack on carbon by a carbon nucleophile, resulting in carbon–carbon bond formation, that makes these such important reactions. Furthermore, as the carbonyl is the most ubiquitous functionality in all of organic chemistry, Grignard reagents have found great utility and widespread use in organic synthesis.

The formation of the organomagnesium halide (Grignard reagent) involves a heterogeneous reaction between magnesium metal and an alkyl, alkenyl, or aryl halide in ether solution. The solvent may be any one of a number of ethers, but diethyl ether and tetrahydrofuran are by far the most popular.

$$\mathrm{R(Ar)}{-}\mathrm{X} + \mathrm{Mg} \xrightarrow{\text{ether}} \mathrm{R(Ar)}{-}\mathrm{Mg}{-}\mathrm{X}$$

R = alkyl, alkenyl
and Ar = aryl

The reaction between an alkyl, alkenyl, or aryl halide and magnesium takes place on the surface of the metal and is an example of a *heterogeneous* (across two phases) reaction. The reactivity of the alkyl halides is in the order Cl < Br < I; fluorides do not generally react. Substituted alkyl halides react in the order 1° > 2° > 3° alkenyl and aromatic halides also form Grignard reagents to varying degrees.

---

[17]See Gordon, N. E. *J. Chem. Educ.* **1930,** *7,* 1487; Rheineoldt, H. *J. Chem. Educ.* **1950,** *27,* 476; Kauffman, G. B. *J. Chem. Educ.* **1990,** *67,* 569; Gilman H. *J. Am. Chem. Soc.* (Proc.) **1937,** *59,* 17; Gibson, C. S.; Pope, W. J. *J. Chem. Soc.* **1937,** 171; *Dictionary of Scientific Biography,* C. C. Gillespie, Ed., Scribner's: New York, 1972, Vol. V, p. 540.

It is important to understand the role of the ether solvent in the formation of the Grignard reagents. The reaction at the surface of the metal is essentially an oxidation–reduction reaction. The metal is partially oxidized to the greater than 1+ state and the organohalide is reduced to a halide ion and a highly polarized carbon–metal bond with the magnesium. The overall reaction can be viewed as forming the species, $R^{\delta-}\text{—}^{\delta+}Mg^{+}X^{-}$as the Grignard reagent. This highly polarized material is insoluble in most nonpolar organic solvents. The reaction will proceed at the surface of the metal until a layer of the insoluble organometallic reagent has formed. At this point, the surface reaction with the magnesium will immediately cease. If protic solvents are used, they would instantly react with the highly basic Grignard reagent, R—MgX, to form the corresponding hydrocarbon, R—H. Thus, the use of either nonpolar or protic solvents does not lead to successful Grignard reagent formation. Why, then, is ether, a relatively nonpolar solvent, essential to the preparation of Grignard reagents?

The magnesium is essentially divalent, and electron deficient, when it reacts with the halide to form the RMgX species. A full octet around the metal atom requires two additional pairs of electrons. It is the energy gained by filling this octet that drives the coordination of the magnesium with two molecules of the ether solvent. This association in turn dramatically increases the solubility of the Grignard reagent in the relatively nonpolar ether solvent, and thus promotes further Grignard reagent formation.

$$R\text{—}X + Mg \xrightarrow{\text{ether}} \begin{array}{c} \overset{\delta+}{CH_3CH_2\ddot{O}CH_2CH_3} \\ \downarrow \\ (R\text{—}Mg\text{—}X)^{\delta-} \\ \uparrow \\ \underset{\delta+}{CH_3CH_2\ddot{O}CH_2CH_3} \end{array}$$

This interaction of RMgX with ether solvent also may be described as a Lewis acid–base interaction in which the coordinating solvent molecules are usually not written. When a Grignard reagent is described, it is important to remember that this vital solvation is always taking place.

A major impurity can be formed by a coupling reaction during the Grignard reagent preparation.

$$2RX + Mg \xrightarrow{\text{ether}} R\text{-}R + MgX_2$$

You will use commercially available phenylmagnesium chloride (Grignard reagent) in today's experiment, thus minimizing the presence of this impurity in your product.

The reactions of Grignard reagents with different types of carbonyl groups yield a number of important functional groups. For example, reaction with formaldehyde yields 1° alcohols; with higher aldehydes, 2° alcohols; with ketones, 3° alcohols; with esters, 3° alcohols; with acyl halides, ketones; with *N,N*-dialkylformamides, aldehydes; and with carbon dioxide, carboxylic acids.

In this experiment you will study the addition of the aryl Grignard reagent (phenylmagnesium chloride) to a diaryl ketone (benzophenone) to yield the corresponding tertiary (3°) alcohol. Because it is possible to vary both the structure of the Grignard reagent and the ketone, a wide variety of 3° alcohols may be obtained by this synthetic route.

The mechanism, as discussed above, can be thought of as involving rapid nucleophilic attack by the Grignard reagent at the carbon of the carbonyl group. Hydrolysis of the resulting alkoxide ion intermediate with dilute acid yields the desired alcohol. The reaction sequence is outlined here:

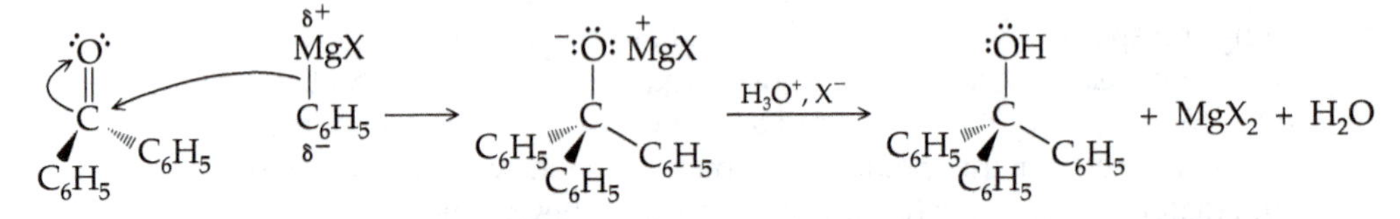

By using Grignard reagents, it is theoretically possible to synthesize a very large number of alcohols. Indeed, there is often more than one synthetic pathway open to a desired product. The choice of route is generally dictated by the availability of starting materials and the associated costs of these compounds.

The Grignard reagent is water sensitive, meaning it will react with water. If water is present, the desired reaction will not go to completion. The reaction of phenylmagnesium chloride with water is shown below.

## EXPERIMENTAL PROCEDURE

Estimated time to complete the experiment: one laboratory period.

---

CAUTION: Ether is a flammable liquid. All flames must be extinguished during the time of this experiment.

---

**Physical Properties of Reactants**

| Compound | MW | Amount | mmol | mp (°C) | bp (°C) | *d* | $n_D$ |
|---|---|---|---|---|---|---|---|
| Diethyl ether | 74.12 | 0.600 mL | | | 34.5 | 0.73 | |
| Phenylmagnesium chloride (2.0 M solution in THF)– 1.5 eq | | | | | | | |
| Benzophenone | 182.21 | 105 mg | 0.58 | 48 | | | |

***Reagents and Equipment.*** In an effort to remove residual water from your glassware, you will rinse your glassware with acetone and then make sure you evaporate away all the acetone. You will use an excess of the Grignard reagent (1.5 equivalents instead of 1.0 equivalents) and accept the loss from the water adsorbed on the glass.

Calculate the amount (mL) of the 2.0 M solution of Grignard reagent required to give a 1.5 equivalent excess relative to benzophenone.

Set up your apparatus as illustrated. Have your TA check your calculation and apparatus set-up and then he or she will dispense the phenyl magnesium chloride into your apparatus.

***The Benzophenone Reagent.*** Prepare a solution of 105 mg (0.58 mmol) of benzophenone in 300 μL of anhydrous diethyl ether in a dry shell vial with a cap. *The ether is measured using a graduated 1-mL syringe and is dispensed in the* ***hood.***

HOOD

Draw the solution immediately into a 1.0-mL syringe, and then insert the syringe needle through the rubber septum on the Claisen head. Place an additional 300 μL of the anhydrous diethyl ether in the empty vial, cap it, and set it aside for later use.

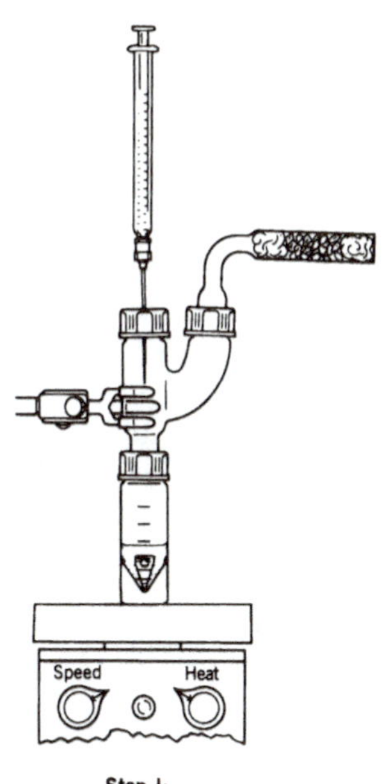


**Step I:**
TA will add PhMgCl

**Step II:**
$(C_6H_5)_2CO$, 105 mg
+ $(CH_3CH_2)_2O$, 600 μL

***Reaction Conditions.*** *Carefully*, with stirring, add the benzophenone solution to the Grignard reagent (1.5 equiv) over a period of approximately 30 s or at a rate that maintains the temperature of the ether solvent at a no more than gentle reflux.

Upon completion of this addition, add the rinse from the capped vial, in like manner, in a single portion.

Stir the reaction mixture for 2–3 min and then allow it to cool to room temperature. Remove the reaction vial from the Claisen head and cap it. During this cooling period the reaction mixture generally solidifies. *Once the reaction vial is detached from the Claisen head, it is recommended that the vial be placed in a 10-mL beaker to prevent loss of product by accidental tipping.*

***Isolation of Product.*** Hydrolyze the magnesium alkoxide salt by the *careful*, dropwise addition of 3 M HCl from a Pasteur pipet, while at the same time using a small stirring rod to break up the solid residue. Continue the addition until the aqueous phase tests acidic with litmus paper. A two-layer reaction mixture forms (ether–water) as the solid gradually dissolves.

---

CAUTION: The addition of the acid may be accompanied by the evolution of heat and some frothing of the reaction mixture. An ice bath should be handy to cool the solution, if necessary. Additional ether may be added, if required, to maintain the volume of the organic phase. Check the acidity of the mixture periodically. The total reaction mixture must be acidic; both insufficient or excess amounts of hydrochloric acid will result in a decreased yield of product during the subsequent workup.

---

Now remove the magnetic spin vane with forceps and set it aside to be rinsed with an ether wash. Cap the vial tightly, shake, carefully vent, and allow the layers to separate.

Using a Pasteur filter pipet, transfer the lower aqueous layer to a clean 5.0-mL conical vial.

*NOTE. Save the ether layer—it should contain your product.*

Wash the acidic aqueous layer with three 0.5-mL portions of diethyl ether (use technical ether, not anhydrous ether; Pasteur pipet). Rinse the spin vane with the first portion of ether as it is added to the vial. Cap the vial, shake (or use a Vortex mixer, if available), vent carefully, and allow the layers to separate. After each extraction, combine the organic phase with the ether solution saved above. The bottom (aqueous) layer is set aside in a 10-mL Erlenmeyer flask until the experiment is completed.

Now extract the combined ether layers with two portions of 0.5 mL of cold water to remove any acidic residue. Combine the aqueous rinse with the previously extracted and stored aqueous layers in a 10-mL Erlenmeyer flask. Dry the ether solution (capped) over 250–300 mg of anhydrous granular magnesium sulfate for approximately 10 min. Stir the drying agent intermittently with a glass rod or swirl the flask. If large clumps of magnesium sulfate begin to develop and the solution remains cloudy, it may be necessary to transfer the ether extracts to another vial for a second treatment with the drying agent, or you may be able to simply add more $MgSO_4$ to the original vial. The ether solution should be *clear* following treatment with the anhydrous magnesium sulfate.

Transfer the dried ether solution to a *previously tared* shell vial containing a boiling stone. Carry out the transfer in 0.5-mL portions, concentrating each ether aliquot by warming the vial in an aluminum block on a hot plate and/or carefully running a stream of air over the solution in in the **hood** between the transfers. Rinse the vial and drying agent with an additional 0.5 mL of ether, add the rinse to the tared shell vial, and, finally, concentrate the solution to dryness to yield the product residue.

HOOD

***Characterization.*** Weigh the crude triphenylmethanol product and calculate the percent yield. Determine the melting point of the material and compare your result to that recorded in the literature.

Characterization of the triphenylmethanol is best done by obtaining the IR spectrum and comparing the spectral data to that of an authentic sample or to the published spectra in *The Aldrich Library of IR Spectra*.

## QUESTIONS

**7-19.** Predict the product formed in each of the following reactions and give each product a suitable name:

**(a)** $CH_3CH_2MgBr + CH_2O \xrightarrow[\text{2. H}^+]{\text{1. ether}}$

**(b)** $p\text{-}CH_3C_6H_4MgBr + CH_3CH_2CHO \xrightarrow[\text{2. H}^+]{\text{1. ether}}$

**(c)** $C_6H_5MgBr + D_2O \xrightarrow{\text{ether}}$

**(d)** 2,4,6-trimethylphenyl–MgBr (ring with $CH_3$ at para position and $CH_3$ groups at both ortho positions) $+ CO_2 \xrightarrow[\text{2. H}^+]{\text{1. ether}}$

**7-20.** Using the Grignard reaction, carry out the following transformations. Any necessary organic or inorganic reagents may be used.

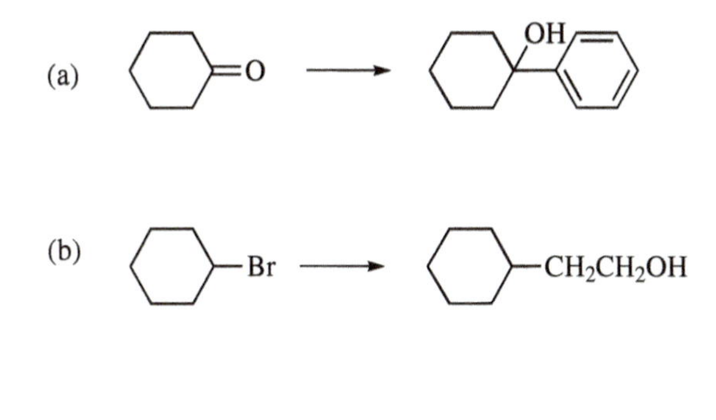


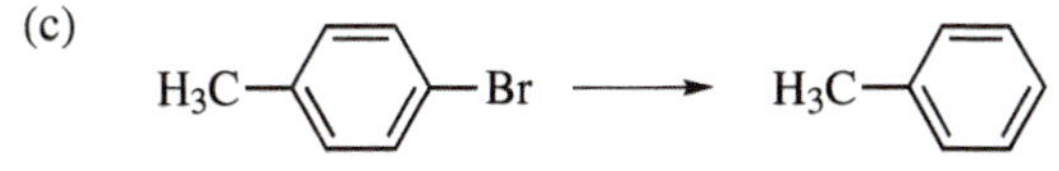


**7-21.** An impurity can be formed during the preparation of the Grignard reagent. For the preparation of phenylmagnesium chloride, that impurity is biphenyl. Ligroin may be used as a solvent for the separation of triphenylmethanol from biphenyl.

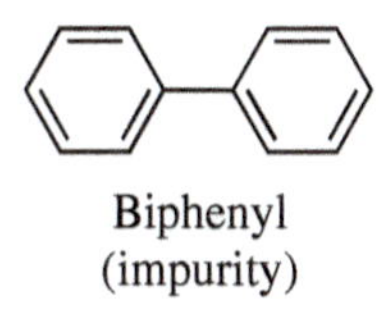

Biphenyl
(impurity)

**(a)** What is ligroin?

**(b)** Can you suggest an alternative solvent that might be used in this step?

## BIBLIOGRAPHY

**The alternative method of preparing the Grignard reagent is adapted from**

Eckert, T. S. *J. Chem. Educ.* **1987,** *64,* 179.

**General references on Grignard reagents:**

Coates, G. E.; Green, M. L. H.; Wade, K. *Organometallic Compounds,* 3rd ed.; Methuen: London; Vol. II, 1968.

Grignard, V. *Compt. Rend.* **1900,** *130,* 1322.

Huryn, D. M. *Comp. Org. Syn.* **1991,** *1,* 49.

Lai, Y. H. *Synthesis* **1981,** 585.

Raston, C. L.; Salem, G. in *The Chemistry of the Metal–Carbon Bond;* Hartley, F. R., Ed.; Wiley: New York, 1987, Vol. 4, p. 159.

Shirley, D. A. *Org. React.* **1954,** *8,* 28.

**A synthesis of triphenylmethanol (triphenylcarbinol) is reported in *Organic Syntheses:***

Bachmann, W. E.; Hetzner, H. P. *Organic Syntheses;* Wiley: New York, 1955; Collect. Vol. III, p. 839.

**The preparation of a series of tertiary alcohols by the addition of Grignard reagents to diethyl carbonate is reported in *Organic Syntheses:***

Moyer, W. W.; Marvel, C. S. *Organic Syntheses;* Wiley: New York, 1943; Collect. Vol. II, pp. 602–603. Also see Braun, M; Gräf, S; Herzog, S. *Organic Syntheses* **1995,** *72,* 32.

# Esterification: An Illustration of Combinatorial Chemistry

EXPERIMENT 3106-11

*Prior Reading*

*Technique 4:* Solvent Extraction

Liquid–Liquid Extraction (p. 72)

Drying of the Wet Organic Layer (pp. 80–83)

***Purpose.*** This exercise explores the classic reactions of carboxylic acids ($RCO_2H$) with alcohols (R′OH), in the presence of acid catalyst, to yield esters ($RCO_2R'$) plus water ($H_2O$, a small stable molecule). A library of esters from 9 acids and 4 alcohols will be synthesized and the one which smells of wintergreen will be identified. This will illustrate the so-called "combinatorial" approach to making large numbers of compounds.

## REACTION

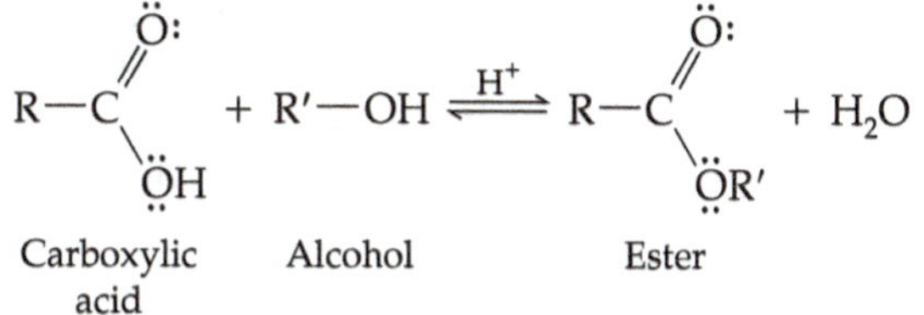


## DISCUSSION

*What is "combinatorial chemistry?"* There are numerous sets of reaction conditions which have been described as combinatorial; they all share as a common goal, the straightforward production of a relatively large number of diverse, but related compounds. In this experiment, you will carry out what is called a "parallel synthesis," in which numerous products will be made at the same time, in separate flasks. For comparison, consider all of the other reactions you have carried out in this laboratory, which were designed and optimized to make a single product.

*Why would you want to make a large number of related compounds?* This is often necessary in the screening process of discovery of new drugs in pharmaceutical companies. A vast number of compounds need to be made and their biological activity measured before a new drug can be discovered. In this experiment, the biological assay will be the odor of the product, and the target will be a compound that smells like wintergreen. (As with all science, you may find something interesting that you were not looking for!) Esters are among the most important of the carboxylic acid (and alcohol) derivatives. Substances possessing this functional group are widely distributed in nature in the form of waxes, essential oils, faty acid esters, and aromas. The ester functionality plays a significant role in biochemistry, both in primary metabolism and in a variety of substances exhibiting remarkable physiological activity in humans (hormones and neurotransmitters). Esters find extensive use in commercial products from fingernail polish remover and artificial sweeteners, to polymeric fibers, plasticizers, and surfactants.

***Biosynthesis of Esters.*** Fatty acids are naturally occurring, long, straight-chain, $C_{12}$–$C_{40}$ carboxylic acids; most contain an even number of carbon atoms.Their biosynthesis provides an important and interesting example of a primary metabolic pathway in which a special type of ester is the essential link between the enzyme and the substrate (acetic acid). The enzyme-bound substrate grows by repeated addition of two-carbon ($C_2$) units and, when eventually released from the enzyme, has undergone an extension of the fatty acid hydrocarbon chain.

The first step in fatty acid biosynthesis involves the formation of a thiol ester, acetyl coenzyme A (acetyl CoA), from acetic acid (present in the primary metabolic pool) and the thiol group (mercapto, or —SH group) of the coenzyme (HSCoA). A thiol ester is an ester in which the single-bonded oxygen (from the alcohol component) is replaced by a sulfur atom,

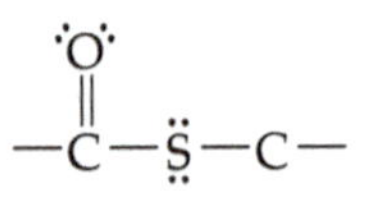

from the coenzyme. Coenzymes are loosely bound, nonprotein factors attached to the enzyme that play an important role in the catalytic function of the enzyme. These coenzymes are distinguished, in an ill-defined manner, from prosthetic groups, which are intimately attached to active sites of enzymes. Part of the role of the CoA is to facilitate the transfer of the substrate (the $C_2$ unit) to a new thiol group of the enzyme (protein), where the next stage of the biosynthesis takes place:

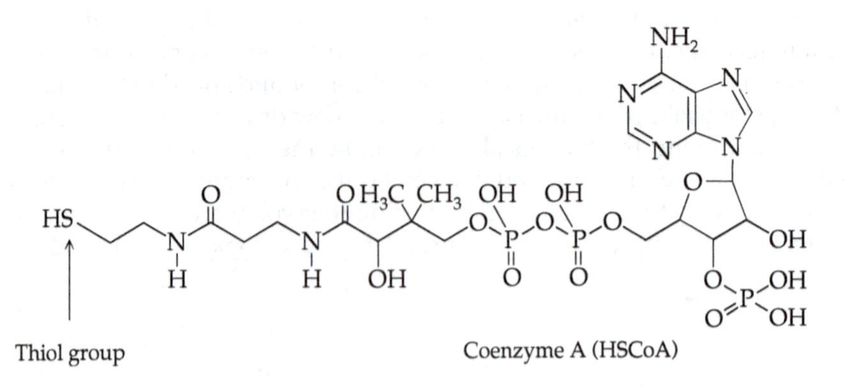

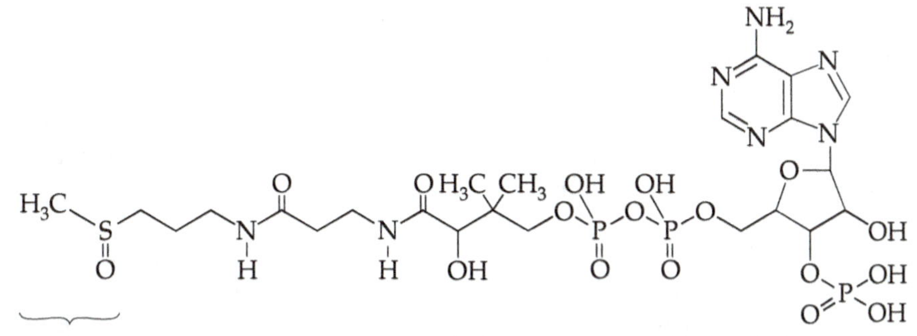

Acetyl thiolester Acetyl coenzyme A ($CH_3COSCoA$)

This reactive thiol ester is capable of undergoing aldol-type condensations under physiological conditions. AcetylCoA is first carboxylated with the help of the enzyme, acetyl CoA-carboxylase, to yield a thiolmalonyl derivative. The resulting intermediate possesses an activated methylene group

$$-\overset{\overset{\large O}{\|}}{C}-\underline{CH_2}-\overset{\overset{\large O}{\|}}{C}-S-$$

which with further enzymatic support undergoes a Claisen condensation with an acetyl group that has been also transferred via acetyl CoA to an appropriate acyl-carrier protein, fatty acid synthetase. Reduction, dehydration, further reduction, and finally hydrolysis of the thiol ester, yields the fatty acid extended by a $C_2$ hydrocarbon group.

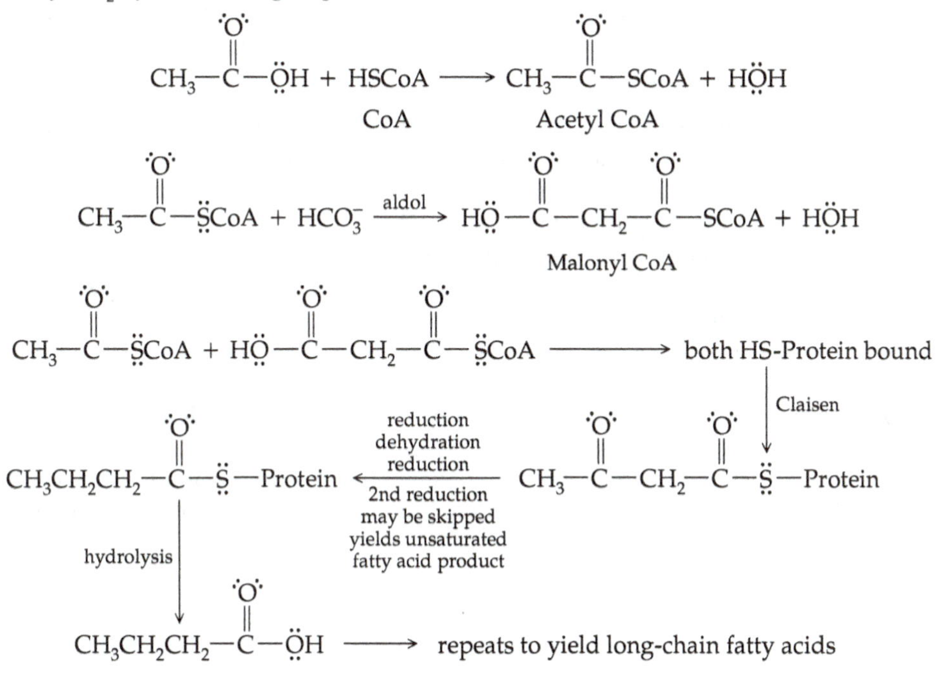


## LIPIDS

***Oils, Fats, Waxes, and Aromas.*** Fatty acids derived from primary metabolism play a key role in the formation of naturally occurring oils, fats, and waxes. Fats and oils are esters of these acids with a triol, glycerol ($HOCH_2CHOHCH_2OH$).

The common fats and oils are formed from mixtures of $C_4$–$C_{26}$ saturated fatty acids with the vast majority derived from $C_{12}$–$C_{18}$ acids. The oils are more likely to include significant contributions from mixtures of unsaturated fatty acids.

Since fats and oils are triesters of glycerol, they are generally called *triglycerides*. In plants and animals, triglycerides function as energy reserves that can be used in primary metabolism when food (energy) is not available to the organism.

Although the fats have high molecular weights, they are generally found to be very low-melting solids, particularly if they contain unsaturated fatty acids. The bent chains, which result from incorporating cis-alkene (C═C) groups into the chain, prevent close packing in a solid and, as a result, such molecules exhibit lower melting points. For example, compare oleic acid ($C_{18}H_{34}O_2$, mp 4 °C) to its saturated analog, stearic acid($C_{18}H_{36}O_2$, mp 69–70 °C). The former melts more than 60 °C lower because of the cis carbon–carbon double bond.

Oleic acid is the simplest of the unsaturated $C_{18}$ fatty acids, because it has a single C═C group located in the middle of the chain. This unsaturated fatty acid is the most widely distributed of all fatty acids. It is the dominant component (76–86%) of the triglycerides in olive oil. Highly saturated fats, on the other hand, are generally solids at room temperature because the straight-chain fatty acids pack together well (see Fig. 6.19 on next page).

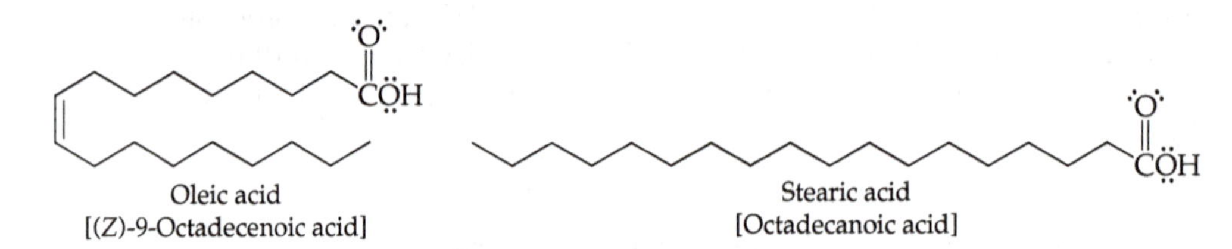


Oleic acid
[(Z)-9-Octadecenoic acid]

Stearic acid
[Octadecanoic acid]

Many vegetable and fish oils are liquid triglycerides. Because these organisms operate at ambient temperatures, evolution dictated that low-melting fats were required to avoid solidification. In warm-blooded animals, higher melting fats can be tolerated and are used.

The cheap and plentiful unsaturated oils can be converted to solid fats by hydrogenation of the alkene groups, which gives straight-chain alkyl groups. As consumers historically have desired to cook with solid, white, and creamy fats (such as lard) derived from animal triglycerides (low in unsaturation), hydrogenation of vegetable oils, such as peanut, soybean, and cottonseed oils, has been carried out on a large scale (this process is referred to as **hardening** the fat).

Unfortunately, the relationship between saturated fats in the human diet and the formation of cholesterol (a simple lipid, see below) plaque and coronary heart disease has been established. The dietary switch to less saturated fats is currently underway.

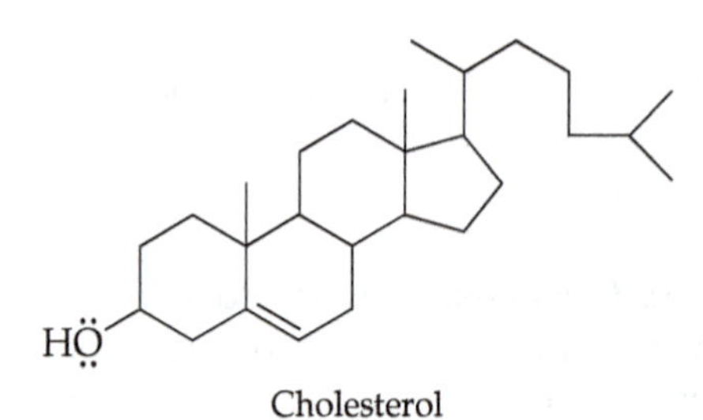


Cholesterol

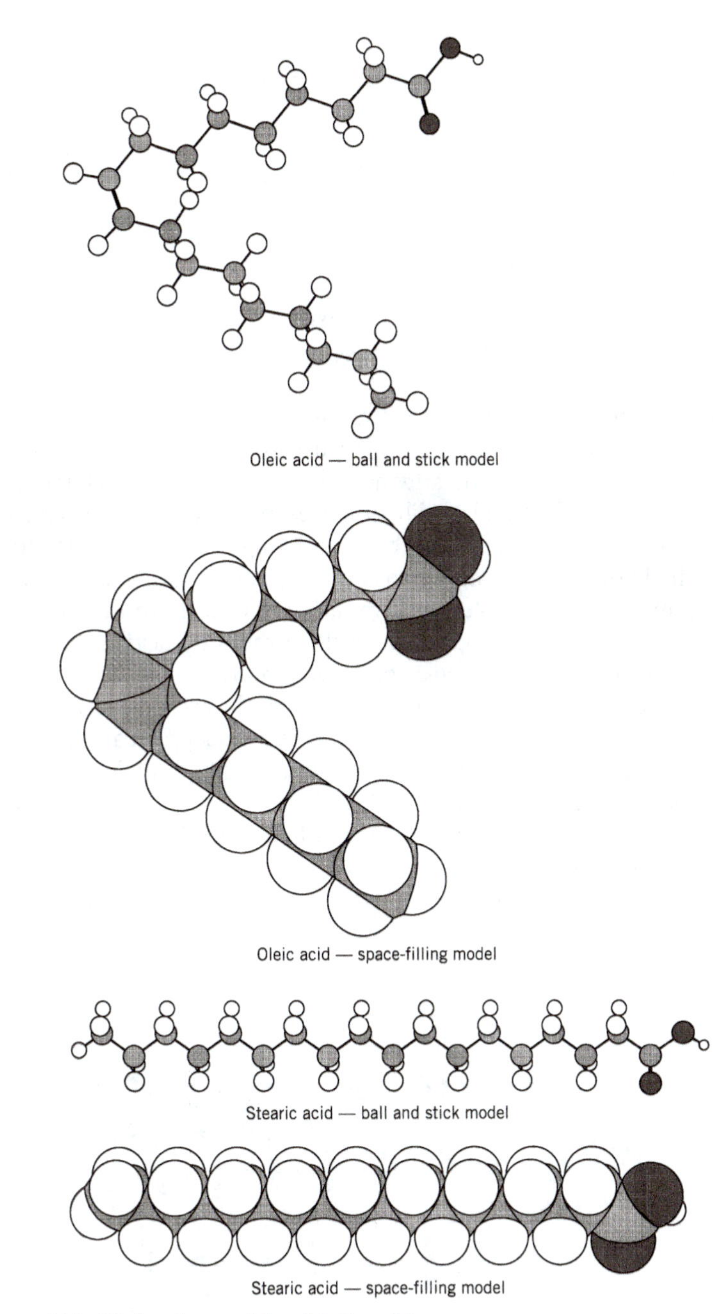


**Figure 6.19 Molecular models of fatty acids.**

The triglycerides obtained from animal fats have been used for a very long time as a source of soap. When fats are boiled with lye (sodium hydroxide) the ester linkages are cleaved by a process known as *saponification* (the term originates from the Latin word for soap, *sapon,* as does the modern French word for soap, *savon*) to yield the sodium salt of the fatty acid and the esterifying alcohol (glycerol).

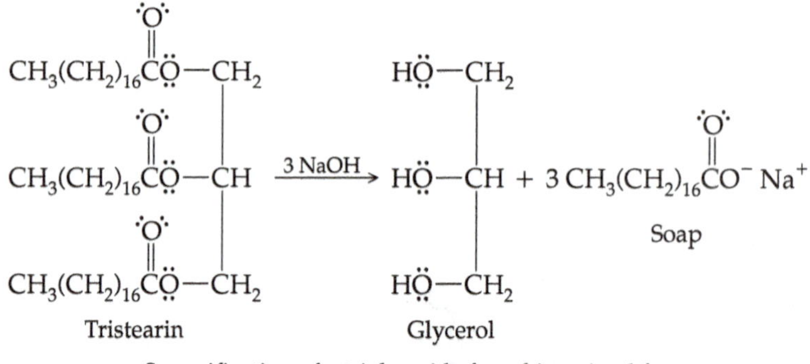


Saponification of a triglyceride found in animal fat

Salts of fatty acids function effectively as soaps because one end of the straight-chain system has the highly polar carboxylate ion and is readily solvated in water. The rest of the fatty acid molecule has all the characteristics of a nonpolar hydrocarbon and readily dissolves in hydrocarbons, such as greases and oils. We refer to the polar end (head) as being hydrophilic (attracted to water) and the hydrocarbon end (tail) as being lipophilic (attracted to oils). When dispersed in an aqueous solution, fatty acids tend to form micelles (spherical clusters of molecules). The lipophilic ends of the fatty acids occupy the interior of the cluster, while the polar ends, which are heavily solvated by water molecules, form the outer surface of the spherical micelle. Micelles absorb the hydrocarbon chains of the triglycerides, and thus soaps break up and help to dissolve the fats and oils that tend to coat skin, clothes, and the surfaces of eating and cooking utensils (see Figs. 6.20 and 6.21).

**Waxes** are naturally occurring esters of fatty acids (in waxes, chain lengths can reach as high as $C_{36}$) and a variety of other alcohols that often possess relatively complicated structures (steroid alcohols) and/or long chains. For example, *n*-octacosanol, $CH_3(CH_2)_{26}CH_2OH$, has been isolated from the esters in wheat

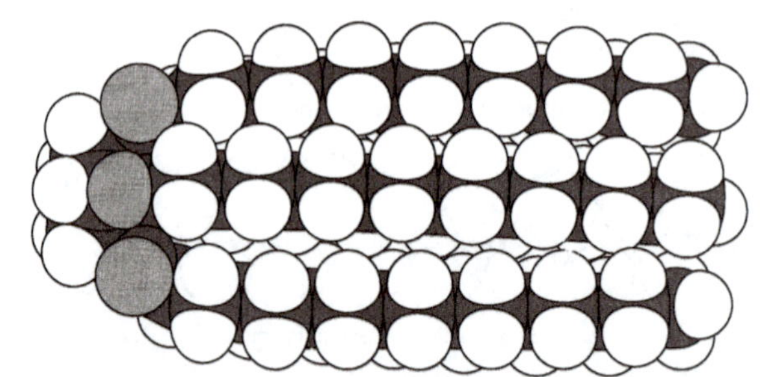

A space-filling model of a saturated triglyceride

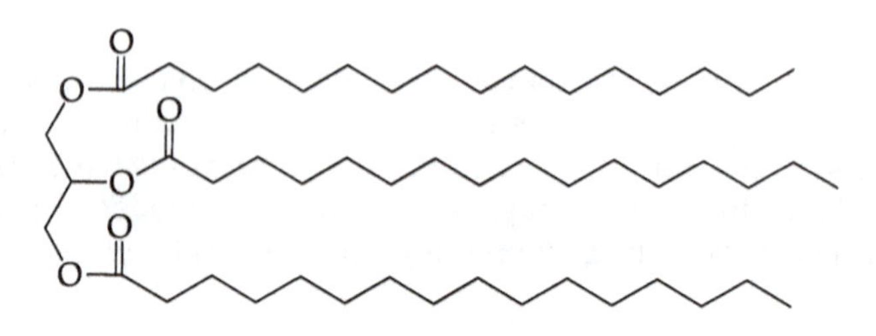

A saturated triglyceride

**Figure 6.20 A space-filling model of a saturated triglyceride.**

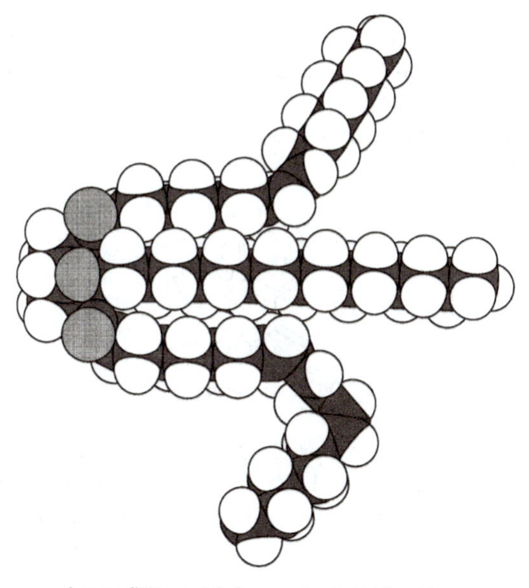

A space-filling model of an unsaturated triglyceride.

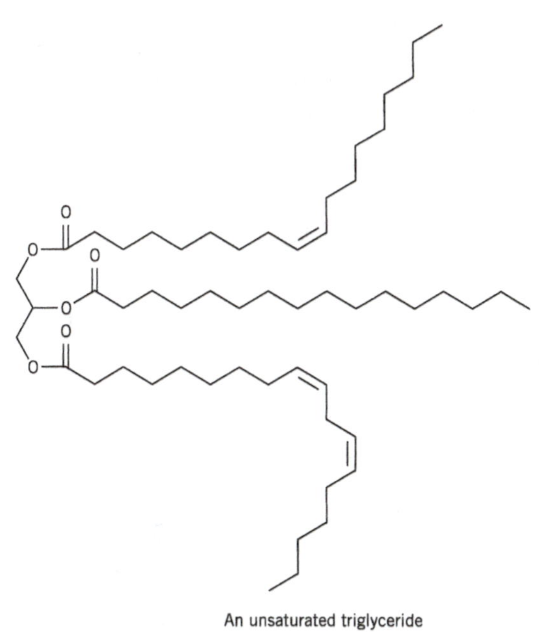

An unsaturated triglyceride

**Figure 6.21 A space-filling model of an unsaturated triglyceride.**

waxes, and a component of carnauba wax (traditionally an automobile wax) has 62 carbons, $CH_3(CH_2)_{33}CO_2(CH_2)_{26}CH_3$. The biological role of carnauba wax is as a leaf coating involved in the conservation of plant moisture. Animal waxes include cetyl palmitate (spermaceti) found in sperm whales and beeswax (one constituent of which has been identified as $CH_3(CH_2)_{29}CO_2(CH_2)_{29}CH_3$, which is used in the construction of the honeycomb).

Lower molecular weight, naturally occurring esters (like what you will synthesize in this experiment) make major contributions to the pleasant aromas of fruits and flowers. These odors have been shown to be composed generally of complex mixtures of materials that have been separated only since the development of modern chemical instrumentation. Single components, however, may play a dominant role in an individual plant or animal. Propyl acetate (pears), ethyl butyrate (pineapples), and 3-methylbutyl acetate (bananas) are examples of simple esters responsible for a particular plant odor. Odors derived from esters are not limited just to esters of straight-chain carboxylic acids, as is demonstrated by oil of wintergreen, methyl salicylate:

Oil of wintergreen

***Phospholipids.*** *Lipid* is a term applied to those natural substances that are more soluble in nonpolar solvents than in water. In its most general sense, it is a broad definition that includes fats, waxes, hydrocarbons, and so on. In biochemistry, lipids are more narrowly defined as substances that yield fatty acids upon hydrolysis.

Another class of glycerides are those substances in which one of the fatty acid groups has been replaced by a phosphoric acid residue: the phospholipids, or more accurately, the phosphoglycerides. The phosphate group is almost always further esterified, usually with a biological amino alcohol, such as choline (the lecithins) or ethanolamine (the cephalins):

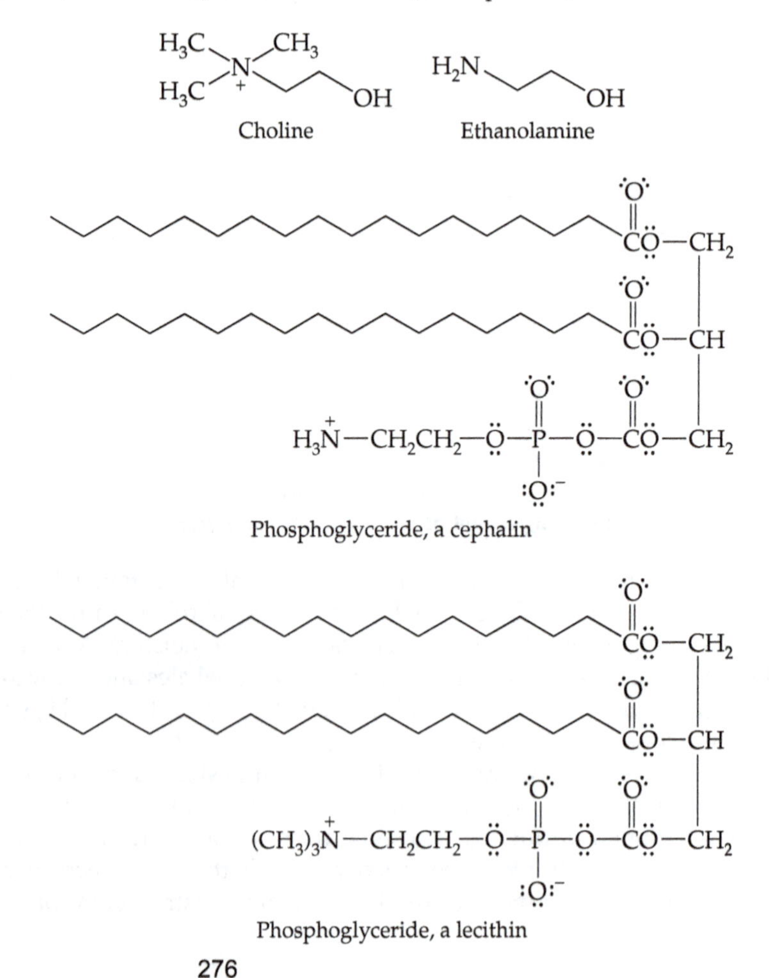

Choline    Ethanolamine

Phosphoglyceride, a cephalin

Phosphoglyceride, a lecithin

These latter groups significantly increase the polarity of the glycerol section of the molecule so that phosphoglycerides undergo strong self-association. In aqueous solutions, this intermolecular attraction can lead to lipid bilayer formation (Fig. 6.22). In a lipid bilayer, the molecules organize themselves to form sheets that contain a double layer of the molecules formed by tail-to-tail association within the interior of the sheet; the outer surface of the lipid bilayer contains the polar heads, which are heavily solvated by water molecules. This association of phosphoglycerides is the key feature in the construction of cell membranes. Thus, esters must have played a vital role at the very earliest stages as cell structures evolved in the development of living systems.

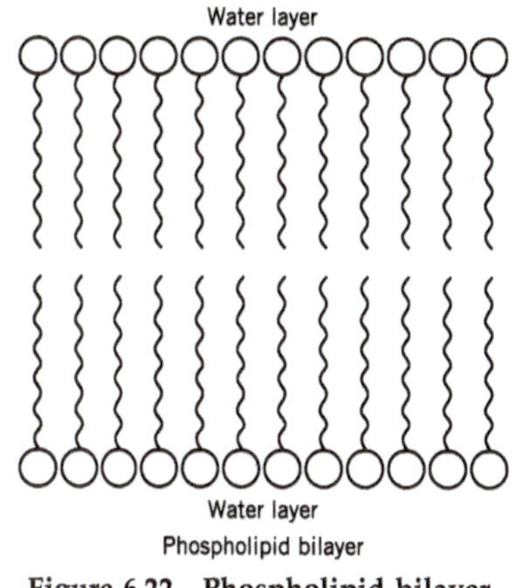


**Figure 6.22 Phospholipid bilayer.**

***Preparation of Esters.*** Esters are generally synthesized by one of four fundamental routes:

1. Esterification of a carboxylic acid with an alcohol in the presence of an acid catalyst
2. Alcoholysis of acid chlorides, anhydrides, or nitriles
3. Reaction of a carboxylate salt with an alkyl halide or sulfate
4. Transesterification reactions

The first of these pathways, known as Fischer esterification, is the method used for the preparation of all the esters in this experiment. The development of this esterification reaction represents just one of a number of major discoveries in organic chemistry by Emil Fischer.

**Emil Fischer (1852–1919)**[7] In 1874 Fischer obtained his Ph.D. from the University of Strasbourg, studying with Adolf von Baeyer. He later had appointments as Professor of Chemistry at Erlangen, Würzburg, and Berlin universities.

In 1875, at the age of 23, and one year after completing his graduate studies, he synthesized phenyl hydrazine ($C_6H_5$—$NHNH_2$) for the first time. This highly reactive reagent later played a key role in Fischer's work on elucidating structures of a large majority of the sugars (carbohydrates), an entire class of important and complex organic molecules. Sugars, or carbohydrates, represent the prime pathway for the storage of radiant energy from the sun, through photosynthesis, as chemical energy. In the short period from 1891 to 1894, Fischer established not only the basic structures, but also the configurations of all the known sugars. In addition, he predicted all the theoretically possible isomers and, in the process, developed a method of representing the three-dimensional molecular structures in two-dimensional drawings that became known as Fisher projection formulas. These representations are still in use today, and have been widely applied beyond sugar chemistry. This work by Fischer led directly to proving the existence of the asymmetric carbon atom, a concept proposed by Vant Hoff and LeBel in 1874.

Fischer was also active in the area of protein chemistry. He demonstrated that amino acids are the basic subunits from which proteins are constructed. Fischer devised methods for the synthesis of many of the known amino acids.

[7]See *Chem. Ind.* **1919,** *42,* 269; Darmstaedter, L.; Oester, R. E. *J. Chem. Educ.* **1928,** *5,* 37; Ratman, C. V. *ibid.* **1942,** *38,* 93; Kauffman, G. B.; Priebe, P. M. *ibid.* **1990,** *67,* 93; *Chem. Eng. News* **1992,** June p. 25. Recommended reading, "The Emil Fischer–William Ramsay Friendship: The Tragedy of Scientists in War." *J. Chem. Educ.* **1990,** *67,* 451.

Perhaps his most ingenious contribution was the "lock and key" hypothesis of how proteins bind with substrates of complementary shapes. This work ultimately led to our understanding of how enzymes, the catalysts of biochemical reactions, function.

Fischer carried out extensive work on the chemistry of purine and on those compounds containing its nucleus. Purine is one of the two nitrogen base ring systems present in DNA. Fischer synthesized approximately 150 members of this class of heterocyclic compounds (including the first synthesis of the alkaloid caffeine (see Experiment 3105-2), uric acid, and the xanthines. He developed a general synthesis of another nitrogen heterocycle, indole, which was so effective that it has become one of the classic synthetic methods of organic chemistry and is known today as the "Fischer indole synthesis":

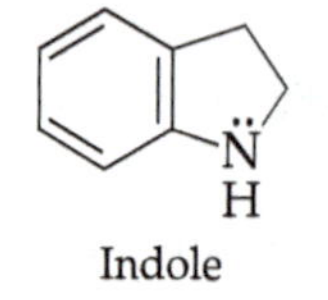

Indole

Fischers work essentially laid the foundation of modern biochemistry. Regarded as the greatest organic chemist of his time, Fischer became the second chemist to receive the Nobel Prize (1902). Depressed by the loss of his young wife at the age of 33, by the loss of two of his three sons (one by suicide, the other in World War I), suffering from the advanced stages of intestinal cancer, and saddened by the socioeconomic conditions of postwar Germany, Fischer committed suicide.

***Mechanism of the Fischer Esterification Reaction.*** The Fischer esterification proceeds by nucleophilic attack of the alcohol on the protonated carbonyl group of the carboxylic acid to form a tetrahedral intermediate. Collapse of the tetrahedral intermediate regenerates the carbonyl group and produces the ester and water. The overall sequence is outlined here:

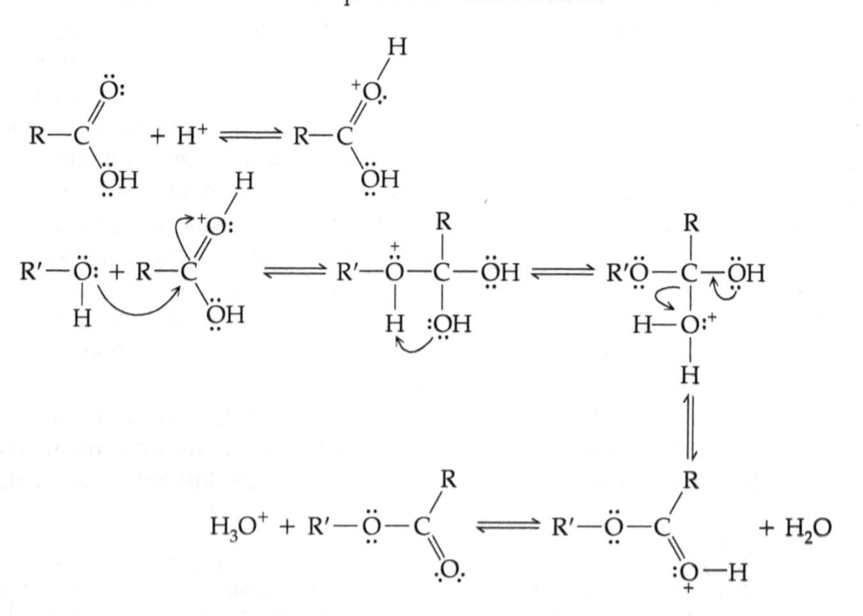

In the Fischer esterification with primary alcohols, the products are only slightly favored by the equilibrium and, therefore, to obtain substantial yields of the ester, the equilibrium must be shifted toward the products. This result can be accomplished in a number of ways. For example, an excess of the starting alcohol can be used to shift the position of equilibrium toward the products. An analogous alternative is to use an excess of the carboxylic acid. A third option to drive the reaction is the removal of one or both of the products (the ester or water) as they are formed during the reaction. The acid catalyst used in Fischer esterifications is generally dry hydrogen chloride, concentrated sulfuric acid, or a strong organic acid, such as *p*-toluenesulfonic acid.

When the carboxyl group and hydroxyl group are present in the same molecule, an intramolecular esterification may occur and a cyclic ester (called a lactone) may be formed. Lactonization requires an acceptable conformation: the two groups must be close and spatially positioned to react. Ring closure is especially favorable if lactone formation yields five- or six-membered (stable and rapidly formed) ring systems.

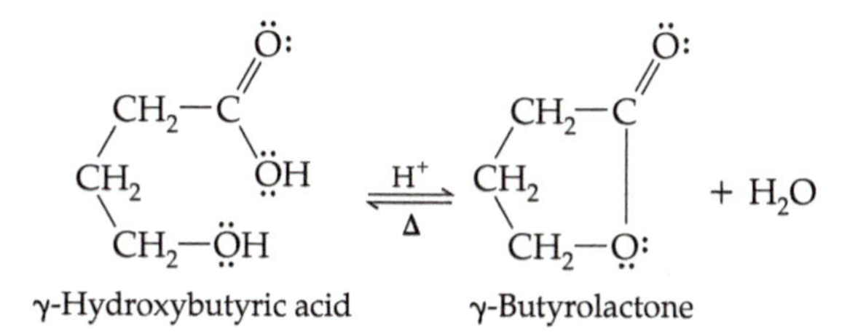

γ-Hydroxybutyric acid γ-Butyrolactone

As noted earlier, Fischer esterification is an equilibrium reaction and is thus reversible. Thus, heating an ester in aqueous solution, in the presence of an acid catalyst, regenerates the corresponding carboxylic acid and alcohol. This latter reaction is called *acid hydrolysis* of an ester. The rate-determining step in both the forward esterification reaction and the reverse reaction, acid hydrolysis, is the formation of the tetrahedral intermediate. It is, therefore, evident that the rate of the reaction will be determined by the ease with which the nucleophile (alcohol on esterification and water on hydrolysis) approaches the carbonyl group. Steric and electronic factors have been shown to have large effects on the rate of esterification. An increase in the number of bulky substituents substituted on the α and β positions of the carbonyl-containing compound decreases the rate (steric effects). Electron-withdrawing groups near the carbonyl group, on the other hand, tend to increase the rate by increasing the electrophilicity (partial positive charge) of the carbonyl carbon (electronic effects). Conversely, electron-donating groups act to retard the rate of esterification (electronic effects).

If you wish to learn more about combinatorial chemistry, there are often articles in Chemical and Engineering New on the subject, (see Borman, S. C. & E. News 1998, 76, 47-67.) There is also an issue of Accounts of Chemical Research (March 1996, 29, 111 - 170) as well as books and a new journal dedicated to the subject. [(1) Combinatorial and Nonpeptide Libraries; Jung, G., Ed.; VCH Publishers: Deerfield Beach, Fl, 1996. (2) Combinatorial Chemistry: Synthesis and Application; Wilson, S. R.; Czarnik, A. W., Ed.; Wiley: New York, 1997. (3) Journal of Combinatorial Chemistry.]

**Prior to lab:** Based on your position in lab, (see Figure 1) determine which alcohol and which acid you will use. Prior to coming to lab to do this experiment, calculate the grams and/or mL of the acid and alcohol that you will use. Bring to lab a table that lists your acid, its molecular weight and the number of grams (or density and mL, if your acid is a liquid; salicylic acid is the only acid that is a solid.) required to give the necessary mmol amount indicated in Figure 1. (The additional quantities of the lower molecular weight materials are helpful to provide an easily manipulated volume.) Your table should also list the identity of the alcohol, its molecular weight, density, and the number of mL required to give 12 mmol of alcohol (all the alcohols are liquids).

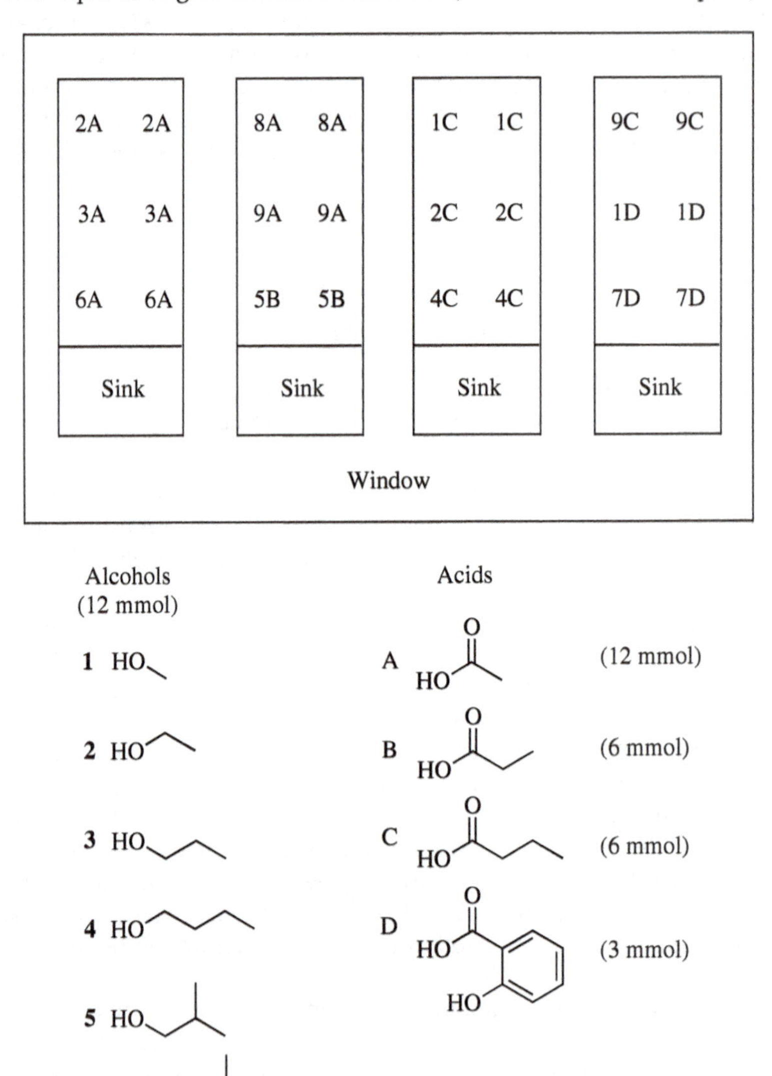


7 HO

8 HO

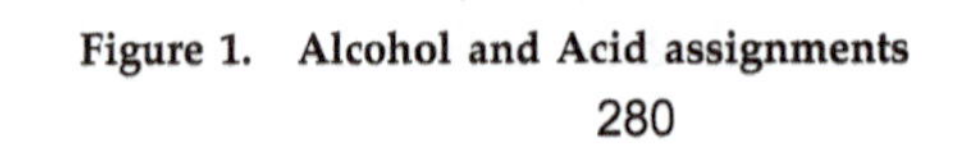


**Figure 1. Alcohol and Acid assignments**

## REACTION

$$\mathrm{RC(=O)OH} + \mathrm{R'OH} \xrightarrow[\Delta]{H_2SO_4} \mathrm{RC(=O)OR'} + H_2O$$

## EXPERIMENTAL PROCEDURE

Estimated time to complete the experient: 3.0 h.

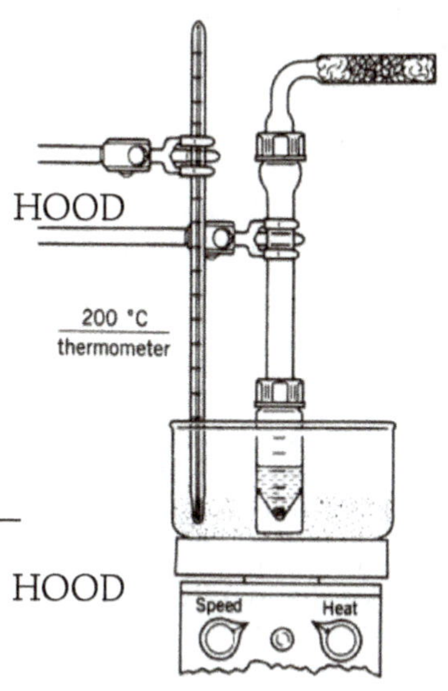


***Reagents and Equipment.*** In the ***hood,*** measure an appropriate amount of your acid and alcohol. Transfer both into a 5 mL conical vial. Add a spin vane and an air condenser, and then ask your TA to dispense about 10 drops of sulfuric acid ($H_2SO_4$) through the air condenser. Reflux gently on a hot plate for 30 minutes. (The hot plate should be on a low setting, for example setting three out of a range of ten.) If you have a solid acid, and it does not completely dissolve on heating, add a bit more alcohol.

---

CAUTION: Cap the vial immediately after addition of each reagent. Dispense the reagents in the ***hood*** using the syringes /pipets provided. Concentrated acetic and sulfuric acids are *corrosive.* Concentrated sulfuric acid is toxic and corrosive and can quickly burn your skin and eyes. Wear your safety goggles. Be particularly careful not to spill any. If you do get acid on your skin, wash it immediately with cold running water and notify your TA. Some of the carboxylic acids and alcohols as well as the esters you will prepare are flammable, irritants and/or toxic. However, the small quantity of ester required for detection by odor will not be hazardous. Nevertheless, do not develop a habit of smelling new or unknown compounds.

---

***Isolation of Product.*** Add about 3 mL of ether to the conical vial and transfer the liquid to a separatory funnel using a Pasteur pipette. Wash the conical vial with about 3 mL water and 3 mL ether and again transfer this to the separatory funnel. You may wish to add a few more mL of ether to make the organic layer easier to handle. Extract the ether three or four times with three to five mL portions of 5% aqueous sodium bicarbonate and then once with distilled water. (Will ether be the top or the bottom layer?) Unreacted acid will extract into the aqueous base, and all the alcohols are water soluble. Separate the ether layer and dry it with $MgSO_4$, filter and carefully evaporate the ether with a gentle stream of air on a hot plate in the hood; the lower molecular weight esters will have relatively low boiling points and will evaporate completely if heated too long.

***Characterization.*** Carefully and gently waft some of the vapor towards your nose and sniff. Only a very small quantity is necessary to detect the odor. Record a description of your odor. You should also smell other groups' esters. Which one has the odor of wintergreen? Can you recognize any other odors? Record all of the odors that you detect, and note which compounds (if any) have no distinct odor. Take an IR spectrum of your ester. If your ester is a solid, obtain its melting point and compare it with that reported in the literature (Aldrich, SciFinder, NIST Webbook, SDBS, etc). Compare the carbonyl peaks of your ester to the other esters synthesized in lab. Can you notice a trend in the C=O peak when the R and R′ groups are varied?

## QUESTIONS

**7-27.** What is the systematic name of the ester you made?

**7-28.** In the preparation of the esters given in this experiment, the reaction product was extracted with 5% sodium bicarbonate solution ($NaHCO_3$) in the isolation step. Why? What gas was, or could have been, evolved during this washing step? Write a balanced equation for the reaction that produced it.

**7-29.** Considering all of the esters that were made in your lab, do you recognize any relationships between the structures of the esters and the odors? Considering all of the esters that were made in your lab, do you recognize any relationships between the structures of the esters and the carbonyl peaks from the IR spectra? (Look at p. 296 for a reminder about factors that affect carbonyl frequencies)

## BIBLIOGRAPHY

**These references are selected from the large number of examples of esterification given in *Organic Syntheses:***

Bailey, D. M.; Johnson, R. E.; Albertson, N. F. *Organic Syntheses;* Wiley: New York, 1988; Collect. Vol. VI, p. 618.

Bowden, E. *Organic Syntheses;* Wiley: New York, 1943; Collect. Vol. II, p. 414.

Eliel, E. L.; Fisk, M. T.; Posser, T. *Organic Syntheses;* Wiley: New York, 1963; Collect. Vol. IV, p. 169.

Emerson, W. S.; Longley, R. I., Jr. *Organic Syntheses;* Wiley: New York, 1963; Collect. Vol. IV, p. 302.

Fuson, R. C.; Wojcik, B. H. *Organic Syntheses;* Wiley: New York, 1943; Collect. Vol. II, p. 260.

McCutcheon, J. W. *Organic Syntheses;* Wiley: New York, 1955; Collect. Vol. III, p. 526.

Mic'ovic', V. M. *Organic Syntheses;* Wiley: New York, 1943; Collect. Vol. II, p. 264.

Peterson, P. E.; Dunham, M. *Organic Syntheses* Wiley: New York, 1988; Collect. Vol. VI, p. 273.

Stevenson, H. B.; Cripps, H. N.; Williams, J. K. *Organic Syntheses;* Wiley: New York, 1973; Collect. Vol. V, p. 459.

**For an overall review of esterification see**

Euranto, E. K. in *The Chemistry of Carboxylic Acids and Esters;* Patai, S. Ed.; Interscience: New York, 1969, p. 505.

# The Stepwise Synthesis of Nylon-6,6

EXPERIMENT 3106-12

***Purpose.*** The important industrial polymer, nylon-6,6, is prepared by the technique of step-growth polymerization.The physical properties of the polymer are examined. One of the two monomers used in the polymerization is synthesized.

***Background of an Industrial Polymer.*** The type of polymerization used in the nylon preparation described in this series of experiments is called "step-growth" polymerization. The technique uses two different difunctional monomers that undergo ordinary organic reactions. In the present case an acid chloride is treated with an amine to produce an amide linkage.

Nylon is a polyamide. In industry it is produced by reaction of two difunctional monomers (or comonomers): a dicarboxylic acid and a diamine. The polymer that you are going to study is of great historical significance in polymer chemistry, because it was the first of the polyamides to be recognized as possessing excellent physical properties for forming very strong fibers. Nylon-6,6 was, in fact, the first commercially produced synthetic polyamide The"6,6" nomenclature refers to the number of carbon atoms in each of the two comonomers. Industrially, nylon-6,6 is prepared from 1,6-hexanediamine (hexamethylenediamine) and hexanedioic acid (adipic acid):

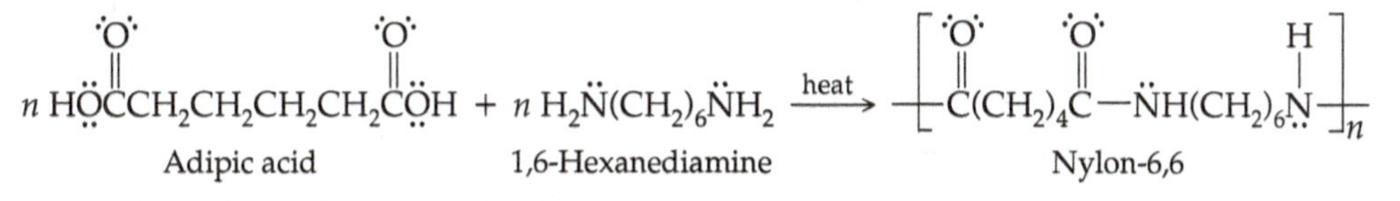


In the industrial process, the diacid and diamine are mixed to form the corresponding amine salt (hexamethylene diammonium adipate), which is then heated under steam pressure (250 psig) at 275° C to form the amide bonds. The resulting polymer has an average molecular weight of about 10,000, with an average of over 400 repeating monomer units in each molecule of polymer and a melting point of about 150° C. Fibers can be drawn from the melted polymer by a "cold-drawing" technique. This method of drawing fibers physically orients the polymer molecules into linear chains that are stabilized by the presence of hydrogen bonds between C=O and the N—H groups of adjacent chains, and the strength of the fiber is thereby increased. The synthetic polyamide linkages in the various forms of nylon are very similar (identical in some cases) to those found in proteins. For example, silk fibers gain their great strength from this type of interaction.

Numerous combinations of diacids and diamines have been evaluated as fiber materials. However, only a few have reached commercial production, which depends on low-cost, easy-to-access intermediates, and satisfactory general and physical properties of the polymer. One such group of materials are the"Aramid"class of fibers, which are prepared from aromatic monomers. One trade name for a material prepared from these type of fibers is Nomex. It has a high degree of heat and flame resistance. Race car driving suits are made from it, and it is also used as an insulator in the space shuttles.

Nylon-6,6 was first synthesized in 1899 by Gabriel and Maas in Germany. It was not until 1929, however, that the substance was shown to possess practical commercial properties. It was Carother's research program on polyamides at DuPont that made the major discoveries that initiated the world's polymer industry. DuPont began production of nylon in October 1939, and the first nylon stockings were manufactured in May 1940. By 1950, 14 chemical plants in 10 countries produced 55,000 metric tons of polyamide fiber. By 1980 worldwide production had expanded to $3.05 \times 10^6$ tons, with about one-third of the polymer synthesized in the United States.

Thus, you should appreciate that the chemical industry carries out organic reactions on massive quantities of material for use in today's highly technological society. The discovery and characterization of these materials all starts in the research laboratory, with many of them initially prepared in microscale quantities. One of the great triumphs of our technology has been the successful scaleup of synthetic organic reactions, but that is a story for another day.

### Preparation of an Acid Chloride: Adipoyl Chloride

Common name: adipoyl chloride
CA number: [111-50-2]
CA name as indexed: hexanedioyl dichloride

***Purpose.*** Adipic acid is converted to its corresponding acid chloride by reaction with oxalyl chloride. The experiment will help you further understand the nucleophilic substitution reaction pathway by which carbonyl-containing compounds undergo reaction.

*Prior Reading*

*Standard Experimental Apparatus: Reflux Apparatus (pp. 23–24)*
*Collection or Control of Gaseous Products: (pp. 105–107)*

## REACTION

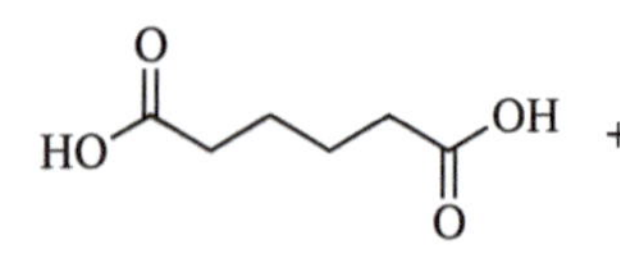


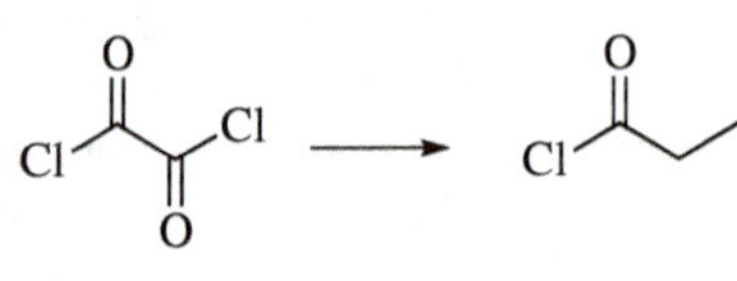


Cl–CO–(CH$_2$)$_4$–CO–Cl $+ 2\,CO_2 + 2\,CO + 2\,HCl$

adipic acid  oxalyl chloride  adipoyl chloride

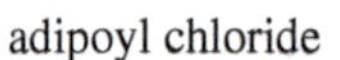

## DISCUSSION

Carboxylic acids react with oxalyl chloride $(COCl)_2$ to produce the corresponding acid chlorides, as shown in the above reaction. Oxalyl chloride is an attractive reagent due to its low cost, and the fact that the byproducts produced in the reaction are gases.Thus, the reaction is driven to completion by the evolution of HCl, $CO_2$ and CO, and a nearly pure acid chloride is obtained. The major drawback to the reaction is that it produces a strong acid (HCl) and thus cannot be used with compounds that are acid sensitive. Thionyl chloride is often used as an alternative reagent.

The reaction proceeds by a nucleophilic acyl substitution pathway. The —OH group is converted into a relatively good leaving group.The intermediate then undergoes attack by the chloride ion at the carbonyl carbon to yield the final product. The sequence is shown here:

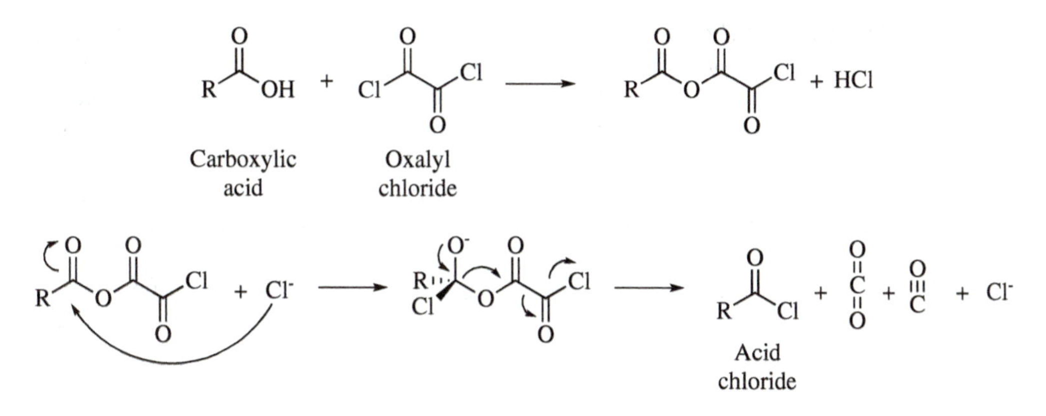


Acid halides are important intermediates and they are used extensively for the conversion of carboxylic acids into other derivatives. For example, acid halides can be used to prepare (in addition to amides): anhydrides, esters, aldehydes, and ketones. Acid halides readily undergo reaction with water (hydrolysis) to form the corresponding carboxylic acid. For this reason the reaction system must be protected from atmospheric moisture when acid halides are formed and/or used.

In the present sequence leading to the formation of nylon, the adipoyl chloride provides a reactive species, which, when treated with a diamine, forms the desired amide linkage inherent to nylon.

## EXPERIMENTAL PROCEDURE

Estimated time to complete the experiment: 2.5 h.

| Physical Properties of Reactants | | | | | | |
|---|---|---|---|---|---|---|
| Compound | MW | Amount | mmol | bp (°C) | mp (°C) | *d* |
| Adipic acid | 146.14 | 500 mg | 3.4 | 153 | 265 | 1.35 |
| Oxalyl chloride | 126.93 | you calculate | 13.6 | −16 | 63-64 | 1.48 |

WARNING: The formation of the acid chloride **must** be performed in the hood. Oxalyl chloride and hydrochloric acid are lachrymators (they make you cry) and carbon monoxide is a colorless, odorless and toxic gas.

***Reagents and Equipment.*** Set up your reactions in the hoods, and follow your TA's instructions. Your glassware should be as dry as possible before starting the reaction. The apparatus consists of a 5 mL conical vial, with a magnetic stir vane inside, and a water cooled condenser. Because there are a limited number of faucets in the hood, the water outlet from one condenser can be connected to the inlet of another. Place 500 mg (3.4 mmol) of adipic acid in the vial. Add the appropriate volume of oxalyl chloride (13.6 mmol) to the apparatus.

***Reaction Conditions.*** With stirring, heat the reaction mixture to reflux within 5 min. Continue to heat the system for 1 h.

***Isolation and Characterization.*** You may check for complete conversion to the acid chloride using the IR spectrometer. The progress of the reaction may be followed by IR analysis. Remove a small sample from the flask and obtain the spectrum of the material. Remove the sample from the instrument sampling compartment immediately following the spectral scan to prevent the HCl gas buildup from damaging the instrument. The reaction is considered incomplete if the IR spectrum displays a weak band on the low-wavenumber side of the acid halide carbonyl peak.

The adipoyl chloride is quite labile, and therefore, it is not purified further, but it is used directly in the preparation of nylon as described in the next section.

### Preparation of a Polyamide: Nylon-6,6

Common names: nylon-6,6, polyhexamethylene adipamide
CA number: [32131-17-2]
CA name as indexed: poly[imino(1,6-dioxo-1,6-hexanediyl)imino-1,6-hexanediyl]

***Purpose.*** The polyamide, nylon, is prepared by the step-growth condensation polymerization of adipoyl chloride with 1,6-hexanediamine. An interfacial (emulsion) polymerization technique is used to generate nylon fibers.

## REACTION

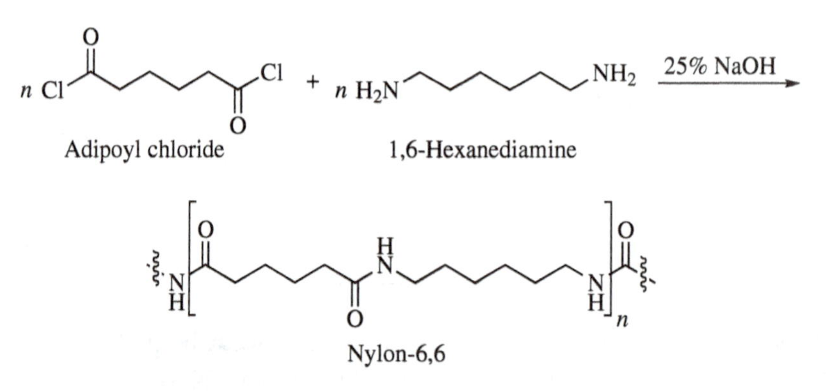


## DISCUSSION

The preparation of nylon outlined in this experiment is not the industrial method (see initial discussion). The use of the reactive diacid chloride reagent allows one to carry out the step-growth polymerization reaction under very mild conditions more convenient to the instructional laboratory.The interfacial (emulsion) polymerization technique used consists of dissolving the adipoyl chloride reagent in a water-immiscible solvent (cyclohexane) and bringing this solution into contact with an aqueous solution of the diamine. A thin film forms at the *interface* of the two solvents as the condensation reaction proceeds. A"rope"of nylon polymer can be pulled from the interface of the two solvents because the film is continuously generated as the reaction occurs.

This polymer has an average molecular weight of ~10,000! In this particular experiment, about 5–7 meter lengths of nylon polymer can be obtained. This particular synthesis of nylon is often used in lecture demonstrations and chemical magic shows.

## EXPERIMENTAL PROCEDURE

**Physical Properties of Reactants**

| Compound | MW | Amount | mmol | mp (°C) | *d* |
|---|---|---|---|---|---|
| Adipoyl chloride | 183.05 | ~622 mg | 3.4 | | |
| 1,6-Hexadediamine (5% aq) | 116.21 | 8 mL | | 41–41 | 1.259 |
| Cyclohexane | 84.16 | 8 mL | | | |

***Reagents and Equipment.*** Dilute the adipoyl chloride with approximately 10 mL of cyclohexane. In a separate 50 mL beaker, combine 8 mL of a 5% aqueous solution of 1,6-hexanediamene and eight drops of 25% NaOH solution. Carefully pour the cyclohexane solution of adipoyl chloride on top of this aqueous solution (see NOTE). **Do not shake or stir the layers.** Two separate layers are desired so that the nylon will form at the interface (boundary) between the two layers. It will be obvious if you have made the polymer.

*NOTE. Add the solution using a Pasteur pipet, taking care to run it down the side of the beaker.*

***Isolation and Characterization.*** Using a copper wire or paper clip bent into a small hook, hook the film in the center of the beaker and draw up the nylon fiber from the solution interface. A slow, steady pull will result in long strands of the polymer. Wash the fibers thoroughly in a beaker of water before handling them. You can take the washed polymer home if you so desire.

## QUESTIONS

**7-22.** Give an explanation of why acid chlorides are more reactive toward nucleophilic substitution than are the corresponding ethyl esters. *Hint:* Consider the nature of the leaving group and the rate-determining step in an addition-elimination sequence.

**7-23.** Acid chlorides are used extensively as electrophiles in the Friedel-Crafts reaction to prepare aromatic ketones. The reaction involves the treatment of an aromatic hydrocarbon with an acyl chloride in the presence of a Lewis acid, such as aluminum chloride. Using this reaction, outline the reaction sequence you would use to prepare benzophenone using benzoic acid and benzene as the starting organic compounds. Hint: Think about the Friedel-Crafts acylation with ferrocene [Experiments 3106-5 and 3106-6].

**7-24.** Explain why the 25% NaOH solution is added to the reaction mixture.

**7-25.** Water, sodium hydroxide and 1,6-hexanediamine are all potential nuclueophiles. Yet only 1,6-hexanediamine reacts with the adipoyl chloride. Offer an explanation as to why.

**7-26.** Predict the structure of the polymer that would result in the condensation of the following reactants. These monomers are used to produce the polyamide Nomex, a high-melting material used as an insulator in space shuttles and as the fire-resistant fabric in clothing worn in race cars.

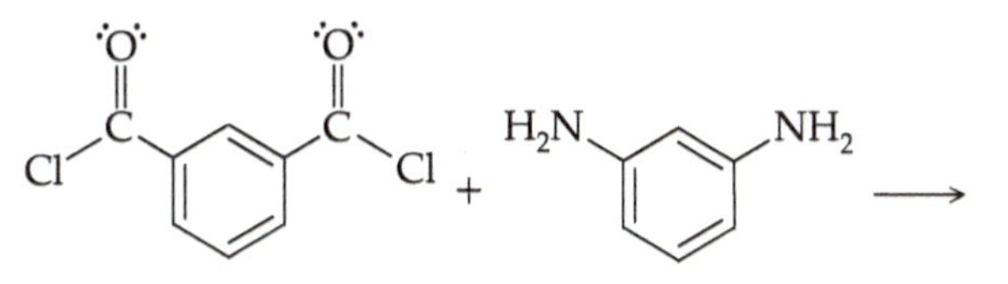

## BIBLIOGRAPHY

**Adipoyl chloride is prepared as an intermediate in several preparations reported in *Organic Syntheses*:**

Fuson, R. C.; Walker, J. T. *Organic Syntheses*; Wiley: NewYork, 1943; Collect.Vol. II, p. 169.
Guha, P. C.; Sankaran, O. K. *Organic Syntheses*; Wiley: NewYork, 1955; Collect.Vol. III, p. 623.

**See your organic textbook for an introduction to this reaction and its scope in synthesis. For example,**

Solomons, T. W. G.; Fryhle, C. B. Organic Chemistry, 9th ed., Wiley: NewYork, 2008, p. 794.

**For a review on the preparation and reactions of acyl halides see**

Ansell, M. F. in The Chemistry of Acyl Halides; Patai, S., Ed.; Wiley: NewYork, 1972, p. 35.

**For detailed information on the production and use of nylon see**

Heckert, W. W. J. *Chem. Educ.* **1953**, 30, 166.
*Kirk–Othmer Encyclopedia of Chemical Technology*, 4th ed.,Vol. 19, Wiley: NewYork, 1996, pp. 454, 470, 485.

**For information on the interfacial polymerization technique see**

Sprague, B. S.; Singleton, *R. W. Text. Res.* J. **1965**, 35, 999.
Morgan, P. W.; Kwolek, S. L. J. *Chem. Educ.* **1959**, 35, 182.
Nikonov,V. Z.; Savinov,V. M. In *Interfacial Synthesis;* Millich, F.; Carraher, C. E., Jr., Eds.; Marcel Dekker: NewYork, 1977,Vol. II, Chap. 15.
Odian, G. *Principles of Polymerization*, 4th ed.; Wiley-Interscience:New York, 2004.

atomic displacements are all related by simple harmonic motion to the overall total vibrational motion (or vibrational energy) of the molecule. There are $3N - 6$ (where $N$ = the number of atoms) *normal modes* (or fundamental vibrations or vibrational degrees of freedom—all these terms are essentially synonymous) present in all nonlinear molecules. Linear molecules have only $3N - 5$ normal modes—in this case there is one more normal mode of vibrational energy present because a rotational degree of freedom has been lost. Rotation around the molecular axis involves no energy because the atomic nuclei are assumed to be point sources of matter. (For an introductory discussion of vibrational energy see Chapter 8W, IR section, Part I B.) ← WWW

Operating under selection rules these normal modes of vibration give rise to absorption bands in the infrared region of the spectrum (see, for example, the infrared spectrum of *n*-hexane, Fig. 8.2, p. 544). In the analysis these modes are often assigned numbers. For example, the 30 modes of benzene (where $N = 12$ in the $3N - 6$ expression) can be assigned 1 through 30 or the numbering can be done using any one of a number of different criteria. Subscripts *a* and *b* are often used to indicate doubly degenerate modes, that is, modes that have identical energies (and thus required to have the same frequency. One of the numbering systems for the benzene ring is used here when the aromatic ring stretching vibrations are identified (see Table 8.6 and Chapter 8W, IR section, Part II D, for a more detailed discussion of normal modes). ← WWW

Many of these vibrational frequencies are associated with small groups of atoms that are essentially uncoupled from the rest of the molecule. The absorption bands that result from these modes, therefore, are characteristic of the small group of atoms regardless of the composition of other parts of the molecule. These vibrations are known as the *group frequencies.* Interpretation of infrared spectra of complex molecules based on group frequency assignments is an extremely powerful aid in the elucidation of molecular structure.

The following four factors make significant contributions to the development of a good group frequency from a molecular vibration:

**1.** The group has a large dipole-moment change during vibrational displacement. This change in moment is formally related to the efficiency of absorption of radiation during the molecular displacement by the expression $I\alpha(\delta\mu/\delta Q)^2$ (where $I$ = intensity, $\mu$ = electric dipole moment, and $Q$ = the normal coordinate [a mathematical description of the vibration]). Thus, if $\delta\mu/\delta Q$ is large, there is a large absorption of infrared radiation which gives rise to very intense bands (the intensity is dependent on the square of the moment change at that vibrational frequency).

**2.** The presence of a large force constant, so that for many of these groups the stretching frequency occurs at high values above the fingerprint region.

**3.** The fundamental mode occurs in a frequency range that is reasonably narrow (little coupling), but sensitive enough to the local environment to allow for considerable interpretation of the surrounding structure.

**4.** The range of frequencies is determined by a number of factors that are now well understood in terms of the mass, geometric, electronic, intramolecular, and intermolecular effects (for an introductory discussion of these effects see Chapter 8W, IR section, Part III A). ← WWW

***Strategies for Interpreting Infrared Spectra***

**1.** Divide the spectrum into two parts at 1350 $cm^{-1}$.

**2.** Above 1350 $cm^{-1}$, absorption bands have a high probability of being good group frequencies. The interpretation is usually reliable and free from ambiguities. We can be much more confident of our assignments in this region even with rather weak bands.

**3.** Because of the reliability of the high-wavenumber region, we always begin the interpretation of a spectrum at this end.

**4.** Bands below 1350 $cm^{-1}$ may be either group frequencies or fingerprint frequencies.

**5.** Below 1350 $cm^{-1}$, group frequencies are less easily assigned. In addition, even if a reliable group frequency occurs in this region, absorption at that frequency is not necessarily a result of that mode. That is, fingerprint bands can also randomly occur in the same location as reliable group frequency bands and the observer cannot usually distinguish which type of band is present.

**6.** To make more confident assignments below 1350 $cm^{-1}$, it is helpful to be able to associate a secondary property, such as band shape, with the particular mode. For example, it helps to know whether the band is very intense, broad, sharp, occurs as a characteristic doublet, gives the correct frequency shift on isotopic substitution, or the like.

**7.** A good rule to remember is that in the fingerprint region the **absence** of a band is more important than the presence of a band. If a band is absent, you can conclude with confidence that a reliable group frequency assigned to this region is absent and therefore the group must be absent from the sample. At the same time you also know that no interfering fingerprint bands occur in the region.

**8.** Before beginning the interpretation, note the sampling conditions and determine as much other information about the sample as possible—such as molecular weight, melting point, boiling point, color, odor, elemental analysis, solubility, and refractive index.

www →

**9.** In the interpretation try to assign the most intense bands first. These bands very often will be associated with a polar functional group.

**10.** Do not try to assign all the bands in the spectrum. Fingerprint bands are unique to a particular system. Occasionally, intense bands will be fingerprint-type absorptions; these bands, generally, will be ignored in the interpretation. Fingerprint bands do, however, play an important role when infrared data are employed for identification purposes.

**11.** The correlation chart (back endpaper) can act as a helpful quick aid for checking potential assignments. It is not a substitute for understanding the theory and operation of group frequency logic. *The use of the correlation chart without a good knowledge of group frequencies is the shortest path to disaster!*[1]

**12.** Try to utilize the so-called *macro group frequency* approach. That is, if the functionality or molecular structural group requires the presence of more than a single group frequency vibrational mode, make sure that all modes are correctly represented. The *macro frequency train* represents a very powerful approach to the interpretation of relatively complex spectral data. This

[1]Bellamy, L. J. *The Infrared Spectra of Complex Molecules,* 3rd ed.; Chapman & Hall: London, 1975, p. 3.

technique is at the core of current work on the automatic computer interpretation of infrared spectra. Contained in the product characterization section of the experiments given in Chapters 6 and 10W (online) are 12 detailed discussions (in Chapter 6, Experiments [5A], [5B], [6], [7], [8], [11], [14], [19], and [20]; in Chapter 10W (online), Experiments [$1_{adv}$], [$4_{adv}$] and [$6_{adv}$]) that demonstrate the operational use of the *macros.* Careful reference to these discussions will be very helpful in the initial stages of learning these interpretation techniques. It should become relatively easy to extend this interpretive approach to other reactions by reference to common infrared library files. Practice using *macro group frequencies* will pay big dividends in the laboratory. This last suggestion is perhaps the most important strategy to master in learning to interpret infrared spectra.

www

## A SURVEY OF GROUP FREQUENCIES IDENTIFIED IN ORGANIC MOLECULES

The useful group frequencies are listed in the following sets of tables.

*NOTE: A detailed description of the associated fundamental vibrational modes, diagrams of the actual displacements of the atoms, along with associated spectra, may be found at Chapter 8W, IR section, Part II. It is highly advisable to study this material.*

www

In the following tables the vibration motion of the localized sections of the molecules assigned to a particular group's frequencies is often described using the following terms:

1. *Symmetric stretch* or *symmetric bend (deformation):* Here the local group retains its symmetry during displacement. The symmetric bend of the methylene group, $CH_2$, is often termed the *scissoring* bend, while the symmetric bend of a methyl group, $CH_3$, is termed an *umbrella* mode—both descriptions imply the type of displacements that are taking place in the vibration.
2. *Antisymmetric stretch* or *bend* (deformation): Here the vibrating system loses its symmetry during the vibration. The displacements involve a reflective (mirror image) displacement during the opposite phase of the simple harmonic vibration and the motion is termed antisymmetric rather than asymmetric. Antisymmetric bends (deformations) are often classified as *twisting, rocking,* and *wagging* vibrations.
3. Some vibrations involving planar sections of the molecules are referred to as *in-plane* or *out-of-plane.* They can be either symmetric or antisymmetric in nature and if they involve the bending of all the displaced bonds of a set of atoms moving together in the same direction they will be termed *all-in-phase.*
4. *Degenerate* vibrations are defined as the case where two or more molecular vibrations are required to occur at the same frequency (see Chapter 8W, IR sections, Part I B and Part I D3 for more details).
5. *Overtones* (integral multiples of the fundamental mode frequency) and *sum tones* (the sum of two different fundamental modes) are forbidden bands that are almost always very weak. Occasionally these bands are good group frequencies.

www

## Group Frequencies of the Hydrocarbons

Alkanes Alkynes
Alkenes Arenes

***Alkanes.*** The C—H vibrational modes of the alkanes (or mixed compounds containing alkyl groups) that are characteristic and reliable group frequencies are summarized in Table 8.1 (also see Chapter 8W, IR section, Part II A).

www

These modes give rise to characteristic bands found in the infrared spectrum of alkanes, such as in the spectrum of *n*-hexane shown in Figure 8.2.

***Alkenes C═C Stretching.*** It is possible to classify open-chain unsaturated systems into two groups, those with C═C stretching modes falling above 1660 $cm^{-1}$ and those with modes falling below 1660 $cm^{-1}$ as shown in Table 8.2 (see also Chapter 8W, IR section, Part II B1).

www

***Alkenes C—H.*** Several fundamental modes associated with the alkene C—H groups are group frequencies and are summarized in Table 8.3.

***Alkynes.*** The group frequencies of the alkynes are summarized in Table 8.4 (see also Chapter 8W, IR section, Part II C).

www

**Table 8.1** Alkane Vibrational Normal Modes

| C—H Vibrational Modes | $\tilde{\nu} \pm 10$ ($cm^{-1}$) |
|---|---|
| **Methyl groups** | |
| Antisymmetric (degenerate) stretch | 2960 |
| Symmetric stretch | 2870 |
| Antisymmetric (degenerate) deformation | 1460 |
| Symmetric (umbrella) deformation | 1375 |
| **Methylene groups** | |
| Antisymmetric stretch | 2925 |
| Symmetric stretch | 2850 |
| Symmetric deformation (scissors) | 1450 |
| Rocking mode (all-in-phase) | 720 |

**Table 8.2** Substitution Classification of C═C Stretching Frequencies

| C═C Normal Modes | $\tilde{\nu}$ ($cm^{-1}$) |
|---|---|
| Trans-, tri-, tetrasubstituted | 1680–1665 |
| Cis-, vinylidene- (terminal 1,1-disubstituted), vinyl-substituted | 1660–1620 |

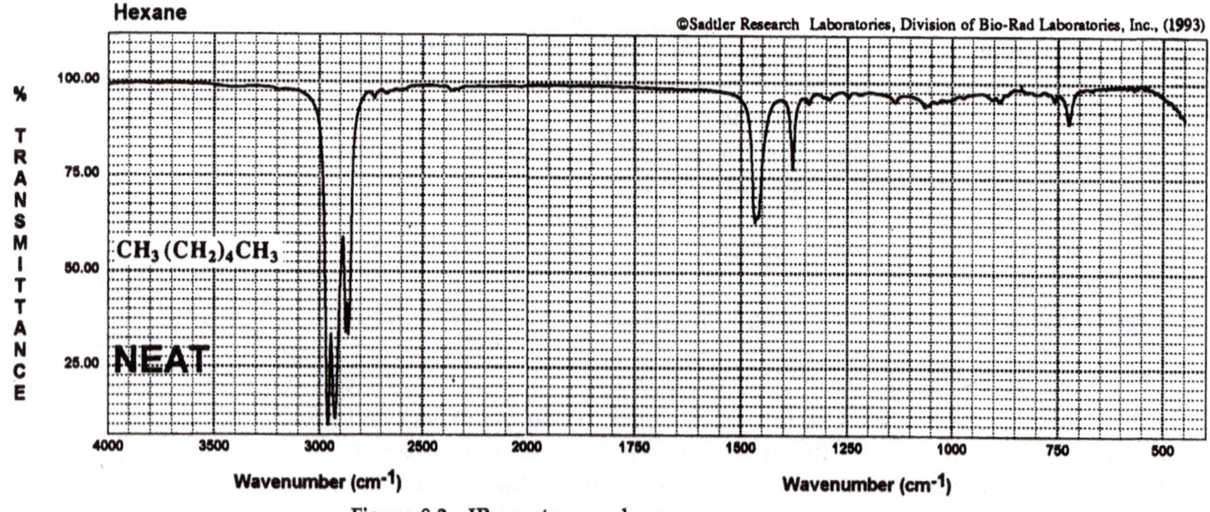


Figure 8.2 IR spectrum: *n*-hexane.

**Table 8.3 Alkene Vibrational Normal Modes**

| C—H Vibrational Modes | $\tilde{\nu} \pm 10$ (cm$^{-1}$) |
|---|---|
| **Stretching modes** | |
| Antisymmetric stretch (=$CH_2$) | 3080 |
| Symmetric stretch (=$CH_2$) | 3020 |
| Uncoupled stretch (=$CH_2$) | 3030 |
| **Out-of-plane bending modes** | |
| *Vinyl group* | |
| Trans hydrogen atoms (in-phase) | 990 |
| Terminal hydrogen atoms (wag) | 910 |
| *Vinylidene group* (=$CH_2$) | |
| Terminal (wag) | 890 |
| *Trans alkene* | |
| Trans hydrogen atoms (in-phase) | 965 |
| *Cis alkene* | |
| Cis hydrogen atoms (in-phase) | ~700 |
| *Trisubstituted alkene* | |
| Uncoupled hydrogen atom | 820 |
| *Tetrasubstituted alkene: no vibrational modes seen in IR* | |

**Table 8.4 Alkyne Vibrational Normal Modes**

| C≡C, C—H Vibrational Modes | $\tilde{\nu} \pm 10$ (cm$^{-1}$) |
|---|---|
| Triple-bond stretch (monosubstituted) | 2120 |
| Triple-bond stretch (disubstituted) | 2225 |
| R—C≡C—H bond stretch (monosubstituted) | 3300 |

***Arenes.*** The group frequencies of the *phenyl* group can be classified as carbon–hydrogen vibrations consisting of stretching and out-of-plane bending modes, plus carbon–carbon ring stretching and out-of-plane bending modes. The in-plane bending modes in both cases are not effective group frequencies.

The wavenumber values for the all-in-phase C—H bending vibrations are presented in Table 8.5.

The generalized group frequencies of the arenes are summarized in Table 8.6 (see also Chapter 8W, IR section, Part II D). ← www

## Group Frequencies of Carbonyl Groups: C=O

The carbonyl group is perhaps the single most important functional group in organic chemistry. It is certainly the most commonly occurring functionality. Infrared spectroscopy can play a powerful role in the characterization of the carbonyl because this group possesses all of the properties that give rise to an excellent group frequency. (Table 8.7; for an in depth discussion see Chapter 8W, IR section, Part III A.) ← www

**Table 8.5 Arene Out-of-Ring-Plane C—H Deformation Modes**

| Arene Fundamentals (C—H bend) (Number of C—H groups directly adjacent) | $\tilde{\nu}$ Range (cm$^{-1}$) |
|---|---|
| 5 | 770–730 |
| 4 | 770–735 |
| 3 | 810–750 |
| 2 | 860–800 |
| 1 | 900–845 |

| *Table 8.6* Arene Group Frequencies | |
|---|---|
| Arene Fundamentals | $\tilde{\nu}$ Range ($cm^{-1}$) |
| C—H stretch | 3100–3000 |
| C=C ring stretch ($\nu_{8a}$) | 1600 ± 10 |
| C=C ring stretch ($\nu_{8b}$) | 1580 ± 10 |
| C=C ring stretch ($\nu_{19a}$) | 1500 ± 10 |
| C=C ring stretch ($\nu_{19b}$) | 1450 ± 10 |
| C—H out-of-plane bend (1H) | 900–860 |
| C—H out-of-plane bend (2H) | 860–800 |
| C—H out-of-plane bend (3H) | 810–750 |
| C—H out-of-plane bend (4H) | 770–735 |
| C—H out-of-plane bend (5H) | 770–730 |
| C—C ring out-of-plane bend (1; 1,3; 1,3,5-substituted) | 690 ± 10 |
| C—H out-of-plane bend sum tones | 2000–1650 |

The major factors perturbing carbonyl frequencies can be summarized as follows:

### *Factors That Raise the C=O Frequency*

1. Substitution with electronegative atoms
2. Decrease in C—CO—C internal bond angle

### *Factors That Lower the C=O Frequency*

1. Conjugation
2. Hydrogen bonding

Several of these factors may be operating simultaneously, so careful judgment as to the contribution of each individual effect must be exercised in

| *Table 8.7* Carbonyl Group Vibrational Frequencies | |
|---|---|
| Compound | $\tilde{\nu}$ ($cm^{-1}$) |
| Ketones, aliphatic, open-chain ($R_2CO$) | 1725–1700 |
| Ketones, conjugated | 1700–1675 |
| Ketones, cyclic | *a* |
| Acyl halides | >1800 |
| Esters, aliphatic | 1755–1735 |
| Esters, conjugated | 1735–1720 |
| Esters (conjugated to oxygen) | 1780–1760 |
| Lactones | *a* |
| Anhydrides: aliphatic, open-chain | 1840–1810 and 1770–1740 |
| Carboxylic acids, aliphatic | 1725–1710 |
| Amides | (see Tables 8.22–8.24) |
| Lactams | *a* |
| Aldehydes | 1735–1720 |

www → [a]See Chapter 8W, IR section, Part III A.

**Table 8.8** Vibrational Normal Modes of the Hydroxyl Group

| $\tilde{\nu}$ (cm$^{-1}$) | Intensity | Mode Description |
|---|---|---|
| 3500–3200 | Very strong | O—H stretch (only strong when hydrogen bonded |
| 1500–1300 | Medium strong | O—H in-plane bend (overlap $CH_2$, $CH_3$ bend) |
| 1260–1000 | Strong | C—C—O antisymmetric stretch |
| 650 | Medium | O—H out-of-plane bend |

**Table 8.9** Substitution Effects on C—O Stretch of Aliphatic Alcohols

| Type of Alcohol | $\tilde{\nu}_{c—o}$ (cm$^{-1}$) |
|---|---|
| $RCH_2$—OH (primary) | 1075–1000 |
| $R_2CH$—OH (secondary) | 1150–1075 |
| $R_3C$—OH (tertiary) | 1200–1100 |
| $C_6H_5$—OH (phenol) | 1260–1180 |

predicting carbonyl frequencies. This judgment develops rapidly with practice at interpretation.

### Group Frequencies of the Heteroatom Functional Groups

(Alkanes)

| | | | |
|---|---|---|---|
| Alcohols | Aldehydes | Ketones | Esters |
| Acyl halides | Carboxylic acids | Anhydrides | Ethers |
| Amines, primary | Nitriles | Amides, primary | Amides, secondary |
| Isocyanates | Thiols | Halogens | Phenyl |

***Hexane.*** Refer to Table 8.1 (see also Chapter 8W, IR section, Part II A, and Fig. 8.2). ← www

***Alcohols.*** A very intense band appears at ~3350 cm$^{-1}$, which is assigned to the stretching mode of the O—H group (Table 8.8; also see Chapter 8W, IR section, Fig. W8.24). ← www

Of particular importance is a strong band in the spectrum of aliphatic alcohols usually located near 1060 cm$^{-1}$. This absorption has been identified as the C—O stretching mode. The vibrational displacements of this fundamental are similar to the antisymmetric stretch of water (see Chapter 8W for a detailed discussion of the vibrational modes of the water molecule). Since the vibration involves significant displacement of the adjacent C—C oscillator, the vibration will be substitution sensitive. These latter shifts can be of value in determining the nature of the alcohol (primary, secondary, or tertiary, see Table 8.9). ← www

***Aldehydes.*** The aldehyde functional groups gives rise to several good group frequencies (Table 8.10; also see Chapter 8W, IR section, Fig. W8.25). ← www

***Ketones.*** The only group frequency mode associated directly with aliphatic ketones is the stretching frequency ($\tilde{\nu}_{C=O}$~1720 cm$^{-1}$), which occurs within the expected region as discussed above. There are, however, several other related bands (Table 8.11; also see Chapter 8W, IR section, Fig. W8.26). ← www

**Table 8.10** Vibrational Normal Modes of the Aliphatic Aldehyde Group

| $\tilde{\nu}$ (cm$^{-1}$) | Intensity | Mode Description |
|---|---|---|
| 2750–2720 | Weak to medium | C(O)—H stretch (see also online Chapter 8W, IR section) |
| 1735–1720 | Very strong | C=O stretch |
| 1420–1405 | Medium | $CH_2$ symmetric bend, —$CH_2$—α to —CHO |
| 1405–1385 | Medium | C—H in-plane bend |

← www

| *Table 8.11* Normal Vibrational Modes of Aliphatic Ketones | | |
|---|---|---|
| $\tilde{\nu}$ (cm$^{-1}$) | Intensity | Mode Description |
| 3430–3410 | Very weak | Overtone of carbonyl stretch |
| 1725–1700 | Very strong | C═O |
| 1430–1415 | Medium | —$CH_2$— symmetric bend, —$CH_2$—α to ketone C═O |

| *Table 8.12* Vibrational Normal Modes of the Aliphatic Ester Group | | |
|---|---|---|
| $\tilde{\nu}$ (cm$^{-1}$) | Intensity | Mode Description |
| 1755–1735 | Very strong | C═O stretch |
| 1370–1360 | Medium | $CH_3$ symmetric bend α to ester C═O |
| 1260–1230 | Very strong | C—CO—O antisymmetric stretch —acetates |
| 1220–1160 | Very strong | C—CO—O antisymmetric stretch —higher esters |
| 1060–1030 | Very strong | O—$CH_2$—C antisymmetric stretch —1° acetates |
| 1100–980 | Very strong | O—$CH_2$—C antisymmetric stretch —higher esters (may overlap with upper band) |

| *Table 8.13* Vibrational Normal Modes of the Acyl Halide Group | | |
|---|---|---|
| $\tilde{\nu}$ (cm$^{-1}$) | Intensity | Mode Description |
| 1810–1800 | Very strong | C═O stretch, acyl chlorides |
| 1415–1405 | Strong | —$CH_2$—symmetric bend, α to —COCl carbonyl |

### Esters

The very strong band found at ~1745 cm$^{-1}$ is typical of the carbonyl frequency of an aliphatic ester, particularly aliphatic acetate esters (Table 8.12; also see Chapter 8W, IR section, Fig. W8.27).

WWW

***Acyl Halides.*** The carbonyl stretching mode dominates the spectrum in aliphatic acyl halides. In acyl chlorides it is an extremely intense band occurring near 1800 cm$^{-1}$ (Table 8.13; also see Chapter 8W, IR section, Fig. W8.28).

WWW

***Carboxylic Acids.*** Acids, observed in the solid or pure liquid states, often possess a very intense band with a width at one-half peak height of about 1000 cm$^{-1}$, which covers the region 3500–2200 cm$^{-1}$. This absorption is characteristic of very strongly hydrogen-bonded carboxylic acid groups (Table 8.14; also see Chapter 8W, IR section, Fig. W8.29).

WWW

***Anhydrides.*** The coupling of the anhydride carbonyls through the ether oxygen splits the carbonyls (in the aliphatic case $\tilde{\nu}_{c=o}$ = ~1830, 1760 cm$^{-1}$) by about 70 cm$^{-1}$ (Table 8.15; also see Chapter 8W, IR section, Fig. W8.30).

WWW

***Ethers.*** The large intensity associated with antisymmetric C—O—C stretching mode relative to the other bands occurring in this part of the fingerprint region, particularly in aliphatic compounds, makes it possible, in most cases, to assign

| *Table 8.14* Vibrational Normal Modes of the Carboxylic Acid Group | | |
|---|---|---|
| $\tilde{\nu}$ (cm$^{-1}$) | Intensity | Mode Description |
| 3500–2500 | Very very strong | O—H stretch intensified by hydrogen bonding |
| 2800–2200 | Very weak | Overtone and sum tones |
| 1725–1710 | Very strong | C=O antisymmetric hydrogen-bonded dimer stretch |
| 1450–1400 | Strong | $CH_2$—CO—O antisymmetric stretch mixed with O—H bend |
| 1300–1200 | Strong | $CH_2$—CO—O antisymmetric stretch mixed with O—H bend |
| 950–920 | Medium | Out-of-plane O—H bend, acid dimer |

| *Table 8.15* Vibrational Normal Modes of the Anhydride Group | | |
|---|---|---|
| $\tilde{\nu}$ (cm$^{-1}$) | Intensity | Mode Description |
| 1840–1810 | Very strong | C=O in-phase stretch |
| 1770–1740 | Very strong | C=O out-of-phase stretch |
| 1420–1410 | Strong | —$CH_2$— symmetric bend α to C=O |
| 1100–1000 | Very strong | C—O stretch, mixed modes |

| *Table 8.16* Vibrational Normal Modes of the Ether Group | | |
|---|---|---|
| $\tilde{\nu}$ (cm$^{-1}$) | Intensity | Mode Description |
| 1150–1050 | Strong | C—O—C antisymmetric stretch, mixed mode |

with confidence the observed strong band (Table 8.16; also see Chapter 8W, IR section, Fig. W8.31). www

***Primary Amines.*** The spectra of these bases usually possess two bands ($\tilde{\nu}_{N-H}$ = ~3380, ~3300 cm$^{-1}$) of medium-to-weak intensity. These bands are assigned to the antisymmetric and symmetric N—H stretching modes, respectively, of the primary amino group (Table 8.17; also see Chapter 8W, IR section, Fig. W8.32). www

***Nitriles.*** The very strong triple bond present in the nitrile group (as in the case of the alkynes) contributes to an unusually high stretching frequency, and the polar character of the group gives rise to very intense bands (Table 8.18; also see Chapter 8W, IR section, Fig. W8.33). www

***Primary Amides.*** The highly polar amide group leads to very strong hydrogen bonding, which in turn leads to greatly intensified N—H antisymmetric and symmetric stretching modes ($\tilde{\nu}_{N-H}$ = ~3375, ~3200 cm$^{-1}$; see Table 8.19; also see Chapter 8W, IR section, Fig. W8.34). www

***Secondary Amides.*** The single N—H group present in secondary amides gives rise to a very strong band near about 3300 cm$^{-1}$, which is indicative of strong hydrogen bonding. The drop in frequency from that of the primary —$NH_2$ scissoring mode near 1600 cm$^{-1}$ allows for confident assignment of substitution on secondary amide groups (Table 8.20; also see Chapter 8W, IR section, Fig. W8.35). www

**Table 8.17 Vibrational Normal Modes of the Primary Amine Group**

| $\tilde{\nu}$ ($cm^{-1}$) | Intensity | Mode Description |
|---|---|---|
| 3400–3200 | Weak to medium | $NH_2$ stretch doublet, (antisymmetric and symmetric modes) |
| 1630–1600 | Medium | $NH_2$ symmetric bend |
| 820–780 | Medium | $NH_2$ wag |

**Table 8.18 Vibrational Normal Modes of the Nitrile Group**

| $\tilde{\nu}$ ($cm^{-1}$) | Intensity | Mode Description |
|---|---|---|
| 2260–2240 | Strong | C≡N stretch, aliphatic |
| 2240–2210 | Strong | C≡N stretch, conjugated |

**Table 8.19 Vibrational Normal Modes of the Primary Amide Group**

| $\tilde{\nu}$ ($cm^{-1}$) | Intensity | Mode Description |
|---|---|---|
| 3400–3150 | Very strong | —$NH_2$ antisymmetric and symmetric stretching modes, hydrogen bonded |
| 1680–1650 | Very strong | C=O stretch, hydrogen bonded |
| 1660–1620 | Strong | —$NH_2$ symmetric bend (overlap with C=O stretch) |
| 1430–1410 | Strong | —$CH_2$—symmetric bend α to amide carbonyl |
| 750–650 | Medium | —$NH_2$ wag |

**Table 8.20 Vibrational Normal Modes of the Secondary Amide Group**

| $\tilde{\nu}$ ($cm^{-1}$) | Intensity | Mode Description |
|---|---|---|
| 3350–3250 | Strong | —NH stretch, intensified by hydrogen bonding |
| 3125–3075 | Medium | Overtone N—H bend (see also) |
| 1670–1645 | Very strong | C=O stretch, hydrogen bonded |
| 1580–1550 | Strong | N—H in-plane bend (see also) |
| 1415–1405 | Strong | —$CH_2$— symmetric bend α to amide C=O |
| 1325–1275 | Medium | C—N stretch mixed with N—H in-plane bend |
| 725–680 | Medium | N—H out-of-plane bend |

www →

www →

**Table 8.21 Vibrational Normal Modes of the Amide Carbonyl: Solution and Solid-Phase Data**

| Amide | Dilute Solution ($cm^{-1}$) | Solid ($cm^{-1}$) |
|---|---|---|
| R—CO—$NH_2$ (primary) | ~1730 | ~1690–1650 |
| R—CO—NHR (secondary) | ~1700 | ~1670–1630 |
| R—CO—$NR_2$ (tertiary) | ~1650 | ~1650 |

Studies of amide carbonyl frequencies in dilute nonpolar solution indicate that hydrogen-bonding effects are largely responsible for the low frequencies observed with primary and secondary amides, but play no role in tertiary amides (see Table 8.21).

**Table 8.22** Vibrational Normal Mode of the Isocyanate Group

| $\tilde{\nu}$ (cm$^{-1}$) | Intensity | Mode Description |
|---|---|---|
| 2280–2260 | Very strong | —N=C=O antisymmetric stretch |

**Table 8.23** Vibrational Normal Mode of the Thiol Group

| $\tilde{\nu}$ (cm$^{-1}$) | Intensity | Mode Description |
|---|---|---|
| 2580–2560 | Weak | S—H stretch |

**Table 8.24** Vibrational Normal Mode of the Alkyl Chloro Group

| $\tilde{\nu}$ (cm$^{-1}$) | Intensity | Mode Description |
|---|---|---|
| 750–650 | Strong | C—Cl stretch (see also) |

WWW

**Table 8.25** Vibrational Normal Modes of the Aryl Chloro Group

| $\tilde{\nu}$ (cm$^{-1}$) | Intensity | Mode Description |
|---|---|---|
| 3080 | Medium | C—H stretch, bonded to ring carbon |
| 1585 | Strong | $\nu_{8a}$ ring stretching |
| 1575 | Weak | $\nu_{8b}$ ring stretching |
| 1475 | Strong | $\nu_{19a}$ ring stretching |
| 1450 | Strong | $\nu_{19b}$ ring stretching |
| 747 | Strong | C—H all-in-phase, out-of-plane bend |
| 700 | Strong | C—Cl stretch |
| 688 | Strong | Ring deformation |
| 1945, 1865, 1788, 1733 | All weak | Sum tones, out-of-plane C—H bends, pattern matches monosubstitution of ring |

***Isocyanates.*** The range of stretching frequencies observed for alkyl-substituted isocyanates is very narrow, $\tilde{\nu} = 2280 - 2260\,\text{cm}^{-1}$, which implies little coupling to the rest of the system (Table 8.22; also see Chapter 8W, IR section, Fig. W8.36). WWW

***Thiols.*** Although weak absorption is associated with the S—H stretching fundamental the band is generally found in a very open region of the infrared spectrum (Table 8.23; also see: Chapter 8W, IR section, Fig. W8.37). WWW

***Alkyl Halides.*** The massive halogen atom is connected to the alkyl section by a fairly weak but highly polarized bond, which dictates that the C—X stretching frequency appears as an intense band at low frequencies (Table 8.24; also see Chapter 8W, IR section, Fig. W8.38). WWW

***Aryl Halides (Chlorobenzene).*** The final system to be considered in this section is the aryl halide, chlorobenzene. Based on the above assignments the group frequencies of the complete hydrocarbon portion and the heteroatom functional group can be assigned as in Table 8.25 (also see Chapter 8W, IR section, Fig. W8.39). WWW

# INFRARED SPECTROSCOPY INSTRUMENTATION AND SAMPLE HANDLING

## Instrumentation

The workhorse infrared instrument used for routine characterization of materials in the undergraduate organic laboratory is the optical-null double-beam grating spectrometer (Fig. 8.3*a*). For a discussion of double-beam spectrometers, see the UV-vis instrumentation discussion (p. 343). Although many

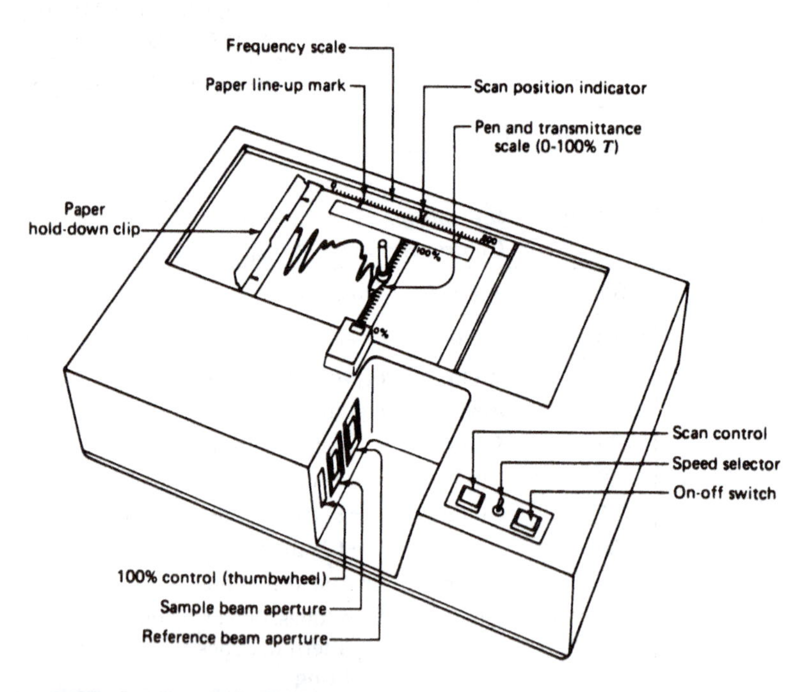


**Figure 8.3*a* The Perkin–Elmer model 710B IR spectrometer.** From Zubric, James W. *The Organic Chem Lab Survival Manual,* 7th ed.; Wiley: New York, 2008. (Reprinted by permission of John Wiley & Sons, Inc., New York.)

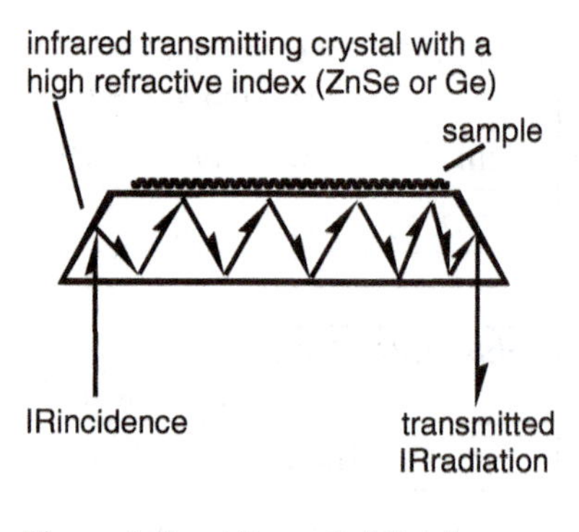

**Figure 8.3*b* Attenuated Total Reflectance (ATR).**

undergraduate instructional laboratories still utilize this type of instrumentation, the winds of change are blowing. ATR (attenuated total reflectance) FT-IRs and lower-cost FT-IR spectrometers, which both depend on computer manipulation of the spectral data, are becoming the infrared instructional instrumentation of choice. Two of the many benefits when using the more expensive ATR FT-IR spectrometer are faster sampling and spectral reproducibility. Even though ATR plates are prone to contamination, one key advantage when working with an ATR crystal (Fig. 8.3*b*) is that no sample preparation is required. This avoids the need to use KBr or mineral oil (Nujol) when working with solids! Many of the spectra utilized in the interpretive discussions (Chapter 8W, IR section) were generated on a prototype of this kind of infrared instrumentation, the Perkin–Elmer model 1600.

This instrument can acquire 16 scans and carry out the required calculations in 42 s. While the spectrum is being printed out(~40 s) the data on a second sample can be acquired. The 42-s acquisition data are significantly superior to those recorded by dispersive instruments that take from 5 to 8 min to scan a sample from 4000 to 600 $cm^{-1}$.

A short description of FT-IR spectrometers is included in the discussion of instrumentation on the website (Chapter 8W, IR section, Part IV).

## Sample Handling in the Infrared

If not working with an ATR FT-IR, the standard techniques of sample preparation employed to obtain infrared spectra of microscale laboratory products are the use of capillary films with liquids on NaCl (or AgCl) plates and the use of KBr disks and melts in solids. This, of course, assumes that for a spectrum to be obtained in the infrared region, the sample must be mounted in a cell that is transparent to the radiation. With an ATR FT-IR spectrometer, liquids and solids are placed in direct contact with the ATR crystal. Since glass and quartz absorb in this spectral region, cells constructed of these materials cannot be

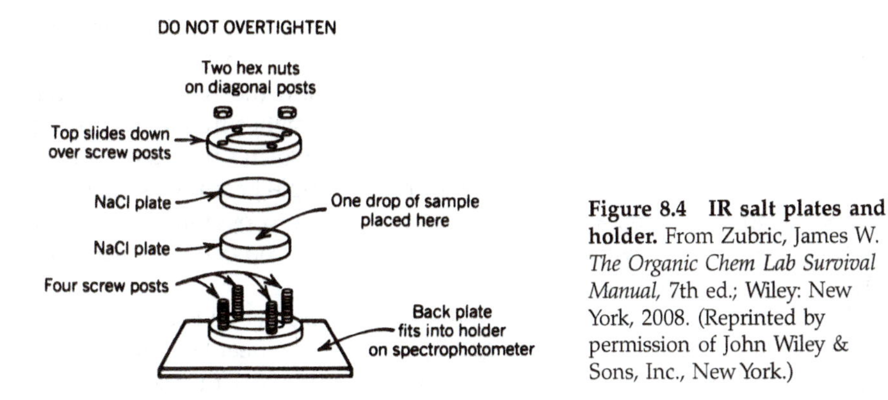


**Figure 8.4 IR salt plates and holder.** From Zubric, James W. *The Organic Chem Lab Survival Manual*, 7th ed.; Wiley: New York, 2008. (Reprinted by permission of John Wiley & Sons, Inc., New York.)

used when working with spectrometers not utilizing ATR technology. Accordingly, alkali metal halides have large spectral regions of transmission in the infrared, as do silver halides. Sodium chloride is the most commonly used material in cell windows in infrared sampling.

***Liquid Samples.*** For materials boiling above 100 °C, the procedure is very simple. Using a syringe or Pasteur pipet, place 3–5 μL of sample on a polished plate of sodium chloride or silver chloride or directly on the ATR crystal. If working with a NaCl or AgCl plate, cover it with a second plate of the same material and clamp it in a holder that can be mounted vertically in the instrument. Be sure that the plates are clean when you start and when you are through! Obviously, the sodium chloride plates cannot be cleaned with water. Silver chloride is very soft and scratches easily; it also must be kept in the dark when not in use because it darkens quickly in direct light. Spectra obtained in this fashion are referred to as *capillary film spectra* (Fig. 8.4).

***Solution Spectra and the Spectra of Materials Boiling Below 100 °C.*** These samples generally require a sealed cell constructed of either sodium chloride or potassium bromide windows. Such cells are expensive and need careful handling and maintenance. They are assembled as shown in Figure 8.5.

***Solid Samples using non-ATR Spectrometers.*** Solid powders could be mounted on horizontal sodium chloride plates, and the beam diverted

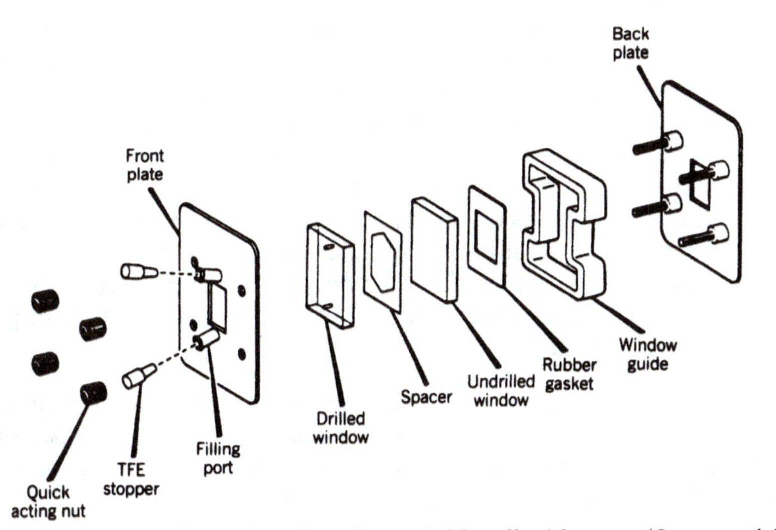


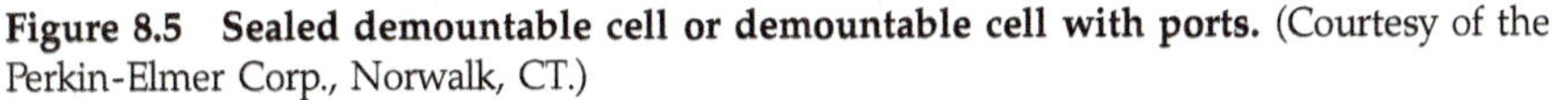

**Figure 8.5 Sealed demountable cell or demountable cell with ports.** (Courtesy of the Perkin-Elmer Corp., Norwalk, CT.)

through the sample by mirrors. This procedure would make sample preparation very easy for solids. Unfortunately, powders tend to scatter the entering radiation very efficiently by reflection, refraction, and molecular scattering. Some of these effects become rapidly magnified at higher frequencies, since they vary with the fourth power of the frequency. Thus, in solid-sample scattering a lot of energy is scattered away from the sample beam. This results in poor absorption spectra, as the instrument is forced to operate at very low energies. The detector cannot differentiate between a drop in energy from absorption or one derived from scattering.

For materials melting below 80 °C the simplest technique is to mount the sample between two salt plates and **gently** warm with a heat lamp until melting occurs. With the fast acquisition times of interferometers, the melting point range is now as high as 100 °C and the spectrum can be obtained so rapidly that the sample does not have time to cool and crystallize. (Heated cells are used in research laboratories, but they are rather expensive and difficult to maintain.) For substances melting above 100 °C, the sampling routine most often employed to avoid scattering problems is the potassium bromide (KBr) disk. Potassium bromide is transparent to infrared radiation in the region of interest. Most important, however, the KBr makes a much better match of the refractive indexes between the sample and its matrix than does air. Thus, reflection and refraction effects at the crystal faces of the sample are greatly suppressed.

In the KBr method the sample (2–3 mg) is finely ground in a mortar, the finer the better for lower reflection or refraction losses. Then, 150 mg of previously ground and dried KBr is added to the mortar and quickly mixed by stirring, not grinding, it with the sample. (Potassium bromide is very hygroscopic and will rapidly pick up water while being ground in an open mortar.) When mixing is complete the mixture is transferred to a die and pressed into a solid disk. Potassium bromide will flow under high pressure and seal the solid sample in a glasslike matrix. Several styles of dies are commercially available. For routine use a die consisting of two stainless steel bolts and a barrel is the simplest to operate (see Fig. 8.6). The ends of the bolts are polished flat to form the die faces. The first bolt is seated to within a turn or two of the head. Then the sample mixture is added (avoid breathing over the die while adding the sample). The second bolt is firmly seated in the barrel, and then the clamped assembly is tightened by a torque wrench to 240 in./lb. After standing for 1.5 min, the two bolts are removed, leaving the KBr disk mounted in the center of the barrel, which can then be mounted in the instrument. After the spectrum of the sample is run, the disk can be retrieved and the sample recovered if necessary (Fig. 8.6). *Always clean the die immediately after use. KBr is highly corrosive to steel.*

When infrared spectra are obtained, it is important to establish that the wavenumber values have been accurately recorded. Successful interpretation of the data often depends on very small shifts in these values. Calibration of the frequency scale is usually accomplished by obtaining the spectrum of a reference compound, such as polystyrene film. To save time, record absorption peaks only in the region of particular interest (this applies only to dispersive instrument derived data).

Turn "top" bolt down to form pellet
Barrel
Sample on this bolt
"Bottom" bolt halfway in

**Figure 8.6 The KBr pellet minipress.** From Zubric, James W. *The Organic Chem Lab Survival Manual*, 7th ed.: Wiley: New York, 2008. (Reprinted by permission of John Wiley & Sons, Inc., New York.)

*NOTE. Most of the infrared spectra found in the experimental sections of the text, which are Fourier-transform derived (Perkin–Elmer 1600), have been plotted on a slightly different scale than the other spectra presented in the text and on the website. The former spectra utilize a 12.5-cm$^{-1}$/mm format below 2000 cm$^{-1}$ and undergo a 2:1 compression above 2000 cm$^{-1}$ (25 cm$^{-1}$/mm).*

# QUESTIONS

*NOTE. Some of the following questions assume that the student is familiar with the infrared material contained on the website available to refer to if needed.* www

**8-1.** The form of the C—H out-of-plane bending vibrations of the vinyl group are shown below:

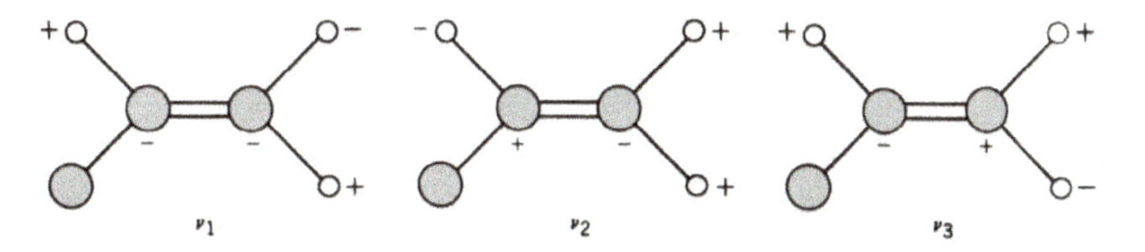


The first two vibrational modes give rise to excellent group frequencies, while the third fundamental does not lend itself to these correlations.

**(a)** Explain the factors that lead to the third vibrational mode being such a poor group frequency.

**(b)** Predict the location in the spectrum of the third fundamental vibration.

**8-2.** In the figure below, the mass of the terminal hydrogen atoms on the acetylene is hypothetically varied from zero to infinity. The response of the C—H symmetric stretching (3374 $cm^{-1}$) and triple-bond stretching (1974 $cm^{-1}$) modes to the change in mass is shown.

**(a)** Calculate the expected deuterium isotope shift for the C—H symmetric stretching mode. Is the hypothetical value close to the calculated value? Explain.

**(b)** Explain why the triple-bond stretching frequency is approximately 100 $cm^{-1}$ higher for high-mass terminal isotopes (>100) than for the low-mass terminal isotopes (<3).

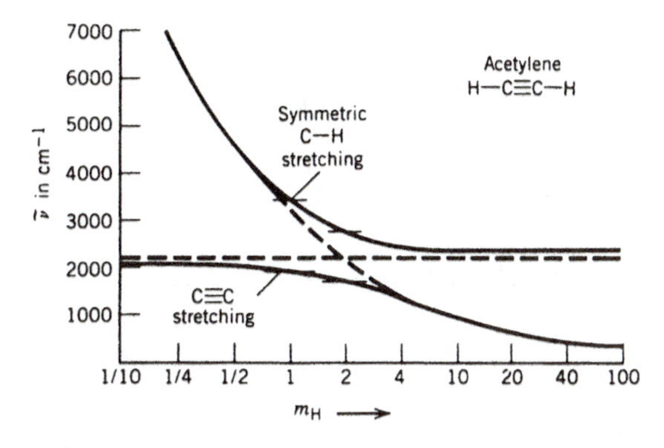


**8-3.** Acetylene has two C—H groups. It will have two C—H stretching frequencies, the in-phase and out-of-phase stretching modes. The in-phase (symmetric) stretch occurs at 3374 $cm^{-1}$ and the out-of-phase stretch at 3333 $cm^{-1}$. Explain why the in-phase vibration is located at a higher frequency than the out-of-phase stretch.

**8-4.** The carbonyl stretching frequencies of a series of benzoyl derivatives are listed below:

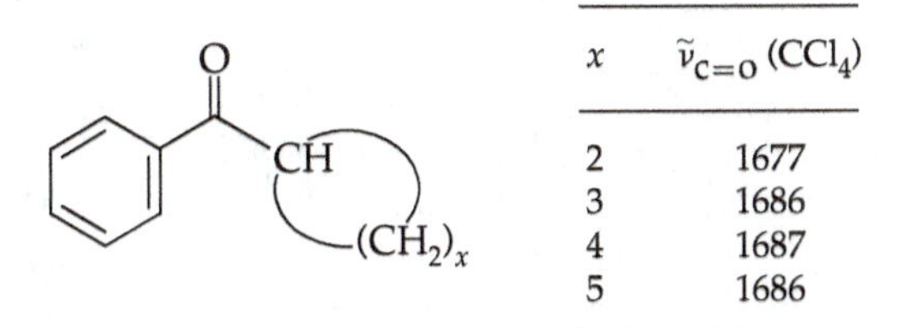


| $x$ | $\tilde{\nu}_{C=O}$ ($CCl_4$) |
|---|---|
| 2 | 1677 |
| 3 | 1686 |
| 4 | 1687 |
| 5 | 1686 |

Consider the $\tilde{\nu}_{C=O}$ of acetone at 1715 $cm^{-1}$ as a reference frequency and identify the factors affecting $\tilde{\nu}_{C=O}$ in the series of compounds listed.

**8-5.** Explain how mass effects act to lower the carbonyl frequency, as well as how inductive and hyperconjugation effects act to raise the carbonyl frequency of aldehydes relative to ketones.

**8-6.** The carbonyl stretching frequency of aliphatic carboxylic acids in dilute solution is located near 1770 $cm^{-1}$. This frequency is much higher than the carbonyl frequency of these substances when measured neat (~1720 $cm^{-1}$). Also, it is considerably higher than the corresponding simple aliphatic ester value (1745 $cm^{-1}$). Explain.

**8-7.** In a number of cases, dipolar interactions control the frequency shifts found in carbonyl stretching vibrations. The table lists wavenumber shifts in going from neat to dilute nonpolar solutions. Explain the observed values.

**Carbonyl Dipolar Interactions**[a]

| Compound | $\Delta\tilde{\nu}$ ($cm^{-1}$) |
|---|---|
| Acetyl chloride | 15 |
| Phosgene | 13 |
| Acetone | 21 |
| Acetaldehyde | 23 |
| *N,N*-Dimethylformamide | 50 |

[a]Shift measured between dilute nonpolar solution and neat sample.

**8-8.** The antisymmetric —$CH_2$—CO—O—stretching vibration in carboxylic acids is heavily mixed with the in-plane bending mode of the O—H group. In alcohols these two vibrations seldom show evidence of mechanical coupling. Explain.

**8-9.** Conjugation of the functional group in alkyl isocyanates has little impact on the antisymmetric —N=C—O stretching vibration located near 2770 $cm^{-1}$ Explain.

**8-10.** In the infrared spectrum of 2-aminoanthraquinone (**I**) two carbonyl stretching frequencies are observed at 1673.5 and 1625 $cm^{-1}$:

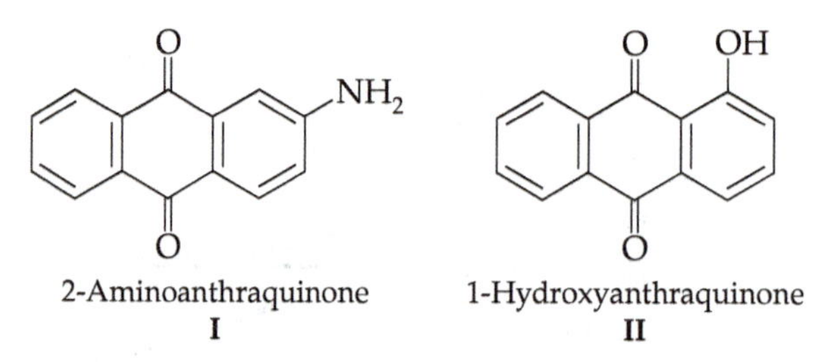


**(a)** Assign carbonyl bands in the infrared spectrum to the carbonyl groups in structure **I** and explain your reasoning.

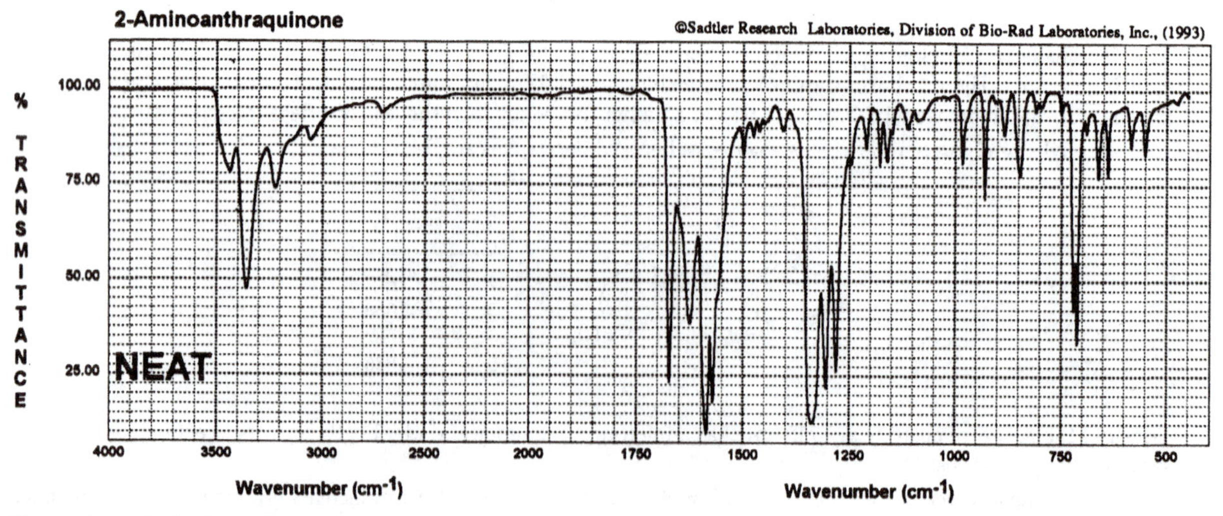

IR spectrum: 2-Aminoanthraquinone.

**(b)** The infrared spectrum of 1-hydroxyanthraquinone (**II**) also exhibits two carbonyl frequencies, which are located at 1675 and 1637 $cm^{-1}$. Assign the carbonyl groups to the related absorption bands. Explain your reasoning.

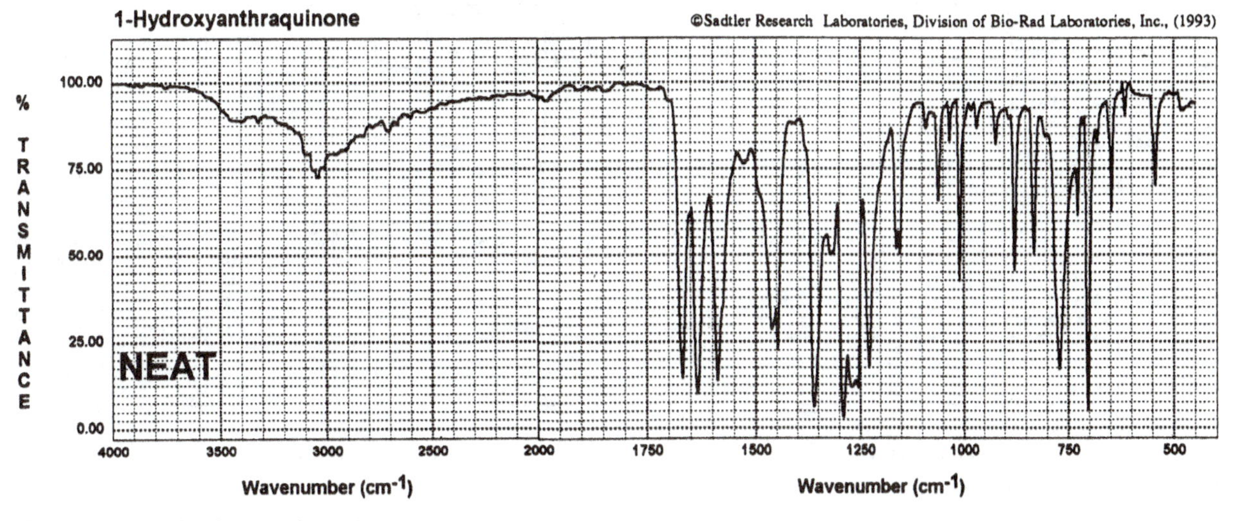


IR spectrum: 1-Hydroxyanthraquinone.

**(c)** The spectrum of 2-hydroxyanthraquinone exhibits a single carbonyl stretching frequency near 1673 $cm^{-1}$. Explain why a single carbonyl band would be expected in the system and why this vibration is located at 1673 $cm^{-1}$.

**8-11.** Suggest a possible structure for the hydrocarbon $C_6H_{14}$, which has the infrared spectrum shown here: Is there more than one correct structure?

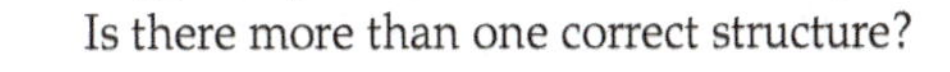

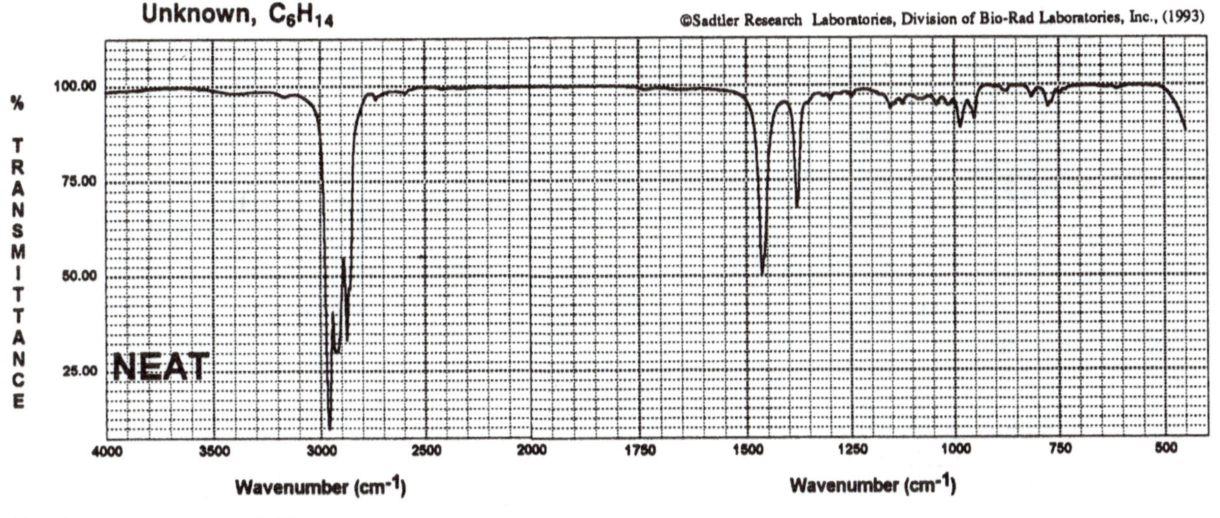


IR unknown spectrum: $C_6H_{14}$.

**8-12.** The hydroxylamine **I** can be oxidized by $MnO_2$ to the amide oxohaemanthidine (**II**). In dilute solution the carbonyl absorption band of **II** occurs at 1702 $cm^{-1}$. Explain this observation.

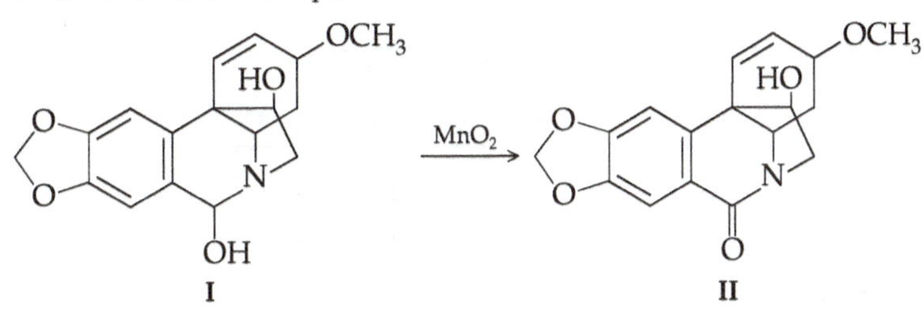

**8-13.** Identify the following alkenes. All samples were obtained from distillation cuts in the $C_6$ boiling range.

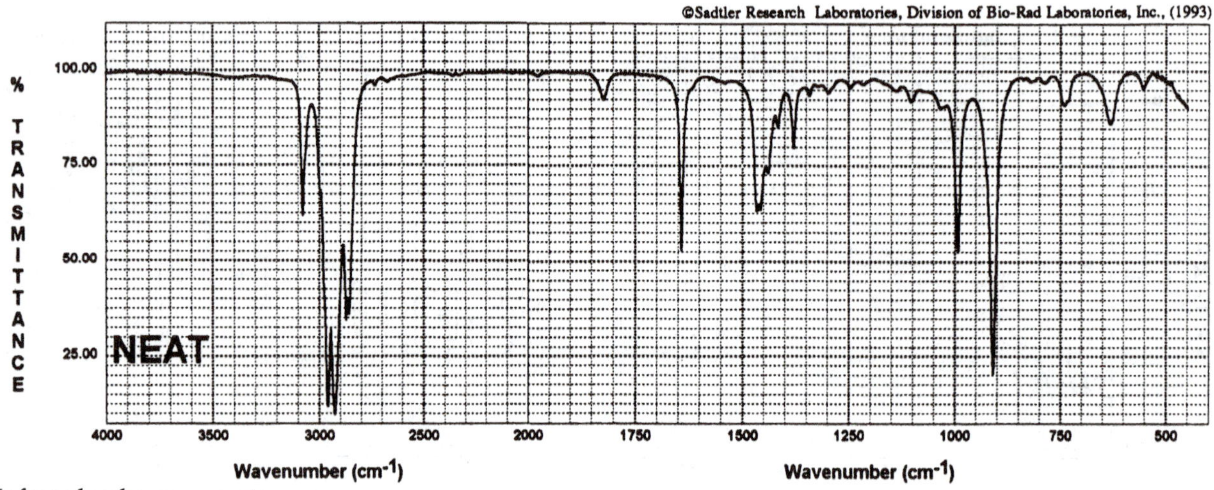


Infrared unknown spectrum *a*.

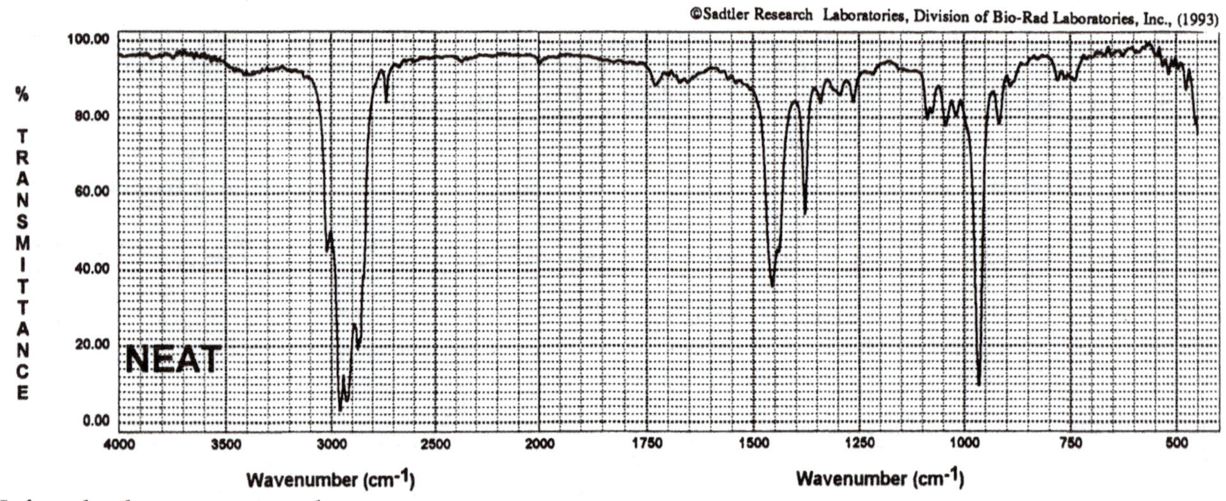


Infrared unknown spectrum *b*.

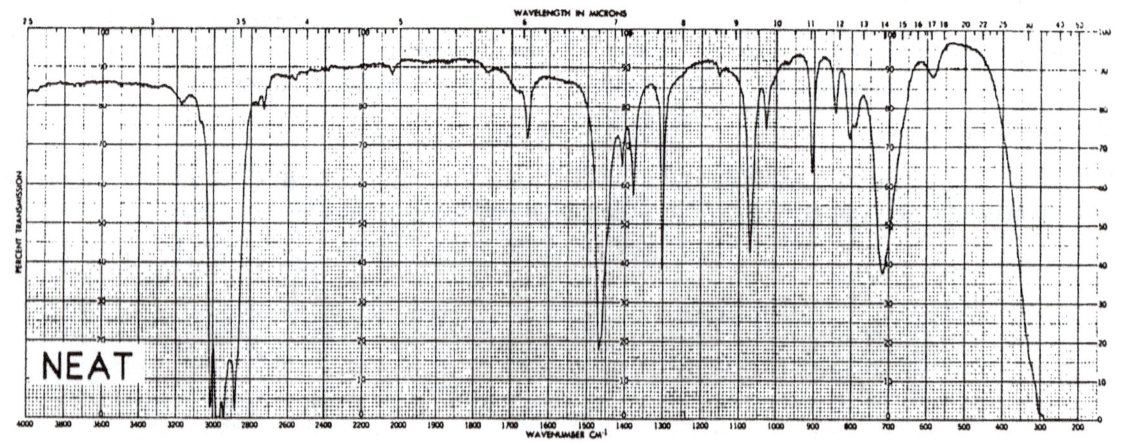


Infrared unknown spectrum *c*. *(Courtesy of Bowdoin College.)*

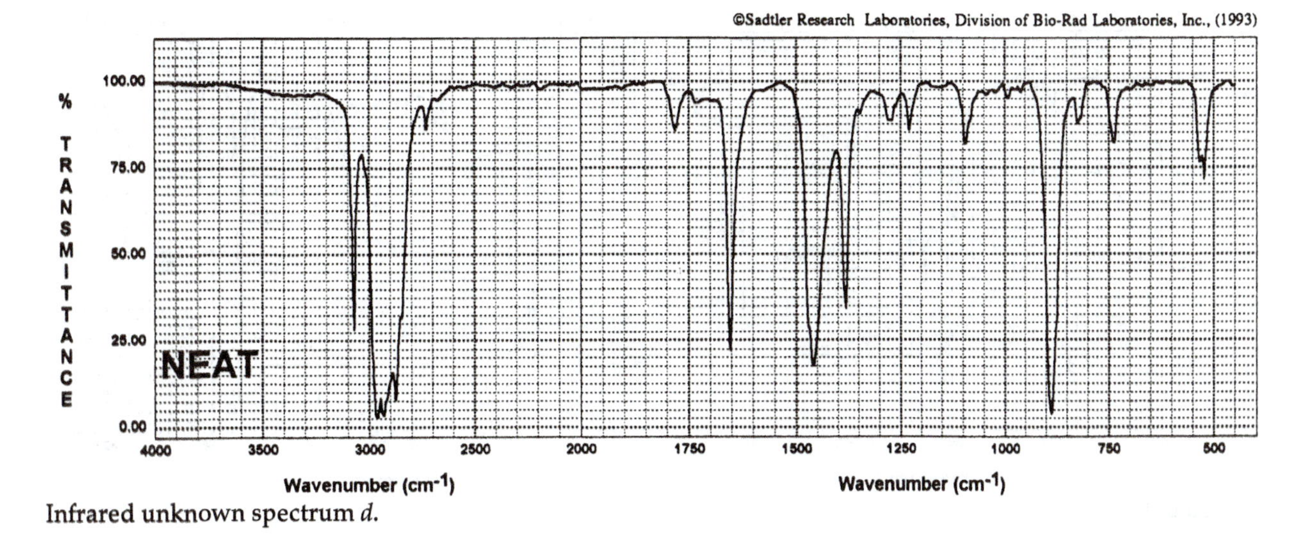


Infrared unknown spectrum *d*.

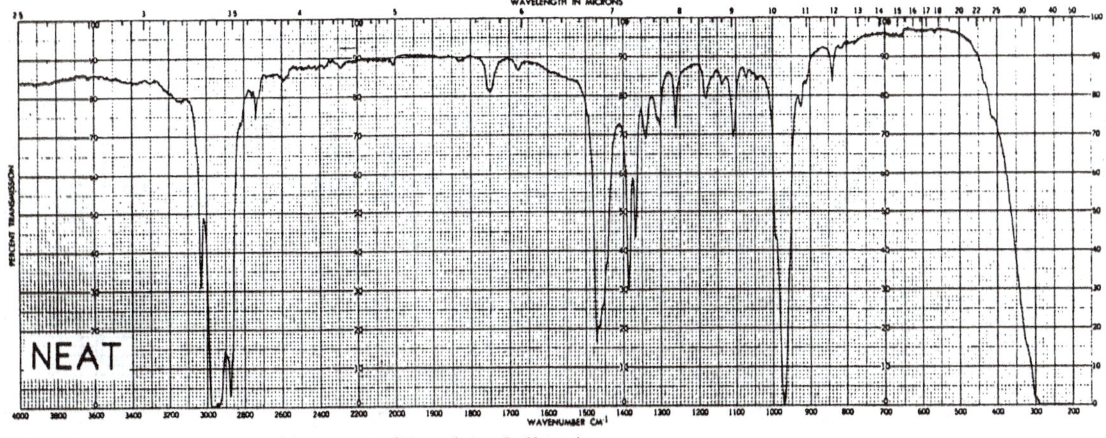


Infrared unknown spectrum *e*. *(Courtesy of Bowdoin College.)*

**8-14.** The infrared spectra of the three xylene (dimethylbenzene) isomers, and an additional aromatic hydrocarbon, are given below. Assign the spectra to the isomers and suggest a potential structure for the remaining unknown substance.

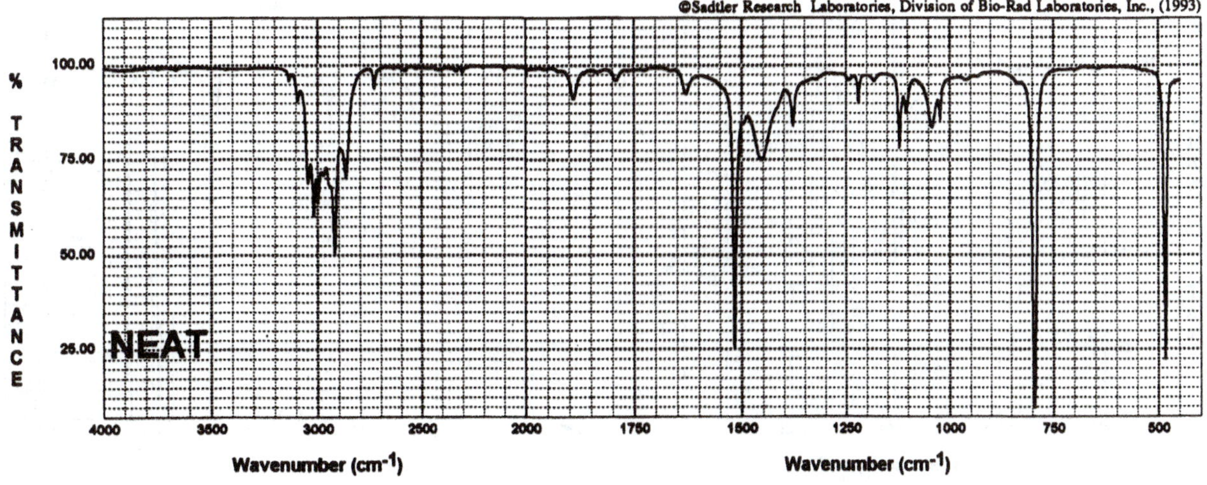


Infrared unknown spectrum *a*.

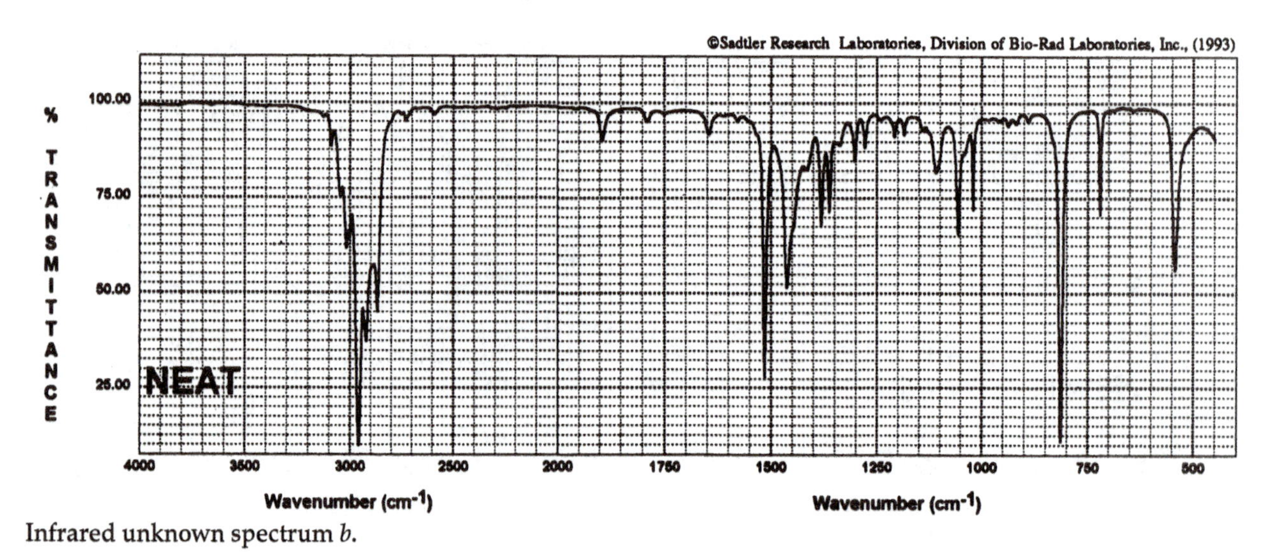


Infrared unknown spectrum *b*.

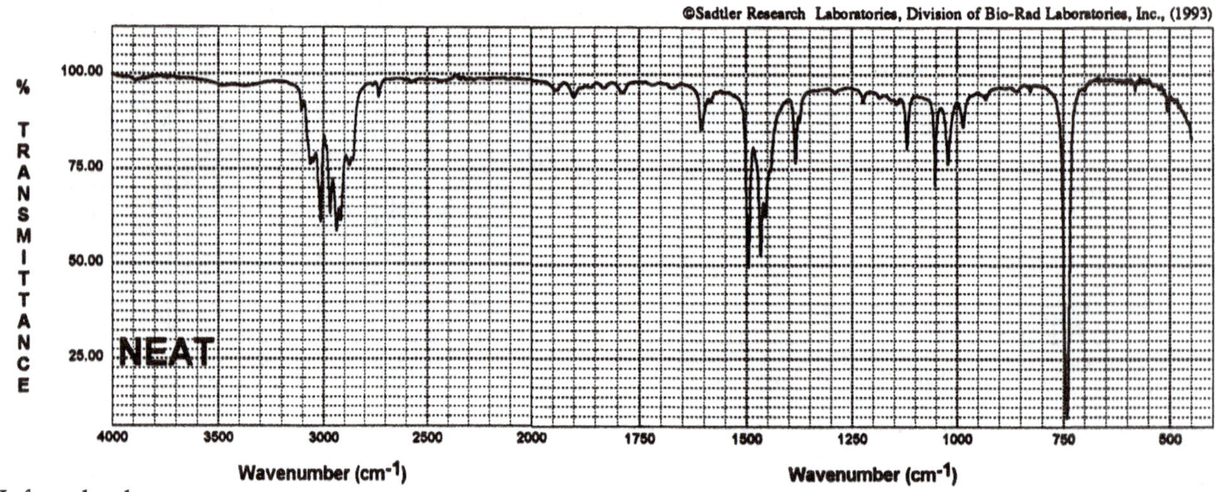


Infrared unknown spectrum *c*.

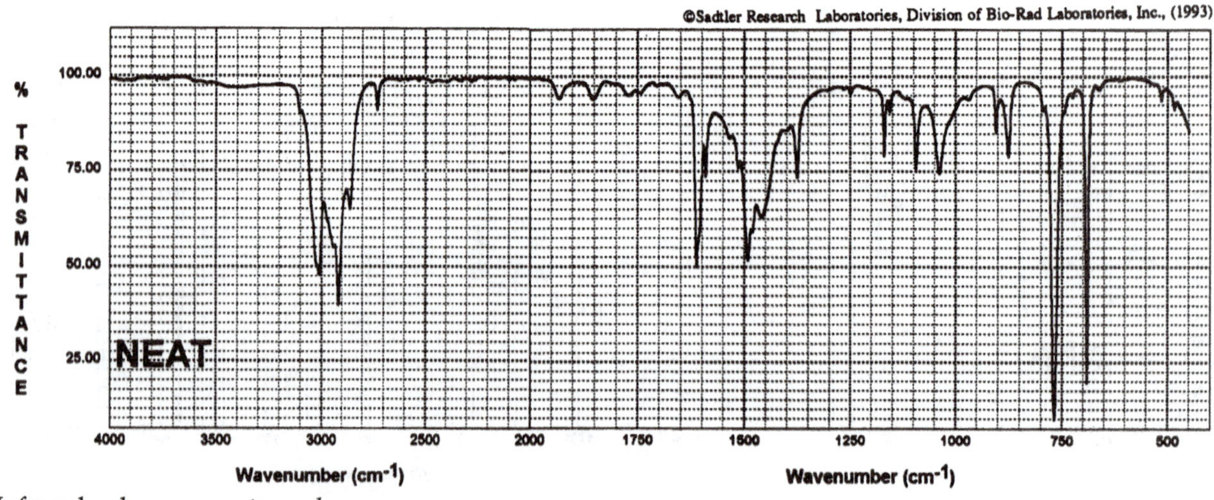


Infrared unknown spectrum *d*.

**8-15.** The C—H stretching mode of chloroform ($CHCl_3$), which occurs at 3022 $cm^{-1}$, is one of the rare exceptions to the 3000-$cm^{-1}$ rule. What is the rule? Suggest an explanation for this exception.

## BIBLIOGRAPHY


*Aldrich Library of Infrared Spectra,* Aldrich Chemical Co., Inc., 940 West Saint Paul Avenue, Milwaukee, WI 53233. 3rd ed., 1981, 12,000 spectra, 8 per page, in one volume arranged by chemical type.

*Aldrich Library of FT-IR Spectra,* Aldrich Chemical Co., Inc., 940 West Saint Paul Avenue, Milwaukee, WI 53233. 2nd ed. 1997, 18,000 spectra, 4 per page, in three volumes arranged by chemical type.

API collection (American Petroleum Institute). About 4500 spectra. M.C.A. collection (Manufacturing Chemists'Association). About 3000 spectra. Chemical Thermodynamics Property Center, Texas A&M University Department of Chemistry, College Station, TX 77843.

Coblentz Society Infrared Spectra Collections for ACS/Labs. (http://www.coblentz.org). An extensive collection of critically evaluated spectra.

Grasselli, J. G.; Ritchey, W. M., Eds., *Atlas of Spectral Data and Physical Constants for Organic Compounds;* CRC Press: 2000 Corporate Blvd. NW, Boca Raton, FL 33431. 2nd ed., 1975.

Sadtler Library. About 95,000 spectra of single compounds; about 12,000 spectra of commercial products. Sadtler Research Labs, 3316 Spring Garden Street, Philadelphia, PA 19014.

**For reviews on ATR FT-IR spectroscopy, see**

Grdadolnik, J. *Acta Chimica Slovenica.* **2002,** *49,* 631.

Hind, A. R.; Bhargava, S. K.: McKinnon, A. *Adv. Colloid Interface Sci.* **2001,** *93,* 91.

Milosevic, M. *Applied Spectroscopy Reviews* **2004,** *39,* 365.

**For reports on the use of ATR FT-IR in the undergraduate laboratory, see**

Schuttlefield, J. D.; Grassian, V. H. *J. Chem. Educ.* **2008,** *85,* 279.

Schuttlefield, J. D.; Larsen, S. C.; Grassian, V. H. *J. Chem. Educ.* **2008,** *85,* 282.


# NUCLEAR MAGNETIC RESONANCE SPECTROSCOPY

## Nuclear spin

Nuclear spin is an energy property intrinsic to a nucleus and analogous to the electron spin that plays such an important role in determining electron configurations. Nuclear spin values are quantized, as are electron spins, and are represented by $I$, the nuclear spin quantum number. Nuclear spin quantum numbers range from 0 through $\frac{7}{2}$ in increments of $\frac{1}{2}$. The nuclei of greatest interest to organic chemists, the $^{1}H$, $^{13}C$, $^{19}F$, and $^{31}P$ nuclei, have spins of $\frac{1}{2}$; the $^{12}C$, $^{16}O$, and $^{32}S$ nuclei have spins of 0 (and thus cannot be observed by nuclear magnetic resonance spectroscopy, NMR); the $^{2}H$ (deuterium, D) and $^{14}N$ nuclei have spins of 1. Since any spinning charged particle (or body) produces a magnetic moment, a nucleus with a nonzero spin quantum number has a magnetic moment, $\mu$.

Nuclear spin values are quantized because the nuclear angular momentum, and thus the nuclear magnetic moment, is quantized. When placed in an external magnetic field, nuclei orient their magnetic moments in certain ways with respect to the magnetic field, which is assumed to be aligned with the $z$ axis of a Cartesian coordinate system. These orientations are referred to as the $z$ components of the nuclear magnetic moment, $\mu_z$. For a nucleus with a spin of $\frac{1}{2}$, $\mu_z$ may be $+\frac{1}{2}$ or $-\frac{1}{2}$. In general, for a nucleus of spin $I$, the $\mu_z$ takes quantized values from $[-I, -I+1, \ldots, I-1, I]$; or $(2I + 1)$ different values in all. For this discussion we will limit ourselves to nuclei with spin $\frac{1}{2}$, since this is easier to describe, and since most nuclei of interest in organic chemistry are of spin $\frac{1}{2}$.

When placed in a static magnetic field of strength $H_0$, the magnetic moment, $\mu_z$, of the spinning nucleus precesses about the magnetic field at a frequency, $\nu$, such that $\nu = \gamma H_0/2\pi$, where $H_0$ is the strength of the applied magnetic field, and $\gamma$ is a characteristic property of the nucleus known as the gyromagnetic ratio. When a nucleus of spin $\frac{1}{2}$ is placed in a magnetic field, the energies of the $\mu_z = +\frac{1}{2}$ and $-\frac{1}{2}$ states are separated, since in one spin state the nuclear magnetic moment is aligned with the applied magnetic field, and in the other spin state the nuclear magnetic moment is opposed to the applied magnetic field.

The amount of separation of the two energy states, $\Delta E$, is proportional to the magnetic field, and is given by the following expression:

$$\Delta E = \frac{h\gamma H_0}{2\pi} = h\nu$$

When nuclei in the magnetic field are exposed to radiation of the proper frequency, transitions between the two energy states are stimulated, and the nucleus is said to be in *resonance,* or to *resonate.* This transition occurs when the frequency and the energy difference are related by the Planck relation, $\Delta E = h\nu$, and thus the sample will absorb energy of frequency $\nu$. The study of these energy changes is known as nuclear magnetic resonance, or NMR, spectroscopy.

## INSTRUMENTATION

In an NMR spectrometer, the magnetic field is provided by a large permanent magnet, electromagnet, or superconducting electromagnet. Commercially available NMR spectrometers have magnets with field strengths that range from 1.4 to 16.3 tesla (the earth's magnetic field at its surface is roughly $5 \times 10^{-5}$ tesla), and thus operate at frequencies from 60 to 700 MHz for protons. In general, most spectrometers with an operating frequency above 100 MHz use a superconducting electromagnet.

Traditionally, NMR spectra were acquired either by holding the applied magnetic field constant and sweeping the radio frequency (rf), or by holding the rf constant and sweeping the applied magnetic field. Energy absorption by the sample was detected, and the result was the NMR spectrum, a plot of intensity (of energy absorption) versus frequency (or field). This instrumental technique is referred to as continuous wave, or CW, spectroscopy (Fig. 8.7). Over the last 20 years, however, it has been commonly replaced by pulsed, or Fourier transform (FT), NMR spectroscopy. Among many other benefits, FT-NMR spectroscopy allows very rapid acquisition of spectral data, which permits analysis of small samples and rare nuclei, such as $^{13}C$.

The basic principles of FT-NMR spectroscopy can be qualitatively explained as follows. Take, for example, an NMR spectrum that contains a single peak at a given frequency. The graph of this spectrum (Fig. 8.8) is a plot of intensity versus frequency. The same information can be conveyed by a plot of intensity versus time that shows a cosine wave at the frequency described by the graph of the usual NMR spectrum. This is shown in Figure 8.9, and for a spectrum with a single frequency, this plot of intensity versus time is almost as easy to interpret as the usual NMR spectrum shown in Figure 8.8.

Of course, it would be very difficult to determine the frequencies of many superimposed cosine waves from this kind of plot, and it would be at best awkward to interpret a complex NMR spectrum presented in such a fashion (Fig. 8.10). The use of the Fourier transform allows us mathematically to

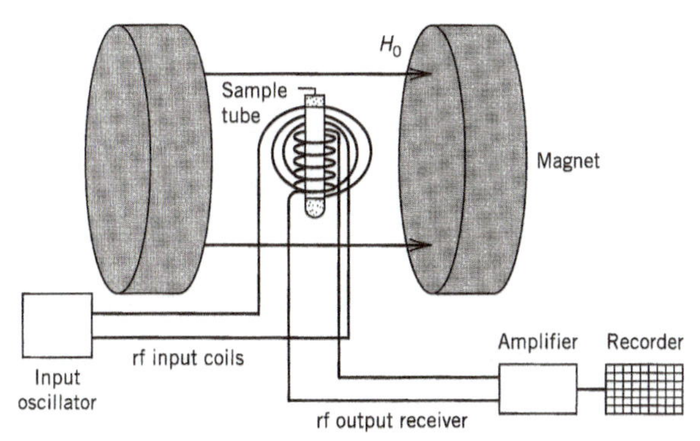


**Figure 8.7 Schematic of NMR spectrometer.** (Reprinted with permission of John Wiley & Sons, New York.)

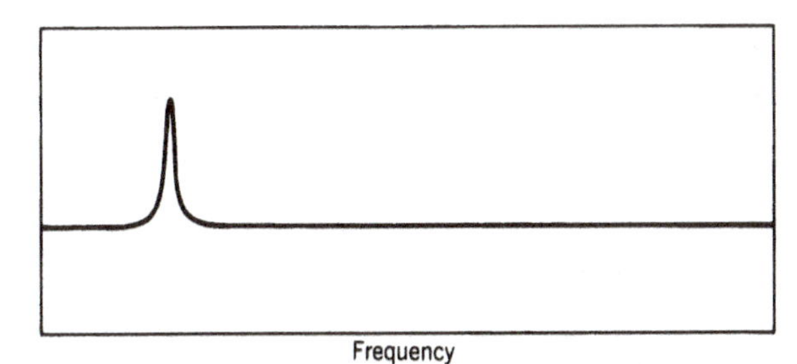


**Figure 8.8 Intensity versus frequency (usual NMR spectrum).**

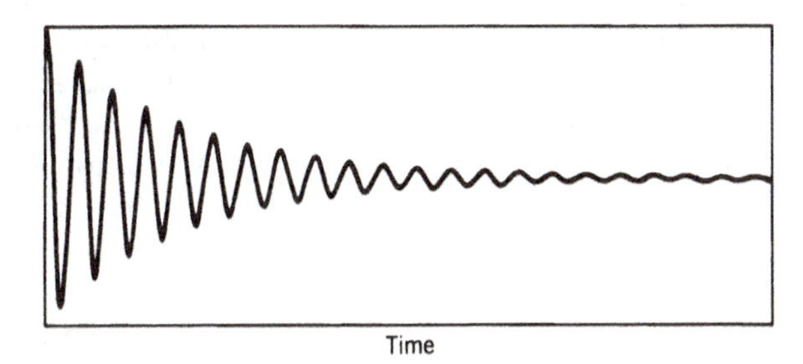


**Figure 8.9 Intensity versus time.**

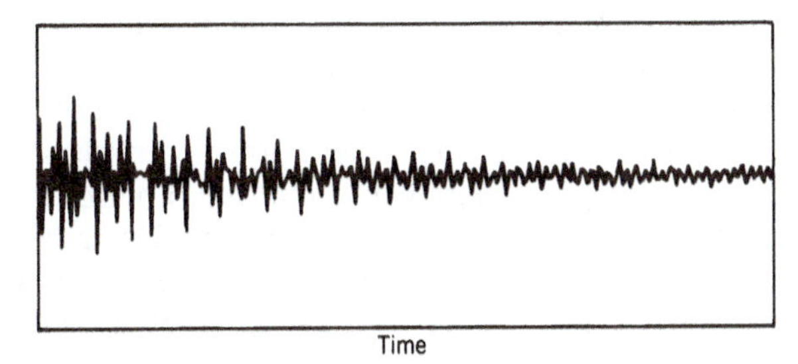


**Figure 8.10 Three-signal NMR spectrum: intensity versus time.**

interconvert these time domain (Fig. 8.10) and frequency domain spectra (Fig. 8.11). Fourier transform of the apparently complex spectrum in Figure 8.10 gives the spectrum in Figure 8.11. It is then easy to see that there are actually only three different resonance signals contained in the time domain of the data of Figure 8.10.

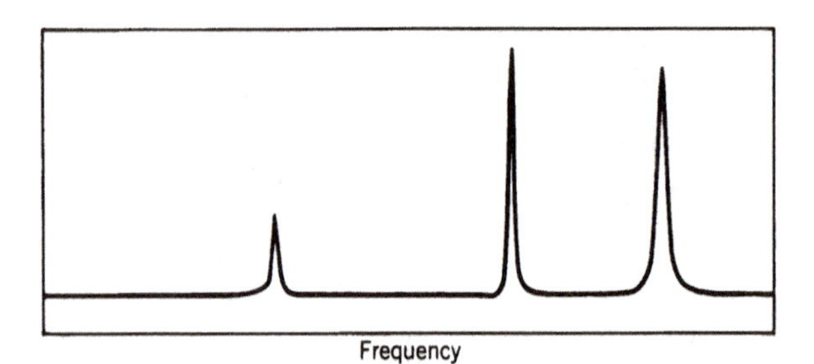


**Figure 8.11 Three-signal NMR spectrum: intensity versus frequency.**

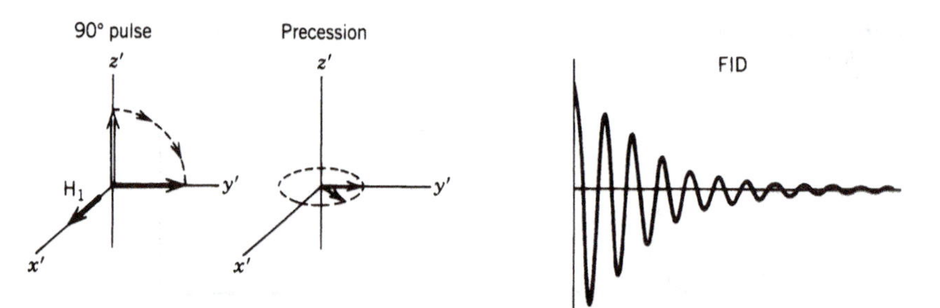


**Figure 8.12 Basic pulsed NMR experiment.**

Fourier transform NMR spectra are obtained by applying a short (~1–10 ms), high-powered pulse of rf energy to the sample (Fig. 8.12). This pulse affects all the nuclei to be observed. Before the pulse is applied, the equilibrium net nuclear magnetization is aligned with the applied magnetic field, along the *z* axis. The coordinate system is presumed to be rotating about the *z* axis at the frequency of the rf pulse. The pulse, applied down the *x* axis, applies a torque to the nuclear magnetic moments and rotates them into the *xy* plane. At this point the pulse is turned off and the nuclear magnetic moments return to their equilibrium alignments. In the process, they precess about the *z* axis (applied magnetic field) in the *xy* plane and induce a current in a detector coil, which can be thought of as being aligned with the *y* axis. This current varies in a sinusoidal manner, and the observed frequency will be the difference between the resonance frequency of the nuclei and the frequency of the rf pulse.

The detected signal, which is called the free induction decay (FID), is digitized and stored. For small organic molecules in a nonviscous solution, the FID will disappear after a few seconds, which corresponds to the time it takes the nuclei to regain equilibrium alignments after the rf pulse. Thus, an entire $^1$H NMR spectrum can be obtained in approximately 2 s, in contrast to the 10–15 min usually needed to obtain a CW spectrum. A major advantage of FT-NMR is that many spectra of a sample can be rapidly obtained and added together to increase the signal-to-noise (*S/N*) ratio. Noise is presumably random about some zero level, so when many spectra are added together, the noise level is reduced, while real signals are reinforced when added. The *S/N* ratio is proportional to the square root of the number of spectra added together. Thus, one can obtain the $^1$H spectrum of a 10-μmol sample in a minute or two. With FT-NMR it is possible to obtain spectra of isotopes that are insensitive and/or of low natural abundance, as well as spectra of large biological molecules in dilute solution. By adding a few hundred spectra together, an adequate $^{13}$C NMR spectrum of a 100-μmol sample can be obtained in about 20–30 min.

## CHEMICAL SHIFT

In a molecule, the magnetic field at a nucleus depends not only on $H_0$, the field generated by the instrument (the external field), but also on the magnetic fields associated with the electron density near the nucleus. Electrons are influenced by the external field in such a way that their motion generates a small magnetic field that opposes the applied field, and reduces the actual field experienced at the nucleus. This reduction is very small (relative to the external field) and is on the order of 0.001%, 10 ppm, for most protons, and about 200 ppm for $^{13}C$ nuclei. Reduction of the external field is known as *shielding*, and it gives rise to differences in the energy separation for nuclei in different electronic environments in a molecule. The differences in the energy separation are known as *chemical* shifts.

The magnitude of the chemical shift depends on the nature of the valence and inner electrons of the nucleus and even on electrons that are not directly associated with the nucleus. Chemical shifts are influenced by inductive effects, which reduce the electron density near the nucleus and reduce the shielding. The orientation of the nucleus relative to $\pi$ electrons also plays an important role in determining the chemical shift. A proton located immediately outside a $\pi$-electron system (as in the case of the protons on benzene rings) will be significantly deshielded. In most molecules the chemical shift is determined by a combination of these factors. Chemical shifts are difficult to predict using theoretical principles, but have been well studied and can usually be easily predicted empirically upon comparison to reference data.

In an NMR spectrum, the absorption of rf energy is detected, as in Figure 8.13, where the energy absorption is shown for increasing frequency. In this example we illustrate the case with two different nuclei, A and X. Since A and X are different, they absorb energy at different frequencies while in the same applied magnetic field.

The spectrum would be displayed as in Figure 8.14. The difference in the resonances is known as a *chemical shift* and is expressed in parts per million (ppm). The use of frequency units is cumbersome and is complicated because

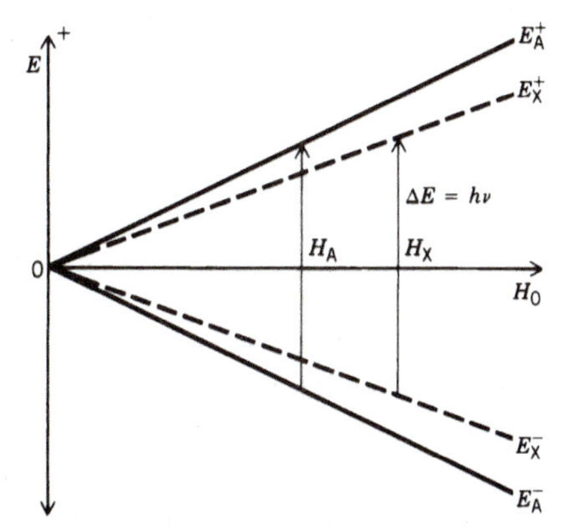


**Figure 8.13** The energy splitting for two chemically different protons. The differences between the *A* energy levels (solid lines) and the *X* levels (dashed lines) have been amplified for illustrative purposes. At 60 MHz, nucleus *A* absorbs energy at field $H_A$ and nucleus *X* absorbs energy at field $H_X$. Nucleus *X* is said to be more strongly shielded than *A*. The resonance for *X* is said to occur upfield of that for *A*.

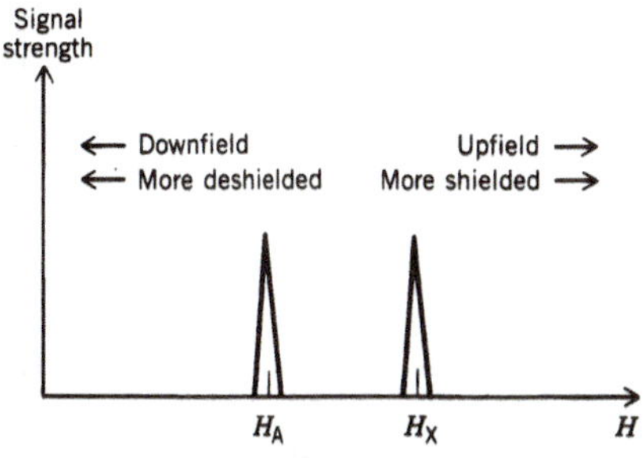


**Figure 8.14** The spectrum for the system in Figure 8.13 as it would be displayed. It is conventional to display the spectrum with magnetic field strength increasing to the right so that upfield (and more strongly shielded) is toward the right and downfield (and deshielded) is toward the left.

NMR spectrometers of different magnetic field strengths (and thus operating frequencies) are used. The use of ppm units allows direct comparisons of spectroscopic data obtained on different instruments, when chemical shifts are referenced relative to a reference compound whose chemical shift is arbitrarily defined as 0 ppm. The accepted reference standard for $^1H$ and $^{13}C$ NMR in organic solutions is tetramethylsilane (TMS), $(CH_3)_4Si$. The chemical shift relative to TMS is symbolized by $\delta$.

Tetramethylsilane is used as a reference substance for a number of reasons. It is more strongly shielded (Si is more electropositive than C) than most other protons and carbon atoms, and its resonance is thus well removed from other areas of interest in the NMR spectrum. Tetramethylsilane is inert and thus unlikely to react with the compound being analyzed, it is volatile (bp 26 °C) and thus easily removed after a sample has been analyzed, and its 12 identical protons per molecule provide a strong signal per molecule of TMS.

## SPIN–SPIN COUPLING

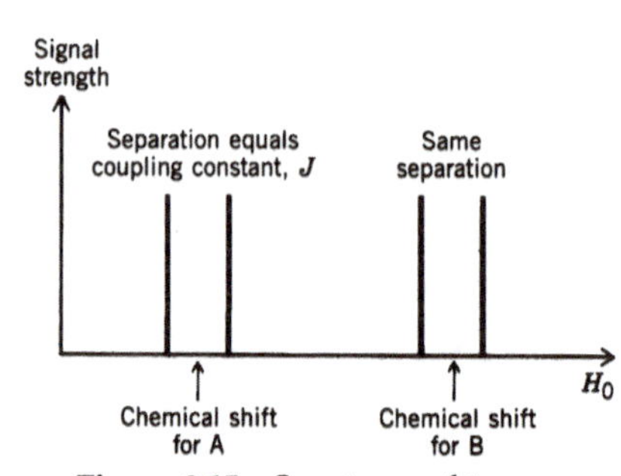


**Figure 8.15 Spectrum of two chemically different protons that are coupled.**

In a molecule with several protons, the exact frequency at which a proton resonates depends not only on the chemical shift of that proton, but also on the spin states of nearby protons. This occurs because the magnetic moments of the nearby protons can either shield or deshield the proton in question from the applied magnetic field, depending on the orientation of the nearby magnetic moments relative to the applied magnetic field. The extent of this perturbation is independent of the applied magnetic field strength. The effect of the spin state of one nucleus on the resonance of another is known as *coupling* or *splitting*.

The spectra resulting from spin–spin coupling depend on the types of nuclei, the distance and geometry between the nuclei, the nature of the bonding, the electronic environment, and the total number of spin states possible. The latter may be illustrated by looking at the spectrum of an imaginary compound that has protons $H_A$ and $H_X$ on adjacent carbons, connected by three bonds: $H_A$—C—C—$H_X$ (Fig. 8.15). In the first approximation we would expect one resonance for $H_A$ and one resonance for $H_X$, and the spectrum would resemble that shown in Figure 8.14. In the presence of coupling, the resonance for $H_A$ splits into two signals, one of which corresponds to $H_X$, having $\mu_z = +\frac{1}{2}$ and the other to $\mu_z = -\frac{1}{2}$. The coupling effect is symmetric in that the $H_X$ resonance also splits into two resonances, one for each spin state of $H_A$. The magnitude of the separation of the $H_A$ pair (a doublet) or the $H_X$ pair (also a doublet) is known as the coupling constant, or $J$. It is usually expressed in frequency units (Hz), since $J$ is independent of the magnetic field strength.

A simple way to explain this is to consider the effect the two possible spin states of $H_X$ have on the resonance frequency of $H_A$. The equilibrium population distribution of the two spin states in $H_X$ is very close to 1:1, since $\Delta E$ is only about $10^{-3}$ cal/mol. Since $H_X$ has a magnetic moment, there are then two slightly different magnetic fields at $H_A$. We thus see two signals for $H_A$, one for those $H_A$ nuclei adjacent to $H_X$ nuclei with $\mu_z$ aligned with the applied magnetic field, and one signal for those $H_A$ nuclei adjacent to $H_X$ nuclei with $\mu_z$ aligned opposed to the applied field. Coupling between protons connected by more than three bonds does occur, but its magnitude, $J$, is usually small and often not directly observed in a usual NMR spectrum.

The splitting becomes more interesting when there are several nuclei of one type. 1,1-Dibromoethane ($CH_3CHBr_2$) has three equivalent protons in the

| Individual $\mu_z$ | SUM |
|---|---|
| (+)(+)(+) | $+\frac{3}{2}$ |
| (+)(+)(−) or (+)(−)(+) or (−)(+)(+) | $+\frac{1}{2}$ |
| (+)(−)(−) or (−)(+)(−) or (−)(−)(+) | $-\frac{1}{2}$ |
| (−)(−)(−) | $-\frac{3}{2}$ |

**Figure 8.16 Possible combinations of spin states for a methyl group.**

methyl group and one proton on the C-1 atom. The methyl group exhibits rapid internal rotation so that its three protons are equivalent. The chemical shift for the C-1 proton is 5.86 ppm and that for the methyl protons is 2.47 ppm. Here we can see an example of decreased shielding resulting from the presence of electronegative substituents. Equivalent protons do not couple with one another (this is an important rule in interpreting spectra), but the methyl protons will affect the proton on C-1, and vice versa.

To analyze the splitting pattern, we need to consider the orientations of the nuclear magnetic moments, with respect to the applied magnetic field, for all three methyl protons. Since each of the three protons may have two spin states that are of nearly equal probability, there are $2^3 = 8$ possible combinations of spin states in all for the methyl protons. The net sums of these may have only four different values, as shown in Figure 8.16. The symbol (+) is used to represent $\mu_z = +\frac{1}{2}$ for a single proton and (−) is used to represent $\mu_z = -\frac{1}{2}$. Thus (+)(+)(−) means that protons 1 and 2 have $\mu_z = +\frac{1}{2}$, while proton 3 has $\mu_z = -\frac{1}{2}$.

The number of different $\mu_z$ states is $(2N + 1)$, where $N$ is the number of equivalent nuclei (of spin $\frac{1}{2}$) Thus the three methyl protons can generate four slightly different magnetic fields, and the proton on the C-1 of $CH_3CHBr_2$ sees (in different molecules) four different magnetic fields. Since these different magnetic fields are not of equal probability, but rather are populated in a ratio of 1:3:3:1, the four signals we see for the proton on C-1 when coupled to the methyl group are of intensities 1:3:3:1, and are referred to as a *quartet*. This is shown schematically in Figure 8.17.

Since the proton on C-1 has two possible spin states of nearly equal probability, the protons of the methyl group experience two slightly different magnetic fields and are observed in the spectrum as two slightly separated signals of equal intensity, or a *doublet*. The separation between each of the C-1 proton signals is the coupling constant, $J$, and will equal the $J$ of the methyl signal. The proposed spectrum is shown in Figure 8.17*b*. The coupling constant in this case is about 7 Hz. The 60-MHz NMR spectrum of 1,1-dibromoethane is shown in Figure 8.17*c*.

*NOTE. The net effect of spin–spin coupling is that a proton (or group of equivalent protons) adjacent to* N *other protons will be observed as a multiplet with (*N *+ 1) lines.*

A proton, or group of equivalent protons, may be coupled to more than one group of nuclei. The spectrum of 1-nitropropane ($CH_3CH_2CH_2NO_2$) is

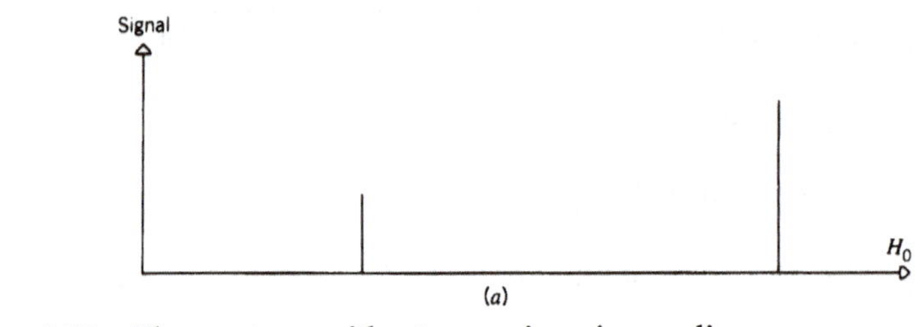


Figure 8.17*a* The spectrum without any spin–spin coupling.

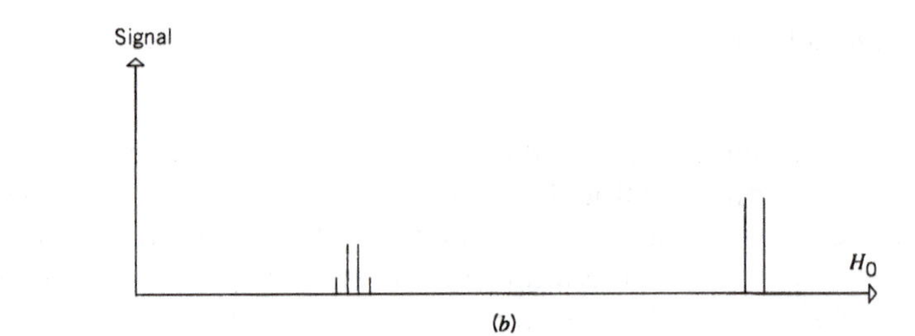


Figure 8.17*b* A "stick figure" spectrum indicating the expected intensities.

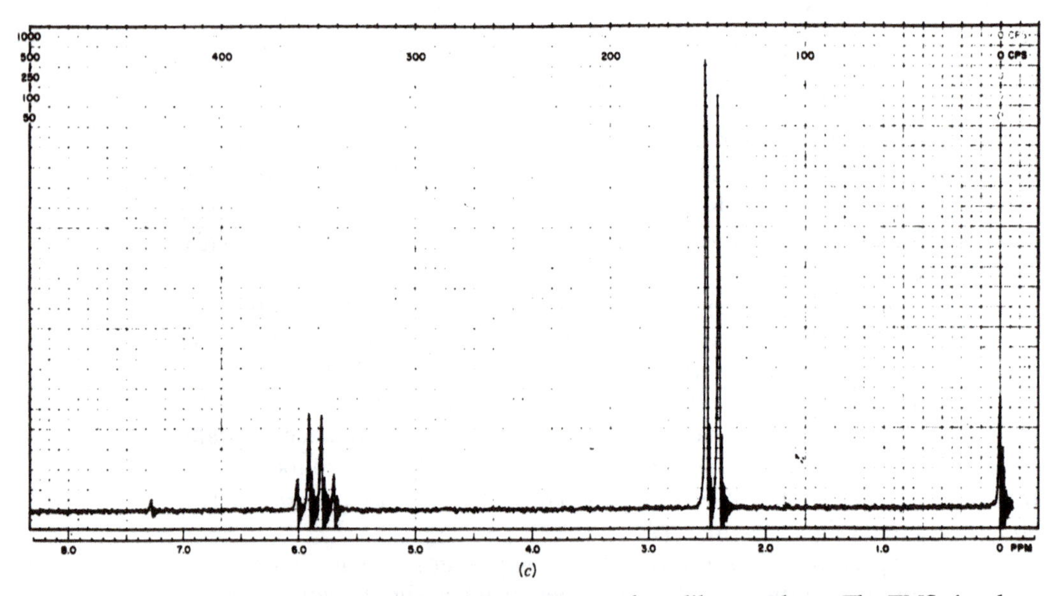


Figure 8.17*c* The actual 60-MHz spectrum of 1,1-dibromoethane. The TMS signal at 0 ppm is seen as well as a weak signal at 7.3 ppm, which is not from this molecule.

shown in Figure 8.18. The signal from the central methylene ($CH_2$) group is seen at about 2.0 ppm. Because the methylene group is adjacent to (and thus coupled to) five protons, its signal is a (5 + 1) or six-line multiplet—a *sextet.*

Nuclei with spins of 1 or greater exhibit more complex spin–spin coupling, since they can exist in more than two different spin states. For a nucleus coupled to *N* nuclei of spin *I*, a multiplet of 2*IN* lines will be observed. Nuclei of spin zero do not couple.

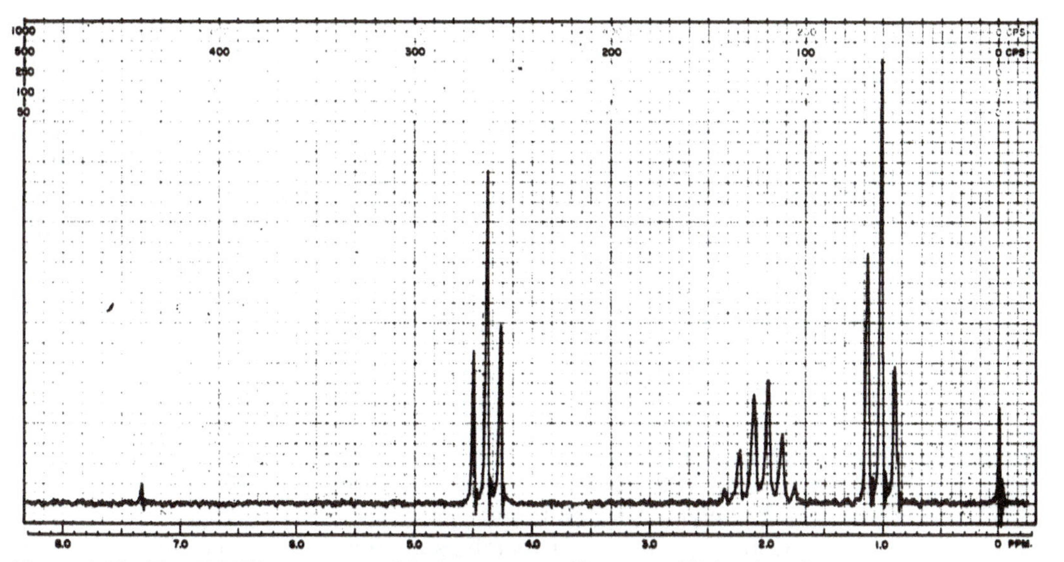


**Figure 8.18 The 60-MHz spectrum of 1-nitropropane.** (Courtesy of Varian Associates, Palo Alto, CA.) **Starting from the right, the TMS signal at 0 ppm is seen. Next is a 1:2:1 triplet at 1.03 ppm. This triplet results from the protons on C-3 and their coupling with the two protons on C-2. Next is the sextet centered at 2.07 ppm. This multiplet is from the protons on C-2 and their coupling with the protons on C-1 and C-3. Finally, we have the signal from the protons closest to the nitro group centered at 4.38 ppm. These protons appear as a 1:2:1 triplet due to their coupling with the protons on C-2.**

## INTENSITIES

The area under an NMR peak is proportional to the number of nuclei giving rise to that signal. The intensity of a resonance is thus best determined by the *integral* of the NMR spectrum over a resonance, or group of resonances. Nuclear magnetic resonance spectrometers can measure the integral, though integration data from an FT spectrometer are less reliable than those from a CW spectrometer. In more complex spectra the intensities are useful as a measure of the number of protons of a given type. For instance, in the above case the integral over both peaks of the methyl group doublet will be three times the integral over the quartet of the proton on C-1. Integration can thus often provide useful information for determining the identity of a compound.

## SECOND-ORDER EFFECTS

So far, all of our examples have consisted of first-order spectra. First-order spectra are those multiplets interpretable through elementary coupling analysis, such as that above; second-order spectra are those that are not interpretable in this manner. These highly symmetric and fairly simple first-order spectra are generally observed when the chemical-shift differences (expressed as a frequency) are much greater than the coupling constant. Second-order effects occur when the coupling constants become comparable to or greater than the chemical-shift differences. Thus, spectra obtained on instruments with higher magnetic fields are more likely to be first order, since the frequency differences between given signals increase with increasing magnetic fields. However, the chemical-shift differences (in ppm) remain the same regardless of magnetic field strength.

Second-order effects may be understood in qualitative terms by considering the limiting cases. Let us consider the hypothetical disubstituted ethylene shown in Figure 8.19, where R and M are substituents that might be identical or may have very different effects on the alkenyl protons. In Figure 8.19*a* the spectrum is shown for the situation in which R and M have very different effects. In this case we will observe a first-order spectrum consisting of two doublets. The coupling constant is the separation in the doublets, and the chemical shift of each nucleus is the geometric midpoint of each doublet.

In Figure 8.19*b*, groups R and M are identical. $H_A$ and $H_B$ are identical in this case and only a single resonance is observed (coupling between equivalent nuclei is not observed).

In Figure 8.19*c* the difference in the chemical environment of $H_A$ and $H_B$ is very slight. The spectrum shown may be seen as intermediate between the limiting cases in Figure 8.19*a* and Figure 8.19*b*. Note that there is a "leaning in" of the doublets as the central members increase in intensity at the expense of the outer members. A full continuum of behavior may be expected with cases observed in which the outer members are lost in the noise and the central members take the appearance of a doublet. This would be one example of a class of spectra known as "deceptively simple spectra."

The second-order spectra of systems with more than two protons are difficult to describe even in qualitative terms. Second-order spectra may well display more lines than one would predict from simple coupling theory. Also, in second-order spectra, the coupling constants and the chemical-shift differences may not be obtainable as simple differences in the positions of spectral lines. Thus, spectra obtained at high frequencies (and magnetic fields) are often more useful. As the operating frequency of the instrument is increased, the chemical-shift differences (in frequency terms) increase while the spin–spin coupling remains constant. Thus, the complicating second-order effects are likely to be less noticeable in high-field spectra. The reader is referred to more extensive treatments of NMR for a discussion of second-order cases.

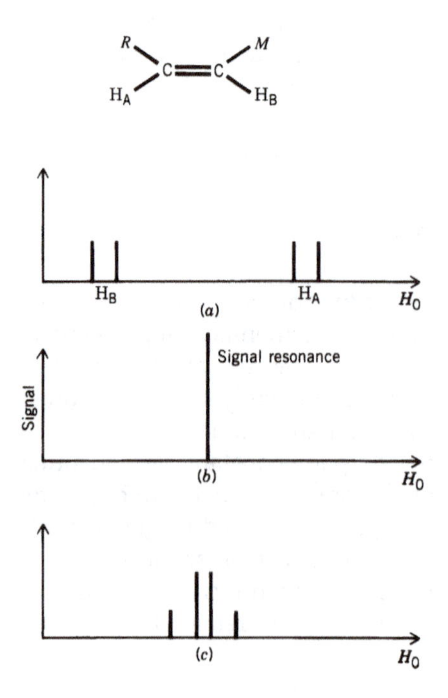


**Figure 8.19 Second-order effects. (a) The chemical-shift difference is much larger than the coupling constant and a first-order spectrum is observed. (b) Protons A and B are equivalent and a single resonance is observed. (c) The chemical-shift difference is of the same order of magnitude or less than the coupling constant. Note the "leaning in" of the peak intensities in this spectrum relative to that in part *a*.**

# INTERPRETATION OF ¹H NMR SPECTRA

The first issues that must be addressed are molecular symmetry and the magnetic equivalence or nonequivalence of protons or other functional groups. Even if two protons, or groups, are chemically equivalent, they may or may not be magnetically equivalent. Although molecular symmetry can often simplify NMR spectra, one must be able to discern which protons or groups are equivalent by symmetry. The two most useful symmetry properties (or symmetry operators) are the plane of symmetry and the axis of symmetry.

A plane of symmetry is simply a mirror plane such that one half of the molecule is the mirror image of the other half, as in *meso*-pentane-2,4-diol:

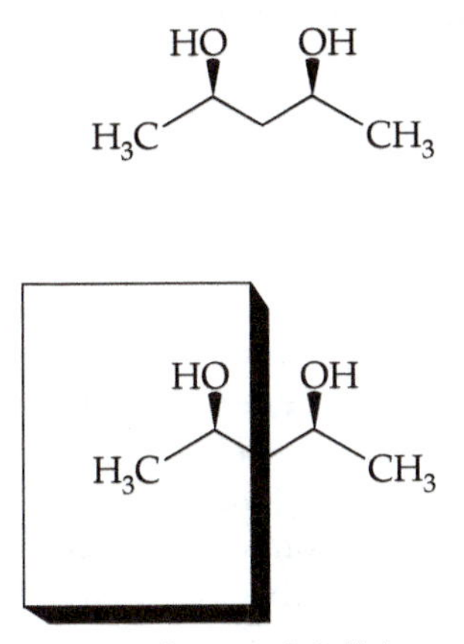


*meso*-Pentane-2,4-diol

The methyl groups are identical by symmetry, and one would expect this stereoisomer to show one methyl doublet in its ¹H NMR spectrum and one signal for a methyl group in its ¹³C spectrum.

Consider the other diastereomer of pentane-2,4-diol, the chiral *d,l* isomer. This isomer has an axis of symmetry. If the molecule is rotated 180° about an axis in the plane of the paper passing through the central carbon, the molecule can be converted into itself:

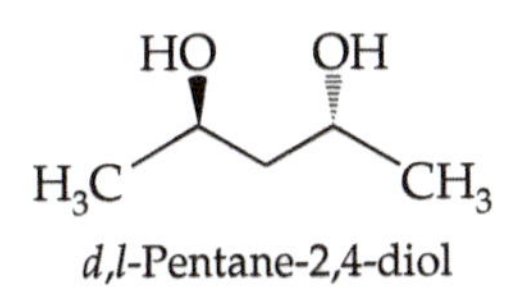


*d,l*-Pentane-2,4-diol

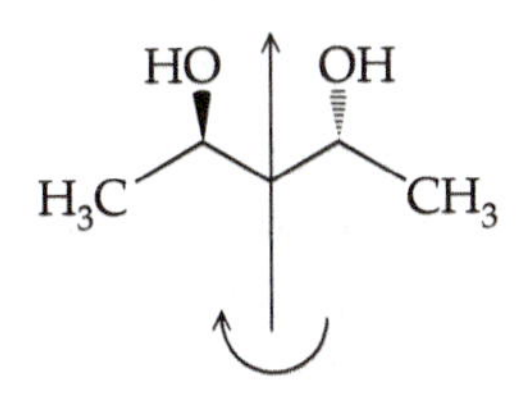


Here, too, the methyl groups are identical by symmetry, and one would expect this stereoisomer to show one methyl doublet in its ¹H NMR spectrum and one signal for a methyl group in its ¹³C spectrum.

It is, however, relatively simple to use NMR spectroscopy to distinguish between these stereoisomers. To do this, look at the two methylene protons on the central carbon, C-3, of each isomer:

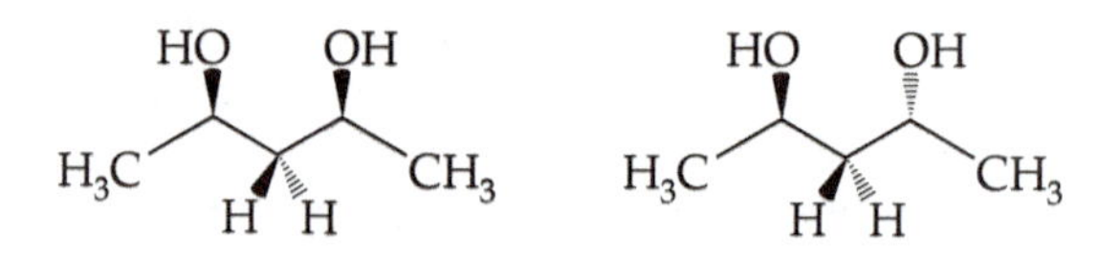


The plane of symmetry in the meso isomer bisects each of the two protons. In the *d,l* isomer, the axis of symmetry interconverts the two protons. Thus, in the *d,l* isomer, the two methylene protons are equivalent by symmetry, but they are not equivalent in the meso isomer. This can also be seen by inspecting the molecule. On the left, one H is syn to both —OH groups and the other is anti to both —OH groups. On the right, each H atom is syn to one —OH group and anti to the other.

The more rigorous way to determine equivalence or nonequivalence is to determine whether the two protons (or groups) are homotopic (identical), diastereotopic, or enantiotopic. To compare two protons, we use the usual Cahn–Ingold–Prelog system for the nomenclature of stereoisomers. We artificially distinguish the relative priority of two protons by a method such as drawing them in different colors or pretending that one is deuterium (as long as the molecule does not contain D). We draw the two possibilities (i.e., the first H as D and then the second H as D) and then determine the stereochemical relationship between the two:

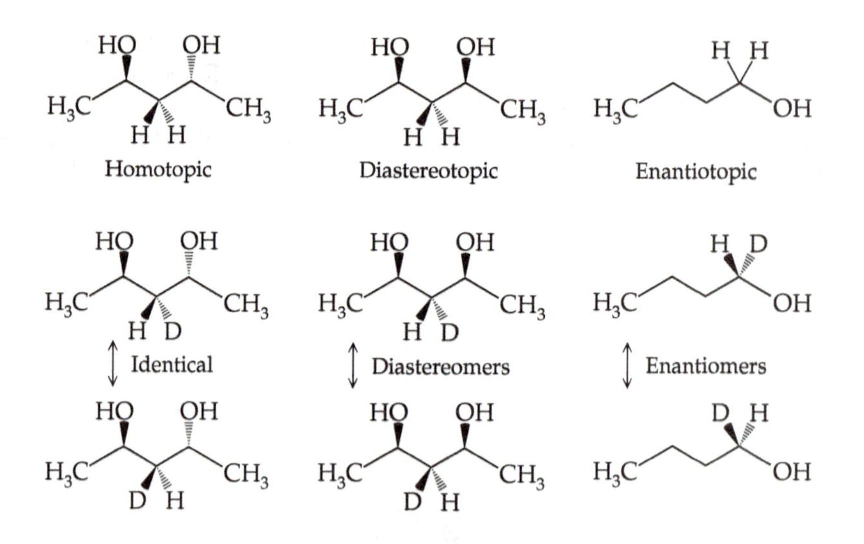


If the two are identical, the two protons are identical, or *homotopic*. If the two structures are diastereomers, the two protons are *diastereotopic*, and if the two structures are enantiomers, the two protons are *enantiotopic*. Diastereotopic protons, or groups, will be magnetically nonequivalent. Enantiotopic protons, or groups, will be magnetically equivalent only in an achiral environment and may appear nonequivalent in a chiral environment, such as a chiral solvent or in a biological sample. Homotopic protons may or may not be magnetically equivalent. Of course, it is possible for magnetically nonequivalent signals to be so close to one another in the NMR spectrum as to overlap (accidentally degenerate).

Homotopic protons may be magnetically nonequivalent if the two protons have different coupling constants to the same third proton. The most common example of this occurs in para-substituted benzenes:

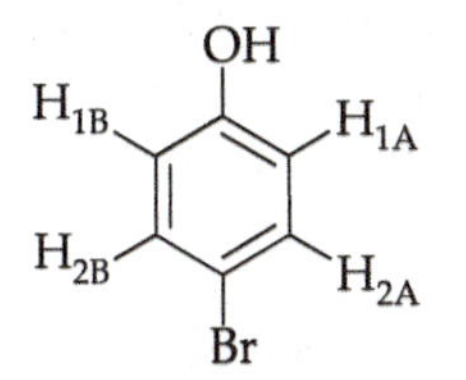


By symmetry, $H_{1A}$ and $H_{1B}$ are equivalent. These protons are not, however, magnetically equivalent because $H_{1A}$ and $H_{1B}$ have different coupling constants to, for example, $H_{2A}$, and the spectrum of this molecule may well be more complex than one would at first expect.

The equivalence or nonequivalence of functional groups, as well as protons, can easily be determined. The $^1$H NMR spectrum of menthol shows three methyl doublets, since the two methyls in the isopropyl group are diastereotopic. The $^{13}$C spectrum of menthol shows three distinct resonances for the three different methyl groups:

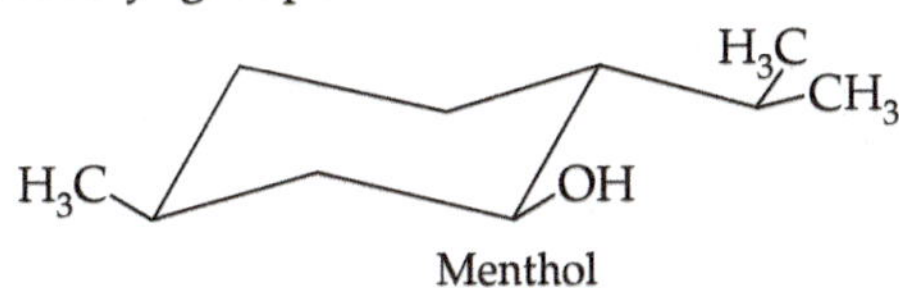


Menthol

## $^1$H CHEMICAL SHIFTS

Figure 8.20 summarizes the chemical shifts of protons in a large range of chemical environments. It is, however, a bit dangerous to use figures such as

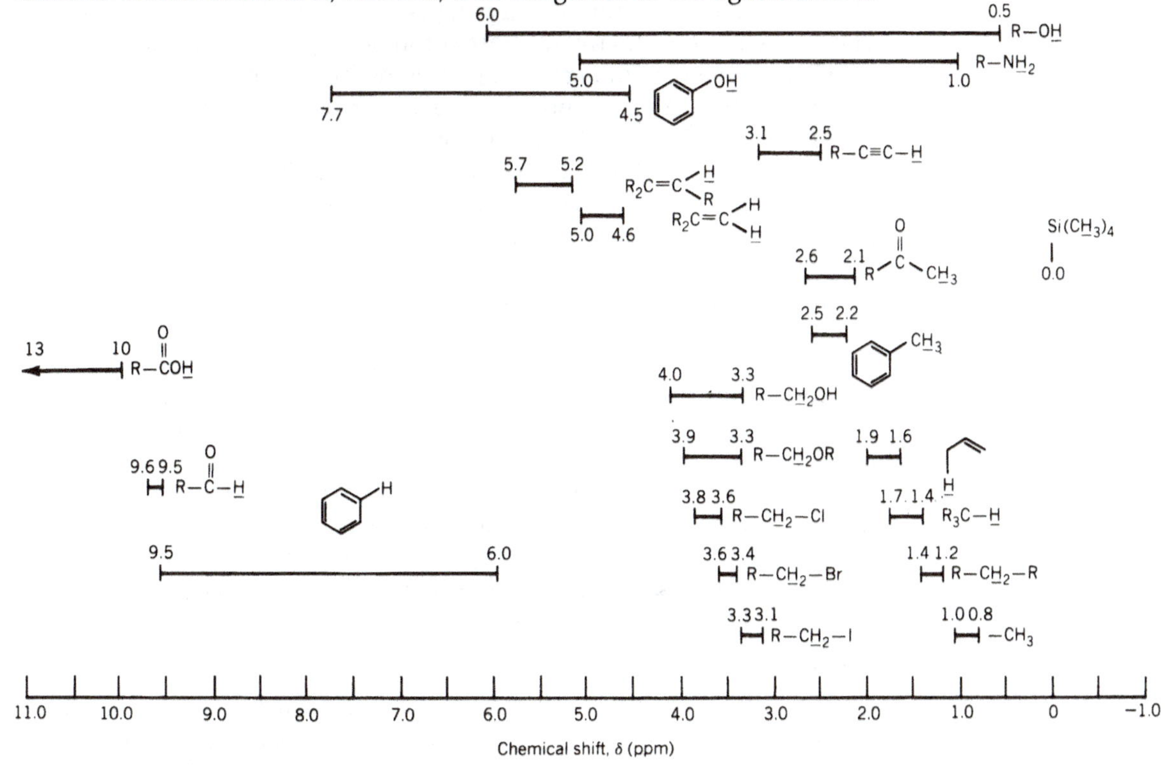


**Figure 8.20 NMR $^1$H chemical shifts.** (From Zubrick, J. W. *The Organic Lab Survival Manual*, 7th ed.; Wiley: New York, 2008. Reprinted by permission of John Wiley & Sons, New York.)

this one without understanding some of the factors that underlie shielding and the chemical shift. To give some flavor of the factors that determine chemical shifts and the range of values observed, we will briefly examine chemical shifts in methyl groups and chemical shifts for protons on $sp^2$ carbon atoms.

Methyl groups bonded to an $sp^3$ carbon generally have chemical shifts in the range 0.8–2.1 ppm as long as there is no more than one electron-withdrawing group attached to the carbon. The shifts generally increase as the strength of the electron withdrawing group increases, or as more electron-withdrawing groups are added. Groups that inductively withdraw electrons reduce the electron density near the methyl group protons. This results in less shielding and a downfield shift of the methyl resonance. This effect is clearly seen in the spectra of 1-nitropropane and 1,1-dibromoethane (Figure 8.17*c* and Figure 8.18), respectively. The chemical shifts for methyl groups bonded to $sp^2$ carbon atoms fall in the range 1.6–2.7 ppm.

In the case of a proton bonded to an $sp^2$ carbon, the location of the proton relative to the $\pi$ cloud plays an important role in determining the chemical shift. In unconjugated alkenes the chemical shifts fall in the range 5–6 ppm. Where more than one proton is bonded to an alkene, complex second-order spectra can be expected at low operating frequencies since the coupling constants are usually fairly large relative to the difference in resonance frequencies. In aldehydes, RC**HO**, the increased electronegativity of the oxygen increases the deshielding and the chemical shift falls in the range 9–10.5 ppm.

The chemical shift in an aromatic system is generally greater than that for alkenes. For example, the chemical shift of benzene is 7.37 ppm, which is substantially greater than the 5.6 ppm for the alkenyl protons of cyclohexene. Much of this difference results from the "ring current" effect and the orientation of the proton relative to the aromatic $\pi$ electrons. If the ring substituents are not strongly electron withdrawing or electron donating, such as alkyl groups, the chemical shift for ring protons will not be shifted greatly from that of benzene itself. Furthermore, these substituents generate only small chemical-shift differences among the ring protons. Thus, the 60-MHz spectra for toluene (methylbenzene) appears to have a single resonance in the aromatic region at about 7.1 ppm. If, on the other hand, the substituents are electron withdrawing, the ortho and para ring protons will be somewhat deshielded relative to benzene. Pi-electron-donating substituents, such as a methoxy group, will increase the shielding of groups ortho and para to it.

## SPIN–SPIN COUPLING

Coupling information is the primary reason that $^1H$ NMR is such a powerful tool for organic structure determination. Since coupling information is transmitted through bonds, coupling provides information about nearby protons and can often be used to deduce stereochemistry.

The sign of the coupling constant (usually symbolized as $J$) may be positive or negative. However, first-order spectra are not sensitive to the sign of the coupling constant. In second-order cases, the sign of $J$ may be determined by a detailed analysis of the spectrum, though the sign of $J$ is generally of little value for organic structure determination.

### Geminal Coupling

Nonequivalent protons attached to the same carbon (geminal protons) will couple with one another. These coupling constants tend to be large (>10 Hz) for $sp^3$ carbon atoms and small (<4 Hz) for $sp^2$ carbon atoms. Geminal coupling constants tend to decrease with decreasing ring size, because of hybridization changes at carbon, and with the increasing electronegativity of the substituents on a given methylene group.

### Vicinal Coupling

Vicinal coupling describes the coupling over three bonds observed between protons attached to two bonded carbon atoms, H—C—C—H. Vicinal coupling constants (*J* values) can range from near 0 to greater than 15 Hz, depending on the stereochemical relationship (dihedral angle) between the coupled protons, the hybridization of the carbon atoms, and the electronegativity of other substituents. For vicinal protons on $sp^3$ carbon atoms, the coupling constant is related to the dihedral angle and is expressed graphically by the Karplus curve (Fig. 8.21).

Though the magnitude of vicinal coupling is very sensitive to the angle of rotation about the central bond, in many simple cases nearly all coupling constants are equal. This situation is often the case if internal rotation about a C—C single bond can occur on a time scale that is very short relative to the NMR time scale, such as in acyclic systems. In these cases, the effect of internal rotation is completely blurred as far as NMR is concerned, and only an average coupling constant is observed. Vicinal coupling constants in freely rotating alkyl groups are usually observed in the 6.5- to 8-Hz range.

When the central C—C bond between two coupled protons is a double bond, rotation is restricted and separate coupling constants for cis and trans protons may be observed. Cis coupling constants fall in the range 5–12 Hz, whereas trans coupling constants range from 12 to 20 Hz. As a result of these large coupling constants, second-order effects are often observed in substituted alkenes in instruments of lower field strengths.

When rotation about carbon–carbon single bonds is restricted, or when stereochemistry dictates significant conformational preferences, nonaveraged coupling constants may be observed that can complicate the appearance of the NMR spectrum. For example, two diastereotopic protons of a methylene ($CH_2$) group may well each couple to a given third proton with different coupling constants. The familiar coupling explained at the elementary level suggests that if one proton is adjacent to, for example, two others, the NMR signal of the first proton will be a triplet. This simplification will be true only if the two coupling constants are identical. A triplet is merely a doublet of doublets with equal coupling constants, which gives rise to the familiar triplet with intensities of 1:2:1 (Fig. 8.22*a*). A doublet of doublets, on the other hand, gives rise to a four-line multiplet with peaks of roughly equal intensity (Figure 8.22*b*).

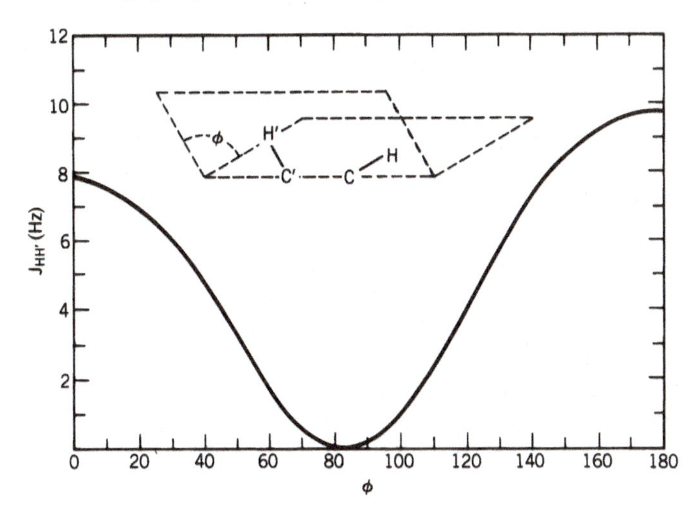


**Figure 8.21** The vicinal Karplus correlation showing the relationship between dihedral angle and coupling constants for vicinal protons.

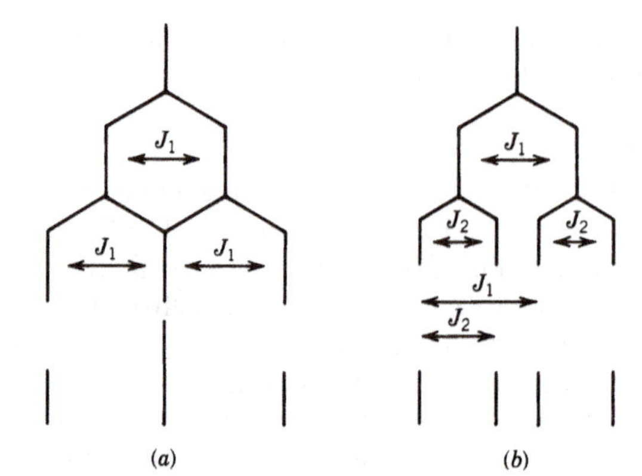


**Figure 8.22** (a) Triplet equals doublet of doublets with equal $J$ values; (b) doublet of doublets.

### Long-Range Coupling

Longer range coupling involving four or more bonds is common in allylic systems and in aromatic rings and other conjugated $\pi$ systems. These coupling constants are generally smaller than the values considered above (i.e., <3 Hz).

## EXAMPLES OF COMPLEX, YET FIRST-ORDER, COUPLING

### Ethyl Vinyl Ether

The coupling constants of even a seemingly complex multiplet can be discerned in a relatively simple manner. First, the total width (outside peak to outside peak) of a first-order multiplet is equal to the sum of all the coupling constants, keeping in mind that, for example, a triplet of $J = 7$ Hz is really a doublet of doublets with both $J$ values equal to 7 Hz. The expansion of the proton spectrum of ethyl vinyl ether is presented as an example in Figure 8.23. Integration data are displayed between the spectrum and the horizontal axis in Figure 8.23*a*.

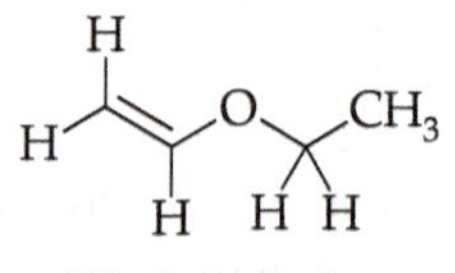


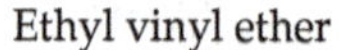
Ethyl vinyl ether

Consider the multiplet centered at 6.45 ppm (Fig. 8.23*b*). By measuring the distance (in Hz) from either outside line to the next inner line, which is 6.8 Hz, the first coupling constant is determined. Then, by measuring from the outside line to the second line in, the second coupling constant is found to be 14.4 Hz. We know that this is the last coupling constant to be found for several reasons. First, if we measure from the outside line to the third line in, we get a value, 21.2 Hz, which is equal to the sum of the previously determined coupling constants. Second, the width of the multiplet (the same measurement in this simple case) is equal to our two coupling constants. Thus, the NMR signal at 6.45 ppm is a doublet of doublets with $J = 14.4$ and 6.8 Hz.

The two doublets of doublets at 4.15 and 3.96 ppm (Fig. 8.23*c*) must be coupled to one another because they both have the coupling constant of 1.9 Hz in common. This geminal coupling constant is typical of the terminal methylene of an alkene.

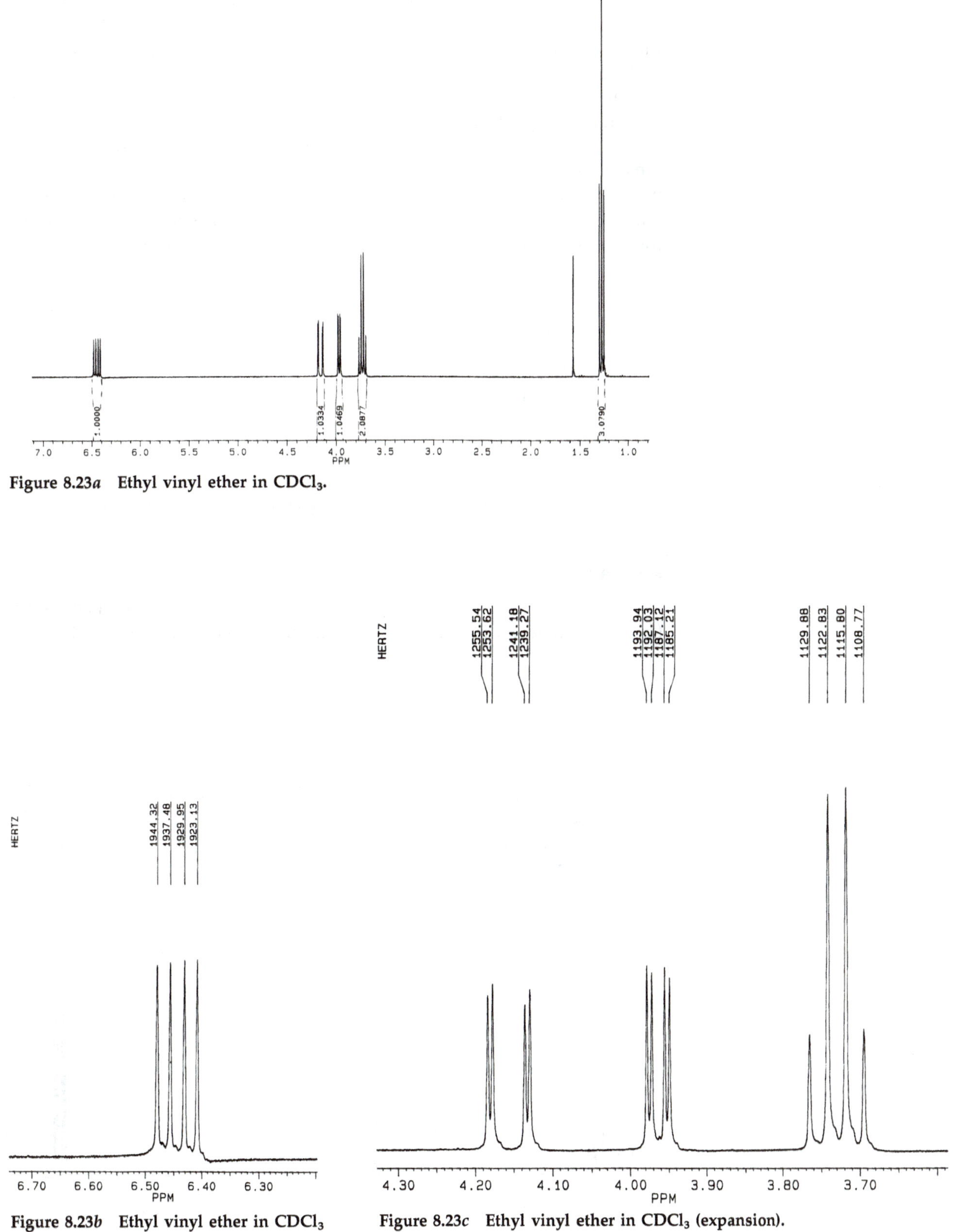


**Figure 8.23*a*** Ethyl vinyl ether in $CDCl_3$.

**Figure 8.23*b*** Ethyl vinyl ether in $CDCl_3$ (expansion).

**Figure 8.23*c*** Ethyl vinyl ether in $CDCl_3$ (expansion).

Since the proton at 3.96 is coupled to the proton at 6.45 ppm by $J$ = 6.8 Hz, and the proton at 4.15 ppm is coupled to the one at 6.45 ppm by $J$ = 14.4 Hz, the proton at 3.96 ppm must be cis, and the proton at 4.15 ppm must be trans, to the alkene proton at 6.45 ppm.

The simple coupling observed for the ethyl group in ethyl vinyl ether can be readily assigned. The triplet at about 1.25 ppm, which integrates for three protons, is due to the methyl group; it is a triplet because the equivalent protons of the methyl group are coupled to the two protons on the adjacent carbon with equal coupling constants. The O—$CH_2$ protons are observed in the NMR spectrum as the quartet at about 3.75 ppm; they are a quartet because they are coupled equally to the three equivalent protons of the methyl group:

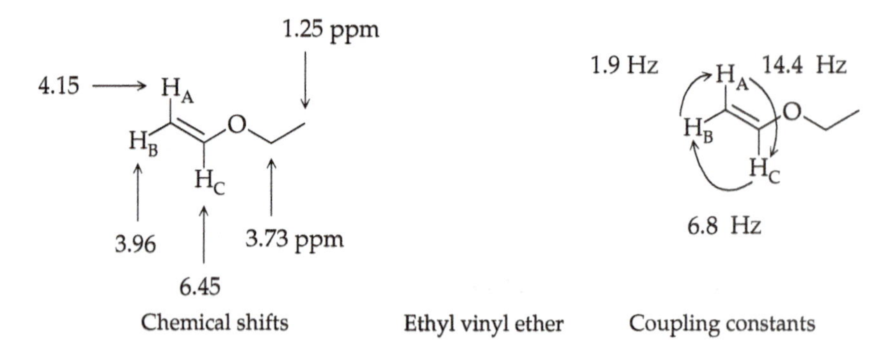


## Allyl Acetate

For a more complex example, refer to the $^1H$ NMR spectrum of allyl acetate (the NMR signal for the methyl group has been omitted) in Figure 8.24*a*.

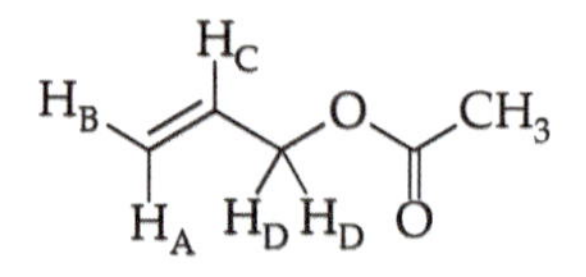


Protons A, B, and C are all chemically distinct, and the two protons labeled D are equivalent to one another by symmetry (the plane of the paper). The multiplet at 4.58 ppm (Fig. 8.24*d*) corresponds to $H_D$ and is a doublet of triplets. The coupling constant for the triplet is 1.4 Hz and the coupling

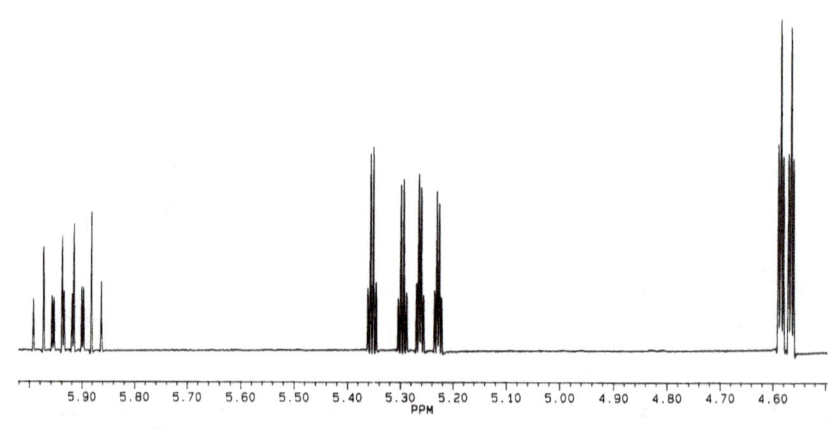


**Figure 8.24*a* Allyl acetate in $CDCl_3$.**

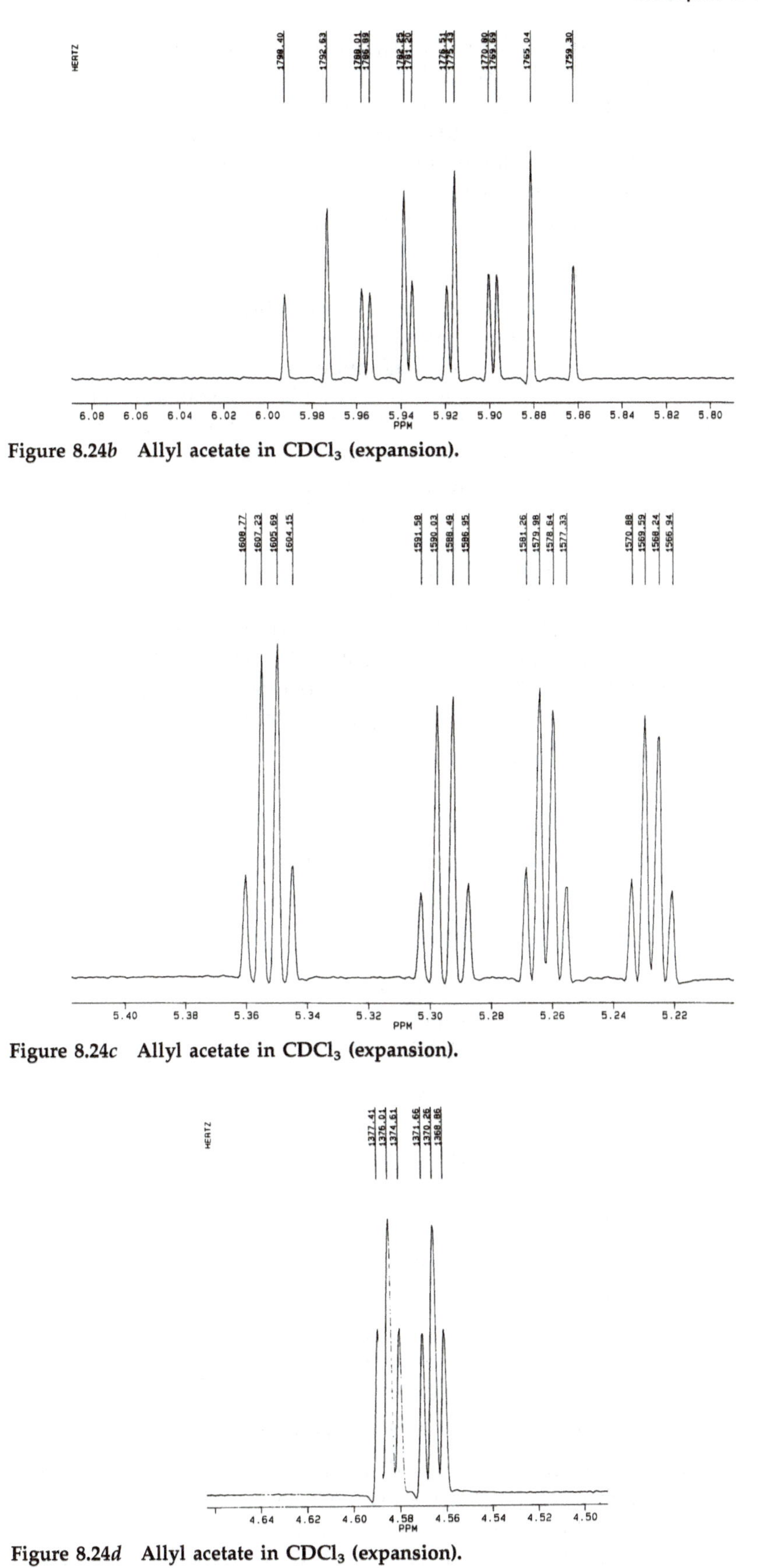


**Figure 8.24*b*** **Allyl acetate in $CDCl_3$ (expansion).**

**Figure 8.24*c*** **Allyl acetate in $CDCl_3$ (expansion).**

**Figure 8.24*d*** **Allyl acetate in $CDCl_3$ (expansion).**

constant for the doublet is 5.8 Hz, which can be measured between any two corresponding peaks in the two triplets.

The four quartets around 5.3 ppm (Fig. 8.24*c*) are actually two doublets of quartets at 5.32 and 5.24 ppm and correspond to $H_A$ and $H_B$ in the structure above. At 5.32 ppm, the multiplet is a doublet of quartets, $J$ = 17.2, 1.5 Hz. At 5.24 ppm, we have another doublet of quartets, $J$ = 10.4, 1.3 Hz. We see quartets because the long-range allylic coupling to the two $H_D$ signals gives a triplet that has a coupling constant $J$ that is approximately equal to the geminal coupling constant (~1.4 Hz) between $H_A$ and $H_B$. Since NMR line widths are naturally several tenths of a hertz, it is not possible to distinguish between coupling constants such as these that differ only by 0.2 Hz. We can unambiguously distinguish $H_A$ and $H_B$ by the magnitudes of their coupling constants to $H_C$, which are 17.2 and 10.4 Hz. Since trans coupling constants are larger than cis coupling constants, $H_A$ must have the 17.2-Hz coupling constant to $H_C$ and is thus assigned to the signal centered at 5.32 ppm. Since $H_B$ is coupled to $H_C$ by $J$ = 10.4 Hz it is assigned to the signal centered at 5.24 ppm.

Finally, we already know what the multiplet for $H_C$ should look like, since we know all of its coupling constants. It is coupled to the two $H_D$ protons with a coupling constant of 5.8 Hz, to $H_A$ with $J$ = 17.2, Hz, and to $H_B$ with $J$ = 10.4 Hz. The multiplet for $H_C$ at 5.93 ppm should be, therefore, a doublet of doublets of triplets with $J$ = 17.2, 10.4, and 5.8 Hz, respectively. There should be $2 \times 2 \times 3 = 12$ lines and the width should be $17.2 + 10.4 + (2 \times 5.8) = 39.2$ Hz. There are indeed 12 lines (Fig. 8.24*b*) and the distance between the outside peaks is 39.1 Hz, which is a perfectly reasonable deviation from the ideal:

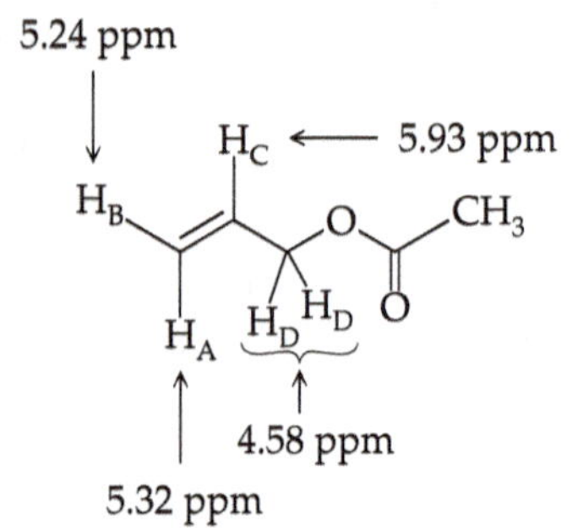

Allyl acetate, chemical shift assignments

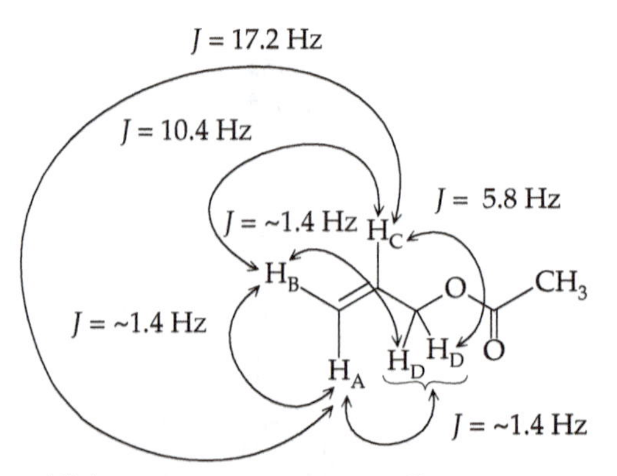

Allyl acetate, proton-proton coupling constants

Since cyclohexane rings are often held in at most two potential conformations (both chairs), coupling constants may allow the determination of relative stereochemistry. On cyclohexane rings in chair conformations, the axial–axial coupling constants for vicinal protons (180° dihedral angle) are on the order of 9–12 Hz.

Equatorial–equatorial and equatorial–axial coupling constants (60° dihedral angles) are on the order of 2–4 Hz. Thus it is often a relatively simple matter to determine stereochemical relationships on a six-membered ring using NMR spectroscopy.

Take, for example, the two diastereomers of 4-*tert*-butylcyclohexanol:

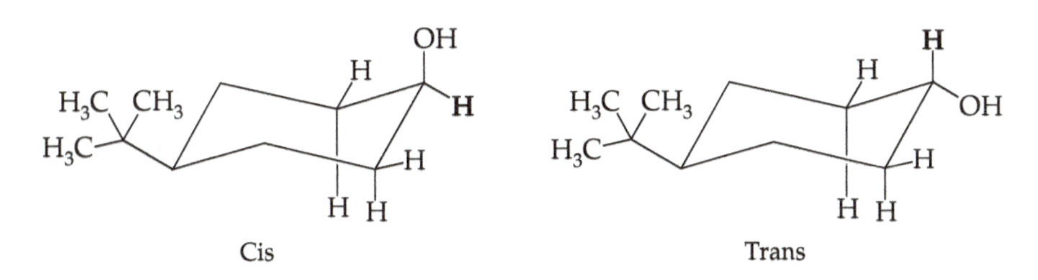


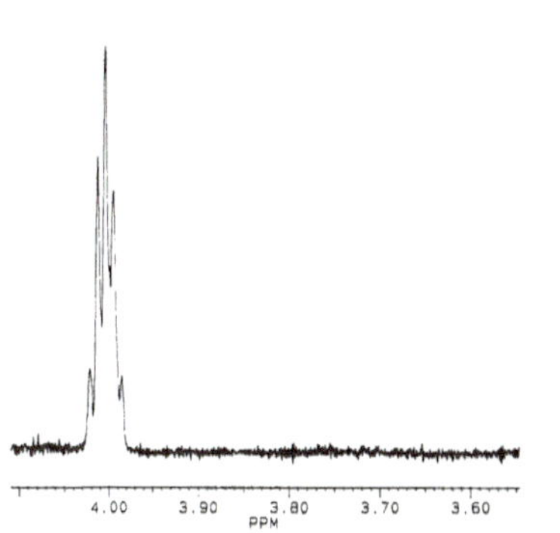


**Figure 8.25** ***cis*-4-*tert*-Butylcyclohexanol.**

We know that the very large *tert*-butyl group will effectively always be equatorial. By examining the coupling to the methine proton of the alcohol, it is simple to determine whether that proton is axial or equatorial. It is also possible to distinguish between these stereoisomers by using chemical-shift information (in one isomer the alcohol methine is seen at ~3.52 ppm and in the other at ~4.04 ppm), but use of coupling information provides a far more definitive and unambiguous determination of stereochemistry.

The alcohol methine proton in the cis isomer is equatorial and thus has a 60° dihedral angle to all four adjacent protons that give rise to a pentet (which is really a doublet of doublets of doublets of doublets with equal coupling constants) with a coupling constant $J$ = ~3 Hz, which is seen in Figure 8.25.

The alcohol methine in the trans isomer is axial and thus has a 180° dihedral angle to each of the two adjacent axial protons and a 60° dihedral angle to each of the two adjacent equatorial protons. This arrangement gives rise to a triplet of triplets with $J$ = ~13 and 3 Hz, which is shown in Figure 8.26.

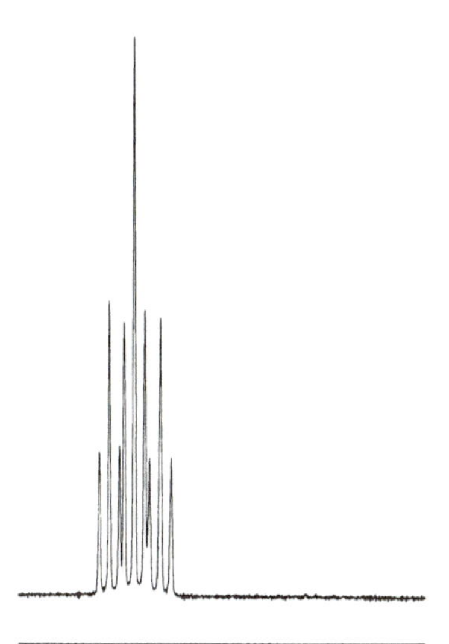

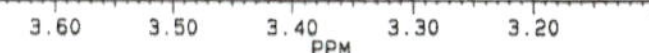


**Figure 8.26** ***trans*-4-*tert*-Butylcyclohexanol.**

## $^{13}$C NMR SPECTROSCOPY

With the advent of Fourier transform (FT) NMR spectrometers, $^{13}$C NMR spectroscopy is now available as a simple and routine tool for the structure determination of organic molecules. Since $^{13}$C is of low natural abundance (1.1%), addition of many spectra is required to obtain acceptable signal-to-noise (*S/N*) levels. With modern spectrometers, $^{13}$C spectra can often be acquired simply by issuing software commands; in some instruments a different probe is inserted into the magnet. Since $^{13}$C resonates at roughly 25% of the proton operating frequency of a spectrometer system, an instrument that acquires $^{1}$H spectra at 300 MHz will be reset to about 75 MHz for $^{13}$C work.

Generally, $^{13}$C NMR spectra are acquired while the entire $^{1}$H frequency range is irradiated by a second rf coil inside the probe assembly. These spectra are referred to as broadband-decoupled $^{13}$C spectra and they do not show the effect of spin–spin coupling to $^{1}$H nuclei. Such decoupling is done because $^{1}$H–$^{13}$C coupling constants can be quite large (a few hundred Hz) relative to chemical-shift differences, which leads to multiplets split over a large portion of the spectrum and subsequent confusion (Fig. 8.27). It is often simpler to see a single line for each distinct carbon atom in a molecule. Furthermore, irradiation of the $^{1}$H spectrum results in signal enhancement of the $^{13}$C signals of the attached carbon atoms. This enhancement is the nuclear Overhauser effect (NOE).

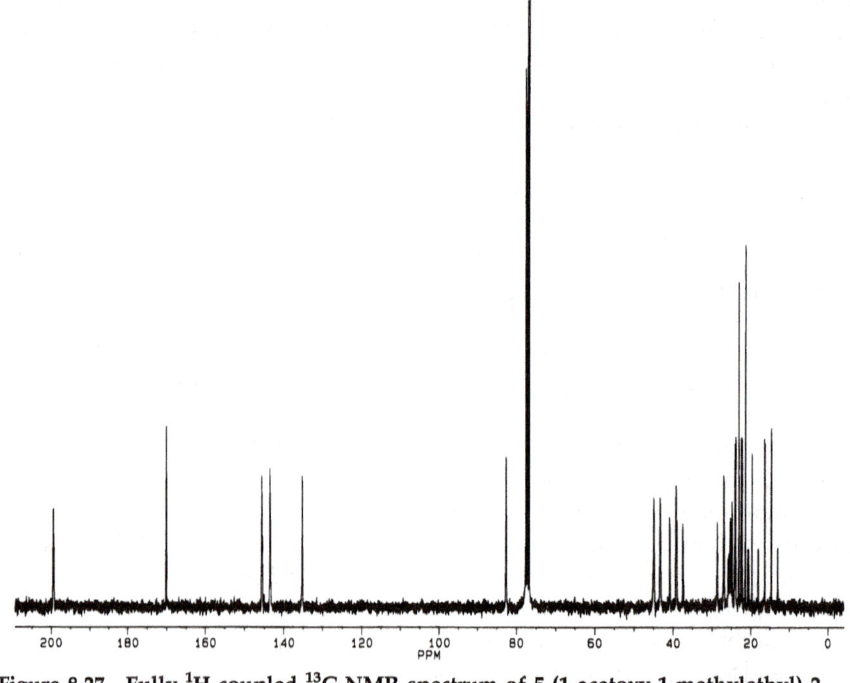


**Figure 8.27 Fully $^{1}H$-coupled $^{13}C$ NMR spectrum of 5-(1-acetoxy-1-methylethyl)-2-methyl-2-cyclohexenone in $CDCl_3$.**

The $^{13}C$ NMR chemical shifts follow the same rough trends as seen in $^{1}H$ chemical shifts. $^{13}C$ chemical shifts, however, are not nearly as amenable to prediction based on the electronegativity of substituents as are $^{1}H$ chemical shifts. The $^{13}C$ chemical shifts are, in general, less sensitive to substituent electronegativities, and are far more sensitive to steric effects than are $^{1}H$ chemical shifts. A brief listing of approximate $^{13}C$ chemical shifts is provided in Table 8.26; a more extensive and thorough listing is available in the Silverstein et al. reference (Bibliography). As in $^{1}H$ NMR spectroscopy, TMS ($Si(CH_3)_4$) is used as the internal reference and the chemical shift of TMS is defined as zero. Except for functional groups such as acetals and ketals, $sp^3$-hybridized carbon atoms appear upfield (to the right) of 100 ppm, and $sp^2$-hybridized carbon atoms appear downfield of 100 ppm. Common carbonyl-containing functional groups appear downfield of 160 ppm. Aldehydes and ketones appear at 195–220 ppm; esters, amides, anhydrides, and carboxylic acids appear at 165–180 ppm.

Typical $^{13}C$ NMR spectroscopy provides an NMR spectrum that is not amenable to integration because of the NOE and insufficient relaxation delays. Therefore, the number of carbon atoms giving rise to a given signal cannot generally be determined by these techniques. It is possible to obtain $^{13}C$ NMR spectra that can be accurately integrated (inverse-gated decoupling), but this experiment requires a great deal of acquisition time to achieve adequate signal-to-noise levels.

Information about C—H coupling can be readily obtained, however. Fully coupled $^{13}C$ NMR spectra are not very useful for structure determination because C—H couplings are large (~120–270 Hz, depending mainly on the hybridization at carbon) and multiplets tend to overlap. Furthermore, when the hydrogen atoms are not irradiated, there is no NOE, and the signal-to-noise ratio suffers significantly. The most common use for coupling information is to

*Table 8.26* **Approximate $^{13}C$ NMR Chemical Shifts**

| Functional Group | Carbon[a] | Chemical Shift/ δ (ppm) |
|---|---|---|
| Alkyl carbon atoms | | ~5–45 |
| | 1° R—**CH**$_3$ | ~5–30 |
| | 2° R—**CH**$_2$—R′ | ~15–35 |
| | 3° R—**C**HR′R″ | ~20–40 |
| | 4° R**C**R′R″R‴ | ~25–45 |
| Alkenyl carbon atoms | | ~110–150 |
| | H$_2$**C**=C | ~100–125 |
| | HR**C**=C | ~125–145 |
| | RR′**C**=C | ~130–150 |
| Aromatic carbon atoms | | ~120–160 |
| Alkynyl carbon atoms | **C**≡C | ~65–90 |
| Nitriles | R—**C**≡N | ~115–125 |
| Alcohols and ethers | **C**—OH(R) | ~50–75 |
| | C—O (epoxides) | ~35–55 |
| Amines | **C**—N | ~30–55 |
| Alkyl halides | **C**—X | ~0–75 |
| Carbonyl groups | **C**=O | ~165–220 |
| Ketones, aldehydes | R**C**OR′, R**C**HO | 195–220 |
| Carboxylic acids, esters | R**C**O$_2$H, R**C**O$_2$R′ | 165–180 |
| Amides, anhydrides | R**C**ON, R**C**O$_2$OCR′ | 160–175 |

[a]R = alkyl group.

determine the number of protons attached to a given carbon atom. This can be done in a variety of ways, some of which do not actually display the carbon signals as multiplets due to coupling to attached protons.

Single-frequency off-resonance decoupling (SFORD) is a useful technique for determining the number of hydrogen atoms attached to a given carbon. The decoupler is tuned off to one side of the proton spectrum and the sample is irradiated at a single frequency giving rise to $^{13}C$ spectra that show C—H couplings as a fraction of their actual values and that show a partial NOE. The apparent C—H coupling is dependent on both the actual coupling constant and the difference between the decoupler frequency and the resonance frequency of the hydrogen in question. The major disadvantage of SFORD is its low signal-to-noise ratio, which is due to two factors. First, there is only a partial NOE. Second, when NMR signals are split into multiplets, the signal intensity becomes distributed among several peaks. Thus, SFORD spectra require significantly more spectral acquisitions than do fully decoupled $^{13}C$ NMR spectra, and to some extent have been replaced with distortionless enhancement by polarization transfer (DEPT) spectra.

DEPT $^{13}C$ NMR spectroscopy provides a rapid way of determining the number of hydrogen atoms attached to a given carbon atom. DEPT spectra result from a multiple-pulse sequence that terminates in a "read pulse," which can be varied according to the spectrum desired. In DEPT spectra, all peaks are singlets; quaternary carbon atoms (without attached hydrogen atoms) are not seen in any DEPT spectra. In DEPT-135° spectra, CH and $CH_3$ groups appear as singlets of positive intensity, and $CH_2$ groups appear as negative peaks. The

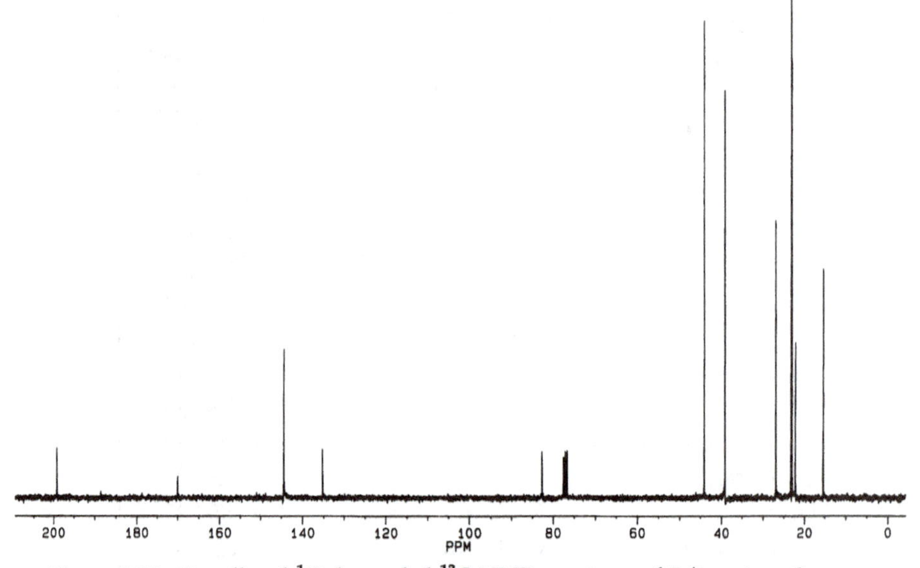


**Figure 8.28 Broadband $^1$H-decoupled $^{13}$C NMR spectrum of 5-(1-acetoxy-1-methylethyl)-2-methyl-2-cyclohexenone in $CDCl_3$.**

DEPT-90° spectra show only CH groups, and thus allow $CH_3$ and CH groups to be distinguished. In combination with a routine fully decoupled spectrum, DEPT spectra allow unambiguous assignment of the number of hydrogen atoms attached to each carbon. In practice, such spectral editing techniques are not perfect, and small residual peaks are often seen where, in principle, there should be none; these are usually small enough to be readily distinguished from the "real" peaks.

The fully coupled $^{13}$C NMR spectrum of the acetoxy-enone (**I**) is shown in Figure 8.27, the broadband decoupled spectrum in Figure 8.28, and the SFORD spectrum in Figure 8.29. The DEPT-135° spectrum is shown in Figure 8.30, and the DEPT-90° spectrum in Figure 8.31. The 1:1:1 triplet centered at 77 ppm is due to the solvent, $CDCl_3$.

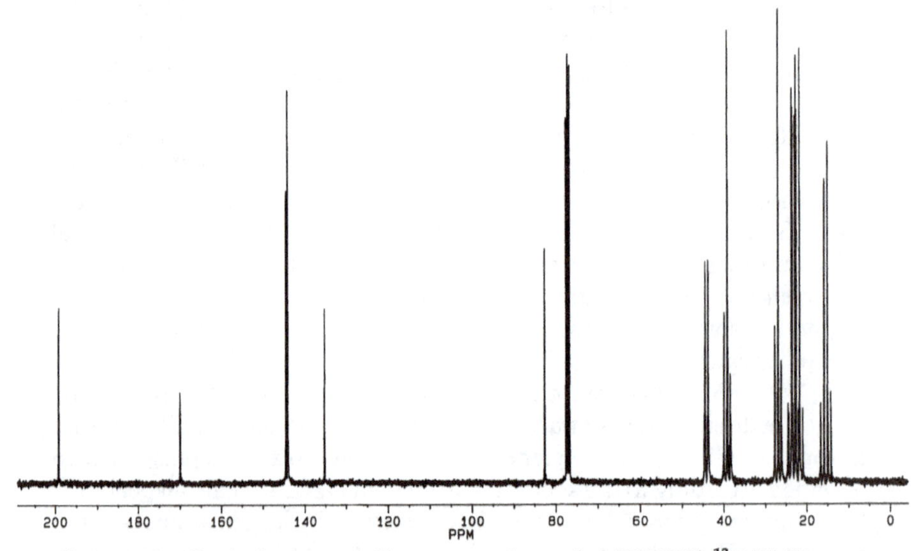


**Figure 8.29 Single-frequency off-resonance decoupled (SFORD) $^{13}$C NMR spectrum of 5-(1-acetoxy-1-methylethyl)-2-methyl-2-cyclohexenone in $CDCl_3$.**

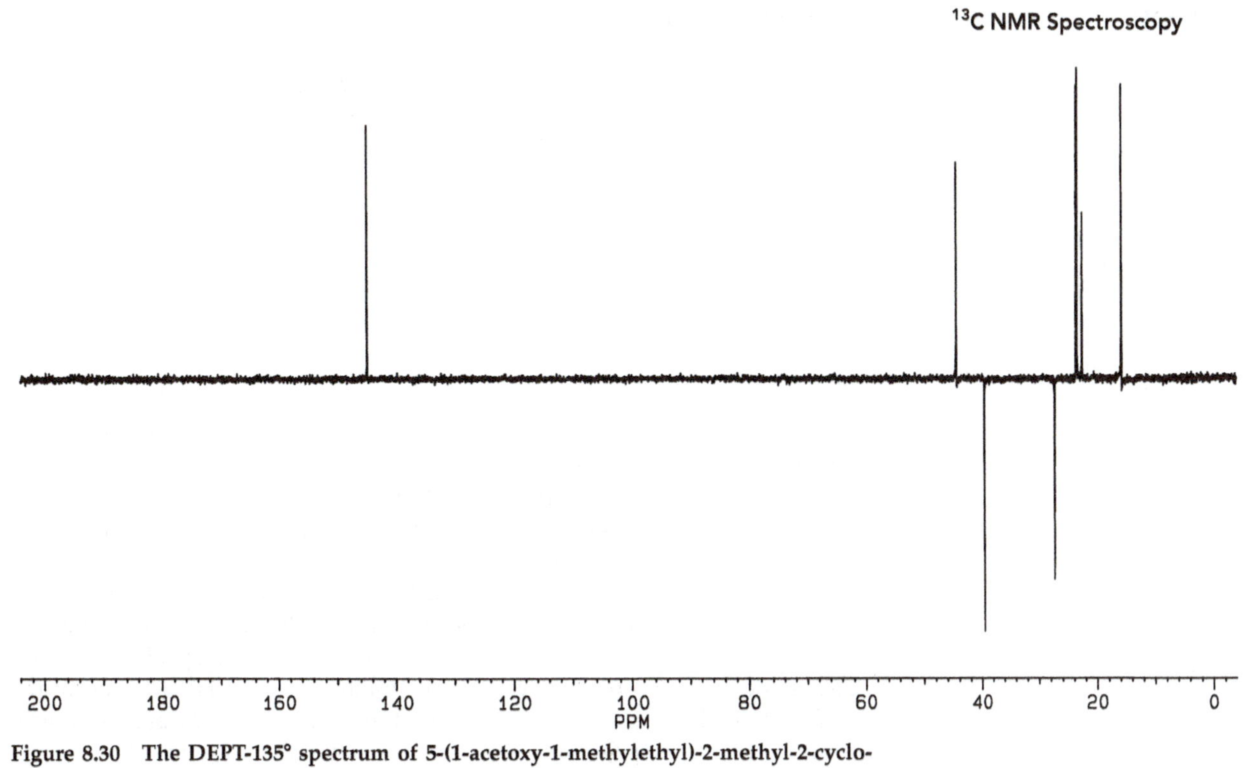


**Figure 8.30** The DEPT-135° spectrum of 5-(1-acetoxy-1-methylethyl)-2-methyl-2-cyclohexenone in $CDCl_3$.

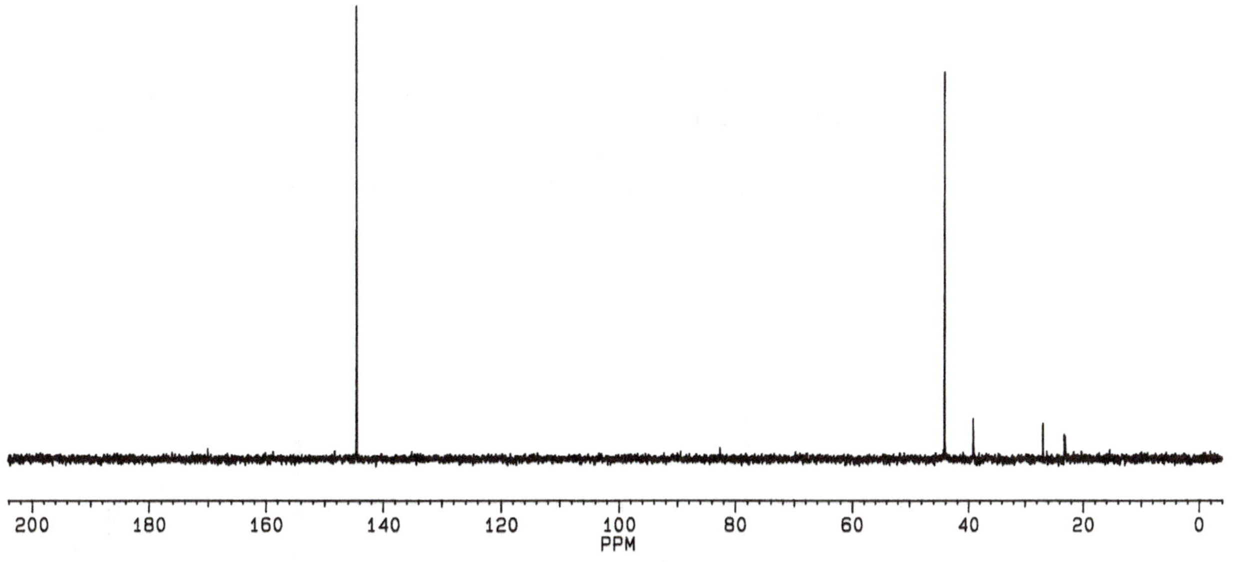


**Figure 8.31** The DEPT-90° spectrum of 5-(1-acetoxy-1-methylethyl)-2-methyl-2-cyclohexenone in $CDCl_3$.

Interpretation and assignment of the $^{13}C$ NMR spectrum are much easier when we unambiguously know how many protons are attached to each carbon.

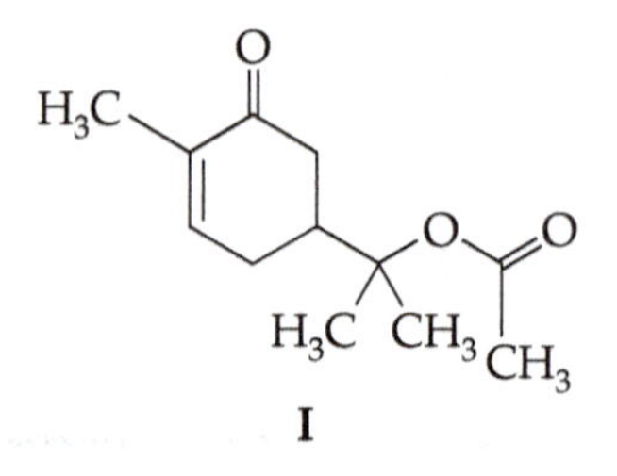


The $^{13}C$ NMR is often better than $^{1}H$ NMR for distinguishing functional groups because typical $^{13}C$ chemical shifts are in the range 0–200 ppm relative to TMS, as compared to 0–10 ppm for proton chemical shifts. Coupling between adjacent $^{13}C$ nuclei is not observed (except in isotopically enriched samples) because the probability of having two rare isotopes adjacent to one another is very small. Because of the absence of decoupling, $^{13}C$ spectra are less complex than $^{1}H$ spectra, and $^{13}C$ spectra are often better suited for the detection and identification of isomeric or other impurities in a sample; it is easy for small peaks to be concealed underneath a complex second-order multiplet in the $^{1}H$ NMR spectrum.

The 300-MHz $^{1}H$ spectrum of 4-cyclohexene-*cis*-1,2-dicarboxylic acid anhydride is shown in Figure 8.32. Owing in part to the presence of two stereocenters, as well as to long-range coupling through the $\pi$ system of the alkene, the entire $^{1}H$ spectrum is second order at this field strength, and no information is available from the coupling constants because the spectrum is too complex. Limited assignments to peaks could be made on the basis of chemical shift, but it would be difficult to make any statements regarding purity of our sample based on the $^{1}H$ NMR spectrum, because an impurity could easily be hidden underneath any of the complex signals.

The $^{13}C$ spectrum of 4-cyclohexene-*cis*-1,2-dicarboxylic acid anhydride (Figure 8.33) is much less complex. Because of the mirror plane of symmetry in the compound, there are only four different carbon atoms and thus only four lines are seen in the fully decoupled $^{13}C$ NMR spectrum. This simplicity makes it easy to detect isomeric or other impurities in this sample. These impurities were not as easy to detect in the $^{1}H$ NMR spectrum. The 1:1:1 triplet centered at 77 ppm is due to the solvent, $CDCl_3$. Although no $^{1}H$–$^{13}C$

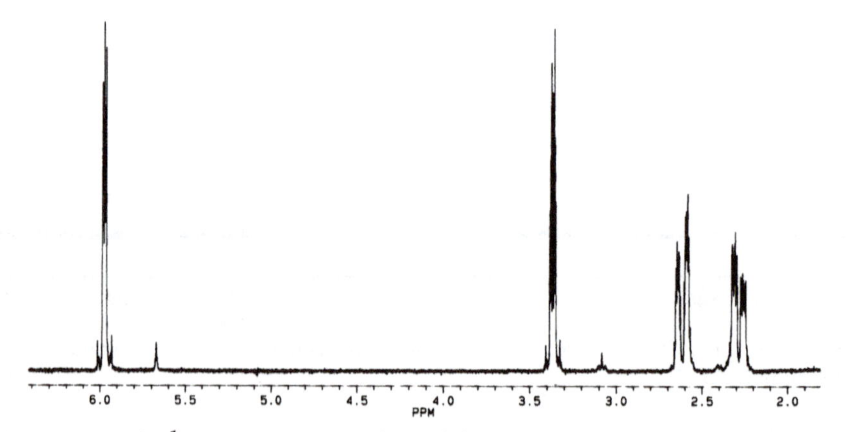


**Figure 8.32 The $^{1}H$ NMR spectrum of 4-cyclohexene-cis-1,2-dicarboxylic acid anhydride in $CDCl_3$.**

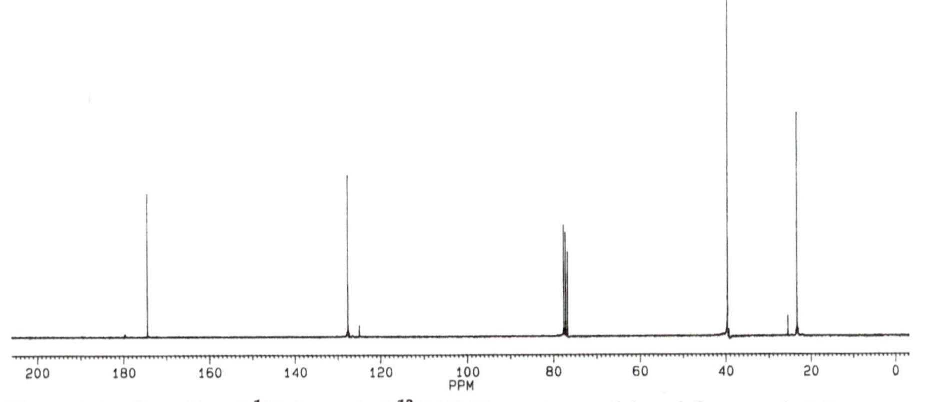


**Figure 8.33 Broadband $^{1}H$-decoupled $^{13}C$ NMR spectrum of 4-cyclohexene-cis-1,2-dicarboxylic acid anhydride in $CDCl_3$.**

coupling is observed because of the $^{1}H$ broadband decoupling, $^{2}H$–$^{13}C$ coupling is observed because $^{1}H$ and $^{2}H$ resonate at different frequencies.

## TWO-DIMENSIONAL NMR SPECTROSCOPY

Two significant developments in NMR spectroscopy are the use of Fourier transform techniques, and the development of two-dimensional (2D) NMR spectroscopy. Two-dimensional spectra are obtained using a sequence of rf pulses that includes a variable delay or delays. A set of FIDs is acquired and stored. The variable delay is incremented by a small amount of time and a new set of FIDs are obtained and stored, and so on. At the end, the resulting matrix of FID data is Fourier transformed twice: once with respect to the acquisition time (as in normal FT-NMR) and second with respect to the time of the variable delay in the pulse sequence. The resulting data represent a surface and are presented as a contour plot of that surface.

The most useful 2D spectra for organic compound identification are called correlation spectra. Correlation Spectroscopy (COSY) spectra are presented as a contour plot with routine proton spectra along both of the axes, as shown in the COSY spectrum of ethyl vinyl ether in Figure 8.34. The spectra along the axes are low-digital-resolution spectra and appear to be a bit different from those generated as usual NMR spectra. Note that the 2D spectrum is symmetric about the diagonal that runs from the lower left corner to the upper right corner. Every peak is represented by a peak along the diagonal and, in fact, the diagonal *is* the normal proton spectrum. Where the contour plot indicates a peak other than on the diagonal, the interpretation is that the corresponding peaks on the two axes represent protons coupled to one another.

For example, draw a line down from the signal at 1.2 ppm on the horizontal axis. This line encounters the diagonal at 1.2 ppm and then there is a peak at 3.7 ppm. By drawing a horizontal line over to the spectrum on the vertical axis, we can see that the peaks at 1.2 and 3.7 ppm are coupled to one another; these are the signals from the methyl triplet and methylene quartet of the ethyl group.

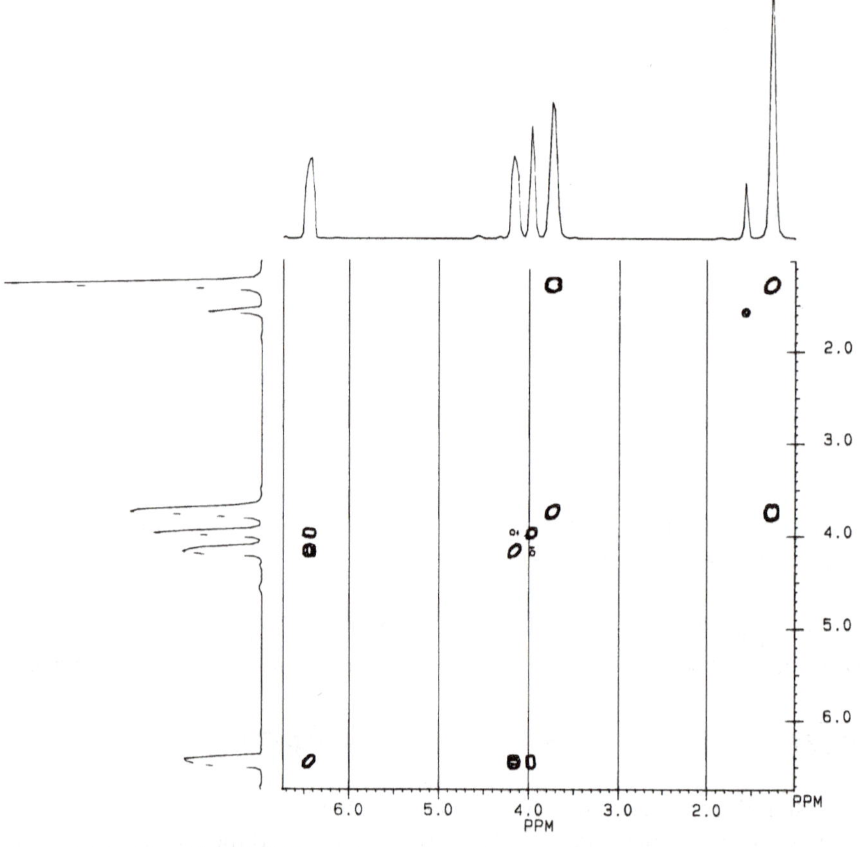


**Figure 8.34 The COSY NMR spectrum of ethyl vinyl ether.**

We can also see that neither of these is coupled to the remainder of the spectrum, which of course is the vinyl group. Each proton in the vinyl group is coupled to each of the others, as we can see in the COSY spectrum. For example, the signal at 4.2 ppm is coupled to both the signal at 4.0 ppm and the one at 6.4 ppm.

It is possible to obtain 2D spectra that correlate the spectra of different nuclei, such as $^{1}H$ and $^{13}C$. The heteronuclear correlation spectrum of ethyl vinyl ether is shown in Figure 8.35. The $^{13}C$ spectrum is along the horizontal axis and the $^{1}H$ spectrum is along the vertical axis. The peaks of the 2D spectrum indicate that the corresponding peaks on the axes represent a carbon and a proton (or protons) that are directly bonded. By using this spectrum we can easily verify the $^{13}C$ and $^{1}H$ chemical-shift assignments below:

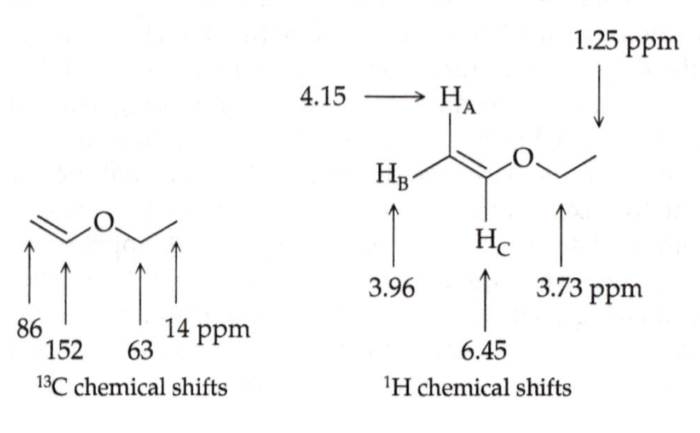

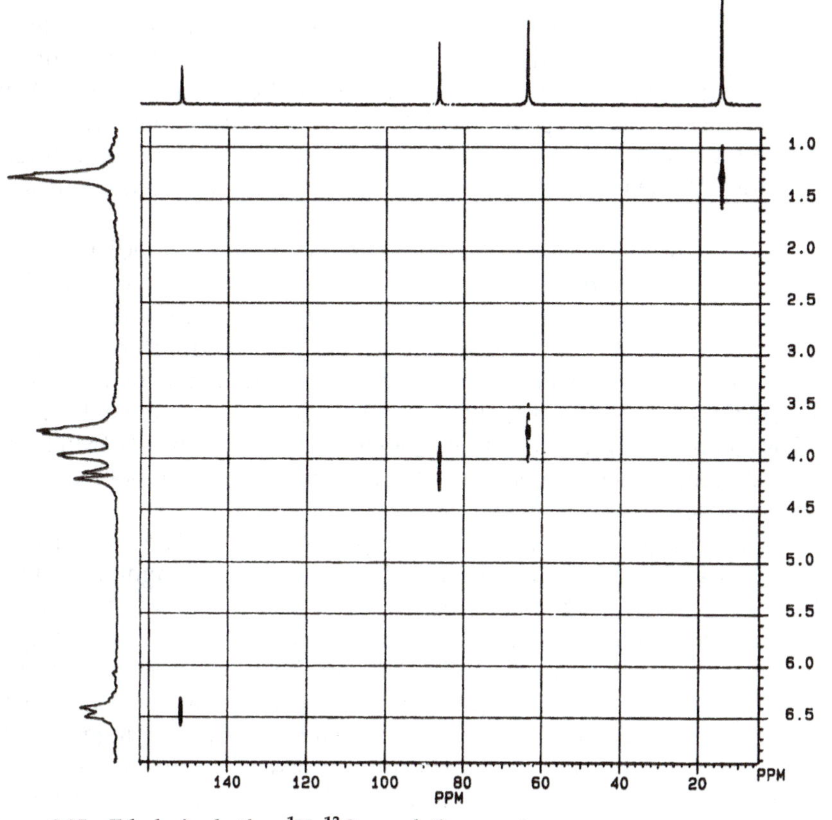


**Figure 8.35** **Ethyl vinyl ether $^{1}H$–$^{13}C$ correlation spectrum.**

There are many other powerful 2D NMR spectroscopic techniques that can provide a wealth of information about molecular structure, even in organic molecules as large as proteins and nucleic acids. A few of the good texts that provide further information about these powerful tools are listed in the Bibliography.

## NUCLEAR MAGNETIC RESONANCE SAMPLING

It is usually simple to prepare a sample of a small organic molecule (MW 500) for NMR analysis. For $^{1}H$ NMR spectroscopy, the sample size compatible with CW spectrometers is in the range 30–50 mg dissolved in about 0.5 mL of solvent. Fourier transform spectrometers require only 2–3 mg of sample in the same volume of solvent. Most samples are measured in solution in thin-walled tubes 5 mm in diameter and 18–20 cm long. NMR sample tubes are expensive and delicate; they must be as perfectly straight, and have as perfectly concentric walls, as possible. The sample tube is filled to a depth of about 3 cm. Filling the tube to this depth maximizes sample concentration in the active part of the NMR probe, and thus the strength of the signal. Adding more solvent just wastes sample by dilution. The sample tube is spun about its axis in the instrument to average out small changes in the tube walls and in the magnetic field strength over the sample volume. There must be enough solvent in the tube to ensure that the vortex, or whirlpool, created when the tube is spun, does not extend down into the portion of the tube where the rf coils in the NMR probe are active. Many instruments use a depth gauge that shows exactly where this area is. Glass microcells are now commercially available. An inexpensive microcell technique for CW-NMR spectrometers using ordinary 5-mm NMR tubes has been described.[2]

The most practical NMR solvent is deuterochloroform ($CDCl_3$); it is relatively cheap and dissolves many different compounds. Handle this solvent with care, in the **hood,** because it is **toxic!** Many other deuterated solvents are commercially available, including acetone, methanol, and water. The universally accepted internal reference compound employed in making these measurements is tetramethylsilane, $Si(CH_3)_4$ (TMS). The most convenient source of TMS is commercially available $CDCl_3$, which contains about 1% TMS for use with CW spectrometers (commercially available 0.03% solutions are more appropriate for FT spectrometers).

HOOD

The most significant problem in sample preparation is the exclusion of small pieces of dust and dirt, because they may contain magnetic material, which will result in poor spectra. Scrupulously clean samples of liquids can often simply be added to the NMR tube, followed by the solvent. Liquids containing visible impurities, and solids, are best prepared by dissolving the sample in about 0.3–0.4 mL of solvent in a small vial. The solution can then be filtered into the NMR tube through a Pasteur pipet plugged with a small piece of cotton, and the pipet can be rinsed with the NMR solvent to achieve the appropriate volume in the NMR tube. Since TMS is volatile (bp = 26.5 °C), the tube should be capped following addition, or the TMS will evaporate. If several people in a laboratory section are going to be obtaining NMR spectra, you will find that the spectrometer will be easier to tune with each new sample if all the sample tubes are filled to exactly the same level.

At this point, all specific instructions are dependent on the NMR spectrometer available to you. In general, the sample is inserted into the magnet and the magnetic field is adjusted very slightly (called shimming or tuning) to obtain a magnetic field that is as homogeneous as possible throughout the sample volume observed by the spectrometer. These adjustments are accomplished by energizing a collection of small electromagnetic coils around the sample, a process that is often done by the spectrometer's computer. Symptoms of a poorly shimmed spectrometer include broad and/or asymmetric peaks. The best place to check for this is on the TMS peak because it is the one peak in the spectrum you *know* should be narrow and symmetric. Once the spectrometer is tuned, the spectrum is obtained and plotted, and the sample is removed from the magnet.

Your NMR sample can be easily recovered by emptying the NMR tube into a small vial, rinsing the tube once or twice with (a nondeuterated) solvent, and then evaporating the solvent in a **hood** under a gentle stream of dry nitrogen.

HOOD

[2]Yu, S. J. *J. Chem. Educ.* **1987,** *64,* 812.

## QUESTIONS

Several 60-MHz $^{1}$H NMR spectra are given below (Figs. 8.36–8.40) along with the molecular formula of the compound.[3] You should be able to account for at least one acceptable structure and for all of the observed resonances.

**8-16.** $C_4H_8O$. Spectrum *a:*

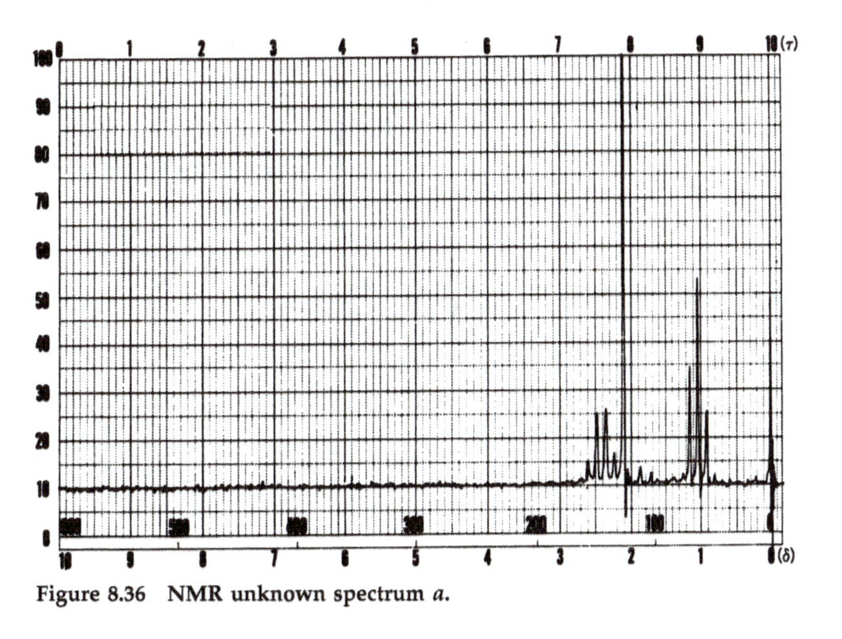


**Figure 8.36 NMR unknown spectrum *a*.**

**8-17.** $C_3H_6O_2$. Spectra *b* and *c*. Two compounds with the same empirical formula:

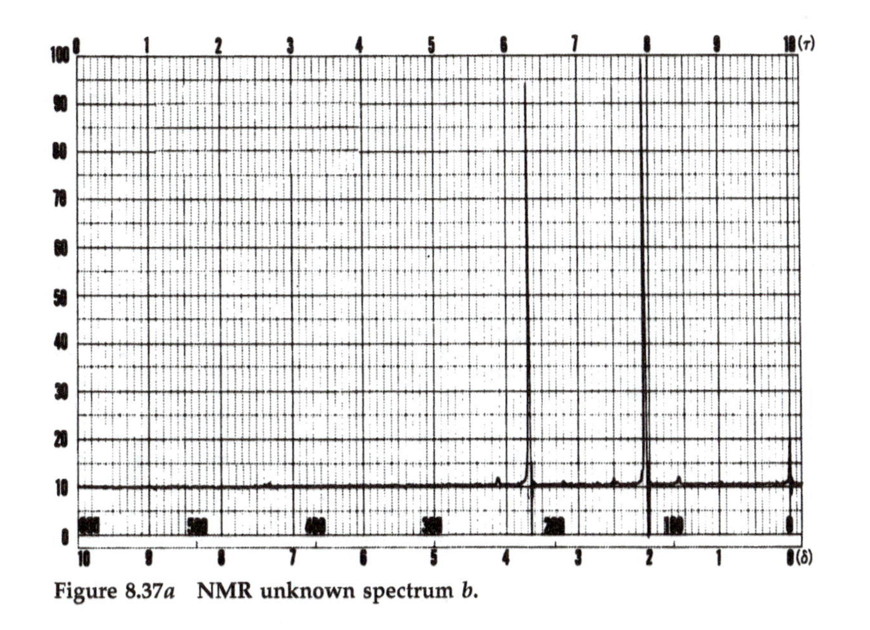


**Figure 8.37*a* NMR unknown spectrum *b*.**

[3]From Pouchert, C. J. *The Aldrich Library of NMR Spectra,* 2nd ed.; Aldrich Chemical Co.: Milwaukee, WI, 1983.

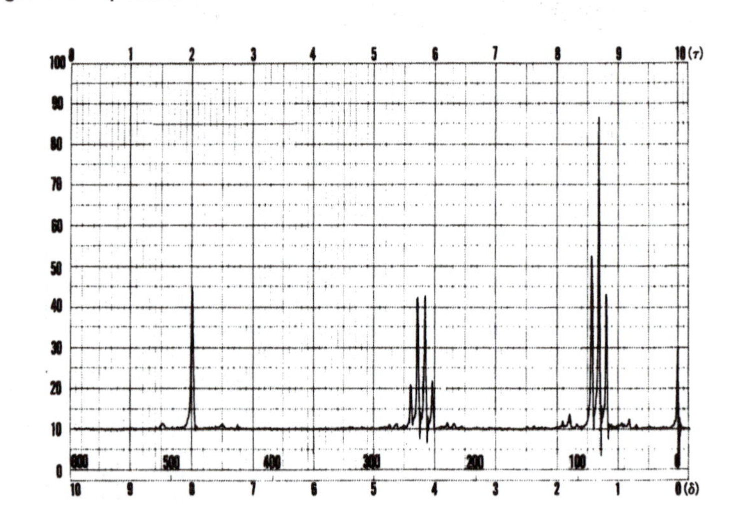

**Figure 8.37*b*** **NMR unknown spectrum *c*.**

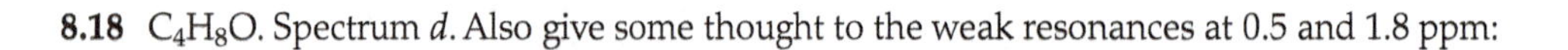

**8.18** $C_4H_8O$. Spectrum *d*. Also give some thought to the weak resonances at 0.5 and 1.8 ppm:

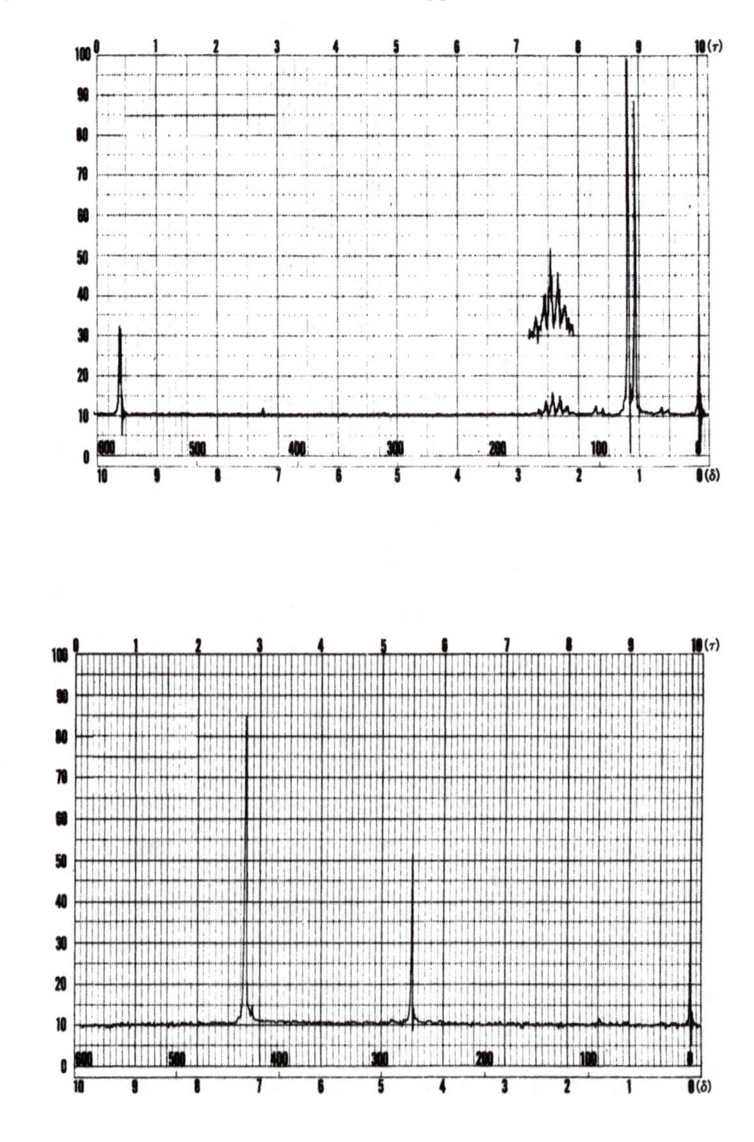

**Figure 8.38** **NMR unknown spectrum *d*.**

**8-19** $C_7H_7Cl$. Spectrum *e:*

**Figure 8.39** **NMR unknown spectrum *e*.**

**8-20.** $C_8H_{10}O$. Spectrum *f*:

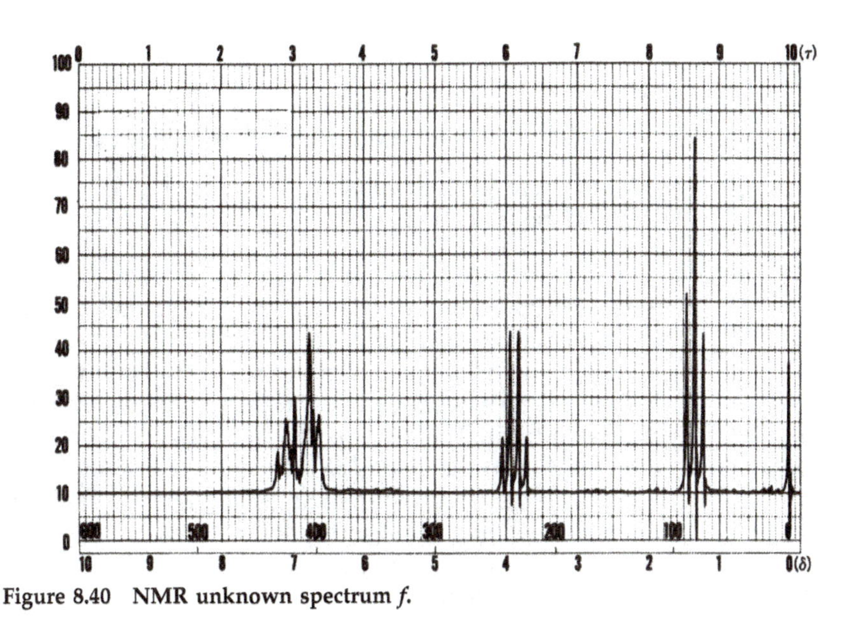


**Figure 8.40 NMR unknown spectrum *f*.**

## BIBLIOGRAPHY

**Nuclear magnetic resonance theory and principles of interpretation:**

Abraham, R. J.; Fisher, J.; Loftus, P. *Introduction to NMR Spectroscopy;* Wiley: London, 1988.

Cooper, J. W. *Spectroscopic Techniques for Organic Chemists;* Wiley: New York, 1980.

Field, L. D.; Sternhell, S.; Kalman, J. R. *Organic Structures from Spectra,* 3rd ed.; Wiley: London, 2002.

Richards, S. A. *Laboratory Guide to Proton NMR Spectroscopy;* Blackwell Scientific Publications: London, 1988.

Silverstein, R. M.; Webster, F. X.; Kiemle, D. J. *Spectrometric Identification of Organic Compounds,* 7th ed.; Wiley: New York, 2005.

Sorrell, T. N. *Interpreting Spectra of Organic Molecules;* University Science Books: Mill Valley, CA, 1988.

**Advanced theory and spectroscopic techniques:**

Atta-ur-Rahman *Nuclear Magnetic Resonance. Basic Principles;* Springer-Verlag: New York, 1986.

Claridge, T. D. W. *High-Resolution NMR Techniques in Organic Chemistry. Tetrahedron Organic Chemistry Series Volume 19;* Pergamon: Oxford, UK, 1999.

Derome, A. E. *Modern NMR Techniques for Chemistry Research;* Pergamon Press: Oxford, UK, 1987.

Duddeck, H.; Dietrich, W.; Toth, G. *Structure Elucidation by Modern NMR. A Workbook,* 3rd ed. Springer-Verlag: New York, 1998.

Sanders, J. K. M.; Hunter, B. K. *Modern NMR Spectroscopy. A Guide for Chemists;* 2nd ed.; Oxford University Press: Oxford, UK, 1993.

Sanders, J. K. M.; Constable, E. C.; Hunter, B. K. *Modern NMR Spectroscopy. A Workbook of Chemical Problems;* 2nd ed.; Oxford University Press: Oxford, UK, 1993.

**Libraries of NMR spectra:**

Bhacca, N. S.; Hollis, D. P.; Johnson, L. F.; Pier, E. A.; Shoolery, J. N. *NMR Spectra Catalog;* Varian Associates: Palo Alto, CA, 1963.

Pouchert, C. J. *The Aldrich Library of NMR Spectra,* 2nd ed.; Aldrich Chemical Co.: Milwaukee, WI, 1983.

Pouchert, C. J.; Behnke, J. *The Aldrich Library of $^{13}C$ and $^{1}H$ FT-NMR Spectra;* Aldrich Chemical Co.: Milwaukee, WI, 1992.

**Online libraries of NMR spectra:**

Online Resource Guide (NMR and Other Spectra (http://www.library.illinois.edu/chx/onlineresources/nmr.html))

SciFinder Scholar (https://scifinder.cas.org/)

## ULTRAVIOLET–VISIBLE SPECTROSCOPY: INTRODUCTION TO ABSORPTION SPECTROSCOPY

In an atom, molecule, or ion, a limited number of electronic energy states are available to the system because of the quantized nature of the energies involved. The absorption of a photon by the system can be interpreted as corresponding to the occupation of a new energy state by an electron. The difference in energy between these two states may be expressed as $\Delta E$:

——— Upper state (excited electronic state, $E_1$)

$\updownarrow$ $\Delta E$

——— Lower state (ground electronic state, $E_0$)

where the energy of the photon, $E$, is related to the frequency of the radiation by the Planck equation,

$$E = h\nu_i$$

where $h$ is Planck's constant, $6.626 \times 10^{-34}$ J s, and $\nu_i$ is the frequency in hertz. In the case above, $\Delta E = E_1 - E_0 = h(\nu_1 - \nu_0) = h\nu_i$.

Thus, when a frequency match between the radiation and an energy gap ($\Delta E$) in the substance occurs, a transition between the two states involved may be induced. The system can either absorb or emit a photon corresponding to $\Delta E$, depending on the state currently occupied (emission would occur if the system relaxed from an upper-level excited state to a lower state). All organic molecules absorb photons with energies corresponding to the visible or ultraviolet regions of the electromagnetic spectrum, but to be absorbed, the incident energy in this frequency range must correspond to an available energy gap between an electronic ground state and an upper-level electronic excited state. The electronic transitions of principal interest to the organic chemist are those that correspond to the excitation of a single electron from the highest occupied molecular orbital (HOMO) to the lowest unoccupied molecular orbital (LUMO). As we will see, this will be the molecule's absorption occurring at the longest wavelength in the electronic absorption spectrum; it is, therefore, the most easily observed.

Electromagnetic radiation can be defined in terms of a frequency $\nu$, which is inversely proportional to a wavelength $\lambda$ times a velocity $c$ ($\nu = c/\lambda$, where $c$ is the velocity of light in a vacuum, $2.998 \times 10^8$ m/s, and $c = \nu\lambda$ is the wave velocity). Thus,

$$\Delta E = h\nu = \frac{hc}{\lambda} = hc\tilde{\nu}$$

where $\tilde{\nu}$ is the wavenumber, defined as the reciprocal of the wavelength $(1/\lambda)\times$ the velocity of light.

Most ultraviolet and visible (UV and vis) spectra are recorded linearly in wavelength, rather than linearly in frequency or in units proportional to frequency (the wavenumber) or in energy values. Wavelength in this spectral region is currently expressed in nanometers (nm, where 1 nm = $10^{-9}$ m) or angstrom units (Å, where 1 Å = $10^{-10}$ m). The older literature is full of UV–vis spectra in which wavelength is plotted in millimicrons (mμ), which are also equivalent to $10^{-9}$ m. For a further discussion of the relationship between frequency, wavelength, wavenumber, and refractive index, see the discussion on infrared spectroscopy.

It is unfortunate that because of instrumentation advantages this region of the spectrum is most often plotted in units that are nonlinear in energy (note the inverse relationship of $E$ to $\lambda$) A convenient formula for expressing the relationship of wavelength and energy in useful values is

$$E = 28{,}635/\lambda \text{ kcal/mol} \qquad (\lambda \text{ in nm})$$

or in terms of wavenumbers

$$E = (28.635 \times 10^{-4})\tilde{\nu} \qquad (\tilde{\nu} \text{ in cm}^{-1})$$

**Table 8.27 Spectroscopic Wavelength Ranges**

| Region | Wavelength (m) | Energy (kJ/mol) | Change Excited |
|---|---|---|---|
| Gamma ray | Less than$10^{-10}$ | $> 10^6$ | Nuclear transformation |
| X-ray | $10^{-8}$–$10^{-10}$ | $10^4$–$10^6$ | Inner shell electron transitions |
| Ultraviolet (UV) | $4 \times 10^{-7}$–$1 \times 10^{-8}$ | $10^3$–$10^4$ | Valence shell electrons |
| Visible (vis) | $8 \times 10^{-7}$–$4 \times 10^{-7}$ | $10^2$–$10^3$ | Electronic transitions |
| Infrared (IR) | $10^{-4}$–$2.5 \times 10^{-6}$ | 1–50 | Bond vibrations |
| Microwave | $10^{-2}$–$10^{-4}$ | 10–1000 | Molecular rotations |
| ESR | $10^{-2}$ | 10 | Electron spin transitions |
| NMR | 0.5–5 | 0.02–0.2 | Nuclear spin transitions |

The electromagnetic spectrum and the wavelength ranges corresponding to a variety of energy-state transitions are listed in Table 8.27. Infrared, UV–vis, and rf are of particular interest to the organic chemist because the excitation of organic substances by radiation from these regions of the spectrum can yield significant structural information about the molecular system being studied.

The absorption of rf energy by organic molecules immersed in strong magnetic fields involves exceedingly small energy transitions (~0.05 cal/mol), which correspond to nuclear spin excitations and result in NMR spectra. When a molecule absorbs microwave radiation, the energy states available for excitation correspond to molecular rotations and involve energies of roughly 1 cal/mol. With relatively simple molecules (in the gas phase) possessing a dipole moment (required for the absorption process) the analysis of the microwave spectrum can yield highly precise measurements of the molecular dimensions (bond lengths and angles). Unfortunately, relatively few organic systems exhibit pure rotational spectra that can be rigorously interpreted.

Absorption of radiation in the infrared region of the spectrum involves the excitation of vibrational energy levels and corresponds to energies in the range of about 1–12 kcal/mol. The excitation of electronic states requires considerably higher energies, from a little below 40 to nearly 300 kcal/mol. The corresponding radiation wavelengths would fall across the visible (400–800 nm), the near-UV (200–400 nm), and the far- (or vacuum) UV (100–200 nm) regions.The long-wavelength visible and near-UV regions of the spectrum hold information of particular value to the organic chemist. Here the energies correspond to the excitation of loosely held bonding ($\pi$) or lone-pair electrons. The far-UV region, however, involves high-energy transitions associated with the inner-shell and $\sigma$-bond electronic energy transitions. This region is difficult to access because atmospheric oxygen begins to absorb UV radiation below 190 nm, which requires working in evacuated or purged instruments (which is why this region is often referred to as the vacuum UV).

## UV–VIS SPECTROSCOPY

As we have seen, the application of electronic absorption spectroscopy in organic chemistry is restricted largely to excitation of ground-state electronic levels in the near-UV and vis regions. When photons of these energies are absorbed, the excited electronic states that result have bond strengths appreciably less than their ground-state values, and the internuclear distances and bond angles will be altered within the region of the molecules where the electronic excitation occurs (see Figure 8.41). It is normally reasonable to assume that nearly all of the molecules are present in the ground vibrational state within the ground electronic state. The upper electronic state also contains a set of vibrational levels and any of these may be open to occupation by the excited electron (see Figure 8.41). Thus, an electronic transition from a particular ground-state level can be to any number of upper-level vibrational states on the excited electronic state.

The shape of an electronic absorption band will be determined to a large extent by the spacing of the vibrational levels and the distribution of band intensity over the vibrational sublevels. In most cases these effects lead to broad absorption bands in the UV–vis region.

The wavelength maximum at which an absorption band occurs in the UV–vis region is generally referred to as the $\lambda_{max}$ of the sample (where wavelength is determined by the band maximum).

The quantitative relationship of absorbance (the intensity of a band) to concentration is expressed by the Beer–Lambert equation:

$$A = \log\frac{I_0}{I} = \varepsilon c l$$

*where*

$A$ = absorbance, expressed as $I_0/I$

$I_0$ = the intensity of the incident light

$I$ = the intensity of the light transmitted through the sample

$\varepsilon$ = molar absorbtivity, or the extinction coefficient (a constant characteristic of the specific molecule being observed); values for conjugated dienes typically range from 10,000 to 25,000

$c$ = concentration (mol/L)

$\lambda$ = length of sample path (cm)

The calculated extinction coefficient and solvent are usually listed with the wavelength at the band maximum. For example, data for methyl vinyl ketone (3-buten-2-one) would be reported as follows:

$\lambda_{max}$ 219 nm ($\varepsilon$ = 3600, ethanol)
$\lambda_{max}$ 324 nm ($\varepsilon$ = 24, ethanol)

Typical UV–vis spectra are shown in Experiments [6], [19D], [A2$_a$], and [A3$_a$]. As part of the characterization data, UV–vis information is also given in Experiments[16], [A1$_a$], [19A], and [33A].

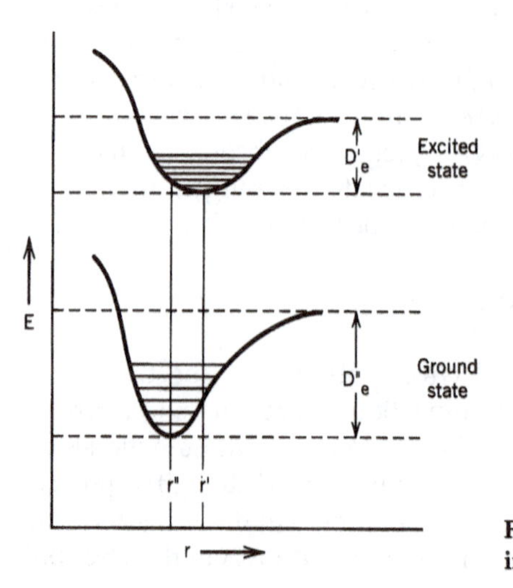


**Figure 8.41 Two electronic energy levels in a diatomic molecule.**

# APPLICATION TO ORGANIC MOLECULES

In organic compounds containing *conjugated* systems of $\pi$ electrons, a particular *chromophore* present can often be identified by the use of UV–vis spectroscopy. A *chromophore* is, in this case, a group of atoms able to absorb light in the UV–vis region of the spectrum. Since the electronic transitions involved are limited primarily to $\pi$-electron (and lone-pair) systems, this type of spectroscopy is less commonly used than the other modern spectroscopic techniques—which, in fact, it predates by several decades. Ultraviolet–visible spectroscopy, however, can play a valuable role in certain situations. For example, if a research problem involves synthesizing a series of derivatives of a complex organic molecule that possesses a strong chromophore, the UV–vis spectrum will be highly sensitive to structural changes involving the arrangement of the $\pi$-electron system.

In a conjugated alkene, such as 1,3-butadiene, the long-wavelength photon absorbed corresponds to the energy required for the excitation of a $\pi$ electron from the HOMO, $\pi_2$ to the LUMO, $\pi^*_3$. For these alkenes, this transition is represented as $\pi \rightarrow \pi^*$; that is, an electron is promoted from a $\pi$ (bonding) molecular orbital to a $\pi^*$ (antibonding) orbital. This type of excitation is depicted below for both ethylene and 1,3-butadiene. Note that as a consequence of extending the chromophore and raising the energy of the highest occupied level in butadiene, the energy gap between the HOMO and LUMO levels of ethylene is larger than that in the conjugated system. Thus, the photon required for excitation of ethylene has a higher energy (higher frequency = shorter wavelength, $\lambda_{max}$ = 171 nm) than the photon absorbed by 1,3-butadiene($\lambda_{max}$ = 217 nm):

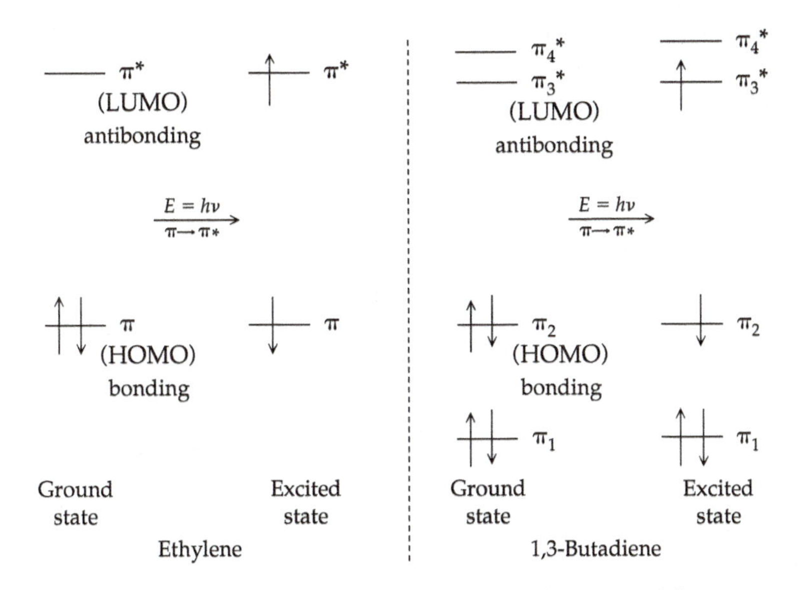


If we continue to extend the chromophore, the decrease of the energy gap between the HOMO and LUMO levels also continues. This drop in $\Delta E$ is then reflected in a drop in energy of the photon required to excite the $\pi \rightarrow \pi^*$ transition. This effect is illustrated in Table 8.28.

As the extension of the chromophore continues, the $\lambda_{max}$ of the $\pi \rightarrow \pi^*$ transition will eventually shift into the visible region. At this point the substance exhibits color. Because the absorbed wavelength is coming from the blue end of the visible spectrum, these compounds will appear yellow. The color will deepen and become red as the energy of the photon required for

*Table 8.28* **Absorption Maxima of Conjugated Alkene**

| Name | Structure | $\lambda_{max}$ (nm) |
|---|---|---|
| Ethylene | $CH_2{=}CH_2$ | 165 |
| 1,3-Butadiene | $CH_2{=}CH{-}CH{=}CH_2$ | 217 |
| 1,3,5-Hexatriene | $CH_2{=}CH{-}CH{=}CH{-}CH{=}CH_2$ | 268 |
| 1,3,5,7-Octatetraene | $CH_2{=}CH{-}CH{=}CH{-}CH{=}CH{-}CH{=}CH_2$ | 290 |

electronic excitation continues to drop. For example, tetraphenylcyclopentadienone is purple (Experiment [A3$_a$]); the dye, methyl red, is deep red (Experiment [26]); and *trans*-9-(2-phenylethenyl)anthracene is golden yellow (Experiment [19D]).

Compounds that contain a carbonyl chromophore $>C{=}\ddot{O}:$ also absorb radiation in the UV region. A $\pi$ electron in this unsaturated system undergoes a $\pi \rightarrow \pi^*$ transition. However, unless the carbonyl is part of a more extended chromophore, such as an $\alpha,\beta$-unsaturated ketone system, the $\pi \rightarrow \pi^*$ transition requires a fairly high-energy photon for excitation, usually below 190 nm in the far-UV and similar to the energy required for excitation of a carbon–carbon double bond. The edge of the $\pi \rightarrow \pi^*$ absorption band may just barely be observed on instrumentation designed for near-UV studies. This partially observed absorption band is generally referred to as *end absorption.* In the case of carbonyls, however, the heteroatom also loosely holds two pairs of nonbonding electrons that are often termed *lone-pair* electrons. These nonbonding electrons reside in orbitals ($n$) that are higher in energy than the bonding $\pi$ orbital, but lower in energy than the antibonding $\pi^*$ orbital. Thus, while a transition from an $n$ level to a $\pi^*$ level is formally forbidden, in fact, weak bands are observed at $\lambda_{max}$ in the near-UV that have their origin in the excitation of a lone-pair electron by an $n \rightarrow \pi^*$ transition. An energy diagram of a typical carbonyl system follows:

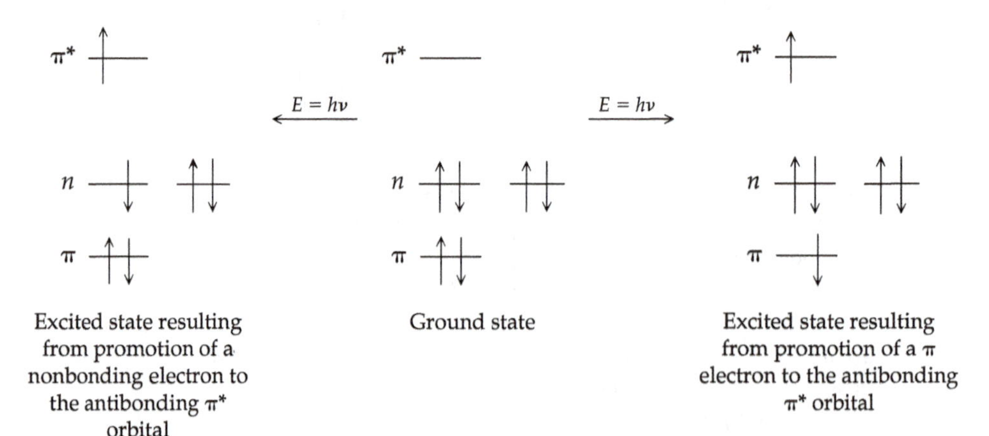


Thus, those substances that contain the carbonyl chromophore absorb radiation of wavelengths that corresponds to both the $n \rightarrow \pi^*$ and the $\pi \rightarrow \pi^*$ transitions. For a simple ketone, such as acetone ($CH_3COCH_3$), the $\pi \rightarrow \pi^*$ transition is found in the far-UV and the $n \rightarrow \pi^*$ in the near-UV. When the carbonyl becomes part of an extended chromophore, such as in methyl vinyl ketone (3-buten-2-one), the spectra reveal that these two transitions have shifted to longer wavelengths—a bathochromic shift (see Fig. 8.42 for the

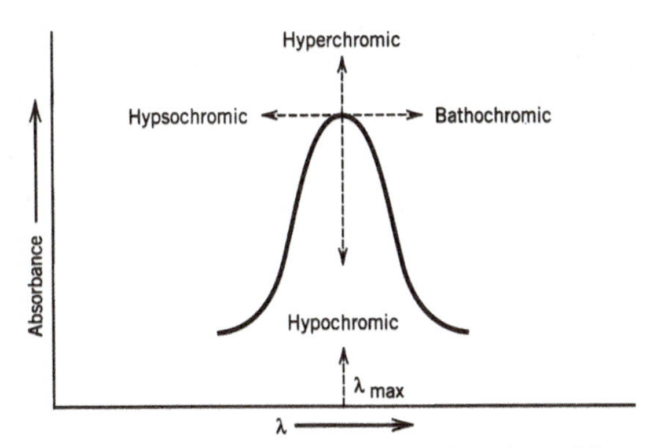


**Figure 8.42 Terms describing direction of wavelengths and intensity shifts.**

definition of terms used in UV–vis spectra to indicate the direction of wavelength and intensity shifts):

$$CH_3-\overset{\overset{\ddot{O}}{\|}}{C}-CH_3 \qquad CH_3-\overset{\overset{\ddot{O}}{\|}}{C}-CH=CH_2$$

| | | |
|---|---|---|
| $n \Rightarrow \pi^*$ | $\lambda_{max}$ 270 nm $\varepsilon_{max}$ 16 | $\lambda_{max}$ 324 nm $\varepsilon_{max}$ 24 |
| $\pi \Rightarrow \pi^*$ | $\lambda_{max}$ 187 nm $\varepsilon_{max}$ 900 | $\lambda_{max}$ 219 nm $\varepsilon_{max}$ 3600 |

Saturated systems containing heteroatoms with nonbonded electrons also exhibit weak absorption bands, often as end absorptions, which have their origin in forbidden $n \rightarrow \sigma^*$ transitions. When these heteroatomic groups are attached to chromophores, both the wavelength and the intensity of the absorption can be altered. These are often referred to as *auxochromes* and *auxochromic shifts.*

Often, model compounds containing a chromophore of interest are referred to as an aid in the interpretation of the UV–vis spectrum of a new structure. Substantial collections of data have been developed for a wide variety of chromophores as an aid to this type of correlation. A number of empirical correlations, such as the Woodward–Fieser rules, of substituent effects on $\lambda_{max}$ values are available. The Woodward–Fieser rules are a set of empirical correlations derived from studies of UV–vis spectral data. Using these rules it is possible to predict with reasonable accuracy the $\lambda_{max}$ for *new* systems containing various substituents on *known* chromophores. The rules are summarized in Table 8.29.

Examples of homoannular and heteroannular dienes are shown below:

A heteroannular diene    A homoannular diene

An example illustrating the use of these rules follows. Calculate the wavelength at which the following steroidal methyl sulfide will absorb:

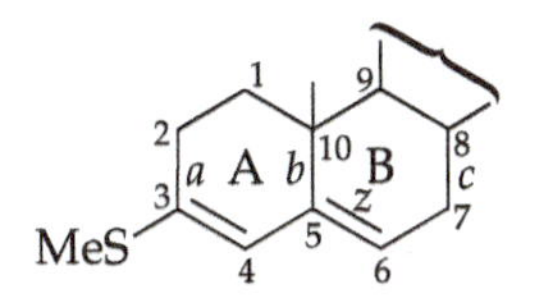

**Table 8.29 Woodward–Fieser Rules for Conjugated Dienes**

| Functionality | Increment (nm) |
|---|---|
| Base value for homoannular diene | 253 |
| Base value for heteroannular diene | 214 |
| Add: | |
| For each double bond extending conjugation | +30 |
| For double bond outside of ring (exocyclic) | +5 |
| For alkoxy groups | +6 |
| For *S*-alkyl groups | +30 |
| For Cl, Br groups | +5 |
| For dialkylamino groups | +60 |
| For parts of rings attached to butadiene fragment | +5 |

The base value for the diene is 214 nm, because the system is heteroannular (if a homoannular diene were present it would take precedence over the heteroannular diene; see the following example). There are three ring residues (or alkyl substituents) attached to the chromophore. Through hyperconjugation, the $\pi$ system is slightly extended by this type of substitution. The residues are labeled *a, b,* and *c.* Each of these substituents is assumed to add 5 nm to the $\lambda_{max}$ of the parent heteroannular diene, for a total of 15 nm. The 5,6-double bond in the B ring marked *z* is exocyclic to the A ring, so empirically we add an additional 5 nm. Finally, for the thiomethyl substituent at the 3 position we add 30 nm. The total is 214 + 15 + 5 + 30 = 264 nm.

Thus we have

| | | |
|---|---|---|
| Predicted value | $\lambda_{max}$ (calcd) = 264 nm | |
| Observed value | $\lambda_{max}$ (obsd) = 268 nm | ($\varepsilon$ = 22,600) |

As another example, consider ergosta-3,5,7,9-tetraene-3-acetate (**I**):

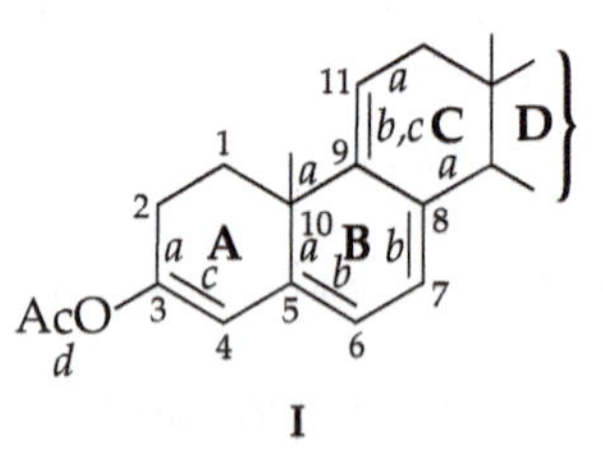


***Prediction of $\lambda_{max}$ for a homoannular diene***

| | | |
|---|---|---|
| Parent homoannular diene in ring **B** | | 253 nm |
| Increments for | | |
| Double bond extending conjugation | *c* [2 × 30] | 60 |
| Alkyl substituent or ring residue | *a* [5 × 5] | 25 |
| Exocyclic double bond | *b* [3 × 5] | 15 |
| Polar substituents | *d* [0] | 0 |
| | $\lambda_{calcd}$ | 353 |

| | |
|---|---|
| Predicted value | $\lambda_{max}$ (calcd) = 353 nm |
| Observed value | $\lambda_{max}$ (obsd) = 355 nm ($\varepsilon$ = 19,700) |

**Table 8.30 Conjugated Carbonyl Systems**

| α,β-Unsaturated Functionality | | | | Base Value (nm) |
|---|---|---|---|---|
| Acyclic or six-membered or higher cyclic ketone | | | | 215 |
| Five-membered ring ketone | | | | 202 |
| Aldehydes | | | | 210 |
| Carboxylic acids and esters | | | | 195 |
| | | | | **Increment (nm)** |
| Extended conjugation | | | | +30 |
| Homoannular diene | | | | +39 |
| Exocyclic double bond | | | | +5 |
| | **Substituent Increment (nm)** | | | |
| **Substituent** | α | β | δ | |
| Alkyl | +10 | +12 | +18 (γ and higher) | |
| Hydroxyl | +35 | +30 | +50 | |
| Alkoxy | +35 | +30 | +31 (γ + 17) | |
| Acetoxy | +6 | +6 | +6 | |
| Dialkylamino | | +95 | | |
| Chloro | +15 | +12 | | |
| Bromo | +25 | +30 | | |
| Alkylthio | | +85 | | |
| **Solvent** | | | | **Solvent Increment (nm)** |
| Water | | | | −8 |
| Ethanol | | | | 0 |
| Methanol | | | | 0 |
| Chloroform | | | | +1 |
| Dioxane | | | | +5 |
| Ether | | | | +7 |
| Hexane | | | | +11 |
| Cyclohexane | | | | +11 |

There are additional rules for carbonyl-containing compounds, such as ketones, aldehydes, carboxylic acids, and so on, and for aromatic compounds. Table 8.30 lists the parameters for conjugated carbonyl systems. Note that in contrast to the conjugated diene compounds, in which we are observing $\pi \rightarrow \pi^*$ transitions, the $n \rightarrow \pi^*$ transitions of the carbonyl $\lambda_{max}$ chromophore are often solvent dependent. Thus, solvent effects will have to be considered when predicting $\lambda_{max}$ values in these systems.

An example of $\lambda_{max}$ (ethanol) calculation for a carbonyl system is presented here:

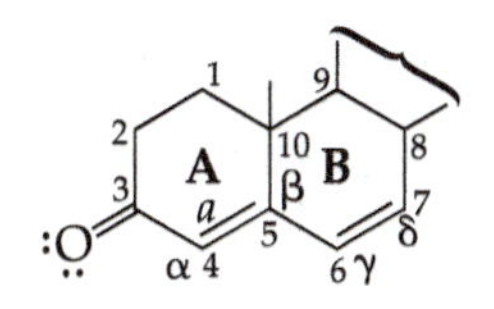

The base value for the α,β-unsaturated six-membered ring ketone system is 215 nm. Extended conjugation adds an additional 30 nm. The presence of an exocyclic double bond, marked *a*, extends the $\lambda_{max}$ another +5 nm. There is asubstituent on the β-carbon atom (+12 nm) and on the δ-carbon atom (+18 nm). There is no solvent effect because the spectrum was obtained in ethanol (0 shift). The total is 215 + 30 + 5 + 12 + 18 = 280 nm:

| | |
|---|---|
| Predicted value | $\lambda_{lmax}$ (calcd) = 280 nm |
| Observed value | $\lambda_{lmax}$ (obsd) = 284 nm |

The Woodward–Fieser rules work well for systems with four or fewer double bonds. For more extensively conjugated systems, $\lambda_{max}$ values are more accurately predicted using the Fieser–Kuhn equation:

$$\text{Wavelength} = 114 + 5\,M + n(48.0 - 1.7n) - 16.5\,R_{endo} - 10R_{exo}$$

*where*

$n$ = number of conjugated double bonds
$M$ = number of alkyl substituents in the conjugated system
$R_{endo}$ = number of rings with endocyclic double bonds in the system
$R_{exo}$ = number of rings with exocyclic double bonds in the system

*Sample calculation:* Find the UV $\lambda_{max}$ of β-carotene:

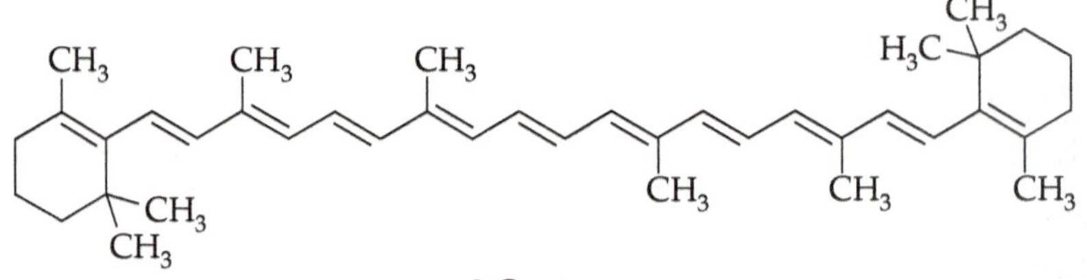


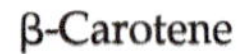
β-Carotene

In the structure there are 11 conjugated double bonds, $n = 11$. There are 6 alkyl groups and 4 ring residues on the conjugated system, $M = 10$. Both rings have an endocyclic double bond, $R_{endo} = 2$ Neither ring has any exocyclic double bonds, therefore $R_{exo} = 0$. Substituting in the equation gives

$$\begin{aligned}\text{Wavelength} &= 114 + 5(10) + 11[48 - 1.7(11)] - 16.5(2) - 10(0)\\ &= 114 + 50 + 322.3 - 33 - 0 = 453\text{ nm}\end{aligned}$$

| | |
|---|---|
| Predicted value | $\lambda_{lmax}$ (calcd) = 453 nm |
| Observed value | $\lambda_{max}$ (obsd) = 455 nm |

Two examples of this correlation scheme (see Table 8.31) are

**1.** 6-Methoxytetralone

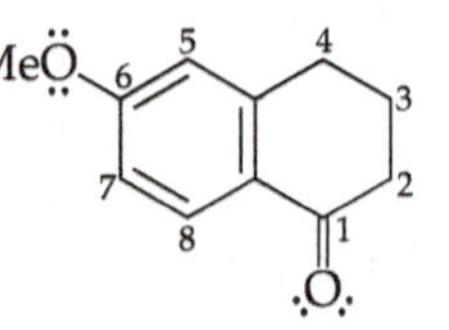


Predicted $\lambda_{max}$ is calculated by taking

Parent value, 246 nm + one *o*-ring residue, 3 + one *p*-OMe, 25 = 274 nm

| | |
|---|---|
| Predicted value | $\lambda_{max}$ (calcd) = 274 nm |
| Observed value | $\lambda_{max}$ (obsd) = 276 nm (ε = 16,500) |

*Table 8.31* **The Benzoyl Chromophore**

| Parent Chromophore $C_6H_5$—CO—R | | | |
|---|---|---|---|
| **Function** | | **Wavelength (nm)** | |
| R = alkyl or ring residue | | 246 | |
| R = H | | 250 | |
| R = OH, *O*-alkyl | | 230 | |
| | **Substituent Increment (nm)** | | |
| **Substituent** | *o*- | *m*- | *p*- |
| Alkyl or ring residue | 3 | 3 | 10 |
| —OH, —$OCH_3$, —*O*-alkyl | 7 | 7 | 25 |
| —$O^-$ (*p*-sensitive to steric effects) | 11 | 20 | 78 |
| —Cl | 0 | 0 | 10 |
| —Br | 2 | 2 | 15 |
| —$NH_2$ | 13 | 13 | 58 |
| —NHAc | 20 | 20 | 45 |
| —$NHCH_3$ | | | 73 |
| —$N(CH_3)_2$ | 20 | 20 | 85 |

*Note.* Spectra obtained in alcohol solvents.

**2.** 3-Carboethoxy-4-methyl-5-chloro-8-hydroxytetralone

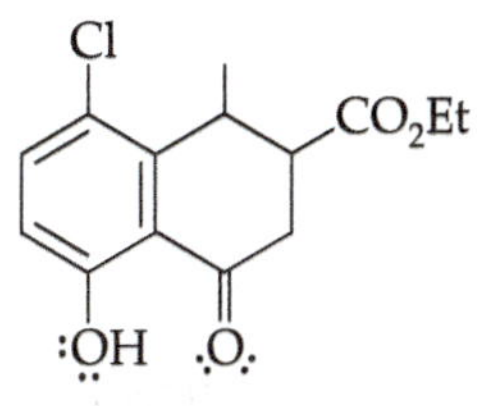


Parent value, 246 nm + one *o*-ring residue, 3 + one *o*-OH, 7 + 0 *m*-Cl = 256 nm

Predicted value $\lambda_{max}$ (calcd) = 256 nm
Observed value $\lambda_{max}$ (obsd) = 257 nm ($\varepsilon$ = 8000)

In summary, UV–vis spectra can make substantial contributions to understanding the molecular structure of organic substances that possess *chromophores:*

**1.** Interpretation of ultraviolet–visible spectra often can be a powerful approach for identifying the molecular structure of that section of a new substance that contains the chromophore.

**2.** The $\lambda_{max}$ increases within a series of compounds that contain a common chromophore that is lengthened (increased conjugation) over the series. The intensity of the absorption ($\varepsilon_{max}$) also generally becomes greater as conjugation increases, but can be very sensitive to steric effects.

**3.** The $\lambda_{max}$ is sensitive to hyperconjugation by alkyl substituents, conformational changes that restrict $\pi$-system overlap, configurational, or geometric isomerization in which $\pi$ systems are perturbed, and structural changes, such as the isomerization of a double bond from an *exocyclic* to an *endocyclic* position and changes in ring size.

*Table 8.32* **Absorption Maxima of Several Unsaturated Molecules**

| Compound | Structure | $\lambda_{max}$ (nm) | $\varepsilon_{max}$ |
|---|---|---|---|
| Ethylene | $CH_2{=}CH_2$ | 171 | 15,530 |
| 1,3-Butadiene | $CH_2{=}CH{-}CH{=}CH_2$ | 217 | 21,000 |
| Cyclopentadiene | (cyclopentadiene ring) | 239 | 3,400 |
| 1-Octene | $CH_3(CH_2)_5CH{=}CH_2$ | 177 | 12,600 |
| *trans*-Stilbene | $(C_6H_5)HC{=}CH(C_6H_5)$ (trans) | 295 | 27,000 |
| *cis*-Stilbene | $(C_6H_5)HC{=}CH(C_6H_5)$ (cis) | 280 | 13,500 |
| Toluene | $C_6H_5CH_3$ | 189<br>208<br>262 | 55,000<br>7,900<br>260 |
| 4-Nitrophenol | $HO{-}C_6H_4{-}NO_2$ | 320 | 9,000 |
| 3-Penten-2-one | $CH_3CH{=}CHCCH_3$ (C=O) | 220<br>311 | 13,000<br>35 |

**4.** In many instances, accurate prediction of the $\lambda_{max}$ of a new molecular system can be made based on empirical correlations of the parent chromophore giving rise to the absorption.

Table 8.32 lists the $\lambda_{max}$ values of a number of common organic molecules.

## INSTRUMENTATION

The acquisition of UV–vis absorption spectra for use in the elucidation of organic molecular structure is now carried out with instrumentation that is typically an automatic-recording photoelectric spectrophotometer. The optical components of one of the classic spectrophotometers is given in Figure 8.43. This system is typical of a high-quality double-beam double-monochromator instrument. The instrument consists of a number of components: the radiation source, monochromator, sample compartment, detector, amplifier, and recorder.

### The Source of Radiation

Radiant energy may be generated by either a deuterium discharge lamp or a tungsten– halogen lamp depending on the spectral region to be observed. Deuterium is generally preferred over hydrogen since the intense radiating ball of plasma is slightly larger in the case of deuterium, and therefore source

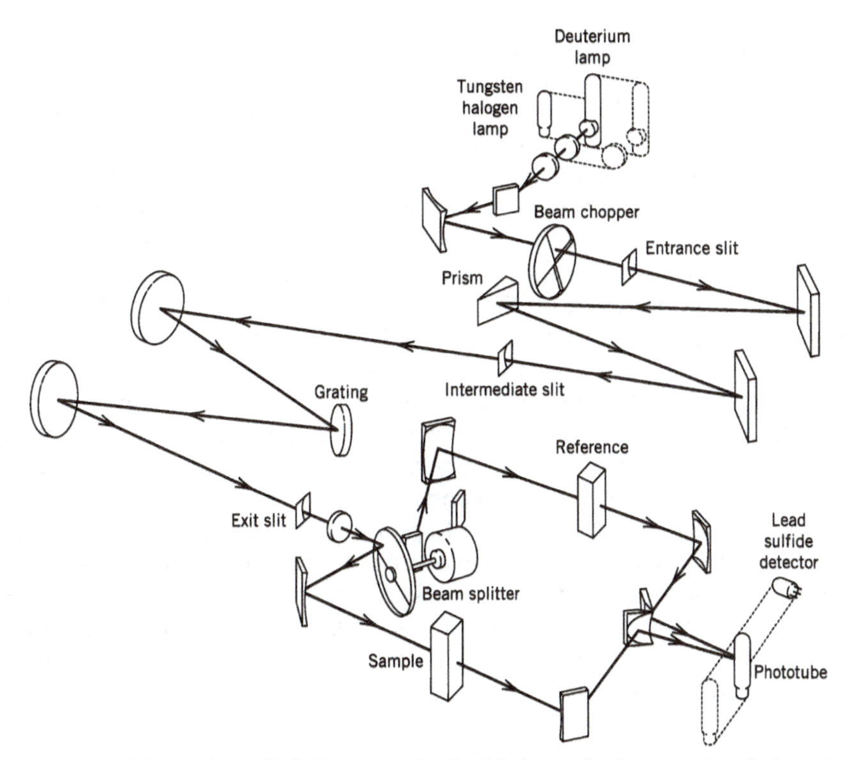


**Figure 8.43 Schematic optical diagram of a double beam-in-time spectrophotometer with double monochromation (Cary Model 17D).** (Courtesy of Varian Associates, Inc.)

brightness is enhanced by a factor of about 4. Below 360 nm, deuterium gas emits an intense continuum band that covers a major portion of the UV region. With special windows the short wavelength cutoff can be extended down to about 160 nm well out into the vacuum-UV. Emission line spectra limit the long wavelength use of these lamps to about 380 nm. The lamps of choice for the region above 350 nm (the visible) are incandescent filament lamps, because they emit a broad band of radiation from 350 nm on the short wavelength end all the way to about 2.5 μm (the near-IR) on the long wavelength side. Most of the radiation emitted falls outside the visible, peaking at about 1 μm in the near-IR. Nevertheless, tungsten lamps are *the* choice for measurements in the visible region, because they are extremely stable light sources.

Thus, radiation sources must possess two basic characteristics: (1) they must emit a sufficient level of radiant energy over the region to be studied so that the instrument detection system can function, and (2) they must maintain constant power during the measurement period. Source power fluctuations can result in spectral distortion.

## The Monochromator

As the name implies, a monochromator (making a single color or hue) functions to isolate a single frequency from the source band of radiation. In practice we settle for isolating a small collection of overlapping frequencies surrounding the monochrome radiation we wish to observe. Thus, the monochromator section of the instrument takes all the source radiation in at one end and releases a very narrow set of bands of radiation at the other end. This function is accomplished, as shown in Figure 8.43, by focusing the entering radiation on an entrance slit that forms a narrow image of the source. After passing through the entrance slit, the spreading radiation is collimated by being reflected off a parabolic mirror, and is converted into parallel light rays (just as in a search light). The collimated radiation is then directed to the dispersing agent, which is usually a quartz prism (quartz is transparent to UV, glass is not) or diffraction grating. The dispersing device spreads the different wavelengths of collimated light out in space. After emerging from the prism the dispersed radiation is redirected to either the same or a new collimator mirror and refocused as an image of the source on the exit slit of the monochromator. The exit slit has only a small fraction of the original radiation focused on it, and allows it to pass through in the image of the source. The remaining frequencies lie at different angles on either side of the exit slit. By mechanically turning the prism or grating, and thus changing the angle of the dispersing device with respect to the exit slit, all of the narrowly dispersed bands of radiation can be passed out of the monochromator in sequential fashion.

Instruments that are designed to reduce unwanted radiation to an absolute minimum will place two monochromators in tandem with an intermediate slit connecting the dispersing systems. In the case illustrated in Figure 8.43 the first monochromator uses a prism, while the second uses a grating. The two monochromators, however, must be in perfect synchronization or no light at all will be transmitted.

### Sample Compartment

After leaving the monochromator the radiation is directed to the sample compartment by a rotating sector mirror, where it is alternately focused on the substance to be examined (which is contained in a cell with quartz windows) and a reference cell (which holds the pure solvent used to dissolve the sample). The system now has two beams, hence the name *double-beam spectrophotometer.* After passing through the sample where the absorption of radiation may occur, the beams are recombined.

The sampling position could be placed either before or after the monochromator. In infrared instruments (such as the PE Model 710B, Fig. 8.3*a*) it was generally found before the monochromator until the introduction of interferometers. In UV systems, the sampling area is most often placed after the monochromator, and for good reasons. If the sample were placed before the monochromator, it would be exposed to the entire band of high-energy UV radiation being emitted by the source over the entire sampling period. By positioning the sample after the monochromator, at any one time the sample sees only the very small fraction of the dispersed radiation passed by the exit slit. Thus, sample stability is greatly protected by this arrangement. Remember that near-UV radiation carries photons with energies that approach those of the bond energies of organic molecules.

### The Detector

The recombined beams are then focused on the detector. Detectors function as transducers because they convert electromagnetic radiation into electrical current. There are a number of radiation-sensitive transducers available as detectors for these instruments. One is the photomultiplier tube. These detectors operate with photocathodes that emit electrons in direct proportion to the number of photons striking the photosensitive surface and possess very large

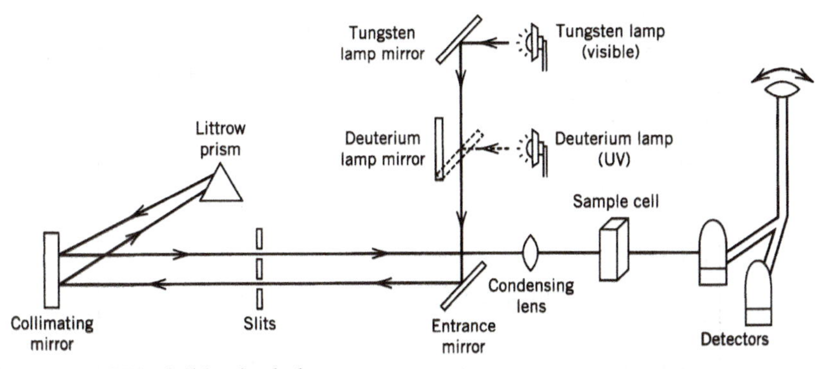


**Figure 8.44 UV–visible single-beam spectrometer.**

internal amplification. Thus, they operate at low power levels. One particular advantage of the photomultiplier is that you can adjust their sensitivity over a wide range simply by adjusting the supply voltage.

### The Electronics: The Amplifier and Recorder

In double-beam instruments, the two signals generated by the sample and reference beams (each referenced against a dark current) in the detector are amplified and the ratio of the sample signal to the reference signal is plotted on a recorder. The simplest of the absorption spectrometers are the single-beam instruments (see Fig. 8.44). These spectrometers are generally employed for problems involving simple one-component analyses. The photometric accuracy of scanned spectra should not be of paramount importance with these systems. Single-beam spectrometers require extremely stable sources and detectors.

## SAMPLE PREPARATION

Ultraviolet spectra are usually obtained on samples in solution using quartz cells. Quartz is used because it is transparent to both UV and visible light. For spectra restricted to the visible region, Pyrex cells are satisfactory (and a good deal less expensive), but because Pyrex absorbs UV radiation, these cells cannot be employed for measurements in this region.

Solution cells usually have a horizontal cross section of 1 $cm^2$ and require about 3 mL of sample solution. Cells must be absolutely clean, and it is advisable to rinse the cell several times with the solvent used to dissolve the sample. A background spectrum of the solvent-filled cell (*without* a reference sample) can easily be obtained at this time and used as a check against contamination of either the cell or the solvent or both.

Because the intensities of electronic transitions vary over a very wide range, the preparation of samples for UV–vis spectra determination is highly concentration dependent. Intense absorption can result from the high molecular extinction coefficients found in many organic chromophores. The sampling of these materials requires very dilute solutions (on the order of $10^{-6}$–$10^{-4}$ M). These solutions can be conveniently obtained by the technique of *serial dilution.* In this method a sample of the material to be analyzed is accurately weighed, dissolved in the chosen solvent, and

*Table 8.33* **Solvents Used in the Near-UV**

| Solvent | Cutoff Wavelength (nm) |
|---|---|
| Acetonitrile | 190 |
| Chloroform | 245 |
| (toxic, substitute $CH_2Cl_2$) | 235 |
| Cyclohexane | 205 |
| 1,4-Dioxane | 215 |
| (toxic, substitute EtOEt) | 218 |
| 95% Ethanol | 205 |
| *n*-Hexane | 195 |
| Methanol | 205 |
| Isooctane | 195 |
| Water | 190 |

*Note.* Since these solvents have no color, they are transparent in the visible.

diluted to volume in a volumetric flask. Sample weights of 4–5 mg in 10-mL volumetric flasks are typical. An aliquot is then taken from this original solution, transferred to a second volumetric flask, and diluted as before. This sequence is repeated until the desired concentration is obtained.

Numerous choices of solvent are available (a list is given in Table 8.33) and most of them are available in "spectral grade." The most commonly used solvents are water, 95% ethanol, methanol, and cyclohexane.

### Criteria for Choosing a Solvent

- The most important factor is solubility of the sample. UV–vis spectra can be very intense, so even low solubility may be quite acceptable in sample preparation.
- The wavelength cutoff for the solvent may be important if the sample absorbs below about 250 nm.
- Sample–solvent molecular interactions must be considered. An example of these effects would be hydrogen bonding of protic solvents with carbonyl systems. Hydrocarbon chromophores are less influenced by solvent character than are the more polar chromophores.

## BIBLIOGRAPHY

American Petroleum Research Institute Project 44 *Selected Ultraviolet Spectral Data,* Vols. I–IV; Thermodynamics Research Center, Texas A&M University: College Station, TX, 1945–1977 (1178 compounds).

Feinstein, K. *Guide to Spectroscopic Identification of Organic Compounds;* CRC Press: Boca Raton, FL, 1995.

Field, L. D.; Sternhell, S.; Kalman, J. R. *Organic Structures from Spectra,* 4th ed.; Wiley: New York, 2008.

Grasselli, J. G.; Ritchey, W. M. *Atlas of Spectral Data and Physical Constants for Organic Compounds,* 2nd ed.; CRC Press: Cleveland, OH, 1975.

Harwood, L. M.; Claridge, T. D. W. *Introduction to Organic Spectroscopy;* Oxford University Press: New York, 1997.

Hesse, M; Meier, H; Zeeh, B. *Spectroscopic Methods in Organic Chemistry,* 2nd ed; Thieme Medical Publishing: New York, 2008.

Lang, L., Ed.; *Absorption Spectra in the Ultraviolet and Visible Region,* Vols. 1–20; Academic Press: New York, 1961–1975; Vols. 21–24; Kreiger: New York, 1977–1984.
Pavia, D. L.; Lampman, G. M.; Kriz, G. S. *Introduction to Spectroscopy: A Guide for Students of Organic Chemistry,* 3rd ed.; Saunders College: Philadelphia, 2000.
Pretsch, E.; Clerc, J. T. *Spectra Interpretation of Organic Compounds;* VCH: Wiley: New York, 1997.
Rouessac, F.; Rouessac, A. *Chemical Analysis: Modern Instrumental Methods and Techniques,* 2nd ed.; John Wiley & Sons; New York, 2007.
*Standard Ultraviolet Spectra;* Sadtler Research Laboratories: Philadelphia.
*UV Atlas of Organic Compounds,* Vols. I–IV; Butterworths: London, 1966–1971.
Williams, D. H.; Fleming, I. *Spectroscopic Methods in Organic Chemistry,* 6th ed.; WCB/McGraw-Hill: New York, 2007.

# MASS SPECTROMETRY*

In comparison with other forms of spectroscopy, such as NMR, IR, or UV–vis, mass spectrometry is unique in terms of how we generate and interpret the spectrum. Instead of monitoring the absorption of electromagnetic radiation in terms of frequency or wavelength, a mass spectrum can be thought of as a snapshot of a rather unconventional organic reaction involving one energetic reactant that decomposes to give a variety of reaction products. By characterizing the composition of this "reaction mixture" we are able to learn the identity of the starting reactant.

The reaction gets started when a molecule in the gas phase is converted to a radical cation by an energetic collision with an electron, as shown below for $N_2O$:

$$N_2O_{(g)} + e^- \longrightarrow N_2O^{+\cdot} + 2e^-$$

The fact that we have formed a positive ion (cation) with an unpaired electron (radical) becomes important for understanding the decomposition reactions. The process of forming the radical cation yields a collection of energized *molecular ions* that contain a range of internal energies. The molecular ion is produced in a low-pressure environment where it is unable to bump into other molecules. Fragmentation (bond breaking) results in the formation of charged and neutral products. Ideally, intramolecular rearrangements, which could complicate determination of the molecules original structure, do not occur. Depending on characteristics of the molecule and the amount of energy deposited, a variety of fragmentation reactions can take place. For example, the molecular ion ($N_2O^{+\cdot}$) may fragment to give the following products through one- or two-step reactions:

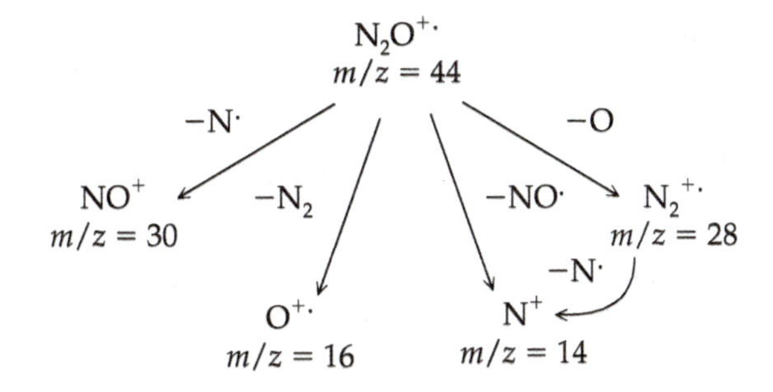


*This section has been written by Elizabeth A. Stemmler, Professor of Chemistry, Bowdoin College.

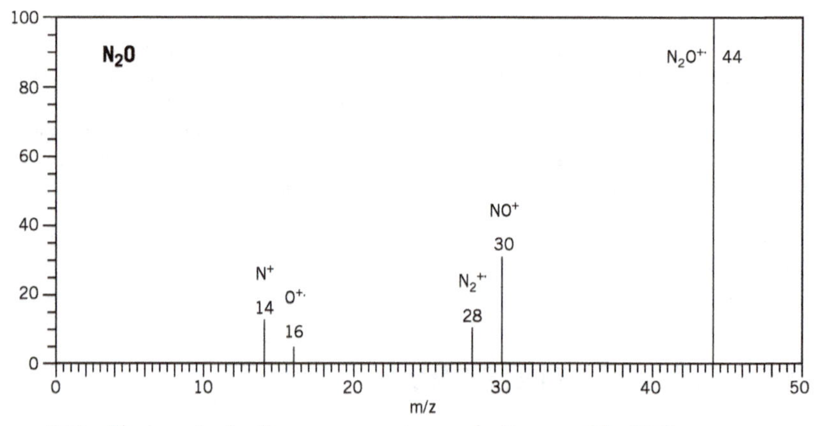


**Figure 8.45 Electron ionization mass spectrum of nitrous oxide, $N_2O$.**

Mass spectrometry derives its name from its ability to distinguish the molecular ion and the different charged reaction products based on the ratio of the ion mass to its charge (*m/z* ratio). In most cases, the charge, *z*, is equal to 1 and we can easily tell the difference between the molecular ion (*m/z* = 44) and products, such as $N_2^+$ (*m/z* 28). In addition, a mass spectrometer can be used to determine the relative amounts of molecular ions and fragment ions present after the reaction has had a little (very little!) time to proceed. A mass spectrum, typically shown in a bar-graph format, is a display of the relative number of each type of ion plotted as a function of the *m/z* ratio (see Fig. 8.45). Instead of reporting the actual number of each type of ion, we normalize the data and give the most abundant ion a value of 100%. Note that only charged species are detected by a mass spectrometer. The neutral products are not observed and their identity must be inferred.

Mass spectrometry is useful to organic chemists because of the information it provides about molecular structure. For example, if the molecular ion is present, that peak can be used to determine the molecular weight (MW) of the neutral molecule. With precise measurement of the *m/z* ratio (to ± 0.0001,for example), the elemental formula of the molecular ion can be determined. For example, $N_2O$ and $CO_2$ both have a molecular weight of 44; however, measurement of their exact masses (44.0011 and 43.9898, respectively) can be used to assign their elemental formula. Mass spectrometry was used to originally determine the exact mass of each element (see Table 8.34), and these exact masses, *not* the atomic weights, are used to calculate mass.

Even when precise mass measurements are not available, the products in the mass spectrum may provide enough information to determine the structure of the neutral molecule. Interpretation of a mass spectrum involves working backward from the observed charged fragments to a proposed molecular structure. For example, $CO_2$ or $N_2O$ (same MW) could be distinguished by an examination of the mass spectrum. The fragment ions at *m/z* 14 and 30 in Figure 8.45 clearly eliminate $CO_2$ as a possible structure. There are no combinations of carbon (mass = 12) and oxygen (mass = 16) that could produce these ions. In addition, the mass spectrum allows us to distinguish between the isomers NNO and NON. What would the mass spectrum of NON show? We would expect to see only a peak at *m/z* 30 ($NO^+$) and no signal at *m/z* 28 ($N_2^+$). The mass spectrum would thus support your chemical intuition that NON is an unlikely and unstable molecular structure.

| *Table 8.34* Exact Masses and the Atomic Weights for Isotopes of Some Common Elements | | | |
|---|---|---|---|
| Element | Nuclide | Mass | Atomic Weight[a] |
| Hydrogen | $^{1}H$ | 1.0078 | 1.0079 |
| | $D(^{2}H)$ | 2.0141 | |
| Carbon | $^{12}C$ | 12.00000 (std) | 12.011 |
| | $^{13}C$ | 13.0034 | |
| Nitrogen | $^{14}N$ | 14.0031 | 14.0067 |
| | $^{15}N$ | 15.0001 | |
| Oxygen | $^{16}O$ | 15.9949 | 15.9994 |
| | $^{17}O$ | 16.9991 | |
| | $^{18}O$ | 17.9992 | |
| Fluorine | $^{19}F$ | 18.9984 | 18.9984 |
| Silicon | $^{28}Si$ | 27.9769 | 28.0855 |
| | $^{29}Si$ | 28.9765 | |
| | $^{30}Si$ | 29.9738 | |
| Phosphorus | $^{31}P$ | 30.9738 | 30.9738 |
| Sulfur | $^{32}S$ | 31.9721 | 32.066 |
| | $^{33}S$ | 32.9715 | |
| | $^{34}S$ | 33.9679 | |
| Chlorine | $^{35}Cl$ | 34.9689 | 35.4527 |
| | $^{37}C1$ | 36.9659 | |
| Bromine | $^{79}Br$ | 78.9183 | 79.904 |
| | $^{81}Br$ | 80.9163 | |
| Iodine | $^{127}I$ | 126.9045 | 126.904 |

[a]Average mass of the naturally occurring isotopes of the element; *not* used for mass calculations in mass spectrometry.

As you will see below, mass spectral interpretation is not always as straightforward as the case given above. Like the outcome of an organic reaction, a mass spectrum will reflect the outcome of competing sequential and simultaneous reaction pathways. For some molecules, very little fragmentation will take place and only the molecular ion is observed. For other less stable molecules, we may have complete conversion of the molecular ion to products, although, because of the high energy required for their formation, we will rarely see complete fragmentation down to products at the atomic level. The interpretation of a mass spectrum requires developing an understanding of important, characteristic reaction pathways and an appreciation of factors influencing ion stability. In many ways, the interpretation of mass spectra provides a place to apply principles of organic chemistry to a unique kind of chemical reaction.

## INSTRUMENTATION

All mass spectrometric instruments contain regions where ionization, mass analysis, and ion detection take place. Mass spectrometry takes place at low pressure; all of the mass spectrometric components are contained in a vacuum

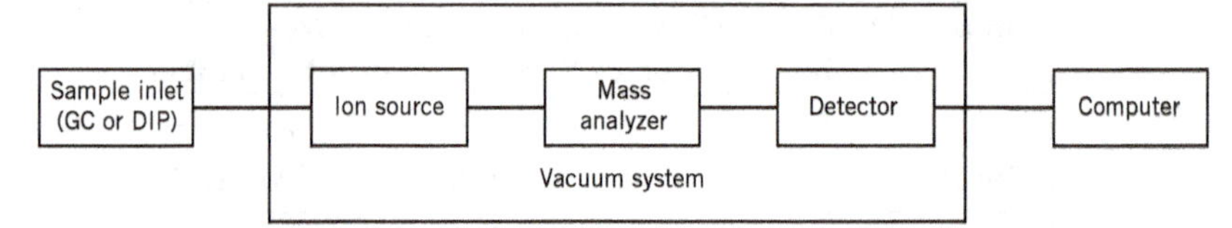


**Figure 8.46 Block diagram of components of a mass spectrometer.**

system at pressures of $10^{-7}$ to $10^{-5}$ torr. Because the instrument must be sealed from the atmosphere to maintain the low pressure, and because samples must be converted to the gas phase prior to ionization, all mass spectrometers have a region devoted to sample introduction. In this region the sample—in the form of a solid, liquid, or gas—is transferred to the low pressure of the mass spectrometer, while preventing the introduction of air. A block diagram of a basic mass spectrometer is shown in Figure 8.46. Ions are generated and fragment in the ion source; the molecular ion and fragments are separated, based upon *m/z* ratios, in the mass analyzer; and the ion signals are converted by the detector into a signal that may be input to a computer.

## Ion Source

The ion source is the region where ions are generated. Mass spectrometrists have many ways of creating ions from different types of samples, including biological materials or the surface of a particle. In our discussions, we will focus only on the most common ionization method, electron ionization (EI). In an EI ion source (Fig. 8.47), we send current through a wire, called a filament. As the filament gets hot, electrons are emitted from the surface. The electrons are produced in an electric field, which results in electron acceleration through

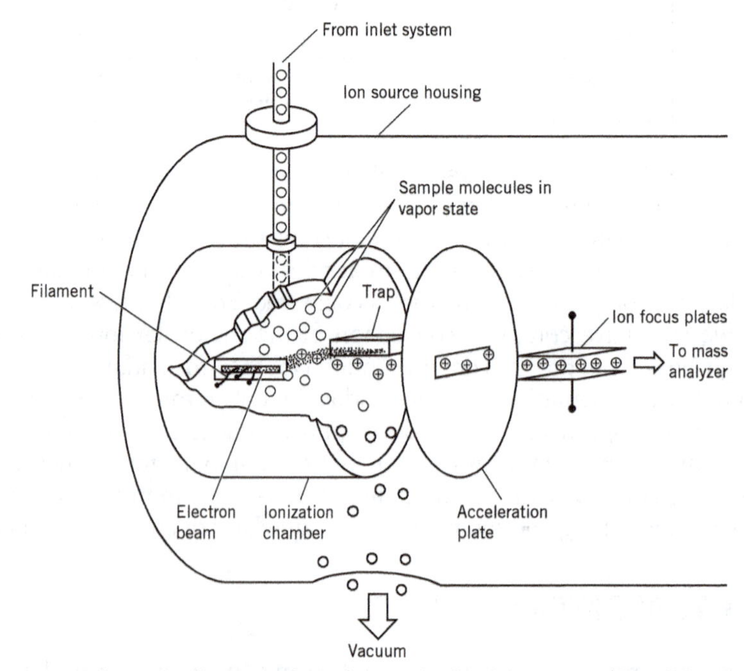


**Figure 8.47 Schematic diagram of an electron ionization source.** (From Watson, J.T. *Introduction to Mass Spectrometry,* 3rd ed.; Lippincott-Raven Publishers: Philadelphia, 1997, p. 140.)

the ion source region where the sample vapor is found. If you work with an electron energy that is too low (below the ionization potential for the molecule), no ions will be produced. As the electron energy increases, the molecular ion ($M^{+\cdot}$) will appear. With further increases in electron energy, fragment ions are observed. Formation of doubly charged ions ($M^{2+\cdot}$) occurs, but the ion intensities are very low:

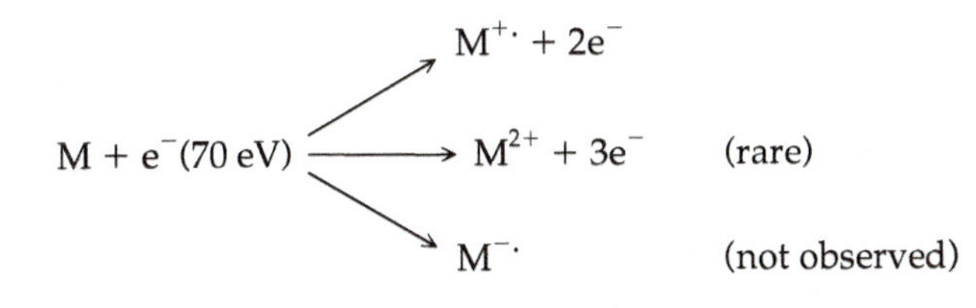


Mass spectra, by convention, are measured with 70-eV electrons. At this energy, the ion intensity is high and the distribution of products remains relatively constant with small changes in electron energy. Formation of negative molecular ions with 70-eV electrons does not occur.

Another important role for the ion source is directing the ions toward the mass analyzer. The ions are pushed and pulled as they pass through one or more metal plates that have a hole in the center for ion transmission. These plates accelerate the ions and keep them directed at the mass analyzer. Depending on the mass analyzer in use, the ions are accelerated toward the analyzer with high or low energy.

## Mass Analyzer

Ion formation and fragmentation in the source is followed by mass analysis. Mass analyzers are used to separate ions based on their mass-to-charge ratios. Organic chemists commonly use two types of mass analyzers: magnetic sector instruments (low- and high-resolution) and quadrupole instruments. Magnetic sectors separate ions based on dispersion of the ions into beams with different *m/z* ratios; quadrupoles are mass filtering devices.

In a magnetic sector instrument, ions are accelerated out of the ion source into a magnetic field with high (kilovolt) kinetic energies. The magnet field, applied perpendicular to the path of the ions, exerts a force that causes the ions to follow a curved path through the magnet (Fig. 8.48). The extent to which the path is bent depends on the mass-to-charge ratio (more specifically, the momentum) of the ions. Light ions are bent more than heavier ions. If the path that the ions must travel is fixed, ions that are too light or too heavy will run into the walls of the mass analyzer, where they are neutralized, and will then be pumped away by the vacuum system. Only the ions with the correct radius (correct *m/z* ratio) will make it to the detector. To measure a complete mass spectrum, the magnetic field strength is varied to bring ions of different *m/z* ratio to focus on the detector.

High-resolution magnetic sector instruments incorporate an additional energy analyzer prior to mass analysis by the magnetic sector. This more precisely defines the kinetic energies of ions entering the magnetic sector, which improves the mass resolution. High-resolution instruments require more expertise to operate and are less common because of their expense, but they can provide the precise and accurate mass measurements needed to determine elemental composition.

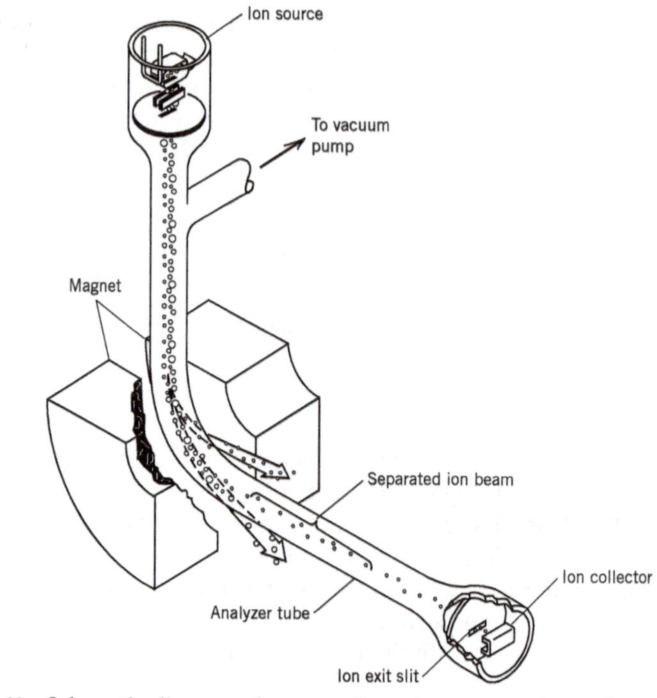


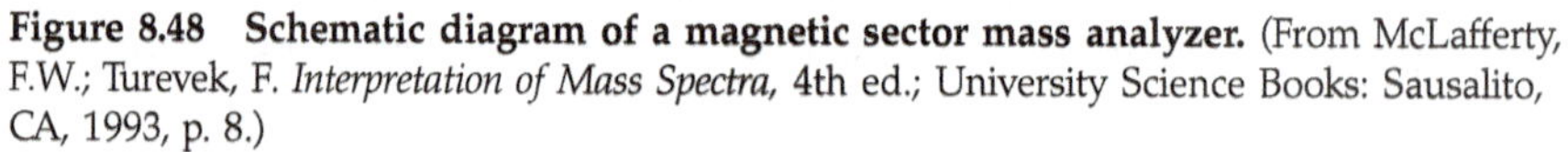

**Figure 8.48 Schematic diagram of a magnetic sector mass analyzer.** (From McLafferty, F.W.; Turevek, F. *Interpretation of Mass Spectra,* 4th ed.; University Science Books: Sausalito, CA, 1993, p. 8.)

A quadrupole is composed of a set of four rods to which (electric) potentials are applied (Fig. 8.49). To allow ions of a particular *m/z* ratio through the rods, a constant positive potential is applied to two opposing rods (the x-rods), while the remaining two rods experience a constant negative potential (the y-rods). In addition, each set of rods experiences a time-varying potential that causes the rod potentials to vary between positive and negative potentials, with the signals 180° out-of-phase. When the x-rods are positive, the y-rods

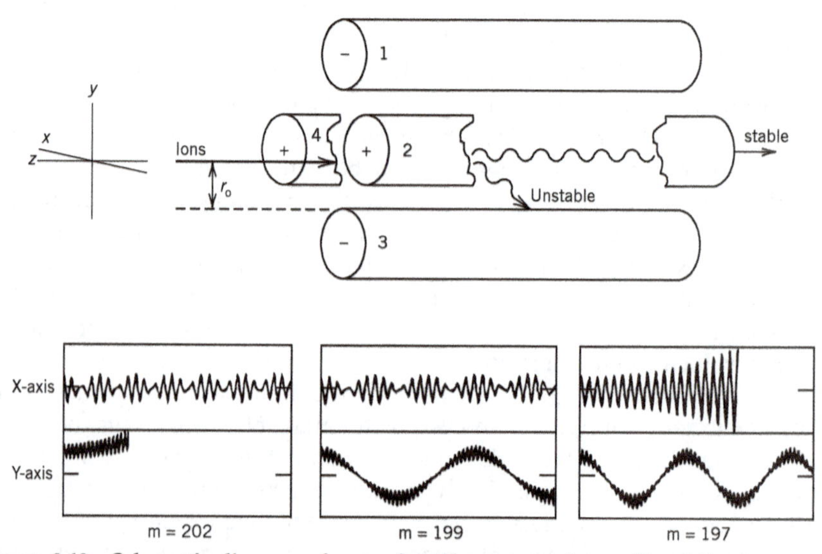


**Figure 8.49 Schematic diagram of a quadrupole mass analyzer. *X* and *Y* axis trajectories for m/z 202, 199, and 197.** (From Steel, C.; Henchman, M. *J. Chem. Educ.* **1998,** *75,* 1049–1054.)

are negative. Mass filtering occurs when a group of ions enters the analyzer. Ions that have *m/z* ratios that are too low or too high will experience unstable trajectories through the rods and will strike a rod, become neutralized, and be pumped out of the system. Only ions with an appropriate *m/z* ratio will have a stable trajectory and will make it through the rods to the detector. Figure 8.49 shows the trajectories of three ions with respect to the x- and y-rods. Only the ion with *m/z* = 199 makes it through the quadrupole. To change the *m/z* ratio of the ions that are transmitted, the magnitude of the constant and time-varying potentials are changed. Mass spectrometers that fit on a laboratory benchtop have a mass range of *m/z* = 10 to 650. With the quadrupole mass analyzer a mass spectrum can be measured rapidly (roughly 1 scan per second), which is important when capillary GC columns are used for sample introduction.

### Detector

Ions can be detected directly through the current produced when they strike a plate; however, we usually make this signal larger through the use of electron multiplier detectors.

### Tuning the Mass Spectrometer

Before the mass spectrometer can be used to collect mass spectra, the instrument must be tuned and calibrated. The tuning procedure involves setting voltages associated with the ion source, lenses, and detector (to optimize sensitivity), and selecting values for potentials applied to the quadrupole (to set the instrument resolution). These tasks are accomplished while a calibration standard is continuously added to the instrument. A common calibration standard is perfluorotributylamine (PFTBA), $(CF_3CF_2CF_2CF_2)_3N$. Usually ions at *m/z* 69, 219, and 502 are monitored (Fig. 8.50).

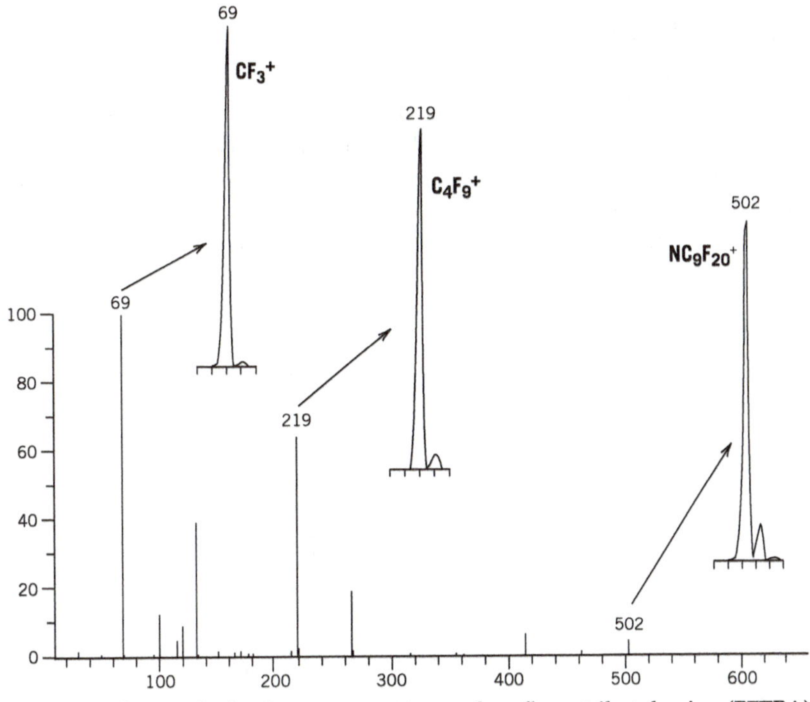


**Figure 8.50 Electron ionization mass spectrum of perfluorotributylamine (PFTBA). Inserts show the peak profiles for *m/z* 69, 219, and 502.**

### Sample Introduction

Samples analyzed by EI mass spectrometry must be converted to gas phase. For pure gases or volatile liquids the samples may be introduced directly through a small orifice that allows an appropriate amount of material into the vacuum chamber. A small amount of a solid sample can be placed in a melting point capillary tube and inserted into the mass spectrometer at the end of a metal rod, called a *direct insertion probe* (DIP). The temperature at the tip of the probe can be varied to promote sublimation of the sample. Another common method of sample introduction is gas chromatography, which is the ideal choice for samples that are impure.

### Gas Chromatography/Mass Spectrometry (GC/MS)

While the goal of a synthetic organic reaction is the production of one pure product in high yield, organic reactions often produce a mixture of reaction products. Chromatographic separation of those products is a useful complement to the mass spectral analysis. The components of the mixture elute from the chromatographic column, ideally, as pure peaks. The mass spectrometer, which is scanning rapidly, is then able to collect a few spectra for each eluting peak. Both the chromatographic retention time and the mass spectrum can be used to help identify components of the mixture. Because the compounds are detected with little bias for one type of compound over another, GC/MS has provided organic chemists with a powerful tool to characterize reaction mixtures and assess product purity.

Sensitivity is another distinguishing feature of mass spectrometry. This sensitivity has allowed mass spectrometers to act as detectors for capillary columns, which can separate mixture containing hundreds of compounds, when less than a nanogram($10^{-9}$ g) of each compound is injected.

### Capillary Columns

Most GC/MS instruments use capillary columns for chromatographic separation. Capillary columns are very long (15- to 30-m), open tubes of fused silica that are coated with a thin coating of the stationary phase (Fig. 8.51). A carrier gas, typically helium, is used as the mobile phase. Capillary column diameters are commonly in the range of 0.25 to 0.53 mm, and the coating of the stationary phase is in the range of 0.25 to 1 μm Thicker coatings are used for the separation of low-boiling compounds.

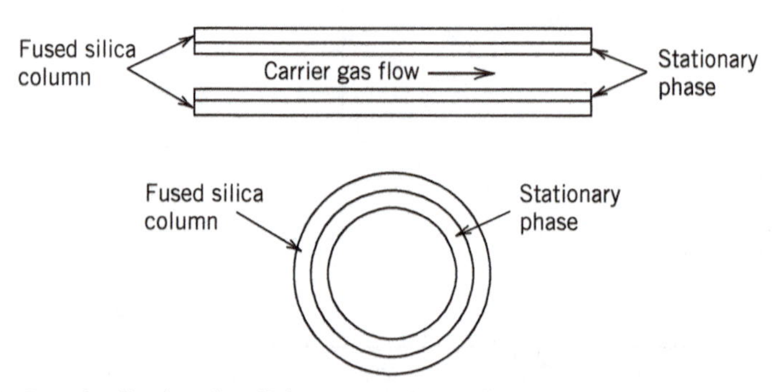


**Figure 8.51 Longitudinal and radial cross sections of a capillary column.**

Chromatographic resolution increases as a function of the square root of the column length, and the extraordinary length of capillary columns means that most simple mixtures are easily resolved on just a few types of stationary phases. One common nonpolar stationary phase is poly(dimethylsiloxane) (R = $CH_3$) which can be made slightly more polar by the incorporation of phenyl groups (typically 5% phenyl) in place of methyl groups:

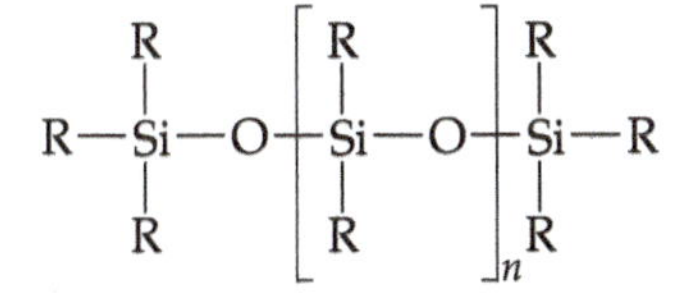

These two stationary phases interact with solutes primarily through dispersion interactions, and compounds elute as a function of boiling point. More polar stationary phases are also available. Because capillary columns are used to separate compounds with a wide range of boiling points, we often make use of a technique called *temperature programming*. This techniques allows you to start with a low oven temperature, to optimize the elution of low boiling components, and then increase the oven temperature at a controlled rate, to decrease stationary phase interactions for high-boiling compounds. The increase in temperature decreases retention times and produces narrower peaks for compounds that would otherwise require a long time to elute as a very broad peak.

While samples may be directly injected onto a packed column, the small diameter of the capillary column presents a problem. In addition, it is easy to overload the capillary column with sample (Table 8.35). Two techniques for getting the sample into the column are split and split/splitless injection.

### Split Injection

In the split injector the sample is injected into the heated injection port and the evaporated sample is mixed with the carrier gas. The sample/carrier gas mixture is then split between the column and a vent, and a fraction of the sample (determined from the column and vent flow) is introduced to the column (Fig. 8.52*a*). This technique is used to introduce concentrated samples.

### Split/Splitless Injection

Splitless injections are used to introduce dilute solutions. The sample is injected into the heated injection port, which is in the "purge off" mode. In this

***Table 8.35*** **Sample Capacity as a Function of Column Diameter and Stationary Phase Thickness**

| Column Diameter (mm) | Stationary Phase Thickness (μm) | Approximate Capacity (ng/component) |
|---|---|---|
| 0.25 | 0.10 (thin) | 25 |
| | 0.25 (most common) | 80 |
| | 1.0 (thick) | 250 |
| 0.53 | 0.10 (thin) | 53 |
| | 1.0 (most common) | 530 |
| | 5.0 (thick) | 2600 |

*Source: Alltech Capillary Instruction Manual*, Bulletin No. 242; Alltech Associates, Inc.: Deerfield, IL, 1991, p. 9.

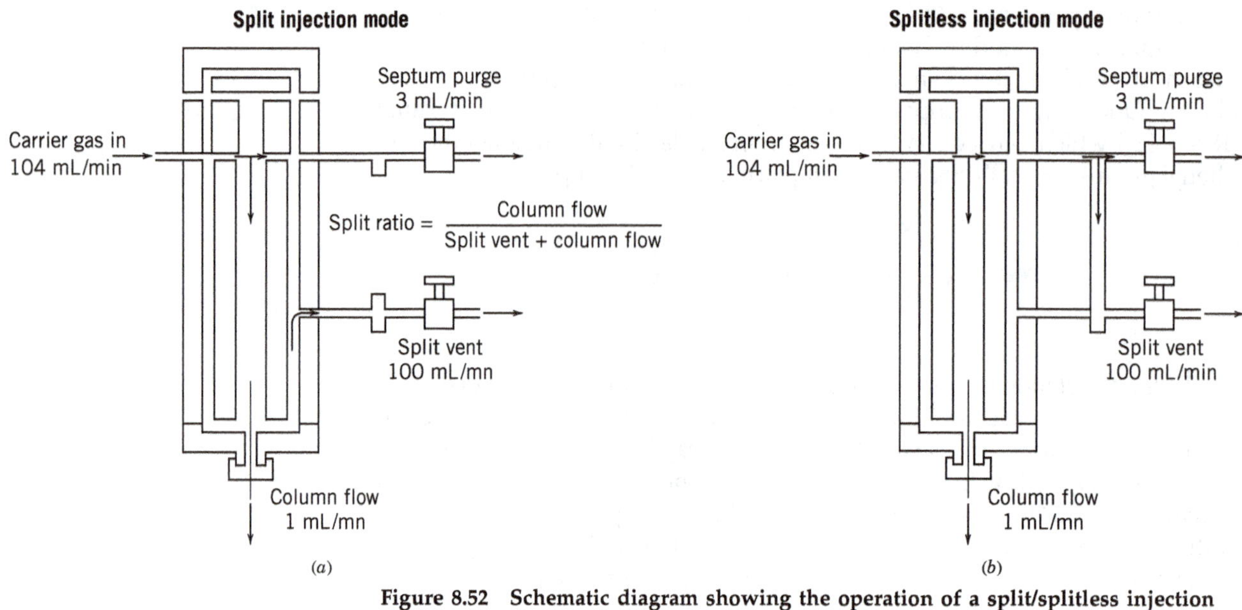


**Figure 8.52 Schematic diagram showing the operation of a split/splitless injection port.**

mode, carrier gas flows through the injector directly to the column (Fig. 8.52*b*). This flow rate is very low (0.5–3 mL/min). Of critical importance to splitless injection is the "solvent effect." The oven temperature is maintained below the solvent boiling point, and the vaporized solvent condenses in the column inlet. This condensed solvent acts like a thick layer of stationary phase and traps sample components. After this concentration period (typically 1 min), the injector is changed to the "purge on" mode. This purge sweeps excess solvent (and other volatile components) out of the injector. Purging too early risks venting volatile components, while purging too late increases interference from the solvent tail.

## FEATURES OF THE MASS SPECTRUM

A low-resolution mass spectrum can provide many pieces of information that help an organic chemist determine the structure of a molecule. One of the most useful pieces of information is the compound's nominal molecular weight, MW, as determined by identification of the molecular ion, $M^{+\cdot}$. In addition, by careful examination of the region around the molecular ion for the presence of isotopes, we can learn more about the elemental formula for the molecule. The mass spectrum also reveals information about the molecular structure through the appearance of groups of ions characteristic of certain compound types. With more experience and an understanding of mass spectral fragmentation pathways, a molecular structure can be proposed by gathering all information from the spectrum and determining if a proposed structure is consistent with the observed ions. Here we will provide a limited introduction to this process with an emphasis on identification of the molecular ion. We will present one case study to show how mass spectrometry, coupled with gas chromatography, can be used to characterize the products of a synthetic organic reaction.

### Terms

The *molecular ion*, represented by $M^{+\cdot}$, is the intact molecule with one electron missing. This should be the peak in the spectrum with the largest mass, but it is not always observed. All spectra have an ion that we call the *base peak*. This is the most abundant peak in the spectrum (*m/z* 44 in the spectrum of $N_2O$ or *m/z* 69 in the spectrum of PFTBA). In the next section you will find that more than one peak may correspond to the $M^{+\cdot}$ or fragment ion when that ion contains elements with different isotopes. We use the term *nominal mass* to describe the mass of the molecule in terms of the most abundant (and, generally, the lightest) isotopes of the element. Relative isotopic abundances for common elements are summarized in Table 8.36.

### Isotope Peaks

The mass spectrum for $N(C_4F_9)_3$ (PFTBA; MW = 671) is shown in Figure 8.50. The peaks at *m/z* 69, 219, and 502 are shown above the spectrum as they were measured by the instrument; the spectrum shows their bar-graph representation. These peaks correspond to $CF_3^+$, $C_4F_9^+$, and $NC_9F_{20}^+$. If you look carefully at the enlarged peaks, you will notice smaller peaks that appear one mass unit above that of the ion. We call these $[A + 1]^+$ peaks. Where do these peaks come from and why does the abundance increase with the mass of the fragment?

If you look at Table 8.36, you will find that fluorine is an isotopically pure element; however, 1.1% of carbon is the $^{13}C$ isotope. You may recall that it is this low abundance of $^{13}C$ that you measure with $^{13}C$ NMR. For nitrogen there is a small amount of $^{15}N$. When a molecule contains more than one atom of an isotopically impure element, you increase the chance of finding the higher mass isotope in the molecule. For example, the $^{13}C$ peaks for *m/z* 69, 219, and 502 of PFTBA (Fig. 8.50) are 1.1, 4.4, and 10.3% of the $^{12}C$ peaks. The relative intensity increases because there is a higher statistical probability of finding one $^{13}C$ when the ion has nine vs. one carbon atom. The $[A + 1]^+$ peak intensity from $^{13}C$ is equal to $n$ times 1.1% the height of the peak $A^+$, where $n$ is the number of carbon atoms. In addition, we need to add contributions from other A + 1 elements, like nitrogen (0.4%). With precise ion intensity measurements, the relative abundance of the [A + 1] peak can be used to determine the number of carbons present in an organic molecule.

**Table 8.36 Relative Isotope Abundances of Common Elements**

| Element | Mass | % | Mass | % | Mass | % |
|---|---|---|---|---|---|---|
| Carbon | 12 | 100 | 13 | 1.1 | | |
| Hydrogen | 1 | 100 | 2 | 0.015 | | |
| Nitrogen | 14 | 100 | 15 | 0.37 | | |
| Oxygen | 16 | 100 | 17 | 0.04 | 18 | 0.2 |
| Fluorine | 19 | 100 | | | | |
| Silicon | 28 | 100 | 29 | 5.10 | 30 | 3.4 |
| Phosphorus | 31 | 100 | | | | |
| Sulfur | 32 | 100 | 33 | 0.79 | 34 | 4.4 |
| Chlorine | 35 | 100 | | | 37 | 32.0 |
| Bromine | 79 | 100 | | | 81 | 97.3 |
| Iodine | 127 | 100 | | | | |

Mass spectra get very interesting when chlorine or bromine is present. Both elements exist as mixtures of the A and A + 2 isotopes (Table 8.36). The characteristic isotope distribution for bromine, with nearly equal abundances for $^{79}$Br and $^{81}$Br, is apparent in the spectrum of bromobenzene (Fig. 8.53*a*). The $M^{+\cdot}$ is observed as a cluster of peaks, with *m/z* 156 containing the lightest isotopes ($^{12}$C and $^{79}$Br). Note that the fragment at *m/z* 77 results from loss of Br, and consequently no Br isotope peaks are observed. When more atoms of A + 2 elements are present, characteristic peak distributions are produced (Fig. 8.54). For example, an ion that contains two chlorine atoms will show three peaks, A, [A + 2], and [A + 4], with a 100:65:10.6 intensity distribution. Working from Figure 8.54, you can use a pattern recognition approach to determine the number of chlorine or bromine atoms present in an ion. For example, the spectrum in Figure 8.53*b* shows two ion clusters that suggest the presence of chlorine. A close examination of the distributions indicates that two chlorines are found in the *m/z* 84 cluster, while the *m/z* 49 cluster contains one chlorine. When looking at the mass difference between these ions we work with the *nominal* mass (mass of the ion that has the lightest isotopes). The mass difference, defined by the nominal mass of each cluster, is 35, which corresponds to the mass of one $^{35}$Cl (Fig. 8.53*b*).

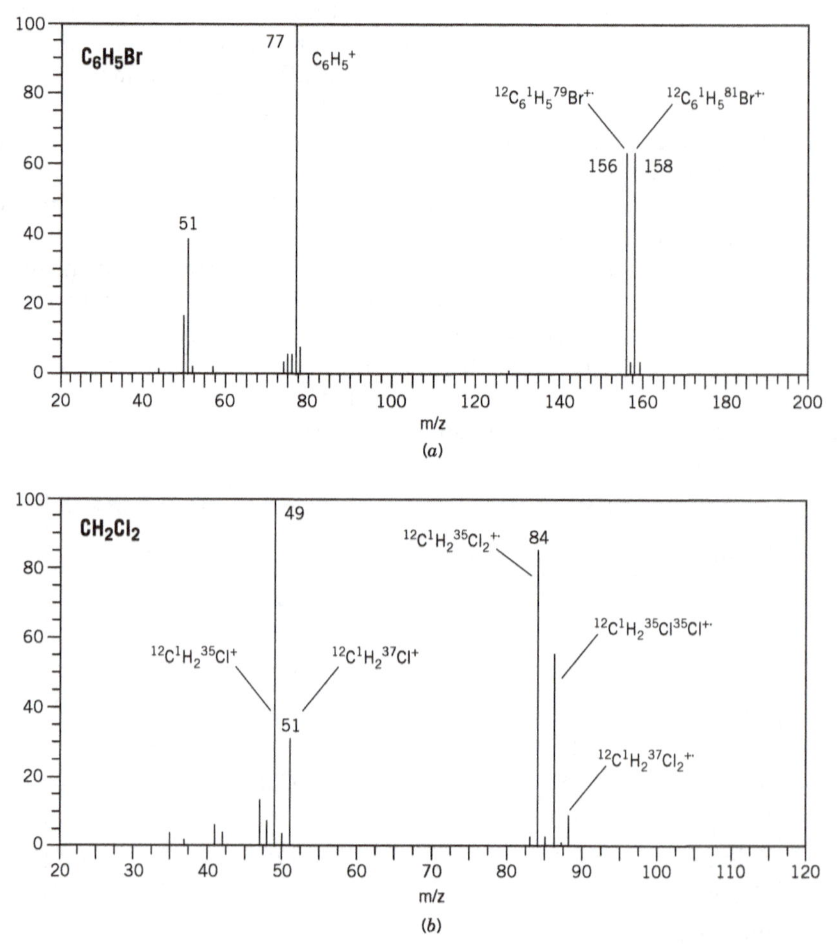


**Figure 8.53 Electron ionization mass spectrum of (a) bromobenzene and (b) dichloromethane.**

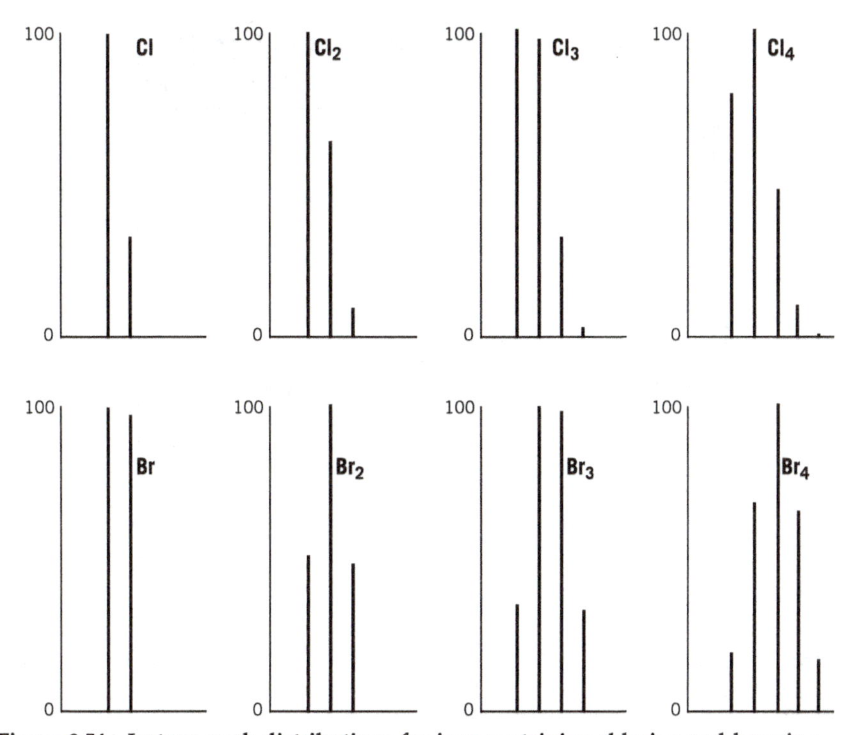


**Figure 8.54 Isotope peak distributions for ions containing chlorine and bromine atoms.**

## Recognizing the Molecular Ion

Compound molecular weight is a valuable piece of information that is not always available from the NMR or IR spectrum. In this section we discuss some things to consider as you examine a mass spectrum and attempt to identify the $M^{+\cdot}$ ion. If you look back at the mass spectra that have appeared above, you will find examples of spectra where the molecular ion is the base peak in the spectrum. In some other cases the molecular ion may be weak or not observed at all! For example, PFTBA fragments extensively and the molecular ion does not appear in the spectrum. The following are some things to consider as you attempt to identify the molecular ion.

The molecular ion should be the highest mass peak in the spectrum. When you have tentatively identified a peak as the molecular ion, you should then determine the masses lost from the molecular ion to give high-mass fragments. For example, in Figure 8.53*b* we found a mass difference of 35 between $M^{+\cdot}$ and the first fragment. A listing of common losses from $M^{+\cdot}$ can be found in Table 8.37. If an observed fragment corresponds to an unreasonable loss, such as M − 12, this strongly suggests that your tentative identification of the molecular ion is incorrect. Remember that it is always possible that no molecular ion is present.

Another useful feature to consider is the *nitrogen rule*. For most elements found in organic molecules, the compound molecular weight will be *even* if the compound has an even number of nitrogen atoms (remember, zero is an even number). In contrast, the mass will be *odd* if the compound contains an odd number of nitrogen atoms. If you are sure that your product could not contain nitrogen, then an odd mass ion could not correspond to the molecular ion.

Mass spectrometrists also use "softer" ionization techniques to obtain MW information. These techniques included measuring EI mass spectra at lower

**Table 8.37 Some Reasonable[a] Losses From $M^{+\cdot}$**

| Fragment[b] | Radical Lost | Neutral Loss |
|---|---|---|
| M-1 | H | |
| M-2 | | $H_2$ |
| M-15 | $CH_3$ | |
| M-18 | | $H_2O$ |
| M-28 | | CO or $C_2H_4$ |
| M-29 | $C_2H_5$ | |
| M-31 | $OCH_3$ | |
| M-32 | | $CH_3OH$ |
| M-43 | $C_3H_7$ | |

[a]Unreasonable losses include [M-4] to [M-14]; [M-21] to [M-25].
[b]For a more complete listing see McLafferty and Turecek (1993).

ionization energies and using a higher pressure ionization technique called *chemical ionization.*

### Mass Spectral Interpretation

The following list contains some factors to consider when interpreting mass spectra. To make best use of this summary, the interested reader should consult the text by McLafferty and Turecek, which is considered by many to be the best resource to learn mass spectral interpretation.

1. Using the considerations described above, identify the molecular ion.
2. If possible, determine the elemental composition for $M^{+\cdot}$ and other important peaks using isotopic abundances. In particular, look for isotope peaks from "M + 2" elements like Cl, Br, S, and Si (Table 8.36 and Fig. 8.54). If you are able to establish a molecular formula, calculate "rings + $\pi$ bonds":

$$\text{For } C_XH_YN_ZO_N \text{ (rings} + \pi \text{ bonds)} = X - \frac{Y}{2} + \frac{Z}{2} + 1$$

*NOTE. An even-electron ion, with no unpaired electrons, will have a fractional value. If halogens are present, they are counted as hydrogens.*

3. Is the molecular weight odd? If so, this indicates an odd number of nitrogen atoms (for organic molecules).
4. Consider the general appearance of the EI mass spectrum: Is it "aliphatic" (lots of fragmentation) or "aromatic" character (minimal fragmentation)?
5. Look for important low-mass ions (Table 8.38).
6. In the region near $M^{+\cdot}$, identify fragments lost from the molecular ion (neutral losses) (Table 8.37). Look for intense high-mass ions that may indicate a characteristic, stable fragment ion.
7. Postulate a structure by assembling the various mass fragments/neutral losses. Do the observed fragment ions make sense in terms of fragment/neutral loss stability considerations? Does the structure make sense in

*Table 8.38* **Some Common Ion Series**

| Ion Series | Compound Type |
|---|---|
| *m/z* 15, 29, 43, 57, 73 | Aliphatic hydrocarbons |
| *m/z* 38, 39, 50–52, 63–65, 75–78 | Aromatic hydrocarbons (not all peaks in ranges will be observed) |
| *m/z* 30, 44, 58 | Amines |
| *m/z* 31, 45, 59 | Alcohols |

*Note.* For a more complete listing see McLafferty and Turecek (1993).

terms of other information, such as the reaction conditions, NMR or IR spectra?

8. Verify a postulated structure by comparing the spectrum with a reference spectrum. The reference spectrum may be found in the literature or it may be measured by purchasing or synthesizing a standard of the postulated structure.

## CASE STUDY: SYNTHESIS OF METHYL BENZOATE*

To illustrate how gas chromatography and mass spectrometry can be used to characterize the products of an organic reaction, we consider the synthesis of methyl benzoate using a base-catalyzed esterification of benzoic acid. The reaction proposed for this synthesis involved deprotonation of benzoic acid by *n*-butyllithium in dry tetrahydrofuran (THF), followed by the addition of methyl iodide, with dimethylformamide (DMF) added to promote the $S_N2$ displacement of iodide by the benzoate anion:

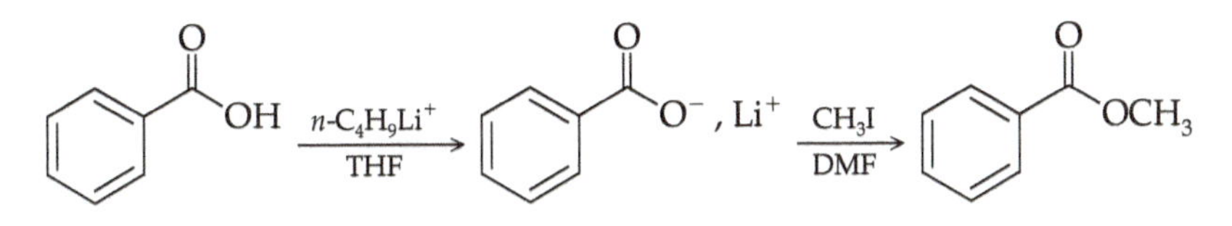


We isolated the reaction products from the reaction mixture by quenching the reaction with water, adding saturated $NaHCO_3$, and extracting the neutral products with diethyl ether. The ethereal solution containing the reaction products was then analyzed by capillary column GC/MS, and the chromatogram shown in Figure 8.55 was produced. The chromatogram displays total ionization as a function of time. The total ionization is a summation of all the ions detected in one scan of the mass spectrometer (one spectrum) plotted as a point as a function of time. The display is often called the TIC (total ionization chromatogram), and the peak areas should reflect the relative amounts of each compound detected by the mass spectrometer.

The chromatogram shown in Figure 8.55 is not quite what we would hope to see. Instead of detecting a single chromatographic peak for our product, we see three peaks. To determine if we made *any* methyl benzoate, and to determine what other components are present in our mixture, we examine the mass

*The synthetic work presented below was carried out by Joshua Pacheco, Bowdoin College Class of 1999.

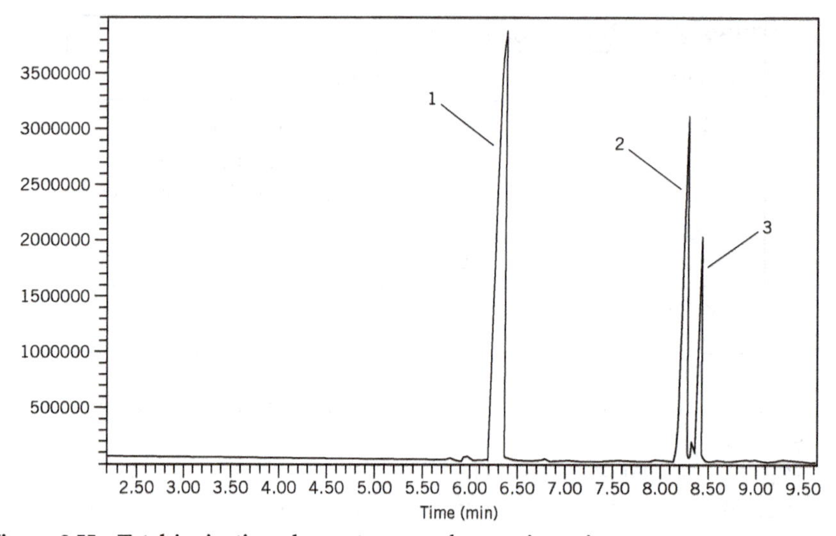


**Figure 8.55 Total ionization chromatogram of a reaction mixture.**

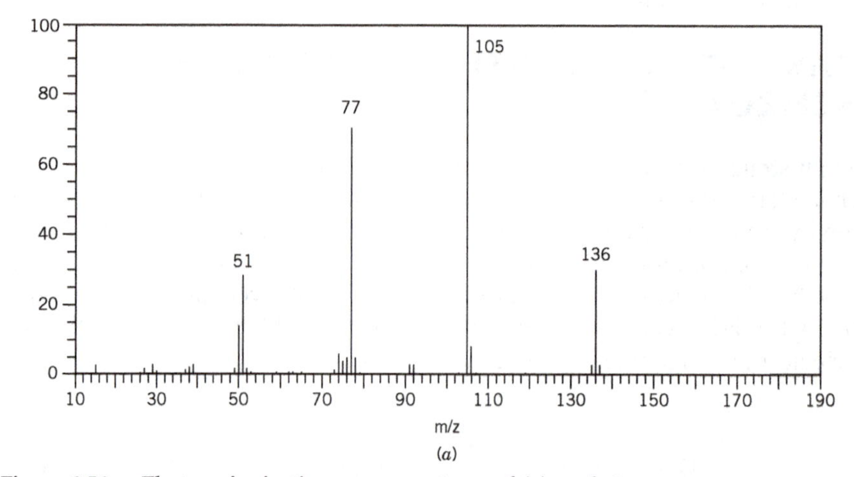


**Figure 8.56*a* Electron ionization mass spectrum of (*a*) peak 1.**

spectrum for each peak. Because we are eager to determine if we made any methyl benzoate, we start by trying to locate a chromatographic peak that has a mass spectrum that corresponds to methyl benzoate. Even if we are not sure what the mass spectrum will look like, we can try to find a spectrum that shows a molecular ion, $M^{+\cdot}$, that corresponds to the molecular weight of methyl benzoate ($C_8H_8O_2$, MW = 136). The mass spectrum for peak **1** (Fig. 8.56*a*) shows an ion at *m/z* 136 that appears as the highest mass peak in the spectrum. Let's now take a closer look at the mass spectrum to see if the fragment ions are consistent with the structure of methyl benzoate.

The base peak in the spectrum appears at *m/z* 105. This intense peak results from a loss of 31 from the molecular ion, which corresponds to loss of $OCH_3$. This is a predicted loss. Upon ionization, we expect one of the nonbonding electrons on the carbonyl oxygen to be lost, and can consider the charge and unpaired electron to be localized on that oxygen. The unpaired electron initiates a cleavage that results in loss of $OCH_3$ radical. The ions at *m/z* 77 and 51, which are characteristic of aromatic rings, may form by cleavage on the other side of the carbonyl group. The ions at *m/z* 105 and 77 may undergo another fragmentation, but the loss of another radical species is not generally observed from ions of this type, where all electrons are now paired up. Instead, even-electron neutrals, such as CO or $C_2H_2$, are lost to give fragments *m/z* 77 and 51, respectively:

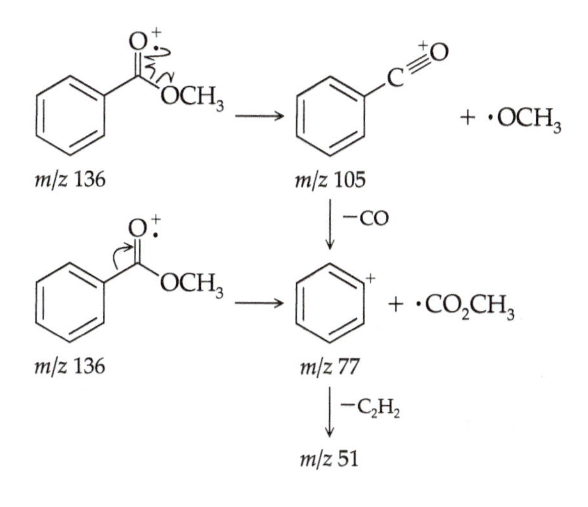


Thus, the mass spectrum for peak **1** is consistent with the structure of methyl benzoate. We could further confirm our identification by consulting a library of mass spectra. If we had any pure methyl benzoate around, we could also prepare a standard and use both the GC retention time and the mass spectrum of the standard as a means of confirming the compound identification.

We can now move on to some other peaks in the chromatogram. The mass spectra for peaks **2** and **3** (Fig. 8.56*b*, *c*) have many similar features to those of methyl benzoate. Both **2** and **3** show ions at *m/z* 51, 77, and 105. We can conclude that both compounds have a carbonyl group attached to an aromatic ring. We can next consider identification of the molecular ion. For peak **2**, the highest mass ion is *m/z* 162. To determine if this is a reasonable assignment for $M^+$, we determine losses from *m/z* 162. The ions at *m/z* 133, 120, and 105 could result from losses of 29, 42, and 57, respectively. None of these losses are unreasonable. We next want to consider two important pieces of information. First, if we assume that the MW is 162, we must add

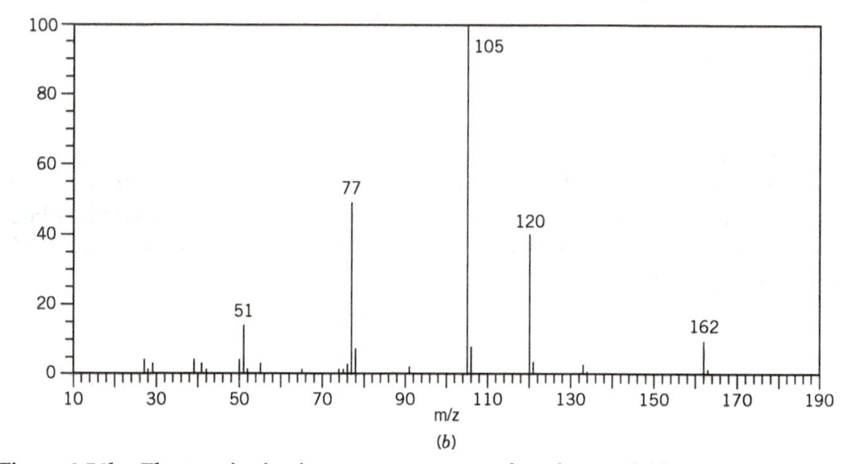


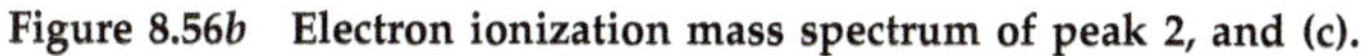


(*b*)

**Figure 8.56*b* Electron ionization mass spectrum of peak 2, and (c).**

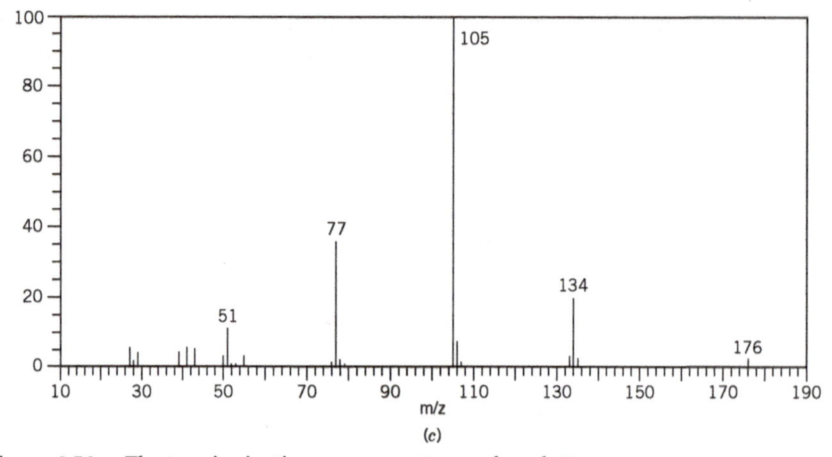


**Figure 8.56*c* Electron ionization mass spectrum of peak 3.**

57 to the carbonyl substituted aromatic ring to make our molecule. Addition of a butyl group is a logical choice:

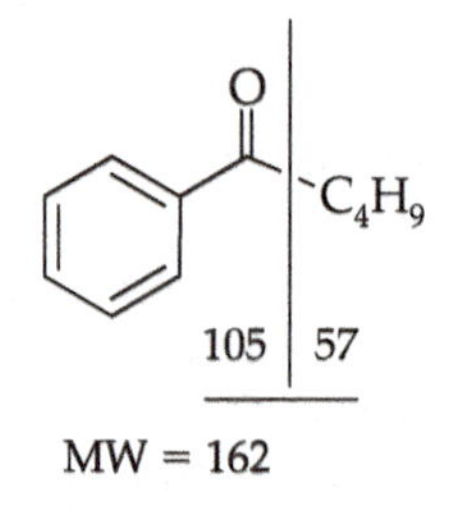


Next we take note of the peak at *m/z* 120. The even mass of this ion, resulting from an even mass molecular ion, indicates that it is a special ion! Even-mass fragments generally result from rearrangement reactions, and rearrangements involving hydrogen transfers to carbonyl groups can produce particularly informative product ions. If a butyl group is attached to the carbonyl, the following fragmentation pathway will occur, which nicely explains the *m/z* 120 peak:

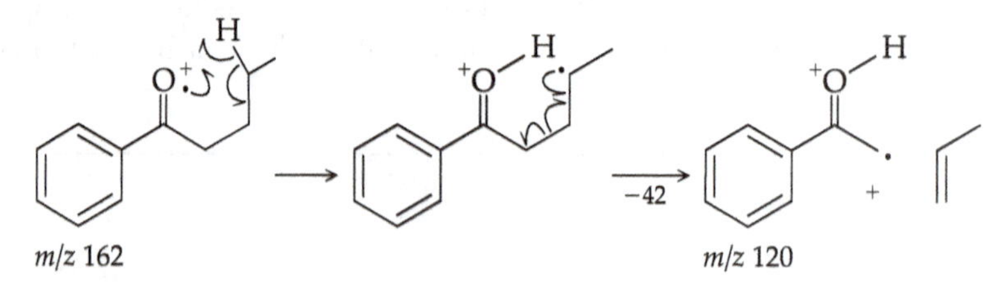


To determine if this assignment makes sense, we go back to consideration of our reaction. How could this product be generated? If we assume that there is some unreacted *n*-butyllithium around after the methyl benzoate has been formed, then the following reaction is possible. Thus, we can feel quite confident in our assignment for peak **2**:

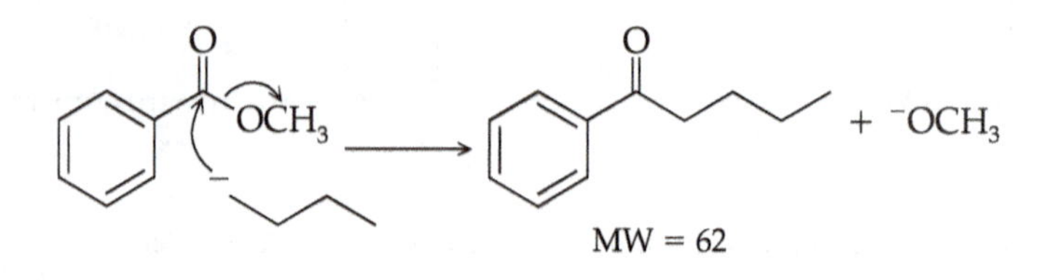

Moving on to peak **3**, we detect an ion at *m/z* 134. Let's start by assuming that this is our molecular ion. The *m/z* 105 peak would result from a loss of 29 ($C_2H_5$) from *m/z* 134, which is a reasonable loss from $M^{+\cdot}$. This *suggests*, erroneously, that ethyl benzoate is our product:

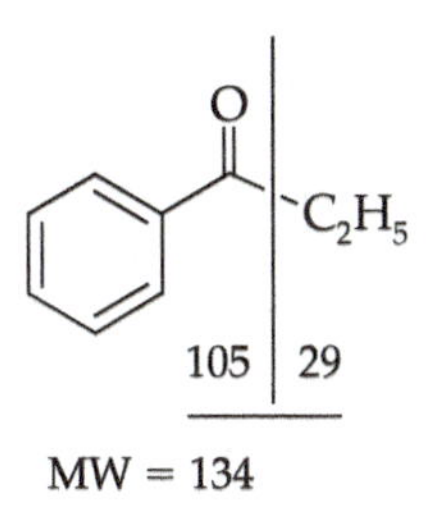


Why is this identification incorrect? If we look back at the chromatogram in Figure 8.55 the chromatographic retention times tell us that something is amiss. Remember that compounds elute from the column in approximate order of increasing boiling point. We would expect that the butyl phenyl ketone would elute *after*, not before, the ethyl phenyl ketone! In addition, you would be hard pressed to propose a mechanism for formation of ethyl benzoate in the context of this reaction. Let's go back and take another look at the mass spectrum. A careful examination shows a small peak at *m/z* 176. We may have incorrectly identified the molecular ion! If $M^{+\cdot}$ is *m/z* 176, this gives losses of 42 and 71 to form *m/z* 134 and 105, respectively. Now it looks like we have a pentyl group attached to the carbonyl, which agrees nicely with the chromatographic retention times. The *m/z* 134 ion becomes one of our special, even mass ions. What does this ion reveal about the pentyl group? The fact that the ion results from loss of 42, and not 58, clearly indicates that this is not an *n*-pentyl group. What makes the most sense is the branching shown below:

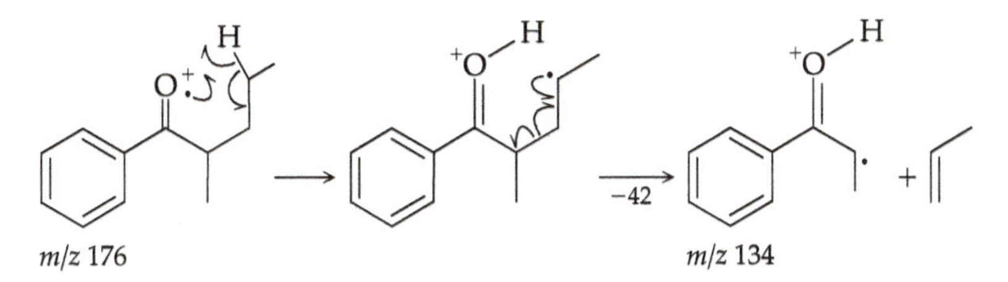


In terms of the chemistry of the reaction, this product also makes sense if deprotonation by *n*-butyllithium occurs α to the carbonyl, followed by reaction with methyl iodide:

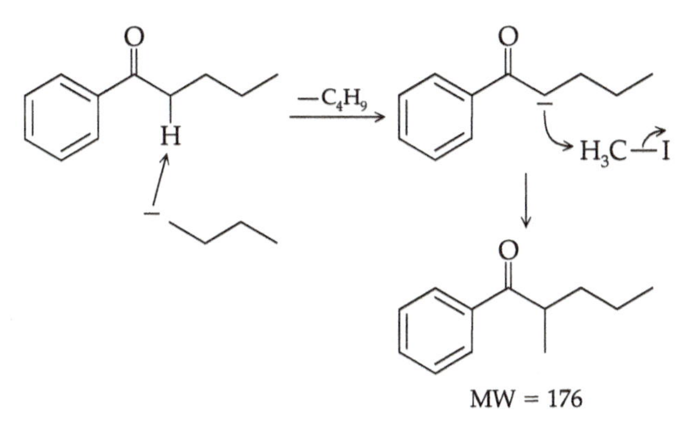

## BIBLIOGRAPHY


**Mass Spectral Interpretation:**

Beynon, J. H. *The Mass Spectra of Organic Molecules;* Elsevier: Amsterdam, 1968.

Biemann, K. *Mass Spectrometry. Organic Chemical Applications;* McGraw-Hill: New York, 1962.

Budzikiewicz, H.; Djerassi, C.; Williams, D. H. *Mass Spectrometry of Organic Compounds;* Holden-Day: San Francisco, 1967.

McLafferty, F. W.; Turecek, F. *Interpretation of Mass Spectra,* 4th ed.; University Science Books: Sausalito, CA, 1993.

Silverstein, R. M.; Webster, F. X.; Kiemle, D. *Spectrometric Identification of Organic Compounds,* 7th ed.; Wiley: New York, 2005.

Watson, J. T.; Sparkman, O. D. *Introduction to Mass Spectrometry: Instrumentation, Applications, and Strategies for Data Interpretation,* 4th ed.; Wiley: New York 2007 (also a good introduction to instrumentation).

**Theory and Instrumentation:**

Dass, C. *Fundamentals of Contemporary Mass Spectrometry;* Wiley: New York, 2007.

De Hoffmann, E.; Stroobant, V. *Mass Spectrometry: Principles and Applications,* 3rd ed.; Wiley: New York, 2007.

Kitson, F. G.; Larsen, B. S.; McEwen, C. N. *Gas Chromatography and Mass Spectrometry: A Practical Guide;* Academic Press: San Diego, 1996.

Message, G. M. *Practical Aspects of Gas Chromatography/Mass Spectrometry;* Wiley: New York, 1984.

*Quadrupole Mass Spectrometry and Its Applications;* Dawson, P. H., Ed., Elsevier: New York, 1976.

**Libraries:**

McLafferty, F. W.; Stauff, D. B.; *The Important Peak Index of the Registry of Mass Spectral Data,* Wiley: New York, 1991.

McLafferty, F. W.; Stauffer, D. B. *Wiley/NBS Registry of Mass Spectral Data;* Wiley: New York, 1989.

*NIST/EPA/MSDC Mass Spectral Library;* National Institute of Standards and Technology. You may also access some mass spectral data through the internet: *WebBook.nist.gov*

# GLOSSARY

**Absorb** To take up matter (to dissolve), or to take up radiant energy.

**Activated complex** An unstable combination of reacting molecules that is intermediate between reactants and products.

**Activation energy** The minimum energy, $\Delta G^{\ddagger}$, necessary to form an activated complex in a reaction. Or the difference in energy levels between the ground state and transition state.

**Active methylene** A methylene group with hydrogen atoms rendered acidic due to the presence of an adjacent ($\alpha$) electron withdrawing group, such as a carbonyl group.

**Activity (of alumina)** A measure of the degree to which alumina adsorbs polar molecules.The activity (adsorbtivity) of alumina may be reduced by the addition of small amounts of water. Thus, the amount of water present in a sample of alumina determines the activity grade. Alumina of a specific activity can be prepared by dehydrating alumina at 360 °C for about 5–6 h and then allowing the dehydrated alumina to absorb a suitable amount of water. The Brockmann scale of alumina activity is based on the amount of water (weight percent) that the alumina contains: grade I = 0%, grade II = 3%, grade III = 6%, grade IV = 10%, and grade V = 15%. For further information, see Brockmann, H.; Schodder, H. *Chem Ber.* **1941,** *74,* 73.

**Adsorb** The process by which molecules or atoms (either gas or liquid) adhere to the surface of a solid.

**Aliphatic** Term used to refer to nonaromatic species, such as alkanes, alkenes, alkynes.

**Aliquot** A portion.

**Alkaloid** A naturally occurring compound that contains a basic amine functional group. They are found particularly in plants.

**Anilide** A compound that contains a $C_6H_5NHCO$ group. An amide formed by acylation of aniline (aminobenzene).

**Bimolecular reaction** The collision and combination of two reactants to give an activated complex in a reaction.

**Capillary action** The action by which the surface of a liquid,where it contacts a solid, is elevated or depressed because of the relative attractions of the molecules of the liquid for each other and for the solid. It is particularly observable in capillary tubes, where it determines the ascent (descent) of the liquid above (below) the level of the liquid in which the capillary tube is immersed.

**Catalyst** A substance that changes the speed of a chemical reaction without affecting the yield or undergoing permanent chemical change itself.

**Characterize** To conclusively identify a compound by the measurement of its physical, spectroscopic, and other properties.

**Condensation reaction** A condensation reaction is an addition reaction that produces water (or another small neutral molecule such as $CH_3OH$ or $NH_3$) as a byproduct.

**Dehydrohalogenation** A reaction that involves loss of HX from a halide by treatment with strong base.

**Deliquescent** Liquefying by the absorption of water from the surrounding atmosphere.

**Dihedral angle** The angle between two intersecting planes. In organic chemistry the term dihedral angle (or torsional angle) is used to describe the angle between two atoms (or groups) bonded to two adjacent atoms, such as H—C—C—H, and can be determined from a molecular model by looking down the axis of the bond between the two central atoms.

**Dipole** The separation of charge in a bond or in a molecule with a positively and negatively charged end.

**Eluant** A mobile phase in chromatography.

**Eluate** The solution that is eluted from a chromatographic system.

**Elute** To cause elution.

**Elution** The flow, in chromatography, of the mobile phase through the stationary phase.

**Emulsion** A suspension composed of immiscible drops of one liquid in another liquid (e.g., oil and vinegar in salad dressing).

**Enol** A functional group composed of a hydroxyl group bonded to an alkene.

**Enolate** The conjugate base of a enol, that is, a negatively charged oxygen atom bonded to an alkene. An enolate results from deprotonation $\alpha$ to a carbonyl group.

**Enthalpy change** ($\Delta H$) The heat lost or absorbed by a system under constant pressure during a reaction.

**Entropy (S)** The randomness, or amount of disorder of a system.

**Entropy change** ($\Delta S$) The change in the amount of disorder.

**Filter cake** The material that is separated from a liquid, and remains on the filter paper, after a filtration.

**Free energy change** ($\Delta G$) A predictor of the spontaneity of a chemical reaction at constant temperature. $\Delta G = \Delta H - T\,\Delta S$

**Glacial acetic acid** Pure acetic acid containing less than 1% water.

**Heterocycle** A cyclic molecule whose ring contains more than one kind of atom.

**Heterolysis** Cleavage of a covalent bond in a manner such that both the bond's electrons end up on one of the formerly bonded atoms.

**Homogeneous** Consisting of a single phase.

**Homolysis** Cleavage of a covalent bond in a manner such that the bond's electrons are evenly distributed to the formerly bonded atoms.

**Hydroboration** Addition of borane ($BH_3$) or an alkyl borane to a multiple bond.

**Hydrogenation** Addition of hydrogen to a multiple bond.

**Hygroscopic** Absorbs moisture.

**In situ** In chemistry, the term usually refers to a reagent or other material generated directly in a reaction vessel and not isolated.

**Kinetics** Referring to the rate of a reaction.

**Lachrymator** A material that causes the flow of tears.

**Ligroin** A solvent composed of a mixture of alkanes.

**Mechanism** A complete description of how a reaction occurs.

**Metabolism** The chemical processes performed by a living cellular organism.

**Metabolites** The compounds consumed and produced by metabolism.

**Methine** A CH group (with no other hydrogen atoms attached to the carbon atom).

**Methylene** A $CH_2$ group (with no other hydrogen atoms attached to the carbon atom).

**Mother liquor** The residual, and often impure, solution remaining from a crystallization.

**Olefin** An older term for an alkene.

**Optical isomers** Enantiomers. Isomers that have a mirror-image relationship.

**Order of reaction** With respect to one of the reactants, the order of a reaction is equal to the power to which the concentration of that reactant is raised in the rate equation.

**Oxonium ion** A trivalent oxygen cation with a full octet of electrons (e.g., $H_3O^+$).

**Paraffins** An older name for alkanes.

**Phase transfer catalysts** Agents that cause the transfer of ionic reagents between phases, thus catalyzing reactions.

**Plasticizer** A substance added to a polymer to make it more flexible or to prevent embrittlement.

**Polymer** A compound of high molecular mass that is built up of a large number of repeating simple molecules, or monomers.

**Racemic** Consisting of an equimolar mixture of two enantiomers.

**Rate equations** Equations giving the relationship between reaction rate and the concentrations of the reactants.

**Reaction mechanism** The stepwise sequence of elementary reactions in an overall reaction.

**Reagent** A chemical or solution used in the laboratory to detect, measure, react with, or otherwise examine other chemicals, solutions, or substances.

**Reflux** The process by which all vapor evaporated or boiled from a vessel is condensed and returned to that vessel.

**Rotamers** Conformational isomers that can be interconverted by rotation about one or more single bonds (e.g., *gauche* and *anti* butane).

**Spontaneous process** A physical or chemical change that occurs without the net addition of energy. $\Delta G < 0$ for a spontaneous process.

**Sublimation** The passing of a solid directly into vapor state without first melting.

**Tare** A tared container is one whose weight has been measured. The term may also refer to the process of zeroing a balance after a container has been placed on the weighing platform.

**Thermodynamics** The chemical science that deals with the energy transfers and transformations that accompany chemical and physical changes.

**Transition state** A combination of reacting molecules that is intermediate between reactants and products.

**Triturate** To grind to a fine powder. (Or, washing solid organic products in a solvent in which the desired product has little solubility.)

**Vapor pressure** The pressure exerted by a vapor in equilibrium with a liquid or solid at a given temperature.

**Ylide** A neutral dipolar molecule in which negative and positive charges are on adjacent atoms.

**Zwitterion** A neutral molecule containing separated opposite formal charges.

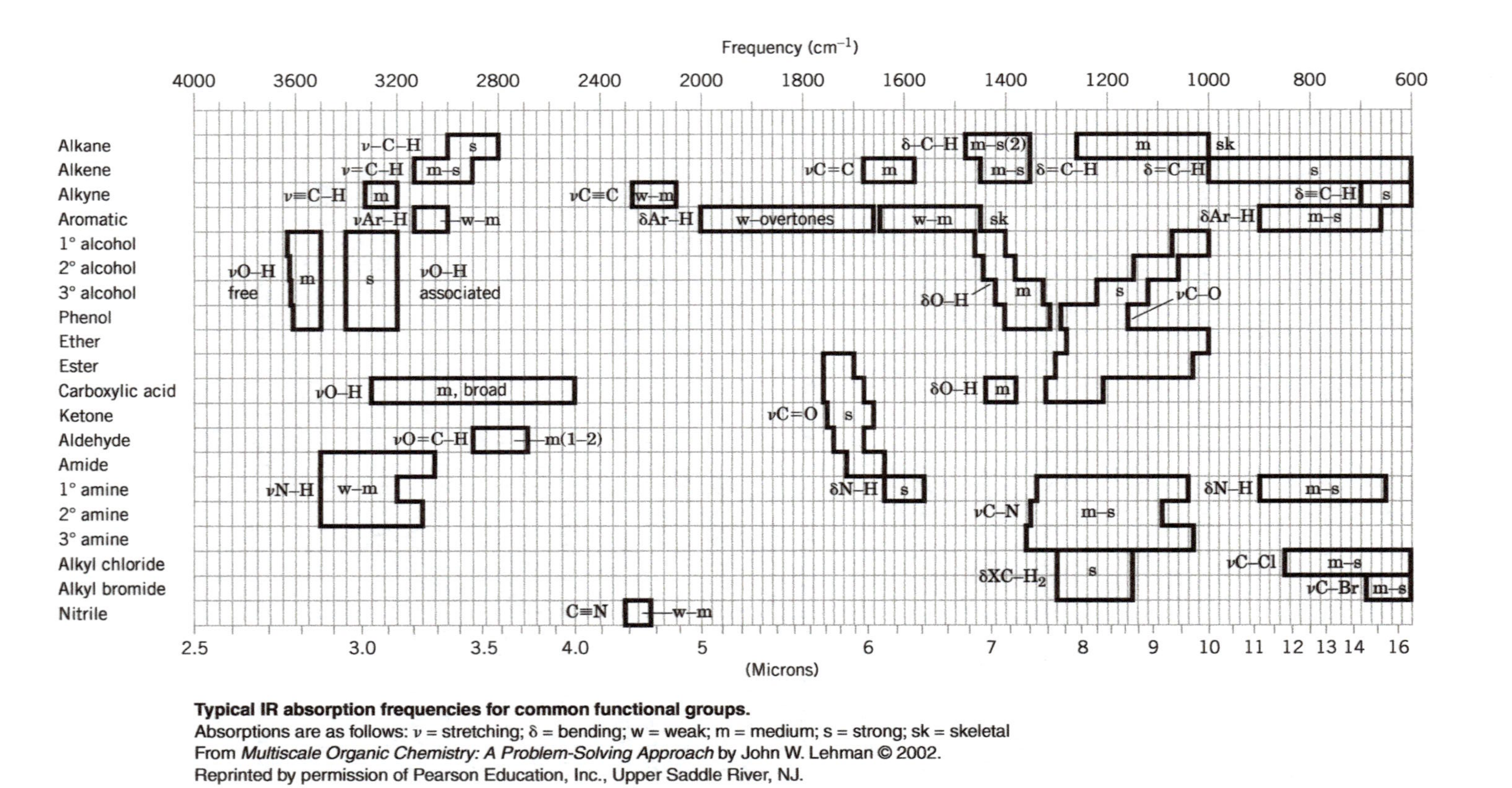


**Typical IR absorption frequencies for common functional groups.**
Absorptions are as follows: ν = stretching; δ = bending; w = weak; m = medium; s = strong; sk = skeletal
From *Multiscale Organic Chemistry: A Problem-Solving Approach* by John W. Lehman © 2002.
Reprinted by permission of Pearson Education, Inc., Upper Saddle River, NJ.

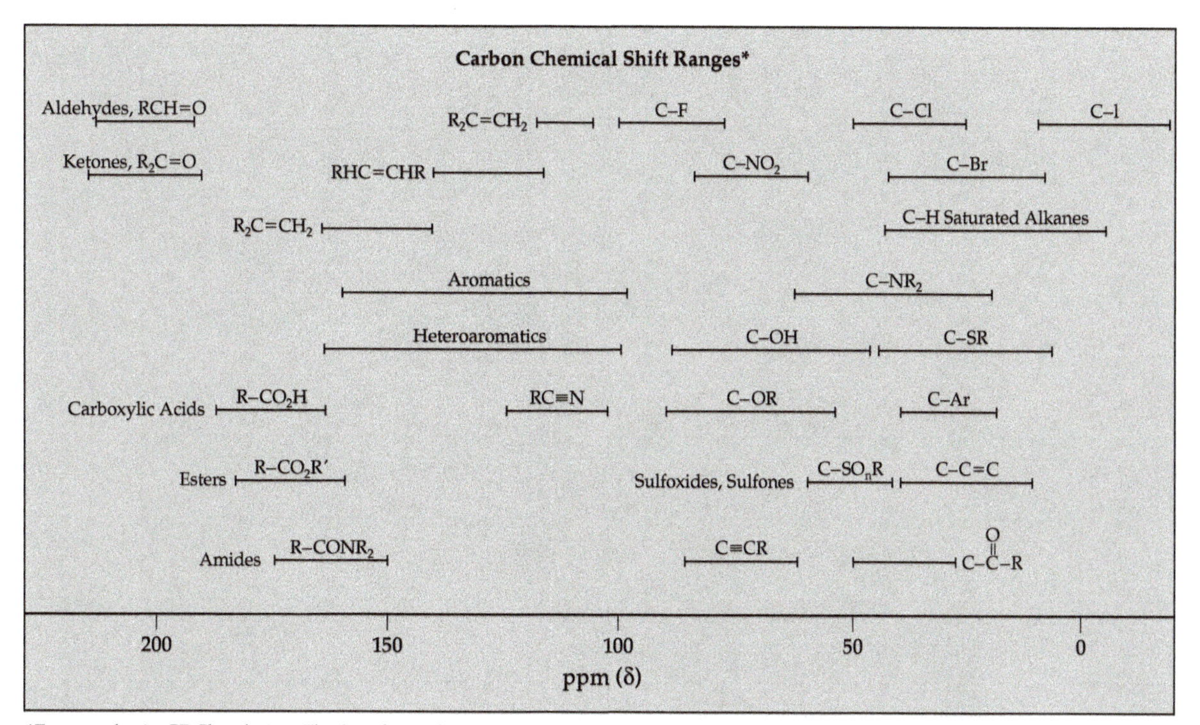


*For samples in $CDCl_3$ solution. The δ scale is relative to TMS at δ=0. Organic Chemistry Michigan State University Source, Dept. of Chemistry, Michigan State University.

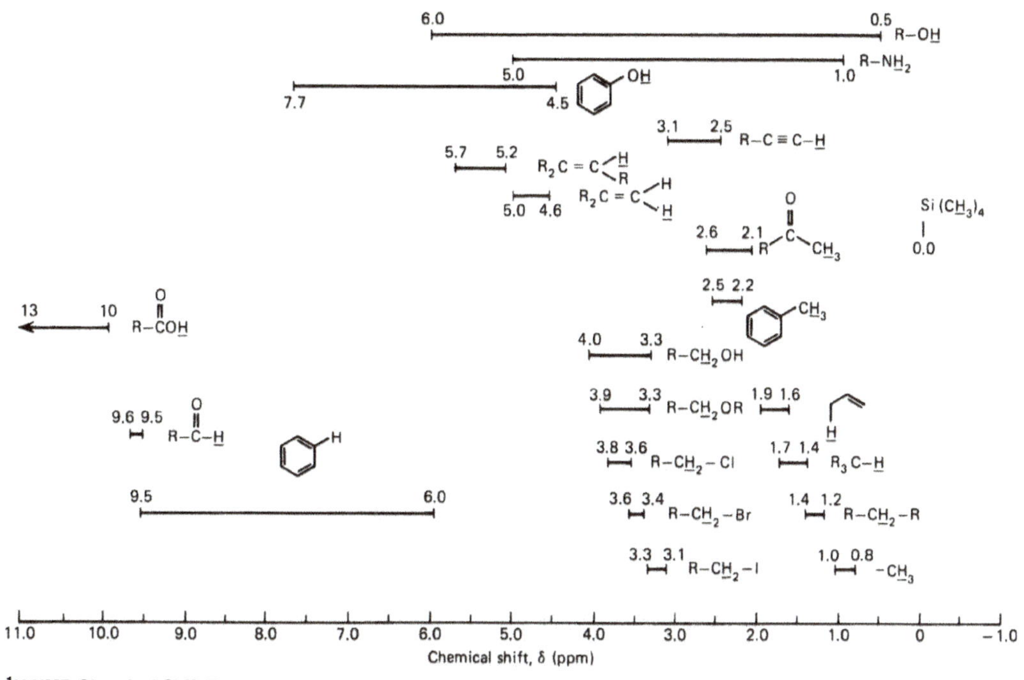


**¹H NMR Chemical Shift Ranges**

NOTES

# NOTES

# NOTES

NOTES

# NOTES

# NOTES

# NOTES

# NOTES

Printed in the USA
K039198SCI112316 01S29053000000001624